OREGON GRADUATE CENTER
LIBRARY

16023

The Environmental Chemistry of Aluminum

Editor

Garrison Sposito, Ph.D.
Department of Soil Science
University of California at Berkeley
Berkeley, California

CRC Press, Inc.
Boca Raton, Florida

Library of Congress Cataloging-in-Publication Data

The environmental chemistry of aluminum / editor, Garrison Sposito.
p. cm.
Bibliography: p.
Includes index.
ISBN 0-8493-4728-9
1. Water--Aluminum content--Environmental aspects. 2. Soils--Aluminum content--Environmental aspects. 3. Acid deposition--Environmental aspects. 4. Environmental chemistry. 5. Biogeochemistry. I. Sposito, Garrison, 1939-
TD427.A45E58 1989
628.1'68—dc19

88-15534
CIP

This book represents information obtained from authentic and highly regarded sources. Reprinted material is quoted with permission, and sources are indicated. A wide variety of references are listed. Every reasonable effort has been made to give reliable data and information, but the author and the publisher cannot assume responsibility for the validity of all materials or for the consequences of their use.

All rights reserved. This book, or any parts thereof, may not be reproduced in any form without written consent from the publisher.

Direct all inquiries to CRC Press, Inc., 2000 Corporate Blvd., N.W., Boca Raton, Florida, 33431.

© 1989 by CRC Press, Inc.

International Standard Book Number 0-8493-4728-9
Library of Congress Card Number 88-15534
Printed in the United States

FOR GERALD LOCKLIN

ἕτερος γὰρ αὐτὸς
ὁ φίλος ἐστίν

PREFACE

naturam expellas furca
tamen usque recurret

The chemistry of aluminum in soils and natural waters has emerged as a scientific problem of critical importance because of widespread public interest in the effects of acidic deposition on terrestrial ecosystems. Acidic deposition is perceived as responsible for major changes in forest productivity and surface water quality in central Europe and northeastern North America over the last half century. Environmental chemists have responded to this public concern in a broad variety of laboratory and field studies which often have substantiated the great complexity of aluminum biogeochemistry in sensitive watersheds.

Aluminum chemistry in aqueous systems, like that of other common elements, is significant environmentally because it has a direct bearing on the mobility, bioavailability, and ecological impact of this metal. Evidence is growing that the specific toxicity effects of aluminum on terrestrial and aquatic organisms are related to its chemical speciation. If this relationship proves to be of general validity, then the production of innocuous chemical species will develop as a major pathway of detoxification of aluminum in forested and agricultural ecosystems. An understanding of the factors that control this pathway will be essential to the management of the biological response to acidic deposition as mediated by surface and subsurface water quality criteria.

This book is intended to provide a comprehensive, fundamental account of the aqueous chemistry of aluminum that is relevant to the environmental context created by the problem of acidic deposition. The objective is to present intelligibly the consensus of experts, not on the unresolved issues at the frontier of research, but on the basic chemical phenomena involving aluminum in natural waters and soils about which some degree of certainty can be expressed. Thus the level of sophistication assumed is on a par with the contents of a standard textbook, like *Aquatic Chemistry* by W. Stumm and J. J. Morgan, but does not require expertise in the particular facets of aluminum reactions in aqueous media. These facets are described in the ten chapters of this book with the hope that environmental chemists will find in them reference material whose value persists despite the changes in understanding that undoubtedly will occur in this rapidly developing subdiscipline.

The first three chapters discuss perhaps the most fundamental aspects of aluminum chemistry: its quantitation in soils and natural waters, including speciation measurements, and its stable chemical forms, both as a dissolved solute and as a solid phase. The emphasis in these chapters is on providing a critical assessment and definitive recommendations for laboratory methodologies and thermodynamic properties relating to aluminum species. The next four chapters build on this foundation to provide details of what may be termed the polymeric chemistry of aluminum: its polynuclear and colloidal hydrolytic species in aqueous solution, its complexes with natural organic ligands including humic substances, and its surface reactions as both adsorptive and adsorbent. These difficult topics are treated firmly in the context of experimental results, as opposed to conceptual modeling, with transience and kinetics playing roles equal in importance to stability and thermodynamics. Areas in which knowledge is not yet well defined are identified throughout the discussion in these chapters as a guide to future research. The final three chapters of the book describe current thinking on the chemistry of aluminum in soil and natural water environments. These chapters attempt to integrate the results of controlled experimentation described in the first seven chapters into pictures of aluminum solubility and speciation in nature. The problem of spatial and temporal variability, metastability, and scale that emerge in the discussion illustrate the challenge that environmental aluminum chemistry presents.

Although the chapters of this book can be read independently, they are in fact more like the bands of color in the visible spectrum, each recognizably distinct but grading into the others with overlapping subtopics and conclusions. A given chapter may be taken, on its principal topics, to be the definitive statement of this book, albeit no attempt has been made to suppress variances among the authors that represent normal vicissitude in the inferences of experts on a difficult subject.

I should like to express my deep gratitude to the following chemists who served as reviewers of the book chapters in draft: Prof. H. L. Bohn, Prof. J. I. Drever, Prof. J. B. Harsh, Prof. P. M. Huang, Prof. J. A. Kittrick, Prof. J. G. McColl, Prof. J. W. Miller, Prof. D. R. Parker, and Prof. L. W. Zelazny. I am also much indebted to Ms. Louise DeHayes for her expert typing of several portions of the final manuscript. Lastly, I must acknowledge the loss of three colleagues in aluminum chemistry who would have contributed to this book as author or reviewer had not fate intervened: Prof. Frank T. Bingham, Dr. Jean-Yves LeBrusq, and Prof. Gordon Pagenkopf. I shall always remember the brightness of their inquiring curiosity and feel keenly the untimely events that deprived our endeavor of their needed contributions.

Garrison Sposito

THE EDITOR

Garrison Sposito received a B.S. degree in soils from the University of Arizona in 1961, then stayed on to complete an M.S. degree in soil physics in 1963 under the direction of Dr. Duwayne M. Anderson. In September 1963 he transferred to the Berkeley campus of the University of California to continue his graduate studies as a University Science Fellow. In July 1965, he received a Ph.D. degree in soil science under the direction of Dr. Kenneth L. Babcock. He then decided to join the faculty at Sonoma State University in northern California, where he initiated and developed the undergraduate program in physics. He remained in physics until 1974, when he joined the Department of Soil and Environmental Sciences at the Riverside campus of the University of California, as an assistant professor in soil science. He was promoted to Associate Professor in 1976, and to Professor in 1978. In 1988 he was appointed Professor Soil Physical Chemistry in the University of California at Berkeley.

Dr. Sposito has made research contributions in the areas of: (1) water, solute, and energy transport through soils; (2) ion exchange and adsorption on soils and clays; (3) computer modeling of soil solutions; and (4) soil organic matter chemistry. He has published more than 160 technical journal articles, 30 book chapters or conference papers, as well as numerous book reviews and technical reports. Seven of his technical journal articles have been reprinted in various hardbound collections of "benchmark papers" relating to soil and water science. He is also the author of a textbook on quantum mechanics (1970), a textbook on classical mechanics (1976), a monograph on chemical thermodynamics applied to soil solutions (1981; Russian edition, 1984), and a monograph on the surface chemistry of soils (1984). He is co-author of a NAS-sponsored syllabus for hydrology courses (1972) and helped to edit a monograph on mineral-organic matter associations (1986).

Dr. Sposito was a NSF Summer Fellow in physics in 1971; a Senior Fulbright Lecturer in theoretical solid-state physics in 1973; a NATO/Heinemann Senior Fellow in soil physical chemistry in 1981, and a John Simon Guggenheim Fellow at the University of Oxford in 1984. In 1970, he received the Award for Distinguished Teaching from the California State University and, in 1980, he received a Distinguished Teaching Award from the University of California. He is believed to be the only person ever to receive distinguished teaching awards from both state university systems in California. In 1982, Dr. Sposito was the recipient of the Soil Science Award, a prize given annually by the Soil Science Society of America "to recognize an individual who has demonstrated unusual creativity, reasoning ability, and technical skill and whose research contributions to basic and applied soil science are distinguished by originality and significance." In 1983 Dr. Sposito was elected a Fellow of the Soil Science Society of America and of the American Society of Agronomy, an honor extended annually to a maximum of 0.3% of the current membership. In 1985 Dr. Sposito was elected a Fellow of the American Geophysical Union, an honor available to a maximum of 0.1% of the current membership, "for his contributions to basic understanding of the physics, chemistry, and thermodynamics of soil-water systems".

CONTRIBUTORS

Paul M. Bertsch
Division of Biogeochemistry
University of Georgia
Savannah River Ecology Laboratory
Drawer E
Aiken, SC 29801

Paul R. Bloom
Department of Soil Science
University of Minnesota
St. Paul, MN 55108

James A. Davis
U.S. Geological Survey
345 Middlefield Road
Menlo Park, CA 94025

Charles T. Driscoll
Department of Civil Engineering
Syracuse University
Syracuse, NY 13244

M. Susan Erich
Department of Plant and Soil Science
University of Maine
Orono, ME 04469

John D. Hem
U.S. Geological Survey
345 Middlefield Road
Menlo Park, CA 94025

Bruce S. Hemingway
U.S. Geological Survey
National Center
Reston, VA 22092

William H. Hendershot
Department of Renewable Resources
Macdonald College
McGill University
Ste-Anne-de-Bellevue
Québec H9X 1C0
Canada

Philip M. Jardine
Environmental Science Division
Oak Ridge National Laboratory
Oak Ridge, TN 37830

Dean S. Jeffries
National Water Research Institute
P.O. Box 5050
Burlington, Ontario L7R 4A6
Canada

W. L. Lindsay
Department of Agronomy
Colorado State University
Fort Collins, CO 80523

Howard M. May
U.S. Geological Survey
345 Middlefield Road
Menlo Park, CA 94025

Darrell Kirk Nordstrom
U.S. Geological Survey
345 Middlefield Road
Menlo Park, CA 94025

Garrison Sposito
Department of Soil Science
University of California at Berkeley
Berkeley, CA 94720

F. J. Stevenson
Department of Agronomy
1102 South Goodwin Avenue
University of Illinois
Urbana, IL 61801

G. R. Vance
Department of Agronomy
1102 South Goodwin Avenue
University of Illinois
Urbana, IL 61801

P. M. Walthall
Department of Agronomy
Louisiana State University
Baton Rouge, Louisiana

Lucian W. Zelazny
Agronomy Department
Virginia Polytechnic Institute
and State University
Blacksburg, VA 24061

TABLE OF CONTENTS

Chapter 1

THE QUANTITATION OF AQUEOUS ALUMINUM

Paul R. Bloom and M. Susan Erich

TABLE OF CONTENTS

I. INTRODUCTION

Determination of aqueous aluminum is important both to scientists interested in toxicity of aluminum to terrestrial and aquatic organisms and to those who study mineral weathering and mineral neoformation. In this chapter, the focus will be on the determination of aluminum in natural waters with an emphasis on fresh-water systems. The definition of natural fresh waters, as used in this chapter, includes lakes and streams, as well as ground waters and interstitial water in soils and sediments.

The analysis of aqueous aluminum is complicated by the low concentrations of aluminum found generally in natural waters (see Chapters 9 and 10) and by the several possible forms of aluminum. Except for the most acidic natural waters, analytical techniques with detection limits of 10 μg/l or less are often needed.

Aluminum forms various monomeric and polymeric complexes (see Chapters 2 and 4), plus many poorly characterized organic complexes (see Chapter 5). Monomeric inorganic complexes of fluorine and PO_4, polymeric hydroxy complexes, and organic complexes can interfere with some analytical techniques for aluminum. In addition, colloidal mineral aluminum can interfere even if samples are filtered through a 0.45- or 0.1-μm membrane. Since many investigators want to be able to determine the thermodynamic activity of aquo Al^{3+}, methods have been developed to estimate the quantity of monomeric inorganic aluminum. These techniques, which discriminate against mineral, polymeric, and organically complexed forms of aluminum, will be discussed in Section IV of this chapter.

II. BASIC METHODOLOGY

A. Atomic Absorption Spectrometry

Direct spectrometric determination of aluminum, i.e., the measurement of light absorbed or emitted by aluminum atoms, provides the basis for the determination of total quantities of aluminum by atomic absorption and emission methods. These methods are based on the fact that atoms in the ground state can absorb or emit light at characteristic wavelengths. In atomic absorption spectrometry (AA), absorbance by ground-state atoms in a flame or furnace is measured. (In emission spectrometry, emission of atoms excited at high temperature is measured.) In these methods most aluminum compounds and complexes are thermally decomposed, thereby allowing the measurement of total aluminum without sample pretreatment. However, in the case of more refractory materials, such as suspended aluminous minerals, the decomposition may not be complete.

Two types of AA analyses are distinguished by whether or not a flame or a furnace is used to vaporize and atomize the sample. Aluminum can be determined using a nitrous oxide-acetylene flame, but for most natural waters this method lacks sufficient sensitivity without preconcentration. Aluminum is more commonly determined using the more sensitive graphite furnace (flameless) techniques. Precise analyses require precise injection of sample into the graphite furnace. Precision is greatly increased with the use of an automatic sampler with mechanical injection.[1]

Interferences which may be present in AA are (1) chemical, (2) ionization, (3) matrix, (4) spectral, and (5) background absorption. Chemical interferences result when species in the sample form compounds with the analyte that are not decomposed in the flame. Aluminum can form refractory compounds, and for this reason the air-acetylene flame, which is cooler than a nitrous oxide flame, is not suitable for aluminum determinations. If the degree of ionization of the analyte significantly reduces the population of nonionized atoms, the absorption signal will be decreased. This interference is overcome by the addition of an alkali salt to suppress ionization, which is recommended for aluminum analyses using the flame.

Matrix components that affect physical properties, such as viscosity and surface tension, can have an effect on the rate of transport of analyte into an AA burner. Also, matrix components may affect the rate and temperature of volatilization of aluminum in the furnace. Matrix effects can be minimized by matching the matrix in the samples and standards as closely as possible.

Spectral interferences occur when other species present in the sample along with the analyte absorb in the spectral region of interest. In general, since absorption spectra consist of transitions from the ground state, the spectra are relatively simple and the occurrence of coincident absorption lines in a given sample is not a frequent problem. The analyst can make accurate determinations in the presence of spectral interferences using the method of standard additions.[2]

Background absorption, which is due to light scattering by molecular species, salt particles, or smoke, or to absorption by molecules in the flame, cannot be overcome by the method of standard additions. Background absorption tends to be especially severe in the graphite furnace, and background correction is generally recommended.

B. Atomic Emission Spectrometry

Measurement of aluminum emission in aqueous samples is accomplished generally by using an inductively coupled plasma (ICP-AES) as the excitation source, and our discussion of emission will be confined to plasma techniques. Flame emission is rarely used for aluminum analysis because of low sensitivity.[3] Another type of plasma, direct current plasma (DCP) is sometimes employed as the source. In principle, its use is very similar to that of the ICP.

The ICP is an excitation source which produces a complex emission spectrum for many sample matrices, so interferences must be carefully investigated and compensated for. However, ionization interferences are not a problem with this technique because of high electron density in the plasma.[4] Both spectral and nonspectral interferences may exist. Spectral interferences are caused by overlapping emission lines. Careful investigation of interferences and application of electronic correction is necessary, especially when working near detection limits where a substantial proportion of the signal may be from the background.[5] The most severe nonspectral interferences are those that are due to the physical properties of the matrix, such as its viscosity.[6] When compositional changes in the samples cause interferences by affecting nebulization and aerosol transport, a known amount of an internal standard element may be added to the samples. Intensity ratios can then be related to concentrations.[4] This problem is minimized when samples and standards are closely matched with regard to acid and salt content.

The aluminum spectrum is given in Table 1.[3] The most commonly used line for absorption is 309.3 nm. Several lines have been investigated for ICP and DCP analyses, including 308.2, 396.2, and 394.4 nm.

C. UV-Visible Spectrophotometry of Organic Aluminum Complexes

1. Absorbance

Determination of aluminum by reacting it with an organic reagent and measuring the light absorbed by the aluminum-organic complex is very common. The most comprehensive discussion of the properties of the many reagents that have been used for aluminum analyses is given by Tikhonov.[7] Table 2 lists nine of the most common reagents and their absorption maxima.

A few considerations are common to all the complexing agents. One is that absorption is sensitive to changes in pH value, so the analytical solutions are buffered generally at the pH value of the maximum absorbance of the aluminum complex. Another is that the reagents are not always specific for aluminum. They may react with other cations as well. Iron is

Table 1
SPECTRAL LINES FOR ALUMINUM

Wavelength (nm)	AA sensitivity (N_2O — C_2H_2 flame) mg/l for 1% abs
226.91	4.0
236.71	3.2
237.31	2.0
256.80	6.4
257.51	4.4
308.21	1.6
309.27	1.0
309.28	
394.49	2.0
396.15	1.2

Table 2
COMMON COMPLEXING AGENTS FOR ALUMINUM

Complexing agent	Wavelength (nm)
Aluminon	535
Eriochrome cyanine R	530—535
Xylenol orange	555
Methylthymol blue	585—590
Haematoxylin	580—620
Ferron	370
8-Hydroxyquinoline	395
Pyrocatechol violet	580—620
Bromopyrogallol red/*n*-tetradecyltrimethyl ammonium bromide	623

probably the most common seriously interfering cation in environmental samples, but other cations may also interfere. Cation interferences are reduced generally by adding another reagent to complex the interfering ion or by extracting interfering ions into an organic phase using a complexing agent.

Anions that complex aluminum may interfere with formation of the analyte complex. Fluoride, PO_4, and organic anions are the three most important interfering anions. These interferences can sometimes be minimized by adjusting the pH value of the reacting solution. Organic compounds may be removed by oxidation before reaction. The fact that naturally occurring anions interfere with the reaction with added complexing agents is the basis for some of the aluminum speciation schemes to be discussed later. Aluminum can also form hydroxy polymers, which react only slowly with organic complexing agents.

2. Fluorescence

The fluorescence of many aluminum-organic complexes forms the basis of several very sensitive analytical techniques. Tikhonov[7] is again a comprehensive source of information about the various organic reagents used to form fluorescing complexes with aluminum. Since these methods are also based on the reaction of aluminum with an organic reagent, the interferences discussed above are also common to these methods. An additional interference for fluorescence analysis is the fluorescence of other components (e.g., organic matter) in the analytical solution.

D. Other Methods

1. Neutron Activation

Neutron activation analysis is a very sensitive, multielement technique that can be used for analyzing untreated natural-water samples. This technique is based on quantitation of the gamma rays emitted by nuclei that have been irradiated with neutrons. A good general discussion of this technique is given by Helmke.[8] Its advantages are sensitivity (10^{-6} to 10^{-9} g), accuracy and precision, lack of chemical interferences, and lack of the need for sample pretreatment. Its major disadvantage is that its use is limited to those investigators who have access to a nuclear reactor. Also, the technique requires attention to detail that can be given only by a well-trained analyst.[8] Radiation safety procedures must be followed because the radioactivity of samples remains high for several hours after irradiation. The aluminum isotope quanititated, ^{28}Al, has a half-life of 2.3 min. Typical analytical conditions for aluminum include a 10-min irradiation, 3-min decay, and 10-min counting. This necessitates custom handling of samples in a nuclear facility.

2. Chromatography

Both high-pressure liquid chromatography (HPLC) and gas chromatography (GC) can be used to separate aluminum from other metals for quantitative determination. In HPLC, chelation complexes of aluminum and other metals can be quantified using a UV-visible detector after separation on a column. For GC analysis of metals, a thermally stable volatile complex must be formed before injection of the metal on a GC column. Complexes of b-diketones are suitable for GC analysis of many metals.[9]

3. Fluoride-Selective Electrode

In potentiometric titrations, the equivalence point can be determined by monitoring the potential that develops between a reference electrode and an indicator electrode that is sensitive to one of the species in the titration.[10] Aluminum can be titrated with fluoride ions and the end point monitored with a fluoride-selective electrode.

III. DETERMINATION OF TOTAL ALUMINUM

A. Sample Handling and Storage

Sample handling and storage are vitally important components of any analytical study. The most careful analyst with the best techniques cannot correct for artifacts introduced during these steps. Unfortunately, preventing artifacts at this stage is not simple and involves consideration of many factors. At the low levels of aluminum characteristic of concentrations in most natural waters, sample contamination is a major analytical problem. Careful acid washing of equipment is a necessity, but may not be sufficient to prevent contamination.[11] Some investigators use Teflon® labware;[12] others advocate EDTA washes (6 g/l) of plastic labware.[13]

Natural-water samples generally contain suspended solids, dissolved and colloidal organic carbon, microbes, and elevated levels of dissolved inorganic carbon compared to water in equilibrium with atmospheric CO_2. Changes in pH value, because of microbiological activity or loss of dissolved CO_2 by degassing, can cause very significant changes in aluminum chemistry. Calculations of the effect of CO_2 degassing on the pH values of natural waters demonstrate that, for samples with pH values greater than about 4.9, degassing can significantly raise the pH value.[14] With increasing pH values, aluminum adsorption by suspended solids and container walls increases, and precipitation of $Al(OH)_3$ can occur.

The problem of pH value change because of CO_2 loss during field sampling of soil solutions can be difficult. The low pressure in a tension lysimeter accentuates the problem of degassing. This problem has been discussed in detail by Suarez.[15] Degassing can also be a problem in zero-tension lysimeters if samples are not processed immediately after collection.

Depending on the purpose of the study, the investigator may or may not want to include colloidal particulate aluminum in an analysis of total aluminum in a natural-water sample. Often the desire is to include only aluminum in "true" solution, i.e., aluminum in compounds of relative molecular mass <1000. These complexes of aluminum would generally have a particle diameter less than 1 nm. The distinction between solution and colloidal species is not absolute. Colloidal and polymeric aluminum are generally included in fractions with particle diameters in the 1- to 100-nm range, which would include aluminum associated with fulvic acid, humic acid, and some clay minerals.[16]

A very common sample-handling technique is to first filter water samples through filters with average pores sizes of 0.1 to 0.45 μm. Therefore, the operational definition of "total aluminum" becomes that aluminum which has passed through the filter and is measured by a particular technique. Some fine colloidal material will pass through these filters. Kennedy et al.[17] found that the use of 0.1- instead of 0.45-μm filters reduces significantly the passage of particulates. They suggested titanium concentrations of >1 μg/l measured by emission spectrometry as an indicator of significant amounts of colloidal clays passing the filter.

Some clay minerals, however, contain little or no titanium.[18] Thus, contamination because of aluminosilicate clays could occur even with very low levels of titanium.

Care must be taken in filtering to avoid contaminating the samples with inorganic and organic materials from the filters, and to avoid sorption of aluminum on the filter material. Jardine et al.[19] found polycarbonate filters to release the lowest amounts of inorganic contaminants of the filters they tested and to have a relatively small sorbing capacity for aluminum. Contamination can be removed by cleaning the filter membranes with dilute acid or by washing carefully with distilled water. Jay[20] found that distilled-water washing in the standard filter holders is inefficient because almost half of the filter area is not directly exposed to leaching. Jay recommended the use of a specially designed filter-washing bath with a wash volume of 4 l for each 47-mm filter disk. This is a very time-consuming procedure. Washing can be avoided by using polycarbonate filters.

Sorption losses may be minimized if the sorption capacity of the filter is satisfied by filtering a large amount of sample and discarding the initial filtrate.[16] However, the amount of colloidal material retained by the filters can change with the amount of solution filtered because the filters are subject to clogging, which changes the pore sizes and filtration rates.[16,17] Therefore, the amounts used for prefiltration rinses and the amount of sample filtered should be kept constant in order to collect the same fraction of aluminum from sample to sample.

In cases where sample volumes are low, as is common for soil solutions or lysimeter samples, high-speed centrifugation may be a better way to exclude large particles. Centrifugation is subject to fewer possible problems than filtration, but it is not a technique that can be easily used in the field. Salbu et al.[21] compared several different means of size fractionation and recommended the analysis of unfractionated samples as well as fractionated samples as a check for possible contamination.

Depending on the analytical technique used, colloidal aluminum can be a problem. Neutron activation quantitatively determines all forms of aluminum. Plasma-emission analysis includes much of the colloidal aluminum. Graphite furnace and flame AA generally determine less of the colloidal aluminum than does ICP. Spectrophotometric techniques with detection by absorbance or fluorescence determine only a portion of the colloidal aluminum. Without acid pretreatment, these techniques generally determine little, if any, of the mineral colloidal aluminum and a portion of the aluminum bound to humic colloids. This point will be discussed in more detail in the following sections.

The determination of total aluminum by any method, with the exception of neutron activation, requires destruction of organic complexes and colloidal mineral aluminum. This can be accomplished easily by heating with an oxidizing acid. For example, James et al.[22] added $5M$ HNO_3 and heated the samples at 130 to 150°C to dryness, followed by heating in a furnace at 450°C for 3 h. The ash was dissolved in 10 ml of 0.1 M HNO_3 plus 1 ml of $NH_2OH \cdot HCl$ to reduce the oxides of manganese and iron.

Some authors prefer to determine acid-reactive aluminum instead of total aluminum. Driscoll[23] recommends adjusting the sample pH value to 1.0 with HCl for 1 h followed by AA analysis after extracting the Al-8-hydroxyquinoline complex into methyl isobutyl ketone (MIBK). The details of the analytical procedure will be discussed in the next section. Driscoll's method determines most of the strongly bound organic aluminum as well as the more reactive polymeric and mineral aluminum. Determination of acid-reactive aluminum depends on the time of contact with the acid, especially if aluminum is determined using a method that involves formation of aluminum complexes. Kennedy et al.[17] used an analytical method similar to that of Driscoll,[23] and found that aluminum analysis of 0.45-μm filtered water samples acidified with 0.04 M HNO_3 and stored for 19 d resulted in a three-fold increase in aluminum, compared to samples from which the aluminum was extracted in the field. The difference, however, was small for 0.1-μm filtered water. The difference caused by acidification was relatively high because the samples were high in sediment and very low in dissolved aluminum. The sample extracted in the field was determined to contain only 13 μ*M* of acid-reactive aluminum.

For the determination of total aluminum or acid-reactive aluminum, the acidification treatment should be done as soon as possible after sampling. This avoids losses because of adsorption on the container or shifts in pH value after degassing.

Some investigators have tried to use freezing as a method of preservation of samples for later analysis. Freezing, however, results in incomplete recoveries after thawing.[17,24] Sample storage at temperatures near freezing may be useful to help preserve samples when acidification is not desired. Røgeberg and Hendrickson,[25] however, found that for samples with pH > 5.0, storage at 4°C for 3 d may result in significant losses. For samples of lower pH values, storage did not affect the results.

B. Flame Atomic Absorption Spectrometry

Flame AA is not very sensitive for aluminum. The sensitivity (the concentration necessary to give a change in absorbance of 0.0044 units) of the most sensitive line, 309.3 nm, is 1 mg/l.[2,3] Although a higher sensitivity is sometimes reported, in actual practice flame AA methods are rarely sensitive enough to directly determine aluminum in natural waters. Several methods involving determination of aluminum after extraction of the 8-hydroxyquinoline complex into a nonaqueous solvent have been developed. Hsu and Pipes[26] found that sensitivity could be increased by a factor of greater than 100 by extracting the Al-8-hydroxyquinoline complex into benzene using 1000 ml of sample and 5 ml of benzene. Fishman[27] used MIBK as the extractant and achieved comparable sensitivity. Barnes[28] employed a method similar to that of Fishman to determine the readily reactive aluminum. She suggested that the optimum range for this method is 2 to 50 μm/l. A similar method is recommended in ''Standard Methods for the Examination of Water and Wastewater'' for samples low in aluminum.[29]

Barnes[28] adjusted the pH value of each sample to 8.3 after the addition of the 8-hydroxyquinoline, but before extraction of the aluminum complex. At this pH value, interference from fluoride and PO_4 is minimized. The use of AA to determine aluminum in the extract instead of spectrophotometry to quantitate the aluminum complex eliminates the possibility of interference from copper. This will be discussed in more detail in Section III.D. For poorly buffered waters, an acetate buffer can be used to control the pH value.[30] Graphite furnace AA as well as flame AA can be used to determine the aluminum in the MIBK phase.[30] The use of the graphite furnace technique increases the sensitivity of the method and allows for the use of smaller samples volumes.

C. Flameless Atomic Absorption Spectrometry

This technique is very sensitive for aluminum; Slavin[31] suggests a detection limit of around 0.1 μg/l. Craney et al.[12] achieved a sensitivity close to this level with $Al(NO_3)_3$ solutions, although McArthur[32] reported a sensitivity ot 50 μg/l for aluminum in seawater.

The literature on aluminum determination by flameless AA is somewhat contradictory. Many inorganic matrix components have been investigated, but not all investigators have found interferences from the same components! These contradictions are probably related to the different experimental conditions employed. For instance, some investigators used pyrolytically coated tubes to hold the sample within the furnace, some used uncoated tubes, and use of a platform or matrix modifiers in the tube is sporadic. It has been shown that the quality of graphite used in the sample tubes influences the degree of interference found. Generally, newer processes producing more impermeable graphite reduce potential interferences.[31]

Matrix modifiers are added to reduce matrix interferences. They work by increasing the volatilization temperature for the element of interest and, thus, allowing a high temperature during the charring phase of furnace heating, which removes more of the unknown matrix components. Another method to reduce interference involves placing a small platform, the L'vov platform, inside the tube. The sample is injected onto the platform, which is effective because its temperature lags behind that of the graphite-tube walls since it is heated by radiation from the walls. Therefore, the volatilization of the sample is slightly delayed and occurs after the gas environment is stabilized at its final temperature. Ideally, under these conditions the integrated absorbance signal is proportional to the quantity of aluminum in the sample, and matrix effects are eliminated. To achieve this effect a fast heating rate of about 1400°C/s is necessary.[31]

One point of agreement in the literature is that perchloric acid and some other chlorine-containing matrices can cause severe interferences because of chlorine binding of aluminum in the vapor phase. This interference is much reduced by the use of a pyrolytically coated graphite tube and a L'vov platform.[33] The pyrolytic coating on the graphite tube prevents entrainment of the chlorine in the otherwise porous graphite. Most investigators recommend the use of a pyrolytically coated tube for aluminum analyses. Because the coating reduces the porosity of the graphite, more symmetrical absorption peaks are generally obtained from coated tubes. The quality of the coating is important, and poorly coated tubes give poor results.[31]

One report recommends the use of uncoated tubes. While determining aluminum in soil leachates, Ross et al.[34] found better precision and slightly higher readings with uncoated tubes than with coated tubes. Interferences were not investigated in this study, and the mechanisms causing the differences in tube performance are not clear.

Tubes which have been used in many firings begin to produce irregular absorption peaks. Changes in analytical sensitivity also occur during this aging process.[13] These changes may be especially pronounced in the first few firings of a new tube. Investigators must be aware of the need to monitor changes in sensitivity during the life of the tube.

Several publications summarize many possible matrix components and the conditions under which they may cause interferences.[13,35,36] The work of Manning et al.,[13] in particular, provides a relatively recent and comprehensive look at aluminum analysis by furnace methods. A few important points are emphasized below. Chloride and fluoride can cause interferences by volatilizing aluminum during the charring phase of heating if the char temperature is too high. Certain chloride salts will not cause this problem because they are removed from the matrix at relatively low temperatures. For example, NaCl is volatilized at 1000°C, below the volatilization temperature for $AlCl_3$, and generally does not interfere with aluminum determination.

Slavin et al.[37] recommend the use of $Mg(NO_3)_2$ as a matrix modifier and an L'vov platform for aluminum analysis. The addition of 0.16% $Mg(NO_3)_2$ causes the aluminum to become embedded in a matrix of magnesium oxide and delays vaporization until the magnesium oxide vaporizes. This delayed vaporization allows the furnace to reach a stable temperature before vaporization from the platform begins. The magnesium oxide also allows a higher charring temperature (1700°C). Without the matrix modifier, loss of aluminum occurs if the sample is charred above 1500°C. The higher charring temperature removes more matrix components and minimizes the difference between charring and atomization temperatures. Manning et al.[13] suggested that charring and atomization should occur no more than 1000°C apart when using the platform.

Craney et al.[12] added 2 and 7 μM H_3PO_4 as a matrix modifier. This improved sensitivity and decreased the degradation in signal with the age of the tube. Craney et al.[12] suggested that H_3PO_4 is superior to $Mg(NO_3)_2$, because the problem of the introduction of contaminant aluminum is less with low concentrations of H_3PO_4 than with the relatively high concentrations of $Mg(NO_3)_2$ used by Slavin et al.[37]

Slavin[31] advocates a system he calls stabilized temperature platform furnace (STPF) for low-level environmental analyses. This system includes the use of a matrix modifier and platform and consists of other components, as well. Analysts should consult this article for general information and recommendations.

An alternative method to obtain results similar to those obtained using a matrix modifier is the use of thorium-treated platforms.[38] Thorium-treated platforms give a narrowed, symmetrical signal for aluminum, which allows for more precision in integrating the signal under the peak.

One investigator has observed the co-volatilization of aluminum and organic material at low temperature (around 500°C), and has investigated the use of a copper compound to suppress this effect (E. Nater, personal communication). The loss of organic aluminum may lead to low recoveries from natural-water samples. This needs further investigation.

One question that arises with the use of furnace methods is how completely they atomize particulate aluminum. Carrondo et al.[39] compared samples of homogenized, dilute sewage sludge injected directly into a furnace with samples that were digested in $HClO_4$–HF–HNO_3 and analyzed in a flame. They found very similar recoveries of aluminum. Samples digested in acid mixtures without HF gave low aluminum recoveries in the flame. Using the graphite furnace and no sample treatment except homogenization, Carrondo et al.[40] found 95 to 100% recovery of aluminum from a zeolite type A with a silicon-to-aluminum ratio of about 1:1. Playle et al.[41] obtained an 85% recovery of aluminum from a very dilute kaolinite suspension using a graphite furnace. However, Salbu et al.[42] found that neutron activation analysis recovered five times as much aluminum from a sample of unfiltered ground water as did a graphite-furnace analysis. After the sample was filtered through a 0.45-μm filter, both techniques found the same (much lower) aluminum level. The lack of recovery in the unfiltered water by the graphite-furnace method may be because of different operating conditions used by Salbu et al.[42] compared to Carrondo et al.[40] and Playle et al.[41] These studies indicate that a graphite furnace can be effective at decomposing particulate inorganic and organic aluminum, while a flame is less effective. However, the furnace does not entirely decompose clays in all samples under all operating conditions.

D. Atomic Emission Spectrometry

The detection limits (the concentration that will produce a signal-to-background intensity ratio equal to twice the standard deviation of the background signal) for aluminum determined by ICP-AES are reported to be around 10 μg/l for 396.15 and 308.21 nm.[5,43] This sensitivity makes the technique very appropriate for most natural-water samples. Elements reported to cause spectral interferences are silver, boron, manganese, and calcium. Of these, manganese and calcium are the most likely to be analytically significant in natural waters. Roura et al.[43] reported that 10 μg/l manganese produces an aluminum signal equivalent to 12 μg/l at 308.21 nm. Nearby high-intensity calcium lines cause an interference on the 394.4 nm aluminum line,[44] and calcium interference also affects the 396.15 line.[43] These interferences generally are corrected simply by assuming a linear relationship between the amount of the interfering ion and the interfering signal produced. The problem of defining background emission in aluminum analyses is one that is best addressed using a scanning monochromator system which can scan across the wavelengths of interest.[45] See Botto[6] for a good discussion of interferences and interference correction procedures.

Roura et al.,[43] working in the range 200 to 800 μg/l, found essentially 100% recoveries for known additions of aluminum to river-water samples. In their analysis of natural-water samples, Janssens et al.[5] measured aluminum in the range 70 to 160 μg/l and found excellent agreement between their ICP and graphite furnace atomic absorption results. In this work, standard deviations were higher for the aluminum analyses than for analyses of several other elements. The authors suggested that this result was because of an insufficiently defined background emission. They emphasize that accurate background correction is essential for aluminum analysis.

Plasma atomic emission techniques are being actively investigated. One potentially important development is the investigation of a reportedly very sensitive aluminum line (167.1 nm) in the vaccum UV.[46] This wavelength can be detected only if the monochromator is purged with dry nitrogen to remove water vapor in the spectrometer.

Although temperatures are very high, aluminum in a plasma is probably not quantitatively released from clay particles. In one study, neutron-activation analysis gave aluminum values that were more than ten times greater than those of ICP for unfiltered ground water samples.[42] Filtering the samples greatly reduced the aluminum detected by neutron activation, indicating the removal of aluminum bound in clay particles.

E. UV-Visible Absorbance Spectrophotometry

Absorbance spectrophotometric analysis of aluminum after reaction with an organic complexing agent can be quite simple and inexpensive, yet very sensitive. Many of these methods discriminate against less reactive forms of aluminum, and can be used to aid in determining chemical species of aluminum. This will be discussed in more detail in Section IV. Of the nine complexing agents listed in Table 2, methylthymol blue, haematoxylin, and xylenol orange are not used commonly to determine aluminum in natural waters. Methylthymol blue has been used to determine aluminum in salt extracts from acid soils,[47] but the method is not sensitive enough for most natural waters. Haematoxylin is much more sensitive and exhibits low interference by iron, manganese, copper, magnesium, PO_4, and fluoride, but formation of the complex takes 60 min.[48] Xylenol orange is moderately sensitive with a molar absorptivity of 1.1×10^4 l/mol/cm.[49] Interference by fluoride and PO_4 are low, but the method requires a heating step to hasten the formation of the complex. This disadvantage can be overcome by automation.[50]

Aluminon was very commonly used for the determination of aluminum prior to the early 1970s. The method is moderately sensitive, since the molar absorptivity of the complex is about 2×10^4 l/mol/cm.[51] Most aluminon methods, however, require a heating step.[52] Also, the absorbance is not always linear with increasing concentration, and sensitivity varies with each batch of reagent.[51] Recently, Bertsch et al.[53] have developed an automated method. This method, however, still requires an initial 30-min heating. More recently, Blamey et al.[54] have used the aluminon method without the heating step to determine monomeric aluminum in hydroponic solutions. Since the solutions contained hydroxy-aluminum polymers, the aluminon method without heating apparently determined monomers, but was not effective in degrading the polymers.

Eriochrome cyanine R is reported to be the most sensitive reagent for aluminum, with a molar absorptivity of 6.75×10^4 l/mol/cm.[7] The sensitivity of this method can even be increased by adding a cationic surfactant to form a ternary complex of aluminum-eriochrome-surfactant.[55] The eriochrome cyanine R method is recommended in "Standard Methods for the Examination of Water and Wastewater" as an alternative to flame AA analysis.[29] However, because of the problem of interferences of PO_4 and fluoride, analysis by AA is suggested as preferable. Also, the eriochrome cyanine R complex with aluminum is stable only for about 10 min.[51] An automated method using eriochrome cyanine R and cetyltrimethylammonium bromide to form a ternary complex with aluminum has been reported by Røyset and Sullivan.[56] Fluoride and PO_4 interference were minimized by raising the pH value of the buffer.

Formation of the ternary complex of bromogallol red and *n*-tetradecyl-trimethyl ammonium bromide provides another very sensitive method for determination of aluminum. The molar absorptivity of the complex is $5.05 \times 10_4$ l/mol/cm, and the absorbance is linear in the range of 0.1 to 0.4 mg/l of aluminum.[57] The method is sensitive to Fe(III) interference, but this can be avoided by reducing the Fe(III) with hydroxylamine hydrochloride and complexing the Fe(II) with *o*-phenanthroline.[57] The method was evaluated for analysis of aluminum in river water and shown to be not sensitive to PO_4 levels found in river water (up to 2 μM). Sensitivity to fluoride was not evaluated. This method may be widely applicable, but it has been evaluated by only one group of investigators.

Ferron, the 7-iodo-5-sulphonic acid derivative of 8-hydroxyquinoline, has been used in many laboratory studies of aluminum-polymer formation[52] and in a few field studies.[58-60] The ferron complex is more hydrophyllic than the 8-hydroxyquinoline complex and is not readily extracted by organic solvents. The sensitivity is approximately similar to aluminon with a reported "limit of usefulness" of 0.05 mg/l. The response is linear up to 1.5 mg/l.[41] Iron interferes, but the interference can be eliminated by adding *o*-phenanthroline and hydroxylamine hydrochloride.[52] Small interferences were reported for manganese and fluoride.[41] The complex forms slowly with hydroxy-aluminum polymers, but rapidly with monomeric aluminum.[61,62]

Bersillon et al.[61] have suggested that the background absorbance of the reagent can be reduced and the complex made more stable if the hydroxylamine hydrochloride and acetate buffer (pH 5.2) containing ferron are mixed at least 5 d, but no more than a month, before use. Jardine and Zelazny[63] investigated the reaction of buffered hydroxylamine hydrochloride with ferron and showed that hydroxylamine hydrochloride reacts with ferron to produce low background absorbance (at 370 nm) after 5 d. They suggest that *o*-phenanthroline, hydroxylamine hydrochloride, and ferron all be premixed at least 5 d, but no more than 25 d, before use. An alternative method for reducing the absorbance of the blank is by the addition of the surfactant, cetyltrimethylammonium chloride.[64]

The pyrocatechol violet method of Dougan and Wilson[51] is nearly as sensitive as eriochrome cyamine R (molar absorptivity = 6.3×10^4 l/mol/cm), but is much less susceptible to interferences by fluoride and PO_4. Iron interferes, but this interference can be eliminated using *o*-phenanthroline and hydroxylamine hydrochloride. Dougan and Wilson[51] made an extensive examination of possible interference in natural and potable water. Their results show that, for most natural waters, interference from dissolved inorganic ions should not be a problem. Orthophosphate does not interfere even at 8.2 mg/l phosphorus. Fluoride does not interfere at 0.5 mg/l and only slightly at l mg/l. They suggested that fluoride interference can be reduced somewhat by the addition of boric acid. Small changes in concentration of pyrocatechol or *o*-phenanthroline do not change absorbance, but careful pH value control is necessary. Dougan and Wilson[51] recommended a final pH value of 6.1 to 6.2 maintained by a hexamine buffer. The color formation is essentially complete in 10 min, but the intensity increases very slowly with time. Absorbance should be measured between 10 and 20 min after the reagents are added.

The pyrocatechol method is precise and accurate even for samples of low concentration. Dougan and Wilson[51] report standard deviation of 4 μg/l for a sample containing 50 μg/l and a recovery of 99.7% for 100 μg/l added to a river-water sample stabilized with 0.1 M HCl. An interlaboratory comparison by Dougan and Wilson[51] has also demonstrated that the method is precise. They compared analyses by six laboratories of tap water spiked with 125 μg/l aluminum and stabilized with 0.1 M HCl. The expected value was 128 μg/l. The reported values ranged from 116 to 133 μg/l with an overall mean of 128 μg/l.

Dale and Hendrickson[65] also conducted an interlaboratory comparison involving nine laboratories using river water spiked with 200 μg/l aluminum, with and without 200 μg/l iron. The solutions were stabilized using 0.04 M H_2SO_4. The coefficient of variation for the unspiked river sample was 40% (mean aluminum = 28 μg/l). For the spiked samples, the coefficients of variation were 13% or less. Without the addition of iron, recovery of aluminum was excellent. With iron, recovery was about 15% low. Dale and Hendrickson[65] recommended the pyrocatechol violet over the eriochrome cyanine R method.

Because the complex forms rapidly at room temperature and no solvent extraction is required, the method of Dougan and Wilson[51] is easily automated.[25,56,66] Grigg and Morrison[66] reported increased precision with automation. Analysis of 0.01 *M* $CaCl_2$ extracts of soils yielded a coefficient of variation of 6.8% for the method as described by Dougan and Wilson[51] and 2.9% for the automated method. The reported detection limit for the automated method was 10 μg/l.

Greater sensitivity can be obtained by extraction of the ternary complex formed by pyrocatechol violet, aluminum, and zephiramine (tetradecyl dimethyl benzyl ammonium chloride) into chloroform or 1,2 dichloroethane.[67] The method is somewhat laborious, involving not only a solvent extraction, but also a washing to remove excess reagent from the solvent phase. The complex, however, is very stable in the organic phase. Analysis of river water with concentrations of 78 to 230 μg/l gave coefficients of variation of 4% or better. The detection limit was 3 μg/l.

The most sensitive of the commonly used spectrophotometric methods for determination of aluminum are those based on the extraction of the 8-hydroxyquinoline (oxine) complex into an organic solvent phase. At one time, the solvent of choice was chloroform.[68-70] This solvent is highly volatile and quite toxic. Also, it is more dense than water, so separations cannot be done without a separatory funnel. More recently, the solvents used have been MIBK,[28,56] toluene,[71] and butyl acetate,[22,72] all of which are less dense than water and less toxic than chloroform. The extraction procedure adds an extra step as compared to the previously discussed methods, but complex formation is rapid (15 s or less)[28,71,72] and extraction generally takes only 10 s to 5 min, depending on the relative volumes of the aqueous and nonaqueous phases.[71] Automation, however, is more difficult than for methods without solvent extraction.

Butyl acetate and MIBK are much more soluble in aqueous solutions than are chloroform or toluene.[71,72] Thus, for these two solvents the aqueous volume of the samples and standards must be controlled. Salt in the samples can also affect the solubility of these organic solvents.[28]

Complete extraction of the aluminum complex is possible in the pH value ranges of 4.5 to 6.5 and 8.0 to 11.5.[73] Commonly used methods use either pH 5.0[71,72] or pH 8.3.[28,56,71] Manganese interference is possible at pH 8.3, but not at pH 5.0. Copper can give a positive interference at both pH values and could be a problem in natural waters where copper is high and aluminum is very low.[71] Phosphate does not interfere at either pH value.[71,72] Fluoride at greater than 0.2 mg/l interferes at pH 5.0 by preventing extracton of aluminum. Increasing the pH value before extraction to 8.3 eliminates interference of fluoride up to about 1 mg/l. Humic and fulvic acids also interfere more strongly at pH 5.0.[56] The aluminum complex is highly stable in the organic phase. Polymeric aluminum and precipitated aluminum are not extracted at either pH 5.0[22,71,72] or pH 8.3.[28,71] Bloom et al.[72] reported that there was no change in absorbance after storage of the extract for 24 h at room temperature.

Both Bloom et al.[72] and May et al.[71] have published methods with extraction at pH 5.0. The concentration of a sodium acetate buffer used was sufficient to allow for the analysis of most natural-water samples without any other adjustment of the pH value. The method of Bloom et al.[72] is appropriate for samples low in fluoride and copper relative to aluminum.[72] To eliminate interference from transition metals and fluoride, May et al.[71] recommend pre-extraction of samples with diethylammonium diethyldithiocarbamate and chloroform at acid pH values followed by extraction of the Al-8-hydroxyquinolate complex at pH 8.3.

Bloom et al.[72] used a ratio of 5 ml of butyl acetate to 25 ml of sample. After extraction in borosilicate glass culture tubes the butyl acetate phase is transferred by pipet. Spectroscopic-grade butyl acetate was used, but the other reagents were reagent grade. With this method, they obtained a detection limit of about 10 μg/l (unpublished data) with a coefficient of variation of 2.5% for a 400-μg/l sample.

By using a wider solution-to-solvent (toluene) ratio and very careful technique, May et al.[71] were able to obtain a detection limit of 0.2 μg/l. They used highly purified reagents, performed the extractions in Teflon® bottles, and extracted 100 ml of sample with 2.0 ml of toluene. This allowed for precise measurements even at very low absorbance (detection of 0.0020 absorbance units above a blank of 0.0040 absorbance units). They demonstrated excellent recoveries of 5.4 and 13.5 μg/l added to surface-water samples and a sample of fluoridated municipal water.[71]

Narayanan and Pantony[74] recommend the removal of transition metals by pre-extraction with 2-isopropyl-8-quinolinol into chloroform. They extracted the Al-8-hydroxyquinolate complex at pH 5.0 with chloroform and reported results superior to those obtained by graphite furnace AA.

The MIBK extraction method that Barnes[28] used to determine atomic aluminum by AA can also be used with spectrophotometric determination of the complex.[56] No published information is avalable concerning sensitivity or precision, but the method should give results in the range of those reported for other solvents. The disadvantage of using spectrophotometric detection of the complex instead of AA analysis of atomic aluminum is that, with samples very low in aluminum positive interference from transition metals, notably copper, is possible.

F. Effect of Natural Organic Ligands on UV-Visible Absorbance Methods

Natural ligands compete with added complexing agents for the binding of aluminum, and can result in negative interference in the spectrophotometric analysis of the complex.[56] This interference has been utilized as a way of discriminating between inorganic monomeric aluminum and organically complexed aluminum (see Section IV.B).[22,75] Because organic ligands in natural-water samples occur in a complex mixture that is difficult to characterize (see Chapter 5), it is more difficult to characterize interferences due to organic ligands than those due to inorganic ligands like fluoride and PO_4. Røyset and Sullivan[56] compared the effects of the addition of 10 to 200 mg/l of humic and fulvic acids on the analysis of 1 mg/l aluminum using three complexation techniques: 8-hydroxyquinoline with extraction at pH 8.3 into MIBK (similar to the AA method of Barnes[28]), eriochrome cyanine R/CTA, and pyrocatechol violet. Analysis of acid-reactive aluminum (acidification to pH 1 for 1 h) showed no interference for the 8-hydroxyquinoline method, but interference for the eriochrome cyanine R method at $C > 50$ mg/l and for pyrocatechol violet at $C > 10$ mg/l. Most natural waters, however, do not have dissolved carbon values greater than 10 mg/l. Determination of "total monomeric aluminum" without prior acidification showed reduced recovery of aluminum with all levels of added carbon. At 10 mg C/l, all the methods recovered 90 to 95% of the aluminum. With 200 mg C/l, recovery was 50 to 70%. The 8-hydroxyquinoline and eriochrome cyanine R/CTA methods had lower recoveries with complex formation at lower pH values compared to complex formation at higher pH values.

Analysis of aqueous soil extracts with aluminum contents ranging from 500 to 800 μg/l showed that with and without acidification to pH 1.0, the highest values were obtained using the 8-hydroxyquinoline method. The eriochrome cyanine R/CTA and pyrocathechol violet results were similar to each other. The 8-hydroxyquinoline results for acid-reactive aluminum compares most closely with the analysis of total aluminum after oxidation of the organic matter using UV light and H_2O_2.[56]

Hoyt and Turner,[76] Bloom et al.,[72] and James et al.[22] showed that the reaction of 8-hydroxyquinoline at pH 5.0 with aluminum organic-matter complexes is time-dependent. Extraction within 15 s gives much lower results than extraction after 30 min if organic ligands were present. James et al.[22] showed that in solutions whose pH value was in the range of 4.5 to 5,.6 and which contained 50 μM aluminum and citrate, less than 10 μM aluminum was determined with 8-hydroxyquinoline at pH 5.0 using a 15-s contract time before extraction.

The interference because of organic matter can be decreased greatly by acidification to pH 1.0 as discussed previously or eliminated by oxidation. Photo-oxidation with or without the addition of small quantities of oxidants like hydrogen peroxide or potassium persulfate is an easy way to remove dissolved organic compounds.[22,77,78] Cronan[78] suggests that a small quantity of 30% H_2O_2 (20 μl in 7 ml of sample) should be added before exposure to UV light. Hydes and Liss[24] in a study of aluminum analysis by the lumagollion fluorescence method, found that the hydrogen peroxide was not necessary.

Campbell et al.[79] found losses of about 20% of the aluminum content (graphite furnace AA) during the photo-oxidation of stream water. They investigated whether sample heating during irradiation was the cause for the loss and found that almost as much loss occurred by heating, without irradiation. Because the pH value changed little during irradiation, the authors ruled out precipitation as a cause of the loss of $Al(OH)_3$. They did not, however, consider the marked decrease in $Al(OH)_3$ solubility with the increase in temperature. The gibbsite dissolution reaction in acid has a ΔH_r° value of −88.1 kJ/mol, which results in a large decrease in solubility with increasing temperature.[80] Very likely, the losses experienced by Campbell et al.[79] were from $Al(OH)_3$ precipitates which were adsorbed on the walls of the quartz tubes. Thus, heating should be avoided during photo-oxidation or acid should be added to prevent precipitation.

Irradiation must be done in tubes that are transparent to UV light. Quartz tubes are ideal but expensive. Our procedure for samples containing 3.6 mg C/l is to add 2 drops concentrated HCl and 6 ml 2% potassium persulfate to 10 ml of sample in a glass beaker. The beakers are covered with Saran Wrap® (which is relatively transparent to UV) and irradiated overnight with low-intensity UV from the top.

G. Fluorescence Spectrophotometry

Fluorimetric methods can be extremely sensitive for the determination of aluminum. Fluorimetry, however, requires very careful laboratory technique. Because of the sensitivity, aluminum contamination is a problem. Double-distilled water is recommended and reagent storage should not be in glass,[24] but care should be taken to avoid containers that may contain fluorophoric compounds, which may leach into solutions. Bloom et al.[72] found it necessary to redistill their reagents to keep the fluorescence of the blank acceptably low. Howard et al.[81] found a significant decrease in blank fluorescence on going from analytical reagent-grade chemicals to a higher grade.

The most common fluorimetric reagent used with aluminum is lumogallion. The standard procedure requires heating for 90 min for complete complex formation and then cooling, because the fluorescence is temperature-dependent. Hydes and Liss[24] suggest a detection limit of about 0.1 μg/l, with a coefficient of variation of 5% for samples containing 1.0 μg/l. Howard et al.[81] improved the detection limit of the procedure to 0.02 μg/l by the addition of a nonionic detergent, Triton® X-100.

Fluoride interferes at concentrations greater than 0.3 mg/l and PO_4 interferes at concentrations greater than 3.0 mg/l.[81] Iron interferes by forming a nonfluorescent complex. This reduces response to aluminum by reducing the effective lumogallion concentration and by absorbing the incident light more effectively than the aluminum complex. Howard et al.[81] used *o*-phenanthroline to reduce iron interference, but Hydes and Liss[24] found that up to 50 μg/l of Fe(III) or Fe(II) could be tolerated, and they suggested that for most natural waters iron interference should not be a problem. Gabriels et al.[82] found interference from copper at 2 mg/l. This is a much higher concentration than is found commonly in natural waters. Other common transition metals were shown to pose even less of a problem for interference.

Hydes and Liss[24] recommended calibration by the method of known addition to compensate for some of the effects of interference. They analyzed five subsamples with aluminum additions to four of the subsamples in their studies of aluminum in an estuary. For samples lower in suspended solids and lower in dissolved organic carbon, distilled-water standards may be sufficient. Fluorescence by dissolved organic matter and competitive aluminum binding by natural organic compounds in colored waters can cause interference.[24] Hydes and Liss[24] removed the organic matter by irradiation with UV light, but the recovery of a known addition of aluminum was low even after the removal of the organic matter, possibly because of resistant organic complexing species not removed by the irradiation, or because of the problem of $Al(OH)_3$ precipitation during photo-oxidation, discussed in Section III.F.

Lumogallion does not react with suspended clay minerals, but the method does seem to determine small quantities of adsorbed aluminum on kaolinite.[24] In river waters with pH values in the range of 8.1 to 8.5, graphite furnace AA determined much higher levels of aluminum even in filtered water (0.1 μm membrane). However, in river water with a pH value of 4.6, the AA and lumogallion results were identical.[24]

Bloom et al.[72] described a fluorimetric method with a similar sensitivity to the lumogallion method with the advantage of not requiring sample heating. The method is similar to the spectrophotometric analysis using 8-hydroxyquinoline and butyl acetate described in Section III.D. The detection limit is 0.3 μg/l and, except near the detection limit, the coefficient of variation was 8% or less. Interference by fluoride, Fe(II), and Fe(III) occurred at concentrations of 800, 44, and 22 μg/l (without the addition of *o*-phenanthroline). With *o*-penanthroline and hydroxylamine hydrochloride, iron did not interfere at concentrations of 8800 μg/l.

A method based on the fluorescence of the aluminum-morin complex has been described.[83] Sensitivity is similar to the two methods discussed above. The morin method is less suitable for routine analysis because it requires extraction into MIBK and two pH value-adjustment steps. Another fluorimetric method using a hydrazone, which does not require heating for complex formation or solvent extraction, has been suggested by Cano-Pavon et al.[84] The sensitivity, however, is not much better than some of the commonly used absorbance spectrophotometric methods. The data of Cano-Pavon et al.[84] suggest a detection limit of about 3 μg/l.

H. Other Methods

1. Neutron Activation

Neutron activation is a very sensitive technique for measuring total aluminum in natural waters. Salbu et al.[85] demonstrated that 0.2 μg/l of aluminum can be detected in waters and that 30 elements can be determined simultaneously. Misra et al.[86] reported the necessity for separation of chlorine and correction for silicon in neutron-activation analysis. They, however, used the NaI(Tl) detector. With the modern Ge(Li) detector and a multichannel analyzer, interferences are no longer a problem in aluminum analysis.

The strength of neutron activation is in studies involving the analysis of many elements, especially the minor elements. Salbu et al,[21] in a study of the association of 20 elements with different particle-size fractions, determined 14 of the elements using neutron activation.

2. Chromatography

HPLC can be used to separate metal chelates for the analysis of several metals at once. Aluminum can be determined simultaneously with iron and copper by elution of 8-hydroxyquinoline complexes in an acetonitrile/water solvent phase.[87] With spectrophotometric detection of the complex, the detection limit is 250 μg/l. Thus, this method is not sensitive enough for the analysis of most natural waters.

GC of volative chelates is much more sensitive and has been used for the analysis of natural waters.[88,89] The detection limit for the analysis of aluminum is <3 μg/l.[88,89] This technique requires the formation of an aluminum-trifluoroacetylacetone complex, in benzene or toluene, and subsequent injection onto a GC column.

3. Fluoride-Selective Electrode

Aluminum in natural waters can be determined by fluoride titration with a fluoride-selective electrode. Trojanowicz and Hulanicki[90] described a method with a detection limit of around

100 μg/l. They used a reductant, hydroxylamine, to prevent iron interference and found an acid digestion necessary for complete aluminum recoveries. They also described an adaptation of their method for continuous-flow analysis.

IV. SPECIATION OF AQUEOUS ALUMINUM

A. Importance of Speciation

The recent interesting the effects of acidic deposition on soils, lakes, and streams has resulted in much effort to develop methods to separate the aluminum species in natural waters. Speciation is important because not all chemical forms of aluminum are equally toxic. Driscoll et al.[91] found that organically complexed aluminum is not toxic to fish and that fluoride-complexed aluminum is more toxic than organic aluminum, but less toxic than the aquo and hydroxy forms. Pavan et al.[92] concluded that toxicity to plants, as measured by the rate of root growth, can best be predicted by the activity of aquo Al^{3+}. Cameron et al.[93] correlated toxicity with the concentration of Al^{3+}. Hue et al.,[94] however, related toxicity to the sum of Al^{3+}, hydroxy aluminum monomers, and $AlSO^{4+}$ concentration. Organically complexed aluminum is apparently much less toxic to plants than inorganic aluminum.[94-95]

Speciation is also important in determining the chemistry of aluminum in acid soils as it relates to the formation or dissolution of minerals. Comparison of ion-activity products for aluminous minerals with solubility product data can be used to evaluate whether a mineral can form.[60,96] Ion-activity product calculations require an estimate of the activity of aquo Al^{3+}.

The soluble aluminum, as defined by filtration through a 0.45- or 0.1-μm membrane filter, includes at least five forms of aluminum. The fine colloidal mineral aluminum and aluminum associated with organic macromolecules, fulvic and humic acids, have been discussed previously in Section III. Additional quantities of aluminum may be sorbed on the surface of colloidal mineral aluminum. Other forms of aluminum include low molecular weight organic complexes and monomeric inorganic complexes. A sixth type of aluminum, polymeric hydroxy aluminum, has also been postulated to exist in natural waters and has been the subject of much recent study. The calculations of Nari and Prenzel[97] suggest that aluminum-hydroxy polymers are stable only in solutions that are highly oversaturated with respect to the precipitation of gibbsite. Comparison of Al^{3+} activities, measured over a range of pH values in soil solutions[60,96] and surface waters,[70,98] with the solubility data of May et al.[71] for gibbsite, suggests that soluble aluminum-hydroxy polymers may not be stable under most conditions in soil and surface waters. Chemical speciation evidence also suggests that little polymeric or amorphous aluminum is found in lake water[30] (see Chapter 4).

If the inorganic monomeric species can be separated from the other aluminum species by chemical or physical means, further speciation is simply a matter of the calculation of the concentration of inorganic complexes. The important complexing inorganic ligands in natural waters are OH^-, fluoride, and SO_4^{2-}. Selected formation constants for complexes with these ligands are given in Chapter 2. The reported interference of PO_4 with some of the spectrophotometric analyses (see Section III.E) suggests that PO_4 ions may complex strongly with aluminum. Data are not readily available for low ionic strength, but the results compiled by Sillén and Martell[99] and those presented in Chapter 2 indicate that complexation with PO_4 should not be a problem at pH values less than about 6.5. However, at high pH values, PO_4 complexation might interfere with the determination of the inorganic species. Calculation of inorganic complexes can readily by accomplished using MINEQL,[100] GEOCHEM,[101] or one of the many other readily available computer programs for speciation.

B. Speciation Based on Rates of Reaction with Complexing Agents

Speciation based on the rate of reaction with a colored organic oigand was first used to separate polymeric from monomeric aluminum species in laboratory studies of hydrolyzed aluminum solutions. Okura et al.[102] and Turner [103] showed that the rate of reaction of 8-hydroxyquinoline (pH 5.0) with hydrolyzed aluminum solutions depends on the quantity of base added and the age of the solutions. They suggested that the quantity of aluminum determined by immediate extraction of the complex into chloroform represents monomeric aluminum. Smith[62] and Bersillon et al.[61] suggested that a 30-s reaction with ferron can also be used to determine monomeric aluminum in hydrolyzed solutions. Jardine and Zelazny[63] further refined the method of Bersillon et al.[61] by determining the rate constants for the monomeric and polymeric species, and were able to better separate polymers from monomers using rate equations.

Hoyt and Turner[76] suggested that in soil extracts the organic complexes of aluminum also react slowly with 8-hydoxyquinoline at pH 5.0. Bloom et al.[72] suggested that a 15-s contact time before extraction could be used in separating monomeric from organic aluminum. James et al.[22] studied the rate of the reaction of 8-hydroxyquinoline (pH 5.0) with solutions containing citrate or fluoride plus added base. They showed that with a 15-s contact time, all of the aquo Al^{3+} and monomeric hydroxy complexes reacted, plus a small fraction of fluoride and citrate complexes. The rate of reaction of SO_4 complexes was not evaluated, but SO_4 does not bind aluminum as strongly as fluoride or citrate. Sulfate complexes probably react rapidly with 8-hydroxyquinoline at pH 5.0.

Barnes et al.[28] suggested that a 10- to 30-s contact time be used before extraction of the 8-hydroxyquinoline complex (pH 8.3) by MIBK and analysis by AA. This procedure determines the "readily reactive aluminum". The evidence discussed in Sections III.E and III.F concerning the reactivity of 8-hydroxyquinoline with solutions concerning fluoride and organic materials suggests that at pH 8.3, the reaction of 8-hydroxyquinoline with fluoride and organic aluminum complexes is almost complete. Thus, the Barnes[28] procedure, unlike the procedure of James et al.,[22] likely determines organic aluminum as well as the fluoride complexes. LaZerte[30] compared the result of using the Barnes method with a 15-s and a 2-h reaction time. He found, for stream and lake waters, less than a 17% increase in aluminum after 2 h, and he concluded that his samples contained little polymeric or amorphous aluminum.

Hodges[75] compared the results of the 15-s 8-hydroxyquinoline method of James et al.[22] and the ferron method of Smith[62] for the measurement of aluminum in artificial solutions containing fulvic acid, SO_4^{2-}, and fluoride. The results show that the aluminum measured by the method of James et al.[22] is similar to the sum of the aquo-, hydroxy-, and SO_4-Al complexes. Apparently, little of the fulvic acid- or fluoride-complexed aluminum was extracted. The ferron-aluminum values were generally higher, suggesting that ferron competes more strongly with fluoride and fulvic acidd than does 8-hydroxyquinoline at pH 5.0.

Hodges[75] also compared the results for an aqueous soil extract. He found both the 8-hydroxyquinoline and the ferron-aluminum values were much greater than the monomeric inorganic aluminum determined by the fluoride electrode method (see Section IV.D). Again, the ferron method determined more aluminum than the 8-hydroxyquinoline method. The fluoride electrode estimates for monomeric aluminum, however, could be low because of fluoride binding by colloidal aluminosilicate minerals (see Section IV.D).

Seip et al.[59] compared the ferron method of Smith[62] with aluminum determined by the pyrocatechol violet method of Dougan and Wilson[51] using a 4-min reaction time. They called the fraction determined by these methods "total monomeric aluminum". Analysis of acidic surface waters gave similar results with the both methods. The authors, however, preferred the pyrocatechol violet method because it is less subject to intereferences.

C. Speciation Using Dialysis and Ultrafiltration Membranes

Dialysis and ultrafiltration can be used to physically separate fine colloidal mineral aluminum and aluminum bound to the macromolecular structures of humic and fulvic acids from the remainder of the soluble aluminum. Dialysis membranes commonly have pore diameters of 1 to 55 nm, whereas ultrafiltration membranes have pore diameters ranging from 1 to 15 nm.[104] Speciation using dialysis and ultrafiltration must be done very carefully to avoid the problems that can be encountered. Membranes may contain both inorganic and organic contaminants and must be washed thoroughly.[30,104] Cleaned membranes may also have a tendency to adsorb metal ions from solutions;[30,104] however, Lazerte[30] found no adsorption losses greater than 5% for the dialysis membrane he used. For solutions of high organic matter content where the predominant anions are macromolecular organic acids that do not diffuse through the membrane, the Donnan potential that develops across the membrane can reduce the quantity of dialyzed aluminum. All of these difficulties have been discussed comprehensively by Florence and Batley[104] and LaZerte.[30]

One difficulty not discussed by either Florence and Batley[104] or LaZerte[30] is the potential problem of changes in pH value during dialysis. Both degassing of CO_2 and the effect of dilution can cause the pH value to rise. If the samples are in equilibrium with solid-phase $Al(OH)_3$, as has been observed both in some streams[58] and lakes, [98] the rise in pH value could result in a loss of aluminum from solution. Many of the problems of dailysis can be avoided by using the method of *in situ* dialysis.[105] wherein a dialysis membrane bag containing distilled water is placed directly into a body of water.

Dialysis and ultrafiltration do not necessarily separate monomeric inorganic aluminum from all forms of organically bound aluminum. It is possible that aluminum is also bound to low molecular weight organic compounds. LaZerte,[30] however, found that only 6 to 12% of the organic matter in stream and lake waters passed through a dialysis membrane. Also, Hodges[75] suggested that most of the organiclly complexed aluminum in soil solutions is because of humic and fulvic acid macromolecules. LaZerte[30] compared the results of dialysis with those of the fluoride electrode technique and found that the monomeric inorganic aluminum calculated from the fluoride electrode data represents most of the dialyzable aluminum. The fluoride electrode results, however, were generally lower. This could be because of some complexation of aluminum by dialyzable organic matter or because of fluoride binding to colloidal mineral matter (see Section IV.D).

D. Speciation Using a Fluoride Ion-Selective Electrode

In most natural waters, the only metal ion that binds a significant fraction of fluoride is aluminum. If aluminum is the only component that binds fluoride, determination of the F^- activity and the total F^- concentration is sufficient for the calculation of aquo Al^{3+} activity.[23,30,75] Fluoride activity can be measured directly with an ion-selective electrode (ISE).[23,30] Total fluoride can be determined with an ISE after the addition of a buffer solution containing metal ion complexing agents.[30]

Because of the low F^- ion activity in most natural waters, a fluoride ISE cannot be calibrated directly using NaF solutions. In solutions containing aluminum, however, stable ISE readings can be obtained to 10^{-8} μM. Both LaZerte[30] and Hodges[75] have checked the electrode response in test solutions, containing known quantities of fluoride and aluminum and found the response to be essentially Nerstian to about 10^{-8} μM. LaZerte,[30] however, found that when he used extrapolated F^- activities for test solutions containing aluminum (pH <5.5), his calculated values of monomeric inorganic aluminum were systematically 7% low.

A sensitivity analysis of the calculations showed that at higher pH values, when hydroxy aluminum exceeds fluoride complexed aluminum, small errors in F^- activity can result in large errors in the calculation of monomeric inorganic aluminum.[30] LaZerte[30] estimated that, for a surface-water sample at pH 5.5, a 10% error in F^- activity can result in a 70% error in monomeric inorganic aluminum!

A potential problem not discussed by LaZerte[30] is that of fluoride binding by components other than aluminum. Fluoride ions can be adsorbed by aluminosilicate clays. Binding of

AlF^{2+} complex ions by fulvic or humic acid may also occur. Further investigation of these sources of fluoride binding is needed.

Results obtained using the fluoride electrode method have been compared to the other speciation methods. Comparisons with the results of reaction with complexing agents and the results of dialysis have been discussed in Sections IV.B and IV.C. Driscoll[23] compared fluoride electrode determinations of monomeric inorganic aluminum with this result from separation using a strong acid resin column (see Section IV.E). As was found for comparison with the other methods, the fluoride method gave lower values. This suggests that fluoride binding other than by aluminum may be a problem with the fluoride electrode method.

E. Speciation Using Morin

Determination of the fluorescence of the aluminum-morin complex after the addition of a small quantity of morin to natural waters can be used to determine Al^{3+} activity.[106,107] Browne et al.[106] determined the stoichiometry, conditional binding constants, and fluorescence efficiency of the chelate complex under conditions typical for acidic natural waters (pH <5.5). This allowed them to calculate Al^{3+} activities from fluorescence data.

The method was evaluated in pure solutions containing aluminum and SO_4 or fluoride and in natural waters.[107] Using the morin method, the calculated binding constants for SO_4^{2-} and F^- aluminum complexes were within 20% of the values reported in the literature. The morin method worked well when tried with samples of acid stream water, but the determined values for Al^{3+} with greater than those obtained with the fluoride ISE method. The reason suggested for this discrepancy is the binding of the fluoride by components other than monomeric inorganic aluminum, as suggested in Section IV.D.

When the recommended quantity of morin is added to acidic natural waters, the morin complexes with only a small fraction of the monomeric aluminum. In the samples determined by Browne et al.[107] <5% of the monomeric aluminum was complexed by morin. This means that the solution equilibrium is not greatly disturbed by the analytical procedure. Also, it is possible to back-calculate to determine the equilibrium of Al^{3+} before the addition of morin.

The morin method, like the fluoride electrode method, offers the advantage of making measurements at equilibrium and does not rely on the differences in rates of reaction as do some other methods discussed. Also, the question of possible adsorption of the complexing agent by other than monomeric aluminum seems to be not as serious as for fluoride. Further evaluation of the application of this method to natural waters is needed before it can be recommended by general use.

F. Speciation Using Ion-Exchange Resins

The reaction of natural waters with cation-exchange resins is the most common method used to separate monomeric inorganic aluminum from other forms of aluminum. Two types of methods have been used.One method utilizes a short-term batch reaction with a chelating resin, Chelex® 100,[75,79,98] whereas the other utilizes the reaction of solutions with a strong acid (sulfonate) resin in anion exchange column.[23,25,91,108] In both methods, monomeric inorganic aluminum reacts much more rapidly with the resin than the other aluminum components, and monomeric aluminum can be estimated from the difference in measured aluminum before and after reaction with the resin.

Chelex® 100, a styrene-divinylbenene copolymer with iminodiacetate functional groups, has been used for the speciation of many metal ions in seawater and fresh waters.[104] Because of the small pore size of the resin (1.5 nm), sorption by the resin is restricted to small cations. The method of Campbell et al.[79] uses 1.0g of resin to react with 400 ml of a well-stirred sample. The resin is pretreated with calcium, magnesium, and NaOH to give an equilibrium pH value of 5.0 and Ca^{2+} and Mg^{2+} concentrations similar to that in the samples of interest. This avoids the large pH value changes that occur if sodium- or hydrogen-

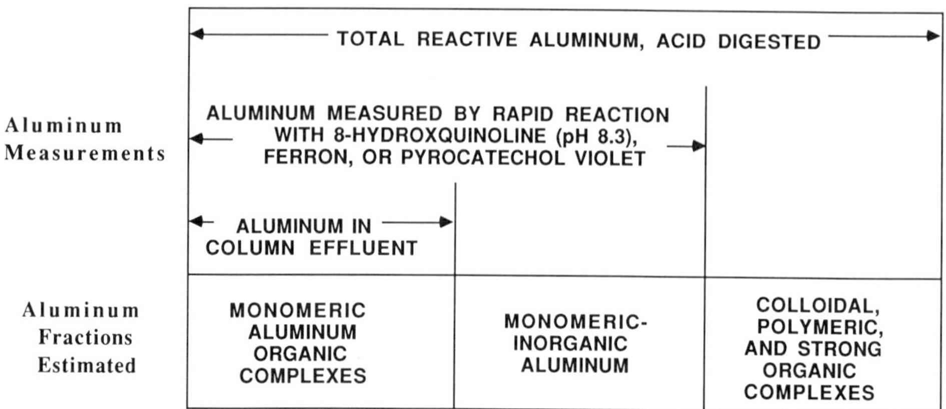

FIGURE 1. Schematic representation of aluminum fractionation of the procedure of Driscoll[107] and similar procedures using a strong acid-resin column to separate monomeric inorganic aluminum.

saturated resins are used. The resin reacts with >85% of the monomeric inorganic aluminum in 30 min and virtually all in 60 min. Aluminum hydroxy polymers in hydrolyzed aluminum solutions are adsorbed at a rate similar to monomeric aluminum.[79] Fluoride has no effect on adsorption, but fulvic and humic acids strongly inhibit adsorption.[79]

Campbell et al.[98] used the chelating resin method to speciate aluminum in 0.4 μm filtered water from 68 Canadian lakes. Aluminum was determined using graphite furnace AA and the contact time with the resin was 30 min. The results show that the nonreactive aluminum fraction greatly exceeded the concentration of resin-reactive aluminum. The activity of aquo Al^{3+} was calculated assuming that the resin-reactive fraction is equivalent to monomeric inorganic aluminum. The calculations suggest that, for samples with pH values between 5.0 and 6.8, Al^{3+} activity can be predicted by the solubility of an $Al(OH)_3$ phase with solubility similar to that of microcrystalline gibbsite. Samples with higher or lower pH values were undersaturated with respect to $Al(OH)_3$.

Cation-exchange column separation using strong acid resin is the more commonly used of the resin procedures.[23,25,59,91,108] Most investigators who use this technique follow the analytical scheme described by Driscoll[109] in 1980 (see Figure 1) with minor variations. Total reactive aluminum is defined by the aluminum determined in 0.1 or 0.4 μm filtered water after a short contact time with HCl added to produce a fixed pH value. Driscoll[23] suggested a pH value of 1.0 for 1 h. This is the most common procedure for determining total acid-reactive aluminum,[23,60,91,108] but Seip et al.[59] used both pH 1.4 for 1 h and pH 1.5 for 24 h. Analytical procedures for determining total aluminum after acid digestion include the ferron procedure of Smith,[59,60,109] the 8-hydroxyquinoline method of Barnes[23] and the pyrocatechol violet method of Dougan and Wilson.[25,59,108]

The procedure of Driscoll[23] requires that a second sample be analyzed for "monomeric aluminum" (see Figure 1). This fraction includes inorganic monomeric aluminum as well as much of the aluminum bound by organic anions. Most of this monomeric aluminum is bound to nondialyzable organic macromolecules (see Section IV.C). This aluminum is considered to be monomeric because each aluminum ion is bound to a separate site on the organic macromolecules. A better term for this fraction of aluminum might be "mononuclear aluminum" (see Chapter 2). The methods used for determining monomeric aluminum include most of the methods discussed in Section IV.B for separation of aluminum species using short-term reactions with complexing reagents. The methods include the ferron method of Smith[59,60,109] (30 s), the pyrocatechol violet method of Dougan and Wilson (4 min),[25,59,108] the 8-hydroxyquinoline/AA method at pH 8.3 of Barnes (immediate extraction),[23] and the 8-hydroxyquinoline method at pH 5.0 of Bloom et al.[96] (15 min). Because of the differences in the reaction rate of these different reagents with the various aluminum fractions in natural waters, these different procedures could be expected to yield somewhat different results. The discussion in Sections III.F and IV.B suggests that the greatest quantity of aluminum should be determined by the 8-hydroxyquinoline/AA method at pH 8.3. The smallest quantity would be found by the 8-hydroxyquinoline method at pH 5.0. The quantity that is represented by the difference between total acid-reactive aluminum and total monomeric aluminum is the acid-reactive aluminum (see Figure 1). This fraction includes strongly bound, organically complexed aluminum and mineral aluminum. Wright and Skogheim,[108] who used the pyrocatechol violet procedure (4 min), showed that after oxidation of the organic matter with potassium persulfate, the measured aluminum increased by 5 to 15%. This represents organically bound aluminum not included in the monomeric aluminum fraction.

A third portion of a sample is passed through a strong acid ion-exchange resin column to separate the monomeric inorganic aluminum from the organic aluminum. Monomeric inorganic complexes, including fluoride and SO_4^{2-} complexes, react rapidly with the resin,[23] but organically complexed aluminum reacts slowly. Driscoll[23] used columns saturated with sodium and hydrogen, such that the pH value of a sample is changed minimally when passing through the columns. Wright and Skogheim,[108] however, used a sodium-saturated column. Driscoll's procedure includes the pumping of a quantity of sample through the column before a sample is taken for aluminum analysis. Between samples, the column is washed with a NaCl solution that has an ionic strength similar to that of the samples. The column effluent is analyzed for aluminum using the same technique used to determine the monomeric aluminum. Monomeric inorganic aluminum is determined by the difference (see Figure 1).

Since the separation of species is dependent on the difference between the rate of reaction of monomeric inorganic aluminum and organic aluminum with the resin column, control of factors that affect contact time with the resin are important. These factors include column size, resin bead size, and flow rate. Unfortunately, most of the reports of column speciation do not state the details of the column methods. Driscoll[23] used a 1-cm diameter column, 9.5 ml long with Amberlite® 120 resin. At flow rates > 2 ml/min/ml of resin the results did not depend on flow rate. Wright and Skogheim[108] used Dowex® 50-X8, 50/100-mesh resin, but no details were given about the column size or flow rate. In the automated procedure of Røgeberg and Hendrickson,[25] a 3-mm diameter column with a flow rate of 2.8 ml/min was used. Because the column will eventually become loaded with aluminum from the samples, columns have a limited useful lifetime. Røgeberg and Hendrickson[25] recommended changing columns after 50 samples are processed.

The method of aluminum analysis of the column effluent must be the same method as determining monomeric aluminum. Because influent and effluent aluminum are determined by the same method, the quantity of momoneric inorganic aluminum calculated from the difference in these two values is not dependent on the analytical method.

Column fractionation can be done in the field as well as in the laboratory. Driscoll[23] fractionated samples in the field using the Barnes[28] method for determining aluminum. With this method, the Al-8-hydroxyquinoline complex was extracted into MIBK for storage and transported to the laboratory for later analysis by AA.

More rapid fractionation can be accomplished using the automated method developed by Røgeberg and Hendrickson.[25] This method utilizes a Technicon Analyzer® to process up to 30 samples/h. Aluminum analysis is by the pyrocathecol violet method.[51] With a two-channel instrument, both monomeric aluminum and column-effluent aluminum can be determined simultaneously.

Hodges[75] compared the results of the strong acid resin-column method of Driscoll[23] with the chelating resin (30 min) method of Campbell et al.[79] using a soil-solution extract and artificial solutions containing fluoride, SO_4^{2-}, and fulvic acid. The chelating resin was shown generally to determine greater quantities of monomeric inorganic aluminum than the strong acid resin. This suggests that the chelating resin may be more aggressive in the reaction with aluminum complexes. Hodges[75] also compared the two resin methods with the fluoride ISE results and found the electrode results to be sometimes greater and sometimes less than the column-resin results. Driscoll,[23] however, reported consistently lower results for monomeric inorganic aluminum with the fluoride ISE compared to the strong acid resin column. LaZerte (unpublished data) compared the results of the automated strong acid resin method of Røgeberg and Hendrickson[25] with dialysis and found both methods gave similar results.

V. SUMMARY

A. Recommendations for the Determination of Total Aluminum

Many factors must be considered in choosing an appropriate analytical technique for total aluminum. Foremost is the nature of the samples. The concentration of aluminum and interfering substances must be considered. Along with those considerations, the analyst must decide just how he or she defines total aluminum. For example, is colloidal aluminum to be included in the analysis? Then the analyst must consider the availability of instrumentation and the cost of analysis.

We recommend filtration using polycarbonate membranes followed by immediate acidification. For total analysis, destruction of organic matter and minerals by heating with an oxidizing acid as described in Section III.A is necessary unless analysis is by neutron activation. For determination of acid-reactive aluminum, acidification to a fixed pH value for a specified time period should be done immediately after filtering (see Section III.A).

Of the analytical methods discussed, we feel that two AA methods and two absorbance spectrophotometric methods merit the most consideration. The AA methods include the Barnes[28] method for flame AA analysis after extraction of the 8-hydroxyquinoline complex at pH 8.3 (see Section III.B), and direct analysis by graphite furnace AA using pyrolytically coated tubes, the L'vov platform, and a matrix modifier (see Section III.C). For high precision, graphite furnace AA should be used with automatic sample injection.

The spectrophotometric methods include the pyrocatechol method of Dougan and Wilson,[51] and the 8-hydroxyquinoline methods of Bloom et al.[72] or May et al.[71] (see Section III.D). The 8-hydroxyquinoline methods require the extra step of solvent extraction, but the ease of formation of the complex and stability of the complex in the organic phase offers the advantage of extraction in the field for later laboratory analysis. This is especially advantageous when a short time between the addition of acid and sample analysis is desired as suggested for the determination of acid-reactive aluminum. The 8-hydroxyquinoline methods also are very precise, which, with the careful analytical techniques suggested by May et al.[71] allows for detection limits similar to that of graphite furnace AA or fluorescence. The disadvantage of absorbance or fluorescence spectrophotometric methods vs. atomic aluminum detection by flame or furnace is that the spectrophotometric methods can be affected by copper. Copper might be a problem for samples low in aluminum that are analyzed by spectrophotometry without pre-extraction of interfering metals as suggested by May et al.[71]

The pyrocatechol method is somewhat less sensitive than the other recommended methods. Also, without automation the method is less precise than the 8-hydroxyquinoline methods. However, for the analysis of soil solutions from acid soils and acid surface waters, the sensitivity of this method is adequate and with automation the method is very precise.

B. Recommendations for the Speciation of Aqueous Aluminum

Three fractions of aqueous aluminum can be defined by existing speciation techniques (see Figure 1). They are acid-soluble aluminum, monomeric aluminum-organic complexes, and monomeric inorganic aluminum. The fraction of interest to most investigators is monomeric inorganic aluminum. Four types of method for estimating monomeric inorganic aluminum were described in Sections IV.B to IV.E.

The most rapid — but least satisfactory — method of determining monomeric inorganic aluminum is to measure the aluminum that reacts rapidly with a complexing reagent. The ferron method of Smith[62] (30 s) overestimates monomeric inorganic aluminum because it also determines some organic aluminum. The pyrocatechol method of Dougan and Wilson[51] with a 4-min contact time determines a similar fraction of aluminum. The 8-hydroxyquinoline method of James et al.[22] (15 s at pH 5.0) determines less aluminum because 8-hydroxyquinoline at pH 5.0 does not compete well with fluoride and organic ligands. The James method provides a rough estimate of monomeric inorganic aluminum, not including fluorine complexes. Rapid extraction of 8-hydroxyquinoline at pH 8.3 using the method of Barnes[28] determines all of the monomeric inorganic aluminum plus most of the organic aluminum. These methods can be used to estimate total monomeric aluminum, but they generally overestimate monomeric inorganic aluminum (see Figure 1).

Because most of the organically complexed aluminum is bound by nondialyzable humic and fulvic acid components, dialysis can be used to estimate monomeric inorganic aluminum. This procedure, however, is time consuming and subject to errors caused by changes in pH value during dialysis.

The fluoride ISE method has the advantage of not disturbing the sample during measurement. Electrode calibration, however, is difficult at the low activities of F^- in natural waters and the assumption of no fluoride binding other than by aluminum may not be true for all samples. The morin method disturbs only minimally the sample equilibrium, but the method needs further evaluation using field samples.

The ion-exchange resin methods are the best methods for general use in estimating monomeric inorganic aluminum. The strong acid resin-column methods and the batch-chelating resin method are quite rapid and are adaptable for field use.[23,79] With the automated resin-column method, samples can be processed very rapidly. The chelating resin may give slightly higher values for monomeric inorganic aluminum than the sulfonate resin method. The sulfonate resin method, however, yields values similar to that of dialysis, but greater than for the fluoride ISE method.

Results of different investigators for determination of monomeric inorganic aluminum should be comparable even with different methods of aluminum analysis if sufficiently rapid column flow rates are used, as suggested by Driscoll.[23] The results for acid-soluble aluminum and monomeric organic aluminum, however, are dependent on the anaylytical method for determining aluminum. Because of the ease of sample handling with the column method and because the wide acceptance of this method allows for direct comparison of results with the results of many investigators, we recommend the method of Driscoll[23] or the automated method of Røgeberg and Hendrickson.[25]

REFERENCES

1. **Welz, P.,** *Atomic Absorption Spectrometry,* 2nd ed., Verlag Chemie, Deerfield Beach, FL, 1985, 65.
2. **Slavin, M.,** *Atomic Absorption Spectroscopy,* John Wiley & Sons, New York, 1978, chap. 3 and 4.
3. **Chakrabarti, C. L.,** Beryllium, boron, aluminum, gallium, indium, and thallium, in *Flame Emission and Atomic Absorption Spectrometry,* Vol. 3, Dean, J. A. and Rains, T. C., Eds., Marcel Dekker, New York, 1975, 95.
4. **Fassel, V. A.,** Quantitative elemental analyses by plasma emission spectroscopy, *Science,* 202, 181, 1978.
5. **Janssens, E., Schutyser, P., and Dams, R.,** ICPAES automated, sequential analysis of Al, Be, Cd, Co, Cr, Cu, Fe, Mn, Ni and Zn in surface water samples, *Environ. Technol. Lett.,* 3, 35, 1982.
6. **Botto, R. I.,** Interference correction for simultaneous multielement determinations by inductively coupled plasma, in *Developments in Atomic Plasma Spectrochemical Analysis,* Barnes, R. M., Ed., Heyden, London, 1981, 141.

7. **Tikhonov, V. N.,** *Analytical Chemistry of Aluminum,* John Wiley & Sons, New York, 1973, chap. 2.
8. **Helmke, R. P.,** Neutron activation analysis, in *Methods of Soil Analysis, Part 2, Chemical and Microbiological Properties,* 2nd ed., Page, A. L., Miller, R. H., and Keeney, D. R., Eds., American Society of Agronomy, Madison, WI, 1982, chap. 4.
9. **Moshier, R. W. and Sievers, R. E.,** *Gas Chromatography of Metal Chelates,* Pergamon Press, Oxford, 1965, chap. 1.
10. **Skoog, D. A. and West, D. M.,** *Fundamentals of Analytical Chemistry,* Holt, Reinhart & Winston, New York, 1969, chap. 24.
11. **Karin, R. W., Buono, J. A., and Fasching, J. L.,** Removal of trace elemental impurities from polyethylene by nitric acid, *Anal. Chem.,* 47, 2296, 1975.
12. **Craney, C. L., Swartout, K., Smith, F. W., and West, C. D.,** Improvement of trace aluminum determination by electrothermal atomic absorption spectrophotometry using phosphoric acid, *Anal. Chem.,* 58, 656, 1986.
13. **Manning, D. C., Walter S., and Carnrick, G. R.,** Investigation of aluminum interferences using the stabilized temperature platform furnace, *Spectrochim. Acta,* 37B, 331, 1982.
14. **Reuss, J. O. and Johnson, D. W.,** Effect of soil processes on the acidification of water by acid deposition, *J. Environ. Qual.,* 14, 26, 1985.
15. **Suarez, D. L.,** Prediction of pH errors in soil-water extractors due to degassing, *Soil Sci. Soc. Am. J.,* 51, 64, 1987.
16. **Salbu, B., Steinnes, E., and Bjornstad, H.,** Use of different physical separation techniques for trace element speciation studies in natural waters, Publ. No. 150, Erickson, E., Ed., International Association of Hydrological Science, Wallingford, Oxfordshire, 1985, 203.
17. **Kennedy, V. C., Zellweger, G. W., and Jones, B. F.,** Filter pore-size effects on the analysis of Al, Fe, Mn, and Ti in water, *Water Resour. Res.,* 10, 785, 1974.
18. **Van Olphen, H. and Fripiat, J. J., Eds.,** *Data Handbook for Clay Materials and Other Non-Metallic Minerals,* Part II, Pergamon Press, Oxford, 1979.
19. **Jardine, P. M., Zelazny, L. W., and Evans, A., Jr.,** Solution aluminum anomalies resulting from various filtering materials, *Soil Sci. Soc. Am. J.,* 50, 891, 1986.
20. **Jay, P. C.,** Anion contamination of water samples introduced by filter media, *Anal. Chem.,* 57, 780, 1985.
21. **Salbu, B., Bjornstad, H. E., Lindstrom, N. S., Lydersen, E., Brevik, E. M., Rambaek, J. P., and Paus, P. E.,** Size fractionation techniques in the determination of elements associated with particulate or colloidal material in natural fresh waters, *Talanta,* 32, 907, 1985.
22. **James, B. R., Clark, C. J., and Riha, S. J.,** An 8-hydroxyquinoline method for labile and total aluminum in soil extracts, *Soil Sci. Soc. Am. J.,* 47, 893, 1983.
23. **Driscoll, C. T.,** A procedure for the fractionation of aqueous aluminum in dilute acidic waters, *Int. J. Environ. Anal. Chem.,* 16, 267, 1984.
24. **Hydes, D. J. and Liss, P. S.,** Fluorimetric method for the determination of low concentrations of dissolved aluminium in natural waters, *Analyst,* 101, 922, 1976.
25. **Røgeberg, E. J. S. and Hendrickson, A.,** An automatic method for fractionation of aluminum species in fresh waters, *Vatten,* 41, 48, 1985.
26. **Hsu, Y. and Pipes, O.,** Modification of technique for determination of aluminum in water by atomic absorption spectrophotometry, *Environ. Sci. Technol.,* 6, 645, 1972.
27. **Fishman, M. J.,** Determination of aluminum in water, *At. Absorpt. Newsl.,* 11, 46, 1972.
28. **Barnes, R. B.,** The determination of specific forms of aluminum in natural water, *Chem. Geol.,* 15, 177, 1975.
29. Standard Methods for the Examination of Water and Wastewater, American Public Health Association, New York, 1980, part 300.
30. **LaZerte, B. D.,** Forms of aqueous aluminum in acidified catchments of Central Ontario: a methodological analysis, *Can. J. Fish. Sci.,* 41, 766, 1984.
31. **Slavin, W.,** Environmental trace analyses with the stabilized temperature platform furnace and Zeeman background correction, *Analytical Techniques in Environmental Chemistry,* (Series on Environmental Science, Vol. 7), Albaiges, J., Ed., Pergamon Press, Elmsford, NY, 1982, 397.
32. **McArthur, J. M.,** The determination of total dissolved iron, manganese, and aluminium in seawater at seawater concentration by direct-injection flameless atomic absorption spectrophotometry, *Univ. Leeds Res. Inst. Afr. Geol. Dept. Earth Sci. Annu. Rep. Sci. Results,* 20, 80, 1976.
33. **Slavin, W., Carnrick, G. R., and Manning, D. C.,** Graphite-tube effects on perchloric acid interferences on aluminum and thallium in the stabilized-temperature platform furnace, *Anal. Chim. Acta,* 138, 103, 1982.
34. **Ross, D. S., Bartlett, R. J., and Magdoff, F. R.,** Graphite furnace determination of aluminum in soil leachates using uncoated graphite tubes, *At. Spectrosc.,* 7, 158, 1986.
35. **Kaiser, M. L., Koirtyohann, S. R., and Hinderberger, E. J.,** Reduction of matrix interferences in furnace atomic absorption with the L'vov platform, *Spectrochim. Acta,* 36B, 773, 1981.

36. **Persson, J., Frech, W., and Cedergren, A.,** Investigations of reactions involved in flameless atomic absorption procedures. V. An experimental study of factors influencing the determination of aluminium, *Anal. Chim. Acta,* 92, 95, 1977.
37. **Slavin, W., Carnrick, G. R., and Manning, D. C.,** Magnesium nitrate as a matrix modifier in the stabilized temperature platform furnace, *Anal. Chem.,* 54, 621, 1982.
38. **Carnrick, G. R. and Slavin, W.,** Use of Th-treated platforms for the determination of Al and Pd, *At. Spectrosc.,* 7, 175, 1986.
39. **Carrondo, M. J. T., Lester, J. N., and Perry, R.,** An investigation of a flameless atomic-absorption method for determination of aluminum, calcium, iron and magnesium in sewage sludge, *Talanta,* 26, 929, 1979.
40. **Carrondo, M. J. T., Lester, J. N., and Perry, R.,** Electrothermal atomic absorption determination of total aluminum (including zeolite type A) in waters and waste waters, *Anal. Chim. Acta,* 111, 291, 1979.
41. **Playle, R., Gleed, J., Jonasson, R., and Kramer, J. R.,** Comparison of atomic absorption spectrophotometric, spectrophotometric and fluorimetric methods for the determination of aluminum in water, *Anal. Chim. Acta,* 134, 369, 1982.
42. **Salbu, B., Bjornstad, H. E., Lindstrom, N. S., Brevik, E. M., Rambaek, J. P., Englund, J. O., Meyer, K. F., Hovind, H., Paus, P. E., Enger, B., and Bjerkelund, E.,** Particle discrimination effects in the determination of trace metals in nautral fresh water, *Anal. Chim. Acta,* 167, 161, 1985.
43. **Roura, M., Baucells, M., Lacort, G., and Rauret, G.,** Determination of aluminium in river water by the inductively coupled plasma-atomic emission spectroscopic (ICP-AES) technique, *Analytical Techniques in Environmental Chemistry,* (Series on Environmental Science, Vol. 7), Albaiges, J., Ed., Pergamon Press, Elmsford, NY, 1982, 377.
44. **Lichte, F. E., Hopper, S., and Osborn, T. W.,** Determination of silicon and aluminum in biological matrices by inductively coupled plasma emission spectrometry, *Anal. Chem.,* 52, 120, 1980.
45. **Floyd, M. A., Fassel, V. A., Winge, R. K., Katzenberger, J. M., and D'Silva, A. P.,** Inductively coupled plasma-atomic emission spectroscopy: a computer controlled, scanning monochromator system for the rapid sequential determination of the elements, *Anal. Chem.,* 52, 431, 1980.
46. **Uehiro, T., Morita, M., and Fuwa, K.,** Vacuum ultraviolet emission line for determination of aluminum by inductively coupled plasma atomic emission spectrometry, *Anal. Chem.,* 56, 2020, 1984.
47. **Arshad, M. A., Huang, P. M., and St. Arnaud, R. J.,** Determination of aluminum in soil extracts by methylthymol blue, *Soil Sci.,* 114(2), 115, 1972.
48. **Dalal, R. C.,** Colorimetric determination of aluminum in soil extracts using haematoxylin, *Plant Soil,* 36, 223, 1972.
49. **Pritchard, D. T.,** Spectrophotometric determination of aluminum in soil extracts with xylenol orange, *Analyst,* 92, 103, 1967.
50. **Edwards, A. C. and Cresser, M. S.,** An improved, automated xylenol orange method for the colorimetric determination of aluminium, *Talanta,* 30, 702, 1983.
51. **Dougan, W. K. and Wilson, A. L.,** The absorptiometric determination of aluminium in water: a comparison of some chromogenic reagents and the development of an improved method, *Analyst,* 99, 413, 1974.
52. **Barnhisel, R. and Bertsch, P. M.,** Aluminum, in *Methods of Soil Analysis, Part 2, Chemical and Microbiological Properties,* 2nd ed., Page, A. L., Miller, R. H., and Keeney, D. R., Eds., American Society of Agronomy, Madison WI, 1982, chap 16.
53. **Bertsch, P. M., Alley, M. M., and Elmore, T. L.,** Automated aluminum analysis with the aluminon method, *Soil Sci. Soc. Am. J.,* 45, 666, 1981.
54. **Blamey, F. P. C., Edwards, D. G., and Asher, C. J.,** Effects of aluminum OH:Al and P:Al molar ratios and ionic strength on soybean root elongation in solution culture, *Soil Sci.,* 136, 197, 1983.
55. **Marczenko, Z. and Jarosz, M.,** Formation of ternary complexes of aluminium with some triphenylmethane reagents and cationic surfactants, *Analyst,* 107, 1431, 1982.
56. **Røyset, O. and Sullivan, T. J.,** Effect of dissolved humic compounds on the determination of aqueous aluminum by three spectrophotometric methods, *Int. J. Environ. Anal. Chem.,* 27, 305, 1986.
57. **Wygenowski, C., Motomizu, S. O., and Toei, K.,** Spectrophotometric determination of aluminum with bromogallol red and a quaternary ammonium salt: determination of aluminum in river water, *Mikrochim. Acta (Wien),* 55, 1983.
58. **Johnson, N. M., Driscoll, C. T., Eaton, J. S., Likens, G. E., and McDowell, W. H.,** Acid rain, dissolved aluminum, and chemical weathering at the Hubbard Brook Experimental Forest New Hampshire, *Geochim. Cosmochim. Acta,* 45, 1421, 1981.
59. **Seip, H. M., Muller, L., and Naas, A.,** Aluminum speciation: comparison of two spectrophotometric analytical methods and observed concentrations in some acidic aquatic systems in southern Norway, *Water Air Soil Pollut.,* 23, 81, 1984.
60. **Driscoll, C. T., Van Breeman, N., and Mulder, J.,** Aluminum chemistry in a forested Spodosol, *Soil Sci. Soc. Am. J.,* 49, 437, 1985.

61. **Bersillon, J. L., Hsu, P. H., and Fiessinger, F.,** Characterization of hydroxy-aluminum solutions, *Soil Sci. Soc. Am. J.*, 44, 630, 1980.
62. **Smith, R. W.,** Relations among equilibrium and nonequilibrium aqueous species of aluminum hydroxy complexes, *Am. Chem. Soc. Adv. Chem.*, 106, 250, 1971.
63. **Jardine, P. M. and Zelazny, L. W.,** Mononuclear and polynuclear aluminum speciation through differential kinetic reactions with ferron, *Soil Sci. Soc. Am. J.*, 50, 895, 1986.
64. **Goto, K., Tamura, H., Onodera, M., and Nagayama, M.,** Spectrophotometric determination of aluminium with ferron and a quaternary ammonium salt, *Talanta*, 21, 183, 1974.
65. **Dale, T. and Hendrickson, A.,** Intercalibration of methods for chemical analysis of water. VI. Results from intercalibration of a pyrocatechol violet method for determining aluminum in water, *Vatten*, 31, 91, 1975.
66. **Grigg, J. L. and Morrison, D.,** An automatic colorimetric determination of aluminium in soil extracts using catechol violet, *Commun. In. Soil Sci. Plant Anal.*, 13, 351, 1982.
67. **Takashi, K., Motomizu, S., and Toei, K.,** Modified extraction procedure for the spectrophotometric determination of trace amounts of aluminum in sea water with pyrocatechol violet and removal of excess reagent, *Analyst*, 105, 328, 1980.
68. **Frink, C. R. and Peech, M.,** Determination of aluminum in soil extracts, *Soil Sci.*, 93, 217, 1962.
69. **Turner, R. C. and Sulaiman, W.,** Kinetics of reactions of 8-quinolinol and acetate with hydroxyaluminum species in aqueous solutions. I. Polynuclear hydroxyaluminum cations, *Can. J. Chem.*, 49, 1683, 1971.
70. **Carter, J. M.,** The determination of aluminum in water by the method involving quinolin-8-ol, *Proc. Analyt. Div. Chem. Soc.*, September, 246, 1975.
71. **May, H. M., Helmke, P. A., and Jackson, M. L.,** Determination of mononuclear dissolved aluminum in near-neutral waters, *Chem. Geol.*, 24, 259, 1979.
72. **Bloom, P. R., Weaver, R. M., and McBride, M. B.,** The spectrophotometric and fluorometric determination of aluminum with 8-hydroxyquinoline and butyl acetate extraction, *Soil Sci. Soc Am. J.*, 42, 713, 1978.
73. **Sandell, E. B.,** *Colorimetric Determination of Traces of Metals*, 3rd ed., Wiley-Interscience, New York, 1959, 231.
74. **Narayanan, A. and Pantony, D. A.,** Spectrophotometric determination of aluminum at parts per billion level in water, *Environ. Sci. Technol. Lett.*, 3, 43, 1982.
75. **Hodges, S. C.,** Aluminum speciation: a comparison of five methods, *Soil Sci. Soc. Am. J.*, 51, 57, 1987.
76. **Hoyt, P. B. and Turner, R. C.,** Effects of organic materials added to very acid soils on pH, aluminum, exchangeable NH_4 and crop yields, *Soil Sci. Soc. Am. J.*, 119, 227, 1975.
77. **Armstrong, F. A. J., Williams, P. M., and Strickland, J. D. H.,** Photo-oxidation of organic matter in sea water by ultraviolet radiation and other applications, *Nature*, 211, 481, 1966.
78. **Cronan, C. S.,** Determination of sulfate in organically colored water samples, *Anal. Chem.*, 51, 1333, 1979.
79. **Campbell, P. G. C., Bisson, M., Bougie, R., Tessier, A., and Villeneuve, J. P.,** Speciation of aluminum in acidic freshwaters, *Anal. Chem.*, 55, 2246, 1983.
80. **Bloom, P. R. and Weaver, R. M.,** Effect of the removal of reactive surface material on the solubility of synthetic gibbsites, *Clays Clay Miner.*, 30, 281, 1982.
81. **Howard, A. G., Coxhead, A. J., Potter, I. A., and Watt, A.,** Determination of dissolved aluminium by the micelle-enhanced fluorescence of its lumogallion complex, *Analyst*, 111, 1379, 1986.
82. **Gabriels, R., Van Keirsbulk, W., and Engels, H.,** Spectrofluorometric determination of aluminum in plants, soils and irrigation waters, *Lab. Pract.*, 30, 123, 1981.
83. **Medina Escriche, J. and Hernandez, F. H.,** Fluorimetric determination of aluminium with morin after extraction with isobutyl methyl ketone. II. Extraction-fluorimetric determination of aluminium in natural and waste waters, *Analyst*, 110, 287, 1985.
84. **Cano-Pavon, J. M., Trujillo, M. L., and Garcia Detorres, A.,** 3-Hydroxypryidine-2-aldehyde 2-pyridylhydrazone as an analytical reagent, *Anal. Chim. Acta*, 117, 319, 1980.
85. **Salbu, B., Steinnes, E., and Pappas, A. C.,** Multielement neutron activation analysis of fresh water using Ge(Li) gamma spectrometry, *Anal. Chem.*, 47, 1011, 1975.
86. **Misra, U. K., Graham, E. R., Upchurch, W. J., and McKown, D. M.,** Activation analysis of aluminum from water extract of soil, *Soil Sci. Soc. Am. Proc.*, 37, 193, 1973.
87. **Bond, A. M. and Nagaosa, Y.,** Determination of aluminum, copper, iron and manganese in biological and other samples as 8-quinolinol complexes by high-performance liquid chromatography with electrochemical and spectrophotometric detection, *Anal. Chim. Acta*, 178, 197, 1985.
88. **Lee, M. and Burrell, C.,** Soluble aluminum in marine and fresh water by gas-liquid chromatography, *Anal. Chim. Acta*, 66, 245, 1973.
89. **Gosink, T. A.,** Rapid simultaneous determination of picogram quantities of aluminum and chromium from water by gas phase chromatography, *Anal. Chem.*, 47, 165, 1975.

90. **Trojanowicz, M. and Hulanicki, A.,** Microdetermination of aluminum with fluoride-selective electrode, *Mikrochim. Acta,* 11, 17, 1981.
91. **Driscoll, C. T., Baker, J. P., Bisogni, J. J., and Schofield, C. L.,** Effect of aluminum speciation on fish in dilute and acidified waters, *Nature,* 284, 161, 1980.
92. **Pavan, M. A., Bingham, F. T., and Pratt, P. F.,** Toxicity of aluminum to coffee in Ultisols and Oxisols amended with $CaCO_3$, $MgCO_3$, and $CaSO_4 \cdot 2H_2O$, *Soil Sci. Soc. Am. J.,* 46, 1201, 1982.
93. **Cameron, R. S., Ritchie, O. S. P., and Robson, A. D.,** Relative toxicities of inorganic aluminum complexes to barley, *Soil Sci. Soc. Am. J.,* 50, 1231, 1986
94. **Hue, N. V., Craddock, G. R., and Adams, F.,** Effects of organic acids on aluminum toxicity in subsoils, *Soil Sci. Soc. Am. J.,* 50, 28, 1986
95. **Bartlett, R. J. and Riego, D. L.,** Effect of chelation on the toxicity of aluminum, *Plant Soil,* 37, 419, 1972.
96. **David, M. B., and Driscoll, C. T.,** Aluminum speciation and equilibrium in soil solutions of a Haplorthod in the Adirondack Mountains (New York, USA), *Geoderma,* 297, 1984.
97. **Nair, V. D. and Prenzel, T.,** Calculations of equilibrium concentrations of mono- and polynuclear hydroxyaluminum species at different pH and total aluminum concentrations, *Z. Pflanzenernaehr. Bodenkd.,* 141, 741, 1978.
98. **Campbell, P. G. C., Bougie, R., Tessier, A., and Villeneuve, J. P.,** Aluminum speciation in surface waters on the Canadian pre-Cambrian shield, *Verh. Int. Ver. Limnol.,* 22, 317, 1984.
99. **Sillén, L. G. and Martell, A. C.,** *Stability Constants of Metal-Ion Complexes,* 2nd ed., Spec. Publ. No. 17, The Chemical Society, London, 1964.
100. **Westall, J. C., Zachary, J. L., and Morel, F. M. M.,** MINEQL, a Computer Program for the Calculation of Chemical Composition of Aqueous Systems, MIT Department of Civil Engineering Tech. Rep. No. 18, 1976.
101. **Mattigod, S. V. and Sposito, G.,** Chemical modeling of trace metal equilibria in contaminated soil solutions using the computer program GEOCHEM, in *Chemical Modeling in Aqueous Systems,* Jenne, E. A., Ed., American Chemical Society, Washington, D.C., 1979, chap. 37.
102. **Okura, T., Goto, K., and Yotsuyanagi, T.,** Forms of aluminum determined by an 8-hydroxyquinoline extraction method, *Anal. Chem.,* 34, 581, 1962.
103. **Turner, R. C.,** Three forms of aluminum in aqueous systems determined by 8-quinolinolate extraction methods, *Can. J. Chem.,* 47, 2521, 1969.
104. **Florence, T. M. and Batley, G. E.,** Chemical speciation in natural waters, *Crit. Rev. Anal. Chem.,* 9(3), 219, 1980.
105. **Benes, P. and Stennes, E.,** *In situ* dialysis for the determination of the state of trace elements in natural waters, *Water Res.,* 8, 947, 1974.
106. **Browne, B. A., McColl, J. G., and Driscoll, C. T.,** Aluminum speciation using morin. I. Morin and its complexes with aluminum, *J. Environ. Qual.,* in press.
107. **Browne, B. A., Driscoll, C. T., and McColl, J. G.,** Aluminum speciation using morin. II. Principles and procedures, *J. Environ. Qual.,* in press.
108. **Wright, R. F. and Skogheim, O. K.,** Aluminum speciation at the interface of an acid stream and a limed lake, *Vatten,* 39, 301, 1983.
109. **Driscoll, C. T.,** Chemistry and Characterization of Some Dilute Acidifed Lakes and Streams in the Adirondack Region of New York State, Ph.D. thesis, Cornell University, Ithaca, NY, 1980.

Chapter 2

AQUEOUS EQUILIBRIUM DATA FOR MONONUCLEAR ALUMINUM SPECIES

Darrell Kirk Nordstrom and Howard M. May

TABLE OF CONTENTS

I. INTRODUCTION

Most reactions involving aluminum in the environment take place in the presence of water. Homogeneous solution reactions involving only mononuclear species of aluminum are usually extremely fast, especially ionic or simple electrostatic reactions such as ion pairing. The high rates of these reactions allow them to be modeled by equilibrium mass-action expressions and the other accoutrements of chemical thermodynamics. Some reactions (e.g., inner-sphere complexation), however, can be slow enough under environmental conditions that their reaction rates must be known and accounted for.

Mass-action expressions can be represented by equilibrium constant equations, which are the basic data used in models for identifying chemical processes in natural waters.[1] Equilibrium data are also applied in analytical chemistry[2] to design and assess procedures and to measure new equilibrium constants. New specialized procedures, such as the determination of dissolved, organically bound aluminum in natural waters, require estimates of the stability of aluminum complexes with a number of different ligands. Consequently, this chapter summarizes information on the hydrolysis of the mononuclear aqueous aluminum ion and on the stability of mononuclear ion pairs or complexes of aluminum with naturally occurring inorganic anions such as fluoride, sulfate, carbonate, and phosphate, and with organic ligands that are used as models for natural organic compounds or as important reagents in the colorimetric analysis of aluminum.

By "mononuclear", we mean species whose stoichiometry indicates that only one aluminum ion is involved in the metal-ligand complex. Polynuclear species are dealt with in Chapter 4.[3] We use the system of symbolic notation described by Sillén and Martell;[4] symbols and definitions used in this chapter are given in Table 1. The temperature range considered is 273.25 to 373.15 K.

II. METHODS OF MEASUREMENT

The use of equilibrium thermodynamics to describe aqueous solution chemistry depends on the availability of data based upon accurate and precise experimental measurements and a reliable electrolyte theory for obtaining standard-state thermodynamic functions. Of these two basic requirements, high-quality experimental measurement is clearly the more fundamental. The quality of thermodynamic data is more closely tied to the reliability of the raw experimental values from which they are derived than to any mathematical method used to interpret those values. Commonly, the results of the most accurate and precise experiments lead to the most appropriate theory. Conversely, thermodynamic data based upon inaccurate experimental results have often misled investigators as to the relative stability of chemical species and the nature of corresponding theories governing those stabilities. Experimental design is thus crucial to the quality of thermodynamic data. "The objective of a well-designed experiment is to obtain data that 'defines' the system as unequivocally as possible."[5] In this section we shall summarize goals, procedures, and other practical issues for the experimental estimation of the thermodynamic properties of aqueous mononuclear aluminum ions and complexes.

The experimental requirements for determining the energetics of aqueous aluminum species are more challenging than those for many other elements because of the limited solubility of aluminum, the stepwise hydrolysis of mononuclear aluminum ions, possible polynuclear aluminum ion formation, strong ligand-complexation affinity, and the effects of polymorphism and surface chemistry variability upon aluminous solid-phase solubilities. The chemical behavior of aluminum severely limits choices among available experimental methods, and it is usually necessary to combine two or more techniques to describe adequately aluminum chemistry in a particular system. Experimental variables should be quantitatively related to

Table 1
SYMBOLS AND DEFINITIONS

C_p°	Standard-state heat capacity per mole at constant pressure
ΔG_f°	Standard-state Gibbs energy of formation from the elements per mole
ΔH_f°	Standard-state enthalpy of formation from the elements per mole
ΔH_r°	Standard-state enthalpy of reaction per mole
I	Ionic strength
K	Equilibrium constant
K	Kelvin temperature
L	Ligand (e.g:, OH^-, F^-, CH_3COO^-, etc.)
R	Molar gas constant
S°	Standard-state entropy per mole
T	Absolute temperature (Kelvin)
V°	Standard-state volume per mole
p	-log
β_n	Cumulative equilibrium-formation constant
[i]	Activity of species i

Notation for ion association constants[4]

Reaction	Equilibrium-constant expression
$AlL_{n-1} + L = AlL_n$	$K_n = \frac{[AlL_n]}{[AlL_{n-1}][L]}$
$Al^{3+} + nL = AlL_n$	$\beta_n = \frac{[AlL_n]}{[Al^{3+}][L]^n}$
$Al^{3+} + H_2O = AlOH^{2+} + H^+$	$*K_1 = \frac{[AlOH^{2+}][H^+]}{[Al^{3+}]}$
$Al^{3+} + nH_2O = Al(OH)_n^{3-n} + nH^+$	$*\beta_n = \frac{[Al(OH)_n^{3-n}][H^+]^n}{[Al^{3+}]}$

either concentrations or activities of reacting species believed to be present at equilibrium. Aluminum hydrolysis and complex formation studies are usually carried out in solutions of an inert supporting electrolyte (e.g., $NaClO_4$, $NaNO_3$, etc.), to fix the relationship between species concentrations and activities over an otherwise broad range of experimental conditions (i.e., the constant ionic medium approach[4]).

A. Potentiometry

Potentiometry is the central technique in virtually all investigations of aqueous aluminum chemistry, through application of readily available ion-selective electrodes (ISE) for measuring pH values and F^- ion activities in studies of aluminum fluoride complexes. Aluminum ion activities cannot be measured directly by ISE because of the lack of a suitably reversible electrode for aluminum.

Potentiometric pH value determinations are easily made using relatively simple procedures and equipment, but are not always correctly performed. It is of critical importance to minimize errors in pH value measurement because H^+ activities appear with large exponent terms in reaction equilibrium expressions. Among the prerequisites for quality determinations of pH value are

1. An appropriate H^+ ISE (pH electrode) that is readily reversible and provides a linear Nernstian response over the range needed; some reference electrode liquid junctions can be adversely affected by perchlorate-background electrolyte
2. Careful electrode calibration, using three (or more, if needed) standard buffers to cover the range of intended use
3. Good temperature control of the electrode/sample environment, to minimize uncertainties arising from the temperature dependence of water dissociation (K_w) and other functions; ± 0.10°C at least, but ± 0.01°C is preferable[5]
4. Use of a supporting electrolyte (often required), especially in studies of low ionic strength solutions of sparingly soluble materials like aluminum minerals or in suspensions of clays or other ion exchangers, so as to normalize electrode response with respect to calibration conditions and to minimize the well-known ''suspension effect''
5. Correction for liquid-junction potentials (often necessary), which can be done with the use of the Henderson equation (e.g., see Plummer and Busenberg[6])

Other recommendations and cautions appropriate for pH value determinations may be found in standard references.[7]

B. Distribution Methods: Solubility, Solvent Extraction, and Ion Exchange

Several classical distribution procedures are major tools in investigations of aluminum chemistry, most especially determinations of equilibrium solubility in aqueous solutions, solvent extractions (for analysis/speciation of dissolved aluminum ions), and ion-exchange procedures. Solubility determinations, carefully executed over a sufficiently broad pH value range, offer the most direct and precise route to thermodynamic constants for the relatively weak $Al(OH)_2^+$ and $Al(OH)_3^\circ$ species that are inaccessible by other techniques. In a study that integrated pH potentiometry and aqueous solubility determination (by means of a solvent extraction analytical technique), May et al.[8] were able to construct an equilibrium-solubility model for gibbsite with sufficient precision and detail to yield a precise estimate of the stability of the $Al(OH)_2^+$ ion, and to limit the possible stability of the $Al(OH)_3^\circ$ species to the extent that it may be rejected as a significant contributor to aluminum hydrolysis and solubility at 298.15 K. Solubility data can be as precise and error-free as potentiometric data are generally accepted to be.

Attainment of equilibrium solubility is best demonstrated by reversibility, i.e., approaching equilibrium from over- and undersaturation with respect to a well-defined solid phase (see also Chapter 3). Ion activity (or solubility) product terms must remain reasonably constant over a reasonable range of solution compositions, and the composition, crystal form, or surface properties of the subject solid must not change significantly. These simple requirements are too often slighted in many reported experiments.

In addition to their classical use for distinguishing complexed species by distributing them between immiscible and therefore separable phases, solvent extraction procedures can be used to distinguish analytically dissolved forms of sparingly soluble metals from polynuclear or suspended, finely divided solid forms of the same element, thereby enabling determination of concentrations of mononuclear dissolved metal ion species. The procedures of Turner,[9] Barnes,[10] Bloom et al.,[11] and May et al.[12] all employ solvent extraction of the 8-hydroxyquinoline complex of aluminum to separate dissolved mononuclear aluminum ion species from aqueous solution, prior to analytical measurement (see Chapter 1).

Ion exchange has been used to quantitate dissolved aluminum complexed by macromolecular organic forms in natural waters, based upon a kinetically slow release of such bound aluminum and its undisturbed passage through short, cation-exchange resin columns that adsorb the more labile aluminum ion species.[13] This technique may find application in studies of the stabilities of such complexes.

C. Spectroscopic Methods

Few spectroscopic techniques are useful in studies of aqueous aluminum ion chemistry. Hydroxy-aluminum ion species cannot be directly determined by UV/visible light spectrophotometry, but such photometry, of absorption by the Al-8-hydroxyquinoline complex at 395 nm, is the quantitation step in three of the four solvent extraction analytical procedures mentioned in the preceding section (see also Chapter 1). Flame and graphite-furnace atomic absorption spectroscopy of aluminum in atomic vapor form are also important analytical tools, but cannot be directly used to assess the properties of aluminum ions in solution. Nuclear magnetic resonance spectroscopy has been used to examine the structural, kinetic, and equilibrium properties of mononuclear, polynuclear, and complexed aluminum ions and of hydroxy-aluminum solids,[3,14] but it lacks sensitivity for the determination of mononuclear stability constants, relative to other methods. Other spectroscopic techniques (IR, Raman, electron spin resonance, Mossbauer, etc.) are similarly ineffective or insensitive owing to their requirements for concentrated sample solutions, nonaqueous samples, or solid samples.

D. Filtration and Centrifugation

Both of these techniques have long been employed to separate suspended solids. Species that will either pass a filter membrane or remain in suspension after centrifugation have been, for reasons of expediency, labeled operationally as ''dissolved''. Neither technique can, with standard available equipment, separate mononuclear species physically from simpler polynuclear ions (e.g., $Al_6(OH)_{15}^{3+}$, $Al_2(OH)_4^{2+}$) that may occur in experimental systems, and neither can contribute directly to the evaluation of the thermodynamic properties of mononuclear aluminum species.

E. Miscellaneous Techniques

Polarography, anode stripping voltammetry, and similar electrochemical techniques are generally inapplicable to the direct study of aqueous aluminum ion chemistry, due to the very negative reduction potential for Al(III). A few recent reports have indicated, however, that cathode stripping voltammetry of aluminum previously reacted with various complex organic ligands may permit electrochemical speciation and determination of dissolved aluminum.[15] Conductivity and dielectric constant determinations measure gross physical properties and cannot directly assess properties of individual weak complexes formed in stepwise reactions and coexisting in dilute solutions. Calorimetry, potentially precise in its classical applications, is insensitive to weak interactions in extremely dilute solutions and, beyond its possible use in estimating the first hydrolysis constant in concentrated solutions, it is generally inapplicable to the study of aluminum ion chemistry in dilute systems at equilibrium. A small but significant group of reports have described the use of kinetic techniques (pressure-, temperature-, or electric field-jump) to characterize aluminum reactions; these methods are noted in Section III below.

III. ALUMINUM HYDROLYSIS AND COMPLEXATION KINETICS IN AQUEOUS SOLUTION

In terms of the effective operational time scales of most experimental procedures, proton-transfer reactions are virtually instantaneous for mononuclear aluminum ion hydrolysis in aqueous solution. Holmes et al.[16] and Fong and Grunwald[17] determined specific rate constants of 4.4×10^9 and 1.1×10^5/s for proton association and dissociation reactions, respectively, in the first step of $Al(H_2O)_6^{3+}$ hydrolysis. In other words, changes in H^+ activity in bulk solution result in simultaneous redistributions of aluminum among hydrolysis species, delayed only by processes such as mixing or diffusion. (Hydrolytic transformations of hydroxy-aluminum polynuclear ions and solids, discussed in Chapter 4, are *much slower*.)

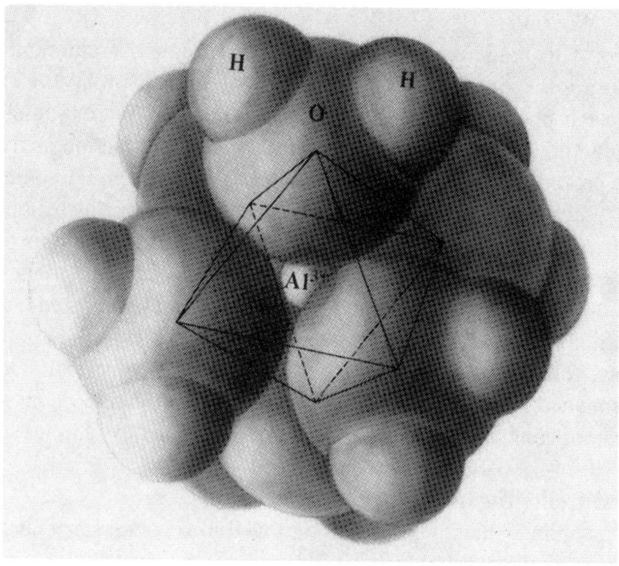

FIGURE 1. Structure of the free aqueous aluminum ion.

Rates of ligand penetration to form inner-sphere complexes with aluminum are orders of magnitude lower than proton transfer rates,[18,19] and similar to the rate of exchange of tightly held coordinating water molecules (for which $t_{1/2} = 4$ s).[5] Fluoride and humic/fulvic acid complexes of mononuclear aluminum ions are of special interest because of the major roles these ligands or complexing macromolecules play in environmental geochemistry. Brosset and Orring[20] and Kleiner[21] first observed that aluminum fluoride complexes were slow to attain equilibrium at low concentrations. Fluoride complexation rates have recently been shown to be pH- and ionic-strength dependent, but invariably high enough to ensure equilibrium within minutes or hours.[22,23] Such reaction time scales demonstrate that equilibrium chemical modeling of natural water aluminum chemistry should be reasonable in most cases, but that laboratory investigations of aluminum complexation must take account of such time requirements. Indeed, in a series of recent reports, Ares[24-27] has described the use of varying rates of F^- complexation to model aluminum ion species chemistry in samples of soil solutions and other natural waters.

IV. AQUEOUS INORGANIC EQUILIBRIUM CONSTANTS AT 298.15 K

A. Hydrolysis Constants

The free aqueous aluminum ion, Al^{3+}, is coordinated by six water molecules in an octahedral configuration, depicted in Figure 1, and can be represented by the formula $Al(H_2O)_6^{3+}$. The hydration formula was estimated from empirical and theoretical considerations and later confirmed by NMR spectra.[28,29] Because of the high positive charge of the trivalent aluminum ion, these water molecules form a tightly bound primary hydration shell. The hydrolysis of aluminum ion is the progressive loss of hydration shell protons to water molecules in the surrounding bulk solution, to maintain dissociation equilibrium. For example, the first hydrolysis reaction can be written

$$Al(H_2O)_6^{3+} + H_2O = Al(H_2O)_5OH^{2+} + H_3O^+ \quad (1)$$

This reaction is essentially the same as that for the ionization of water, except for the stabilizing presence of the aluminum cation, and the rate of aluminum hydrolysis is nearly

Table 2
SUMMARY OF ALUMINUM HYDROLYSIS CONSTANTS DATA AT 298.15 K AND ZERO IONIC STRENGTH

	p^*K_1	$p^*\beta_2$	$p^*\beta_3$	$p^*\beta_4$
Range	4.3—5.86	8.56—10.65	15.0—16.76	19.43—27.98
No. of values	20	8	4	22
Mean	4.985(±0.28)[a]	9.52(±0.85)[a]	15.9	22.97(±1.94)[a]
Median	4.984	9.5	—	22.59
Mode	4.99	—	—	—
Selected value	5.00(±0.04)[a]	10.1	16.8	22.7(±0.6)[b]

[a] These parenthetical values represent one standard deviation.
[b] This parenthetical value represents the range of reliable reported values.

as fast as that for the hydrolysis of water.[16] Such a relationship can be useful in the evaluation of the temperature dependence of hydrolysis data.

For the sake of convenience, Reaction 1 will be abbreviated to the more common expression:

$$Al^{3+} + H_2O = AlOH^{2+} + H^+ \tag{2}$$

for which the mass-action expression is

$$^*K_1 = \frac{[AlOH^{2+}][H^+]}{[Al^{3+}]} \tag{3}$$

taking the activity of water to be unity and designating activities with brackets.

Hydrolysis constants for Al^{3+} have been measured using a variety of techniques, including potentiometry, spectroscopy, spectrophotometry, distribution methods, calorimetry, and electrochemisty (see Chapter 4 for a comprehensive discussion). The agreement among studies using different techniques is quite good for $^*K_1$ and attests to the accuracy of the results. A search of the literature revealed 45 investigations of the first hydrolysis constant, for which 75 values have been reported over a range of ionic strength up to three molal and at temperatures between 283 and 373 K. The range of values reported for 298.15 K and zero ionic strength is shown in Table 2. Some measurements have been made or estimated at higher temperatures, but these have been omitted to simplify the evaluation.

Parks[30] made a careful evaluation of solubilities and hydrolysis constants for aluminum hydroxide species and concluded that the most reliable studies for p^*K_1 were those of Schofield and Taylor,[31] Frink and Peech,[32] and Raupach,[33] based on criteria involving the attainment of equilibrium, absence of complexing anions, awareness of polymer formation, and zero ionic strength corrections. He averaged these results and obtained $p^*K_1 = 5.01 \pm 0.04$. Baes and Mesmer[34] concluded that the investigation of Kubota[35] was the most reliable and they reanalyzed his results to obtain $p^*K_1 = 4.97 \pm 0.02$. We have examined these same studies and included the results of Volokhov et al.,[36] May et al.,[8] Couturier et al.,[37] Nishide and Tsuchiya,[38] and Nikolayeva and Tolpygina[39] to calculate a mean value of 5.00 ± 0.04 (one standard deviation). This mean value was calculated from 12 separate determinations and, since it is no different from that reported by Parks,[30] it is not likely that any further measurements or refinements of the data will lead to a significant decrease in the uncertainty. This constant has been determined accurately and precisely enough for the purposes of modeling the chemistry of aluminum in natural waters or most other applications to aqueous aluminum chemistry. Table 2 summarizes the statistics for the mean, median, and mode for p^*K_1 and the other hydrolysis constants.

The second hydrolysis reaction,

$$Al^{3+} + 2H_2O = Al(OH)_2^+ + 2H^+ \quad (4)$$

for which the mass-action expression is

$$^*\beta_2 = \frac{[Al(OH)_2^+][H^+]^2}{[Al^{3+}]} \quad (5)$$

is very difficult to characterize quantitatively because it (and the third hydrolysis reaction) occurs under conditions of very low aluminum solubility, beyond reach of most analytical techniques. The experimental difficulty in determining the second and third hydrolysis constants, combined with the interference from polymerization reactions at elevated aluminum concentrations, has resulted in fewer experimental values for these constants, accompanied by greater uncertainties. Our literature search uncovered 18 reports, of which only 8 described investigations carried out at 298.15 K and corrected for ionic strength effects (see Table 2). We have selected $p^*\beta_2 = 10.1$ from the study by May et al.,[8] who have reported the only solubility investigation, executed over a broad range of pH values, that demonstrates reversibility and is consistent with the selected value for p^*K_1 and with other measurements of p^*K_2; i.e., it is the most internally and externally consistent set of data available. We believe that greater accuracy is gained by selecting the single most precise and consistent experimental result instead of an average value, when few data are available.

The third hydrolysis reaction,

$$Al^{3+} + 3H_2O = Al(OH)_3^\circ + 3H^+ \quad (6)$$

for which

$$^*\beta_3 = \frac{[Al(OH)_3^\circ][H^+]^3}{[Al^{3+}]} \quad (7)$$

is even more difficult to characterize experimentally than the second because of the vanishingly small concentrations of aluminum present in most experimental systems near pH 6. All indications are that the value of $^*\beta_3$ must be less than 10^{-15} and the estimate made by Baes and Mesmer[34] of $p^*\beta_3 = 16.5$ is consistent with the values reported by May et al.[8] of $p^*\beta_3 = 16.76$ and by Nazarenko and Biryuk[40] of $p^*\beta_3 = 16.75$. We have selected the value from May et al.[8] to maintain consistency.

The fourth hydrolysis reaction is

$$Al^{3+} + 4H_2O = Al(OH)_4^- + 4H^+ \quad (8)$$

for which

$$^*\beta_4 = \frac{[Al(OH)_4^-][H^+]^4}{[Al^{3+}]} \quad (9)$$

The fourth hydrolysis constant, $^*\beta_4$, is usually obtained from solubility studies of aluminum hydroxides or oxide hydroxides in basic solutions ($pH > 8$). Solubility determinations are very sensitive to such factors as particle size and surface area, and the structural integrity and phase composition of the reacting solid surface. These factors have rarely received adequate attention in reported studies. Thus, when $*\beta_4$ values are directly obtained from solubility data based upon uncertain knowledge of surface properties, and when the solubility product constant for a simple mineral such as gibbsite can vary over three orders of magnitude,[8,41,42] there must be considerable uncertainty attached to reported estimates of $*\beta_4$. It has also been suggested that gibbsite is not stable in alkaline solutions and that bayerite or boehmite may have been the solubility controlling phase.[8] Hemingway et al.[43] observed no significant particle size effect in the enthalpy of gibbsite particles of mean diameter from about 0.05 μm to more than 50 μm. Hence, one must conclude that either (1) the differing reported solubility behavior of gibbsite is controlled by the properties of surface layers that are only a small fraction of the total mass of the investigated samples, or (2) particles less than 0.05 μm were reacting in the solubility studies.

From 26 references, we have found 31 values of $*\beta_4$ determined at temperatures below 373 K. The 22 values for $p*\beta_4$ determined at 298.15 K (see Table 2) tend to form a bimodal distribution around either 22.2 or 23.3. Valid arguments can be made why either set of data may be inaccurate. Careful reviews by Parks[30] and Hemingway[44] recommend $p*\beta_4 = 23.3$, based upon evaluations. The compilation and evaluation of hydrolysis data in the Baes and Mesmer[34] book recommends $p*\beta_4 = 23.0$. Several other experimental solubility and complexation studies[8,37,45,46] all suggest a value close to 22.2.

Dyrssen[47] has recalculated $*\beta_3$ and $*\beta_4$ from the data of May et al.[8] because their last two hydrolysis constants disobey the general principle that $*K_1 > *K_2 > *K_3 > *K_4$. However, this principle only applies to a series of constants for which the reacting species have no specific electronic effects or steric interferences.[5] The aluminate ion, $Al(OH)_4^-$, is complexed by a tetrahedrally coordinated shell of hydroxide ions[48,49] and would not be expected to follow the usual succession of hydrolysis constants for which octahedral coordination has been demonstrated.[50,51] Therefore, it is difficult to justify the modification by Dyrssen.[47]

Apps et al.[52] have compiled and evaluated all available solubility and hydrolysis data for gibbsite, diaspore, and boehmite in basic solutions over a wide range of temperatures and ionic strengths. They have determined best fits of the data and computed the results in terms of pK_{s4} and free energies and enthalpies of formation. They adopted the solubility product constant of Bloom and Weaver[42] for gibbsite and calculated $\Delta G_f^\circ(Al^{3+})$ from that using the Hemingway et al.[43] redetermination of the standard Gibbs energy of formation for gibbsite. From their results we calculated $p*\beta_4 = 22.7\ (\pm 0.7)$ which is nearly at the midpoint between the bimodal distribution of $p*\beta_4$ values. We recommend this value as a reasonable estimate until the cause of the bimodal distribution is resolved.

B. Standard Gibbs Energies of Formation for Hydrolysis Species

In transforming equilibrium constants into standard Gibbs energies of formation, one often introduces considerable uncertainties. However, there is usually no need for such transformations, so we generally recommend against using Gibbs energies of formation (for individual species) when making thermodynamic calculations. It will be demonstrated that, with reference to aqueous reactions involving aluminum, it makes little difference whether one uses Gibbs energies or equilibrium constants. The same unresolved uncertainties will remain in either case.

Prior to the new derivation of the entropy and Gibbs energy of formation of $Al^{3+}_{(aq)}$ by Hemingway and Robie,[53] it was common practice to calculate the Gibbs energy of formation of a mineral from a solubility study by employing the Gibbs energies of the aqueous species taken from some readily available source, such as NBS Circular 500.[54] Parks[30] noted considerable uncertainty in the reported values for $Al^{3+}_{(aq)}$ and chose $\Delta G_f^\circ = -116 \pm 1$ kcal/

mol (-485 ± 4 kJ/mol) to be consistent with the NBS tabulation.[55] Hemingway and Robie[53] carefully reviewed the historical evolution of investigations leading to the free energy of $Al^{3+}_{(aq)}$ and concluded that there were too many errors involved. They proposed a more direct thermochemical pathway, using their more recent calorimetric determination of ΔG°_f for gibbsite, together with the solubility product constant for gibbsite from Kittrick[56] and Singh.[57] The calculation is simply:

$$\Delta G^\circ_f(Al^{3+}) = \Delta G^\circ_r + \Delta G^\circ_f(\text{gibbsite}) - 3\Delta G^\circ_f(OH^-) \qquad (10)$$

where $\Delta G^\circ_r = -RT \ln K_{so}$ for gibbsite solubility. This pathway, yielding $\Delta G^\circ_f(Al^{3+}) = -489.4$ kJ/mol, is much more precise and less subject to error, *as long as the solubility product constant is reliable*. Unfortunately, more recent results from May et al.[8] and Bloom and Weaver[42] reflect a fourfold variation in K_{so} for crystalline gibbsite, depending upon the nature of the solid phase and how it is prepared. There are some problems with the procedures by which both of these investigations were carried out, but the conclusion appears inescapable that variations in sample preparation, structural order, crystallinity, and/or surface properties of the solid phases can generate this range in solubility product constants (see Chapter 3). The best value for $\Delta G^\circ_f(Al^{3+})$ should be obtained using the K_{so} derived from the most crystalline and carefully prepared gibbsite, i.e., that of Bloom and Weaver,[42] which would result in $\Delta G^\circ_f(Al^{3+}) = -486.2$ kJ/mol (-116.2 kcal/mol). However, Bloom and Weaver[42] do not appear to have used inert supporting electrolytes in their solubility runs, except for one sample whose surface had not been cleaned (and it gave consistently different results from the others). It is not clear whether the variation in K_{so} was because of the absence of a supporting electrolyte or the improved sample-preparation technique.

Our conclusion is that $\Delta G^\circ_f(Al^{3+})$ lies in the range -486 to -490 kJ/mol (-116 to -117 kcal/mol), because of the range of log K_{so} (about ± 0.5 log unit). Until further solubility studies are completed, this ambiguity cannot be resolved. Once a more accurate value is found, it becomes a simple matter to calculate the free energies for each hydrolysis species. In theory, solubility data for other reactions could be used to check the value of $\Delta G^\circ_f(Al^{3+})$, but for the aluminum-water system, other pathways (e.g., kaolinite or cryolite solubility) are less reliable than that of gibbsite solubility. Consequently, it makes no difference whether one chooses to use Gibbs energies or equilibrium constants in their calculations; the uncertainty is the same. The same conclusion applies to the Gibbs energies for the other dissolved mononuclear aluminum species.

C. The Formation of Other Complexes

When the hydrated aluminum ion approaches a hydrated anion, both hydration sheaths are attenuated.[58] A new species is formed if no more than the inner hydration shell of each ion is retained. If both inner shells are retained, then for an anion or ligand, e.g., L^-, an ion pair, $[AlOH_2OH_2L]^{2+}$, forms, where the intermediate water molecules are shown in preferred orientation. Ion pairs are often poorly defined by experimental measurement because of the weakness of interaction. When only one water molecule separates the aluminum ion from the ligand an outer-sphere complex is formed, $[AlOH_2L]^{2+}$. When the ligand bonds directly with the aluminum ion an inner-sphere complex is formed, $[AlL]^{2+}$. Inner- and outer-sphere complexes are easier to measure reliably. Sometimes, outer-sphere complexes are also designated ion pairs; ion pairs and complexes are then distinguished only on the basis of whether the bonding between is dominantly coulombic (= ion pair) or covalent (= complex).[59] The first, more rigorous definition of inner- and outer-sphere complexes was chosen because it is preferred by inorganic and analytical chemists and by the International Union of Pure and Applied Chemistry (IUPAC).[58] Needless to say, these definitions are simplifying constructs; there is a continuous transition from the directly bonded species

Table 3
SUMMARY OF ALUMINUM FLUORIDE ASSOCIATION CONSTANTS AT 298.15 K AND ZERO IONIC STRENGTH

log K_1	log β_2	log β_3	log β_4	log β_5	log β_6	Source
7.00	—	—	—	—	—	42B/55P
6.13	11.15	15.00	17.74	19.37	19.84	43B/42B[a]
6.61	11.97	16.03	18.71	20.04	(20.44)[b]	43B/59K
7.01	12.75	17.02	19.72	20.91	20.86	59K/68H
6.98	12.60	16.65	19.03	—	—	69B
6.89	12.04	15.72	18.47	—	—	71A
7.0(±0.1)	12.7(±0.2)	16.8(±0.3)	19.4(±0.4)	20.6(±0 .5)	20.6(±1)	Selected Value

[a] These values are for 0.53 *M* KNO_3 solutions.
[b] This value was estimated by 68H based on the trend reported in 59K.

REFERENCES

42B **Brosset, C.,** Ph.D. thesis, University of Lund, Sweden, 1942.
43B **Brosset, C. and Orring, J.,** *Sven. Kem. Tidskr.*, 55, 101, 1943.
55P **Paul, A. D.,** Univ. Calif. Rad. Lab Rep. UCRL-2926, 1955.
59K **King, E. L. and Gallagher, P. K.,** *J. Phys. Chem.*, 63, 1073, 1959.
68H **Hem, J. D.,** *U.S. Geol. Surv. Water-Supply Pap.*, 1827-B, 1968.
69B **Baumann, E. W.,** *J. Inorg. Nucl. Chem.*, 31, 3155, 1969.
71A **Agarwal, R. P. and Moreno, E. C.,** *Talanta,* 18, 873, 1971.

to the completely separated ions. There is often a mixture of different forms present at any given moment. Even for the same two ions, the complex can change from ion-pair to outer-sphere to inner-sphere dominance depending on the concentration or temperature.

D. Fluoro Complexes

The fluoride ion, F^-, has the same charge and nearly the same ionic radius as the hydroxide ion, OH^-, and fluorine is the most electronegative element. Hence, it forms inner-sphere complexes with the aluminum ion up to the maximum of octahedral coordination (AlF_6^{3-}). The fluoro and hydroxo complexes are the two strongest groups of inorganic ion associations with aluminum in natural waters.

Bond and Hefter[60] have made a critical survey of aqueous fluoride stability constants from the literature, through 1977, as part of the IUPAC Chemical Data Series. They evaluated aluminum fluoride stability constants taken from 13 papers and divided them into the 4 categories: Recommended (±0.05), Tentative (±0.2), Doubtful (±1), and Rejected, based on the estimated reliability of the logarithmic values for the stability constants (shown in parentheses). They found no Recommended values, but did give two Tentative sets of stability constants, K_2 through K_3, and Doubtful K_4 at 298.15 K and zero ionic strength. We considered two additional reports for K_1 and one additional report for K_2 through K_6. These values are summarized in Table 3. Bond and Hefter[60] categorized the results from Baumann[61] and Agarwal and Moreno[62] for K_1 through K_3 as Tentative because of the good agreement between them. The value of log K_1 is slightly less than 7.0, which agrees well with Paul,[63] who corrected Brosset's[64] results. King and Gallagher[65] also corrected Brosset's results[20] but only to I = 0.07 M. Hem[66] corrected the values from King and Gallagher to zero ionic strength and found log K_1 = 7.01. Hence, we have selected the value of 7.0 with an estimated reliability of ±0.1. This stability constant is the most important one for aluminum fluoride complexes in natural waters. Based on good agreement between Hem, Paul, and Baumann for log K_1, we have selected the means between Hem's correction of earlier data and Baumann's data for log β_2, log β_3, and log β_4. For log β_5 and log β_6 we extrapolated the general trend from the previous constants, noting that there is no meaningful difference between these two constants and that AlF_6^{3-} is negligible for any natural water. These values are also consistent with the ordering principle $K_1 > K_2 > K_3 > K_4 > K_5 > K_6$. Other available reports on aluminum fluoride complexing have not been corrected for ionic strength effects and, in some instances, the investigations were so poorly done that corrections are not even warranted. Such reports were not considered in the present review.

Evidence has been obtained[22] for mixed-ligand or outer-sphere complexes, such as $AlOHF^+$, as intermediates during the formation of aluminum fluoride inner-sphere complexes, but the data suggest that they are rather short-lived and that the equilibrium constants representing the complete inner-sphere formation should be adequate for most environments. However, as the pH value of a natural water drops below 3.5 and the temperature falls below 15°C, reaction half-lives can increase to the range of minutes to hours, and such complexes may become important.

E. Sulfato Complexes

Although the complexing of aluminum with sulfate is significantly weaker than with fluoride, it is strong enough to have an effect on conductance and spectroscopic measurements of aluminum sulfate solutions. Aluminum-sulfate associations have been reported as ion pairs, outer-sphere complexes, and inner-sphere complexes. Such variable speciation is a common characteristic for many aqueous polyvalent metal-sulfate associations. The distribution of these three forms is strongly temperature-dependent and, to a lesser degree, dependent on ligand concentration.[67] Kalidas et al.[68] and Akitt et al.[67] have made measurements of K_1 for the inner-sphere complex with comparable results, but most measurements have been made of the over-all formation constants, i.e., a composite of the three types.

Garrels and Christ[69] have found that the sulfato complexes for metal ions of the same ionic charge have nearly the same stability constant, in contrast to hydroxo and carbonato complexes for which the stability constants differ greatly among different cations. They suggested that this consistency implied the predominant association in aqueous solution was an outer-sphere complex, since the hydrated radii are all similar and the bond strengths between such hydrated cations and sulfate might be similar in magnitude. A survey of trivalent metal sulfate formation constants revealed that, at 298.15 K and zero ionic strength, the log K_1 values rarely fell outside the range 3.5 to 4.0. This result was used to discard two reported measurements of about 2.0 for log K_1. The remaining five values from our literature search range from 3.01 to 3.89 (see Table 4), with a bimodal distribution suggesting that the value should be close to either 3.2 or 3.8. It is difficult to justify choosing one number over the other. Consequently, we have selected the midpoint value, log $K_1 = 3.5 \pm 0.5$, with an uncertainty that covers the range of better literature values. Although this falls between the IUPAC definitions of Tentative and Doubtful, we consider it to be adequate for most chemical-equilibrium calculations.

Only two measurements of log β_2 were found for 298.15 K and zero ionic strength and they are from two investigations that gave nearly identical values. Both studies are not of high quality and need confirmation. We have adopted the mean value and suggest that it should be adequate for most purposes of chemical modeling.

F. Phosphato Complexes

Aqueous systems involving aluminum-phosphate interactions are notoriously difficult to characterize experimentally and aluminum-phosphate complexes have not often been studied. Bjerrum and Dahm[70] have estimated log $K \approx 3$ ($I = 0.1$, $T = 18°C$) for the reaction

$$Al^{3+} + H_2PO_4^- = AlH_2PO_4^{2+} \tag{11}$$

Table 4
SUMMARY OF ALUMINUM SULFATE ASSOCIATION CONSTANTS AT 298.15 K AND ZERO IONIC STRENGTH

$\log K_1$	$\log \beta_2$	Source
3.20	5.10	62B
3.73		65N
3.01	4.90	69I
3.2		70R
3.89		71K
3.5(±0.5)	5.0(±0.5)	Selected Value

REFERENCES

62B **Behr, B. and Wendt, H.,** *Z. Elektrochem.*, 66, 223, 1962.
65N **Nishide, T. and Tsuchiya, R.,** *Bull. Chem. Soc. Jpn.*, 38, 1398, 1965.
69I **Izatt, R. M., Eatough, D., Christensen, J. J., and Bartholomew, C. H.,** *J. Chem. Soc. A*, 47, 1969.
70R **Richburg, J. S. and Adams, F.,** *Soil Sci. Soc. Am. Proc.*, 34, 728, 1970.
71K **Kryzhanovskii, M. M., Volokhov, Yu. A., Pavlov, L. N., Eremin, N. I., and Mironov, V. E.,** *Zh. Prikl. Khim.*, 44, 476, 1971.

and the investigation of Bohn and Peech[71] supported that value. Similarly, the same two investigations estimated log K ≈ 7 for

$$Al^{3+} + HPO_4^{2-} = AlHPO_4^{+} \quad (12)$$

These values are entirely reasonable based upon the correlation plot of Langmuir,[72] who independently estimated values of 3.1 and 7.4, respectively.

G. Miscellaneous Complexes

No values for aluminum carbonate complexes can be found in the extensive compilations by Sillén and Martell,[4,73] Smith and Martell,[74] and Högfeldt.[75] Only two estimates of the strength of these interactions were found in the literature. Turner et al.[76] estimated the formation constant of $AlCO_3^+$ from a correlation plot of metal-carbonate with metal-oxalate complexes. Such an estimate may not be reliable because this correlation applies only to divalent metal carbonates. The only measurement known to us is that of Ohman and Forsling,[77] who determined aluminum carbonate complexation in 0.6 *M* NaCl at 9.5 to 100 kPa CO_2 pressure and pH = 3.95 to 4.35. The best fit of their data suggested the formation of outer-sphere or ion-pair associations with tentative stoichiometries of $Al_2(OH)_2CO_3^{2+}$ and $Al_3(OH)_4HCO_3^{4+}$. The high P_{CO_2} and the difficulty of interpreting these results indicates that aluminum and carbonate or bicarbonate ions have very weak interactions that can be neglected for most natural waters.

Other possible inorganic complexes with aluminum in natural waters appear to be negligible. Ohman and Sjoberg[78] have found that aluminum borate complexes are too weak to be considered and Ohman[79] used arguments involving chemical properties and availability

Table 5
TENTATIVE THERMODYNAMIC PROPERTIES OF ALUMINUM HYDROLYSIS SPECIES AT 298.15 K[a]

	ΔG_f°(kJ/mol)	ΔH_f°(kJ/mol)	S°(J/mol/K)	C_p°(J/mol/K)	V°(cm³/mol)
Al^{3+}	−487.8(±2)	−531(±4)	−312(±15)	−119	−45
$Al(OH)^{2+}$	−696.5(±2)	−769(±4)	−177(±15)	75	—
$Al(OH)_2^+$	−904.3	−990	11.95	−194	—
$Al(OH)_3^\circ$	−1104	−1222	133.3	−505	—
$Al(OH)_4^-$	−1307(±10)	−1478	127	58.5	—

[a] Values for ΔG_f° are based on the average solubility product constants for gibbsite from Kittrick,[56] Singh,[57] May et al.,[12] and Bloom and Weaver,[42] the calorimetric value for ΔG_f° (gibbsite) from Hemingway and Robie,[53] and the hydrolysis constants given in Table 2. $\Delta H_f^\circ(Al^{3+})$ is from Hemingway and Robie,[53] $S^\circ(Al^{3+})$ is calculated from ΔG_f° and ΔH_f°, and $C_p^\circ(Al^{3+})$ and $V^\circ(Al^{3+})$ are from Hovey and Tremaine.[81] All other ΔH°, S°, and C_p° values are based on temperature-dependent equations for log K as described in the text (see Table 6); the given values for Al^{3+} above and values for $H_2O(1)$ are from the revised JANAF Tables.[80]

in natural waters to deduce that no other associations are strong enough to be amenable to measurement.

V. STANDARD-STATE THERMODYNAMIC PROPERTIES OF AQUEOUS INORGANIC ALUMINUM SPECIES

The Gibbs energy of formation of Al^{3+} is not precisely known because of the difficulty of defining the solubility product constant of gibbsite. Consequently, only a range of $\Delta G_f^\circ(Al^{3+}) = -486.2$ to 489.4 kJ/mol could be derived and we recommend the midpoint value of −487.8(±2) kJ/mol until further investigations resolve the imprecision. Using this value, we have computed the free energies for the other hydrolysis products and tabulated them in Table 5. These values were obtained from their respective stability constants (see Table 2) and the Gibbs energy of water taken from the revised JANAF Tables.[80] The $\Delta H_f^\circ(Al^{3+})$ is from Hemingway and Robie[53] and $S^\circ(Al^{3+})$ is calculated from ΔG_f° and ΔH_f°. The values for $C_p^\circ(Al^{3+})$ and $V^\circ(Al^{3+})$ are from Hovey and Tremaine.[81] The partial molal volume is consistent with previously reported data (e.g., Millero[82]).

The enthalpies, entropies, and heat capacities for the hydrolysis products, shown in Table 5, were derived from nonlinear, least-squares regressions of log K(T) data using the general equation

$$\log K = A + B/T + C(\log T)$$

and the following constraints: (1) the temperature range was restricted to 273.15 to 373.15 K; (2) the fitted curves were forced through the selected 298.15 K values from Table 2; (3) the fitted curves were considered acceptable if the differences between actual and fitted data points were within the estimated error for the points (<0.5%); and (4) the fitted curves were smooth and well-behaved including the constraint that they not show any anomalous departures from a trend parallel to the temperature-dependent ionization constant of water, $K_w(T)$. The above form of log K (T) assumes ΔC_p = constant (see Nordstrom and Munoz,[84] p. 382). The program FITIT[83] was used to compute the regression parameters. The temperature-dependent equations for the equilibrium constants and the enthalpies of hydrolysis are given in Table 6.

Although some 35 data points for $\log {}^*K_1$ were found, from the literature search for $T \neq 298.15$ K, only ten data points met the criteria listed above. These points were taken from

Table 6
ANALYTICAL FUNCTIONS FOR THE TEMPERATURE DEPENDENCE OF ALUMINUM HYDROLYSIS REACTIONS (SEE TEXT FOR SOURCES OF DATA)

Species	Equation	Temperature range (°C)
Al^{3+}	$C_p^\circ = 566.2 - 1.452T - (27338)/(T - 190)$	10—55
$AlOH^{2+}$	$\log{}^*K_1 = -38.253 - 656.27/T + 14.327 \log T$	0—100
	$\Delta H_r^\circ(kJ/mol) = 12.564 + 0.11912T$	0—100
$Al(OH)^{2+}$	$\log{}^*\beta_2 = 88.500 - 9391.6/T - 27.121 \log T$	0—100
	$\Delta H_r^\circ(kJ/mol) = 179.80 - 0.2255T$	0—100
$Al(OH)_3^\circ$	$\log{}^*\beta_3 = 226.374 - 18247.8/T - 73.597 \log T$	0—100
	$\Delta H_r^\circ(kJ/mol) = 349.35 - 0.6119T$	0—100
$Al(OH)_4^-$	$\log{}^*\beta_4 = 51.578 - 11168.9/T - 14.865 \log T$	0—100
	$\Delta H_r^\circ(kJ/mol) = 213.82 - 0.0585T$	0—100

the data of Schofield and Taylor,[31] Nikolayeva and Tolpygina,[39] Volokhov et al.,[36] and Couturier et al.[37] The differences between observed and calculated values were within 0.5%. This agrees well with the differences between results of separate studies at the same temperature, using the general equation. These results also agree quite closely with the data of Ley[85] and of Kahlenberg et al.,[86] as corrected by Kubota[35] for temperatures of 323.15 to 373.15 K.

There are almost no data for the second and third hydrolysis constants at temperatures other than 298.15 K. Hence, we assumed that their temperature dependence closely parallels the temperature dependence of the first hydrolysis constant, with the constraint that the curves go through the selected values at 298.15 K.

The fourth hydrolysis constant has been studied in detail by Apps et al.[52] in conjunction with an evaluation of gibbsite, boehmite, and diaspore solubilities in basic solutions. We agree with their interpretation that an average value, $\log{}^*\beta_4 = 22.7$, should be used at the present time and we accept their evaluation of the temperature dependence. We fitted their data for 273.15 to 373.15 K and note that our $C_p^\circ(Al(OH)_4^-)$differs from theirs because of the difference in temperature range and the uncertainties of the data. The resultant curve deviates only slightly from parallelism with $\log K_w(T)$ at higher temperatures.

The thermodynamic properties of the fluoro and sulfato complexes can be easily calculated from the data presented in the previous tables. The lack of measurements at $T \neq 298.15$ K precludes retrieving reliable heat-capacity data for these additional species.

Table 7 lists the association enthalpies for fluoro and sulfato complexes of aluminum. The main source of data on aluminum fluoride enthalpies of complexation is that of Latimer and Jolly,[87] who made measurements over a range of ionic strength. King and Gallagher[65] converted these data to $I = 0.07\ M$ and these values are given in Table 7 for the fluoro complexes. The first and second complexation enthalpies for aluminum sulfate have been measured by Izatt et al.,[88] and these have been selected for Table 7. The first sulfato complex has an enthalpy that agrees well with the value reported by Kryzhanovskii et al.,[89] but neither of these sulfate-association enthalpies are well defined in comparison with those for the fluoro or hydroxo complexes.

VI. ALUMINUM-ORGANIC STABILITY CONSTANTS

The association constants for complexing of aluminum by organic ligands are shown in Table 8. This table includes all of the suggested compounds that might serve as models for complexing behavior between aluminum and natural organic substances. Only data from

Table 7
CUMULATIVE ENTHALPIES OF ASSOCIATION FOR FLUORIDE AND SULFATE COMPLEXES OF ALUMINUM AT 298.15 K

$Al^{3+} + nF^- = AlF_n^{3-n}$	$Al^{3+} + nSO_4^{2-} = Al(SO_4)_n^{3-2n}$
ΔH_1 = 4.44 kJ/mol	ΔH_1 = 9.6 kJ/mol
ΔH_2 = 8.28 kJ/mol	ΔH_2 = 13kJ/mol
ΔH_3 = 9.04 kJ/mol	
ΔH_4 = 9.20 kJ/mol	
ΔH_5 = 7.70 kJ/mol	
ΔH_6 = 7.00 kJ/mol	

Table 8
ALUMINUM-ORGANIC ASSOCIATION CONSTANTS[a]

	Reaction	log K	Source	Remarks
		I. Monocarboxylic Acids		
A.	Methanoic acid (formic acid), HCO_2H			
	K_1	0.56/1.36	73T/75K	I = 1.0 *M* (NO_3^-/ClO_4^-), 25°C
	β_2	1.76/2.02	73T/75K	I = 1.0 *M* ($(NO)_3^-/ClO_4^-$), 25°C
B.	Ethanoic acid (acetic acid), CH_3CO_2H			
	K_1	1.51	75K	I = 1.0 *M* (ClO_4^-), 25°C
C.	Propanoic acid, $CH_3CH_2CO_2H$			
	K_1	1.69/1.78	75K/75T	I = 1.0 *M* (ClO_4^-), 25°C
		1.81	75T	I = 1.0 *M* (ClO_4^-), 35°C
	β_2	3.42	75T	I = 1.0 *M* (ClO_4^-), 25°C
		4.04	75T	I = 1.0 *M* (ClO_4^-), 35°C
D.	2-Hydroxypropanoic acid, $CH_3CHOHCO_2H$			
	K_1	4.26	75T	I = 1.0 *M* (ClO_4^-), 25°C
		2.28	75T	I = 1.0 *M* (ClO_4^-), 25°C
	β_2	4.88	75T	I = 1.0 *M* (ClO_4^-), 25°C
		4.94	75T	I = 1.0 *M* (ClO_4^-), 25°C
E.	Butanoic acid, $CH_3CH_2CH_2CO_2H$			
	K_1	1.58	75K	I= 1.0 *M* (NO_3^-), 25°C
F.	Benzoic acid, $C_6H_5CO_2H$			
	K for $Al^{3+} + L^- + OH^- = AlOHL^+$			
		12.1	70Nb	I = 0.5 *M* (ClO_4^-), 25°C
G.	Nitroacetic acid, $O_2NCH_2CO_2H$			
	K_1	0.48	49P	I = 0.4 *M*, 18°C
H.	D-2-Hydroxypropanoic acid (lactic acid), $CH_3C(OH)HCO_2H$			
	K_1	2.38	71H	I = 0.2 *M* (NO_3^-), 20°C
	β_2	4.56	71H	I = 0.2 *M* (NO_3^-), 20°C
	β_3	6.66	71H	I = 0.2 *M* (NO_3^-), 20°C
I.	D-2,3,4,5,6-Pentahydroxyhexanoic acid (D-gluconic acid), $C_6H_{12}O_7$			
	K_1	1.98	84M	I = 0.1 *M* (NO_3^-), 25°C
		II. Dicarboxylic Acids		
A.	Ethanedioic acid (oxalic acid), HO_2CCO_2H			
	K_1	6.1/4.9	68B/70Gc	I = 1.0 *M* (ClO_4^-), 25°C
	β_2	11.1	68B	I = 1.0 *M* (ClO_4^-), 25°C
	β_3	15.1	68B	I = 1.0 *M* (ClO_4^-), 25°C

Table 8 (continued)
ALUMINUM-ORGANIC ASSOCIATION CONSTANTS[a]

	Reaction	log K	Source	Remarks
B.	Propanedioic acid (malonic acid), $HO_2CCH_2CO_2H$			
	K_1	3.4	74T	I = 1.0 *M* (ClO_4^-), 25°C
	β_2	6.0	74T	I = 1.0 *M* (ClO_4^-), 25°C
	β_3	8.6	74T	I = 1.0 *M* (ClO_4^-), 25°C
C.	Butanedioic acid (succinic acid), $HO_2CCH_2CH_2CO_2H$			
	K_1	3.83	74T	I = 1.0 *M* (ClO_4^-), 25°C
	β_2	5.92	74T	I = 1.0 *M* (ClO_4^-), 25°C
	β_3	8.97	74T	I = 1.0 *M* (ClO_4^-), 25°C
D.	Benzene-1,2-dicarboxylic acid (phthalic acid), $C_6H_4(CO_2H)_2$			
	K_1	3.18	70Nb	I = 0.5 *M* (ClO_4^-), 25°C
	β_2	6.32	70Nb	I = 0.5 *M* (ClO_4^-), 25°C
E.	L-Hydroxybutanedioic acid (malic acid), $HO_2CCH_2C(OH)HCO_2H$			
	K_1	5.34	69P	I = 0.2 *M* (NO_3^-), 20°C
	β_2	9.32	69P	I = 0.2 *M* (NO_3^-), 20°C
F.	*meso*-2,3-Dihydroxybutanedioic acid (*meso*-tartaric acid), $HO_2CC(OH)HC(OH)HCO_2H$			
	K_1	5.62	72M	I = 0.1 *M* (ClO_4^-), 25°C
	β_2	9.95	72M	I = 0.1 *M* (ClO_4^-), 25°C
G.	D,L-2,3-Dihydroxybutanedioic acid (D,L-tartaric acid), $HO_2CC(OH)HC(OH)HCO_2H$			
	K_1	5.32	72M	I = 0.1 *M* (ClO_4^-), 25°C
	β_2	7.65	84M	I = 0.1 *M* (NO_3^-), 25°C
H.	Oxydiacetic acid (diglycollic acid), $HO_2CCH_2OCH_2CO_2H$			
	K_1	3.16	72N	I = 0.5 *M* (ClO_4^-), 25°C
	β_2	5.25	72N	I = 0.5 *M* (ClO_4^-), 25°C
I.	Thiodiacetic acid, (thiodiglycollic acid) $HO_2CCH_2SCH_2CO_2H$			
	K_1	1.93	72N	I = 0.5 *M* (ClO_4^-), 25°C
	β_2	6.80	79D	I = 0 *M*, 25°C
J.	Dithiodiacetic acid, (dithiodiglycollic acid) $HO_2CCH_2SSCH_2CO_2H$			
	β_2	6.36	79D	I = 0 *M*, 25°C
K.	D-2,3,4,5-Tetrahydroxyhexanedioic acid (saccharic acid), $HO_2CCH(OH)CH(OH)CH(OH)CH(OH)CO_2H$			
	K for $Al^{3+} + L^{2-} = AlH_{-1}L + H^+$			
		1.57	84M	I = 0.1 *M* (NO_3^-), 25°C

III. Tricarboxylic Acids

	Reaction	log K	Source	Remarks
A.	2-Hydroxypropane-1,2,3-tricarboxylic acid (citric acid), $HOC(CO_2H)(CH_2CO_2H)_2$			
	K_1	7.98	84M	I = 0.1 *M* (NO_3^-), 25°C
		7.14	83Oa	I = 0.1 *M* (Cl^-), 25°C
	K for $Al^{3+} + H_3L = AlHL^+ + 2H^+$			
		−2.20	84M	I = 0.1 *M* (NO_3^-), 25°C
		−2.68	83Oa	I = 0.1 *M* (Cl^-), 25°C

IV. Monohydroxy Phenols

	Reaction	log K	Source	Remarks
A.	2-Hydroxybenzoic acid (salicylic acid), $C_6H_4OHCO_2H$			
	K_1	14.5	75S	I = 0.1 *M* ClO_4^-), 25°C
		12.9	69H	I = 0.1 *M* (NO_3^-), 20°C
		13.0	83Ob	I = 0.6 *M* (Cl^-), 25°C

Table 8 (continued)
ALUMINUM-ORGANIC ASSOCIATION CONSTANTS[a]

	Reaction	log K	Source	Remarks
	β_2	23.2	69H	I = 0.1 *M* (NO_3^-), 20°C
		23.6	83Ob	I = 0.6 *M* (Cl^-), 25°C
	β_3	29.8	69H	I = 0.1 *M* (NO_3^-), 20°C
B.	2-Hydroxy-5-sulfobenzoic (5-sulfosalicylic acid), $HO_3SC_6H_3(OH)CO_2H$			
	K_1	12.3	69H	I = 0.1 *M* (NO_3^-), 20°C
	β_2	20.0	69H	I = 0.1 *M* (NO_3^-), 20°C
	β_3	25.8	69H	I = 0.1 *M* (NO_3^-), 20°C
	V. Dihydroxy Phenols			
A.	1,2-Dihydroxybenzene (catechol), $C_6H_4(OH)OH$			
	K_1	16.1	84M	I = 0.1 *M* (NO_3^-), 25°C
	β_2	29.1	84M	I = 0.1 *M* (NO_3^-), 25°C
	β_3	37.8	84M	I = 0.1 *M* (NO_3^-), 25°C
B.	1,2-Dihydroxybenzene-4-sulfonic acid (catechol-5-sulfonic acid), $HO_3SC_6H_3(OH)_2$			
	K_1	16.6	69H	I = 0.1 *M* (NO_3^-), 20°C
	β_2	29.9	69H	I = 0.1 *M* (NO_3^-), 20°C
	β_3	39.2	69H	I = 0.1 *M* (NO_3^-), 20°C
C.	1,2-Dihydroxybenzene-3,5-disulfonic acid (tiron), $(HO_3S)_2C_6H_2(OH)_2$			
	K_1	19.06	77M	I = 0 *M*, 25°C
	β_2	31.10	77M	I = 0 *M*, 25°C
	β_3	33.5	77M	I = 0 *M*, 25°C
	VI. Trihydroxy Phenols			
A.	1,2,3-Trihydroxybenzene (pyrogallol), $C_6H_3(OH)_3$			
	K_1	24.50	70Gb	I = 0.2 *M* (Cl^-), 22°C
	β_2	44.55	70Gb	I = 0.2 *M* (Cl^-), 22°C
	β_3	57.95	70Gb	I = 0.2 *M* (Cl^-), 22°C
B.	3,4,5-Trihydroxybenzoic acid (gallic acid), $C_6H_2(OH)_3CO_2H$			
	K_1	14.24	81O	I = 0.6 *M* (Cl^-), 25°C
	β_2	25.37	81O	I = 0.6 *M* (Cl^-), 25°C
	β_3	33.31	81O	I = 0.6 *M* (Cl^-), 25°C
	K for $Al^{3+} + H_3L = AlHL^+ + H^+$	−4.3	81O	I = 0.6 *M* (Cl^-), 25°C
	VII. Napthols			
A.	1-(2-Hydroxy-5-sulfophenylazo)-2-napthol (solochrome violet R) $HO_3SC_6H_3(OH)N_2C_{10}OH$			
	K_1	18.4	63C	I = 0 *M*, 25°C
	β_2	31.6	63C	I = 0 *M*, 25°C
B.	3-Hydroxy-2-naphthoic acid, $C_{10}H_6OHCO_2H$			
	K_1	13.38	66M	I = 0 *M*, 25°C
C.	1,8-Dihydroxynaphthalene-3,6-disulfonic acid (chromotropic acid), $C_{10}H_4(OH)_2(SO_3H)_2$			
	K_1	17.1	69H	I = 0.1 *M* (NO_3^-), 20°C
	β_2	29.8	69H	I = 0.1 *M* (NO_3^-), 20°C

Table 8 (continued)
ALUMINUM-ORGANIC ASSOCIATION CONSTANTS[a]

	Reaction	log K	Source	Remarks
		VIII. Amino Acids		
A.	L-Aminobutanedioic acid (aspartic acid), $HO_2CCH_2C(NH_2)HCO_2H$			
	K_1	16.29	72S	I = 0.1 *M* (ClO_4^-), 25°C
	β_2	30.69	72S	I = 0.1 *M* (ClO_4^-), 25°C
	β_3	42.19	72S	I = 0.1 *M* (ClO_4^-), 25°C
B.	Iminodiacetic acid (IDA acid), $NH(CH_2CO_2H)_2$			
	K_1	8.10	71La	I = 0.5 *M* (ClO_4^-), 25°C
	β_2	15.07	71La	I = 0.5 *M* (ClO_4^-), 25°C
C.	L-2-Aminopentanedioic acid (glutamic acid), $HO_2CCH_2CH_2C(NH_2)HCO_2H$			
	K_1	15.12	72S	I = 0.1 *M* (ClO_4^-), 25°C
	β_2	29.40	72S	I = 0.1 *M* (ClO_4^-), 25°C
	β_3	38.60	72S	I = 0.1 *M* (ClO_4^-), 25°C
D.	Nitrilotriacetic acid (NTA acid), $HO_2CCH_2N(CH_2CO_2H)_2$			
	K_1	11.4	67B	I = 0.2 *M*, 25°C
E.	*N*-(2-Hydroxyethyl)iminodiacetic acid (HIDA acid), $HOCH_2CH_2N(CH_2CO_2H)_2$			
	K_1	7.74	74M	I = 0.1 *M*, 25°C
F.	*N*-(2-Hydroxyethyl)ethylenedinitrilo-*N*,*N'*,*N*-triacetic acid (HEDTA), $C_{10}H_{18}O_7N_2$			
	K_1	14.4	74M	I = 0.1 *M*, 25°C
G.	Ethylenedinitrilotetraacetic acid (EDTA), $(HO_2CCH_2)_2NCH_2CH_2N(CH_2CO_2H)_2$			
	K_1	16.5	74M	I = 0.1 *M*, 25°C
H.	*trans*-1,2-cyclohexylenedinitrilotetraacetic acid (CDTA), $C_{14}H_{22}O_8N_2$			
	K_1	19.6	74M	I = 0.1 *M*, 25°C
I.	Triethylenetetranitrilohexaacetic acid (TTHA), $C_{18}H_{30}O_{12}N_4$			
	K_1	21.0	70H	I = 0.1 *M*, 25°C
J.	Pyridine-2,5-dicarboxylic acid (isocinchomeronic acid), $C_7H_5O_4N$			
	K_1	3.95	70Na	I = 0.5 *M*, 25°C
	β_2	7.24	70Na	I = 0.5 *M*, 25°C
K.	Pyridine-2,6-dicarboxylic $acid_2$ (dipicolinic acid), $C_7H_5O_4N$			
	K_1	4.87	68N	I = 0.5 *M*, 25°C
	β_2	8.32	68N	I = 0.5 *M*, 25°C
L.	*N*,*N*-bis(2-hydroxyethyl)glycine (bicine), $HO_2CCH_2N(CH_2CH_2OH)_2$			
	K for $Al^{3+} + HL = AlHL^{3+}$	3.38	84M	I = 0.1 *M* NO_3^-), 25°C
		IX. Amines and Other Complexes		
A.	7-Iodo-8-hydroxyquinoline-5-sulfonic acid (ferron), $C_9H_6O_4NIS$			
	K_1	7.6/6.8	61L/71Lb	I = 0.1 *M*, 25°C/28°C
	β_2	14.7/13.8	61L/71LB	I = 0.1 *M*, 25°C/28°C
	β_3	20.3	61L	I = 0.1 *M*, 25°C
B.	8-Hydroxyquinoline (oxine), C_9H_7ON			
	β_3	33.42	71B	20 Vol.% EtOH, ambient temp.
		33.75	73S	I = 0.1 *M* (ClO_4^-), 60% dioxane, 25°C
C.	2-Amino-2-propylphosphonic acid, $NH_2C(CH_3)_2PO_3H_2$			
	K_1	11.42	69D	I = 0.1 *M* (Cl^-), 25°C
	β_2	19.53	69D	I = 0.1 *M* (Cl^-), 25°C
D.	Amino(phenyl)methylenediphosphonic acid, $NH_2C(PO_3H_2)_2C_6H_5$			
	K_1	18.79	69D	I = 0.1 *M* (Cl^-), 25°C
	β_2	26.31	69D	I = 0.1 *M* (Cl^-), 25°C
E.	3,4-Dihydroxy-3-cyclobutene-1,2-dione (squaric acid), $C_4H_2O_4$			
	K_1	2.83	69T	I = 0.3 *M* (ClO_4^-), 25°C

Table 8 (continued)
ALUMINUM-ORGANIC ASSOCIATION CONSTANTS[a]

	Reaction	log K	Source	Remarks
F.	3-Hydroxy-2-methyl-4-pyrone (maltol acid), $C_6H_6O_3$			
	K_1	7.7	69C	I = 0.1 *M* NO_3^-), 25°C
	β_2	15.25	69C	I = 0.1 *M* NO_3^-), 25°C
	β_3	21.90	69C	I = 0.1 *M* NO_3^-), 25°C
G.	Acetohydroxamic acid, $CH_3C(O)NHOH$			
	K_1	7.95	63A	I = 0.1 *M*, 20°C
H.	L-Ascorbic acid (vitamin C), $C_6H_8O_6$			
	K_1	1.89	66V	I = 0.1 *M*, 25°C
	β_2	3.7	66V	I = 0.1 *M*, 25°C
I.	Folic acid (vitamin B_{10}), $C_{19}H_{19}O_6N_7$			
	K_1	5.80	70Nc	I = 0 *M*, 30°C
	β_2	10.50	70Nc	I = 0 *M*, 30°C
	β_3	15.15	70Nc	I = 0 *M*, 30°C
J.	Pentane-2,4-dione (acetylacetone), $CH_3C(O)CH_2C(O)CH_3$			
	K_1	8.6	55I	I = 0 *M*, 30°C
	β_2	16.5	55I	I = 0 *M*, 30°C
	β_3	22.3	55I	I = 0 *M*, 30°C
K.	3,3′,4′-Trihydroxyfuchsone-2″-sulfonic acid (pyrocatechol violet), $HO_3SC_6H_4C(C_6H_3(OH)O)C_6H_3(OH)_2$			
	K_1	25.12	70Ga	I = 0.2 *M* (Cl^-), ambient temp.
	β_2	47.39	70Ga	I = 0.2 *M* (Cl^-), ambient temp.
	β_3	68.13	70Ga	I = 0.2 *M* (Cl^-), ambient temp.
L.	3,3′,4′,5,7-Pentahydroxyflavone(quercetin), $C_{15}H_{10}O_7$			
	β_2	12	75L	I = 0 *M*, 25°C
M.	5-Hydroxy-2-hydroxymethyl-4-pyrone (kojic acid), $C_6H_6O_4$			
	K_1	7.7	59O	I = 0.1 *M*, 25°C
	β_2	14.2	59O	I = 0.1 *M*, 25°C
	β_3	19.5	59O	I = 0.1 *M*, 25°C

[a] The primary sources of data for this table are the IUPAC volume on metal-ion complexes,[100] the three-volume set on organic stability constants by Martell and Smith,[101-103] and journal publications.

REFERENCES

49P **Pedersen, K. J.,** *Acta Chem. Scand.*, 3, 676, 1949.
55I **Izatt, R. M., Fernelius, W. C., Haas, C. G., Jr., and Block, B. P.,** *J. Phys. Chem.*, 59, 170, 1955.
59O **Okac, A. and Kolarik, Z.,** *Collect. Czech. Chem. Commun.*, 24, 266, 1959.
61L **Langmyhr, F. J. and Storm, A. R.,** *Acta Chem. Scand.*, 15, 1461, 1961.
63A **Anderegg, G., L'Eplattenier, F., and Schwarzenbach, G.,** *Helv. Chim. Acta*, 46, 1400, 1963.
63C **Coates, E., Evans, J. R., and Rigg, B.,** *Trans. Faraday Soc.*, 59, 2369, 1963.
66M **Makitie, O.,** *Suom. Kemistil. B*, 39, 175, 1966.
66V **Veselinovic, D. S. and Susic, M. V.,** *Bull. Chem. Soc. Belgrade*, 31, 37, 1966.
67B **Bhat, T. R., Das, R. R., and Shankar, J.,** *Indian J. Chem.*, 5, 324, 1967.
68B **Bottari, E. and Ciavatta, L.,** *Gazz. Chim. Ital.*, 98, 1004, 1968.
68N **Napoli, A.,** *Talanta*, 15, 189, 1968.
69C **Chiacchierini, E. and Bartusek, M.,** *Collect. Czech. Chem. Commun.*, 34, 530, 1969.
69D **Dyatlova, N. M., Medyntsev, V. V., Balashova, T. M., Medved, T. Ya., and Kabachnik, N. I.,** *Zh. Obshch. Khim.*, 39, 329, 1969.
69H **Havelkova, L. and Bartusek, M.,** *Collect. Czech. Chem. Commun.*, 34, 3722, 1969.
69P **Pavlinova, A. V. and Vysotskaya, T. D.,** *Ukr. Khim. Zh.*, 34(4), 402, 1968.
69T **Tedesco, P. H. and Walton, H. F.,** *Inorg. Chem.*, 8, 932, 1969.
70Ga **Goina, T., Olariu, M., and Bocanicio, L.,** *Rev. Roum. Chim.*, 15, 1049, 1970.
70Gb **Goina, T. and Olariu, M.,** *Rev. Roum. Chim.*, 15, 1547, 1970.
70Gc **Gordienko, V. I. and Mikhailyuk, Y. I.,** *Zh. Anal. Khim.*, 25(12), 2267, 1970.

70H **Harju, L.,** *Anal. Chim. Acta,* 50, 475, 1970.
70Na **Napoli, A.,** *J. Inorg. Nucl. Chem.,* 32, 1907, 1970.
70Nb **Napoli, A. and Liberti, A.,** *Gazz. Chim. Ital.,* 100, 906, 1970.
70Nc **Nayan, R. and Dey, A. K.,** *Z. Naturforsch. Teil B,* 25, 1453, 1970.
71B **Bryuk, E. A. and Ravitskaya, R. V.,** *Zh. Anal. Khim.,* 26(9), 1752, 1971.
71H **Hurnik, B.,** *Rocz. Chem.,* 45, 147, 1971.
71La **Liberti, A. and Napoli, A.,** *J. Inorg. Nucl. Chem.,* 33, 89, 1971.
71Lb **Lingaiah, P. and Sandarom, E. V.,** *J. Indian Chem. Soc.,* 48, 961, 1971.
72M **Manning, P. G. and Ramamoorthy, S.,** *J. Inorg. Nucl. Chem.,* 34, 1997, 1972.
72N **Napoli, A.,** *J. Inorg. Nucl. Chem.,* 34, 1225, 1972.
72S **Singh, M. K. and Srivastava, M. N.,** *J. Inorg. Nucl. Chem.,* 35, 1621, 1973.
73S **Steger, H. F. and Corsini, A.,** *J. Inorg. Nucl. Chem.,* 35, 1621, 1973.
73T **Tedesco, P. H., de Rumi, V. B., and Quintana, J. A. G.,** *J. Inorg. Nucl. Chem.,* 35, 285, 1973.
74M **Martell, A. E. and Smith, R. M.,** *Critical Stability Constants,* Vol. 1, Plenum Press, New York, 1974, 469.
74T **Tedesco, P. H. and Quintana, J. A. G.,** *J. Inorg. Nucl. Chem.,* 36, 2628, 1974.
75K **Kereichuk, A. S. and Ilicheva, L. M.,** *Russ. J. Inorg. Chem.,* 20, 291, 1975.
75L **Lind, C. and Hem, J. D.,** *U.S. Geological Survey, Water-Supply Paper,* 1827-G, 1975.
75S **Secco, F. and Venturini, M.,** *Inorg. Chem.,* 14, 1978, 1975.
75T **Tedesco, P. H. and de Rumi, V. B.,** *J. Inorg. Nucl. Chem.,* 37, 1833, 1975.
77M **Martell A. E. and Smith, A. E.,** *Critical Stability Constants,* Vol. 3, Plenum Press, New York, 1977.
79D **Dubey, S. N., Singh, A., Kalra, H. L., and Puri, D. M.,** *J. Indian Chem. Soc.,* 56, 451, 1979.
81O **Ohman, L.-O. and Sjoberg, S.,** *Acta Chem. Scand.,* A35, 201, 1981.
83Oa **Ohman, L.-O. and Sjoberg, S.,** *J. Chem. Soc. Dalton Trans.,* 2513, 1983.
83Ob **Ohman, L.-O. and Sjoberg, S.,** *Acta Chem. Scand.,* A37, 875, 1983.
84M **Motekaitis, R. J. and Martell, A. E.,** *Inorg. Chem.,* 23, 18, 1984.

those studies which recorded both temperature and ionic strength are provided. Unfortunately, very few studies attempted to make ionic strength corrections to the common reference state of infinite dilution. However, such corrections could be made for much of this data so that it would be more amenable to chemical equilibrium computations. There are simply not enough data to make a proper evaluation or selection of best values. There were only four stability constants for which more than one reliable measurement was found and listed in Table 8. In these four instances the agreement is excellent.

Natural organic substances dissolved in surface waters have been found to complex a significant proportion of the dissolved aluminum content of such waters.[90,91] These organic ligands have not been well characterized but some of their properties are known or can be postulated (see Chapter 5). About 90% of all dissolved organic carbon in natural waters consists of carboxylic acids (primarily in the form of fulvic and humic acids[92]) and the most appropriate model compounds for trace-metal complexation in soils are thought by some investigators to be salicylic and phthalic acids.[93] This hypothesis seems to be valid for natural organic ligands reacting with aluminum, based upon the study of Plankey and Patterson,[23] who showed that two kinetically distinguishable components of a fulvic acid were bound to aluminum; one site compared well with the kinetic and equilibrium data for aluminum-salicyclic acid binding and the other with data for aluminum-phthalic acid binding. Lind and Hem[94] argued that flavanoids are also good candidates because they are the largest group of naturally occurring phenols and they may polymerize to form strongly colored compounds observed in many natural waters. Hence, they chose quercetin as a model compound and measured its aluminum complexation constant. Ohman[79] pointed out that among the low relative molecular mass organic compounds, aluminum complexation ability seems to be greatest for polycarboxylic acids, so he investigated aluminum-citrate complexing.[95] For analogues to high relative molecular mass organic compounds he chose gallic acid,[96,97] salicylic acid,[98] and catechol.[99] He also suggested that polynuclear aluminum ion formation occurred with some of these organic compounds.

ACKNOWLEDGMENTS

We should like to acknowledge the assistance of Jim Ball in fitting data, Jeanne Dileo-Stevens for drafting the figure, and especially Jo Sains for her courageous and masterful efforts to conquer our PC and produce a flawless manuscript. We are also very grateful to John Hem, George Parks, Owen Bricker, and John Apps for their critical comments and discussions during their review of our manuscript.

REFERENCES

1. **Jenne, E. A., Ed.,** *Chemical Modeling in Aqueous Systems: Speciation Sorption, Solubility, and Kinetics,* ACS Symp. Ser. No. 93, American Chemical Society, Washington, D.C., 1979.
2. **Perrin, D. D.,** Recent applications of digital computers in analytical chemistry, *Talanta,* 24, 339, 1977.
3. **Bertsch, P. M.,** Aqueous polynuclear aluminum species, in *The Enironmental Chemistry of Aluminum,* Sposito, G., Ed., CRC Press, Boca Raton, FL, 1989, chap. 4.
4. **Sillén, L. G. and Martell, A. E.,** *Stability Constants of Metal-Ion Complexes,* Spec. Publ. No. 17, The Chemical Society, London, 1964.
5. **Hartley, F. R., Burgess, C., and Alcock, R.,** *Solution Equilibria,* Ellis Harwood, New York, 1980.
6. **Plummer, L. N. and Busenberg, E.,** The solubilities of calcite, aragonite and vaterite in CO_2-H_2O solutions between 0 and 90°C, and an evaluation of the aqueous model for the system $CaCO_3$-CO_2-H_2O, *Geochim. Cosmochim. Acta,* 46, 1011, 1982.
7. **Bates, R. G.,** *Determination of pH: Theory and Practice,* 2nd ed., John Wiley & Sons, New York, 1973.
8. **May, H. M., Helmke, P. A., and Jackson, M. L.,** Gibbsite solubility and thermodynamic properties of hydroxy-aluminum ions in aqueous solution, *Geochim. Cosmochim. Acta,* 43, 861, 1979.
9. **Turner, D. R.,** Three forms of aluminum in aqueous systems determined by 8-quinolinolate extraction methods, *Can. J. Chem.,* 47, 2521, 1969.
10. **Barnes, R. B.,** The determination of specific forms of aluminum in natural water, *Chem. Geol.,* 15, 177, 1975.
11. **Bloom, P. R., Weaver, R. M., and McBride, M. B.,** The spectrophotometric and fluorometric determination of aluminum with 8-hydroxyquinoline and butylacetate extraction, *Soil Sci. Soc. Am. J.,* 41, 1068, 1978.
12. **May, H. M., Helmke, P. A., and Jackson, M. L.,** Determination of mononuclear dissolved aluminum in near-neutral waters, *Chem. Geol.,* 24, 259, 1979.
13. **Driscoll, C. T.,** A procedure for the fractionation of aqueous aluminum in dilute acidic waters, *Int. J. Environ. Anal. Chem.,* 16, 267, 1984.
14. **Akitt, J. W., Farthing, A., and Howarth, O. W.,** Aluminum-27 nuclear magnetic resonance studies of the hydrolysis of aluminum (III). III. Stopped-flow kinetic studies, *J. Chem. Soc. Dalton Trans.,* p. 1609, 1981.
15. **Van den Berg, C. M. G., Murphy, K., and Riley, J. P.,** The determination of aluminum in seawater and freshwater by cathodic stripping voltammetry, *Anal. Chim. Acta,* 188, 177, 1986.
16. **Holmes, L. P., Cole, D. L., and Eyring, E. M.,** Kinetics of aluminum ion hydrolysis in dilute solutions, *J. Phys. Chem.,* 72, 301, 1968.
17. **Fong, D.-W. and Grunwald, E.,** Kinetic study of proton exchange between the $Al(OH_2)_6^{3+}$ ion and water in dilute acid. Participation of water molecules in proton transfer, *J. Am. Chem. Soc.,* 91, 2413, 1969.
18. **Miceli, J. and Stuehr, J.,** Ligand penetration rates into metal ion coordination spheres. Aluminum (III), gallium (III), and indium (III) sulfates, *J. Am. Chem. Soc.,* 90, 6967, 1968.
19. **Secco, F. and Venturini, M.,** Mechanism of complex formation. Reaction between aluminum and salicylate ions, *Inorg. Chem.,* 14, 1978, 1975.
20. **Brosset, C. and Orring, J.,** Studies on the consecutive formation of aluminum fluoride complexes, *Sven Kem. Tidskr.,* 55, 101, 1943.
21. **Kleiner, K. E.,** Interaction between aluminum and fluorine ions. The system $Al(NO_3)_3$-NaF-$Fe(NO_3)_3$-KSCN in nitric acid solution, *Zh. Obshch. Khim.,* 20, 221, 1950.
22. **Plankey, B. J., Patterson, H. H., and Cronan, C. S.,** Kinetics of aluminum fluoride complexation in acidic waters, *Environ. Sci. Technol.,* 20, 160, 1986.
23. **Plankey, B. J. and Patterson, H. H.,** Kinetics of aluminum-fulvic acid complexation in acidic waters, *Environ. Sci. Technol.,* 21, 595, 1987.

24. **Ares, J.,** Identification of aluminum species in acid forest soil solutions on the basis of Al:F reaction kinetics. I. Reaction paths in pure solutions, *Soil Sci.,* 141, 399, 1986.
25. **Ares, J.,** Identification of aluminum species in acid forest soil solutions on the basis of Al:F reaction kinetics. II. An example at the Solling area, *Soil Sci.,* 142, 13, 1986.
26. **Ares, J.,** Identification of hydrolysis products of aluminum in natural waters. I. n-Dimensional calibration of Al/F kinetic pathways, *Anal. Chim. Acta,* 187, 181, 1986.
27. **Ares, J.,** Identification of hydrolysis products of aluminum in natural waters. II. ALSPEC, a computerized procedure for quantifying equilibria with inorganic and organic ligands, *Anal. Chim. Acta,* 187, 195, 1986.
28. **Connick, R. E. and Poulson, R. E.,** Nuclear magnetic resonance studies of the aluminum fluoride complexes, *J. Am. Chem. Soc.,* 79, 5153, 1957.
29. **Akitt, J. W., Greenwood, N. N., Khandelwal, B. L., and Lester, G. D.,** ^{27}Al nuclear magnetic resonance studies of the hydrolysis and polymerisation of the hexa-aquo-aluminum (III) cation, *J. Chem. Soc. Dalton Trans.,* p.604, 1972.
30. **Parks, G. A.,** Free energies of formation and aqueous solubilities of aluminum hydroxides and oxide hydroxides at 25°C, *Am. Mineral.,* 57, 1163, 1972.
31. **Schofield, R. K. and Taylor, A. W.,** The hydrolysis of aluminum salt solutions, *J. Chem. Soc. (London),* p. 4445, 1954.
32. **Frink, C. R. and Peech, M.,** Hydrolysis of the aluminum ion in dilute aqueous solutions, *Inorg. Chem.,* 2, 473, 1963.
33. **Raupach, M.,** Solubility of simple aluminum compounds expected in soils. I. Hydroxides and oxyhydroxides, *Aust. J. Soil Res.,* 1, 28, 1963.
34. **Baes, C. F. and Mesmer, R. M.,** *The Hydrolysis of Cations,* Wiley-Interscience, New York, 1976, 112.
35. **Kubota, H.,** Properties and Volumetric Determination of Aluminum Ion, Ph.D. thesis, University of Wisconsin, Madison, 1956.
36. **Volokhov, Y. A., Pavlov, L. N., Eremin, N. I., and Mironov, V. E.,** Hydrolysis of aluminum salts, *Zh. Prikl. Khim.,* 44, 246, 1971.
37. **Couturier, Y., Michard, G., and Sarazin, G.,** Constantes de formation des complexes hydroxydes de l'aluminum en solution aqueuse de 20 a 70°C, *Geochim. Cosmochim. Acta,* 48, 649, 1984.
38. **Nishide, T. and Tsuchiya, R.,** The formation of $Al^{3+}(SO_4)^{2-}$ ion pair in an aqueous solution of potassium aluminum alum, *Bull. Chem. Soc. Jpn.,* 38, 1398. 1965.
39. **Nikolayeva, N. M. and Tolpygina, L. N.,** Hydrolysis of aluminum salts at increased temperatures, *Izv. Sib. Otd. Akad. Nauk SSSR Ser. Khim. Nauk,* 3, 308, 1969.
40. **Nazarenko, V. A. and Biryuk, E. A.,** The hydrolysis constant of aluminum ions in solutions with a variable ionic strength; the correlation of such constants for group III ions with the position of the element in the periodic system, *Russ. J. Inorg. Chem.,* 19, 341, 1974.
41. **Feitknecht, W. and Schindler, P.,** *Solubility Constants of Metal Oxides, Metal Hydroxides, and Metal Hydroxide Salts in Aqueous Solution,* Butterworths, London, 1963, 199.
42. **Bloom, P. R. and Weaver, R. M.,** Effect of the removal of reactive surface material on the solubility of synthetic gibbsites, *Clays Clay Miner.,* 30, 281, 1982.
43. **Hemingway, B. S., Robie, R. A., and Kittrick, J. A.,** Revised values for the Gibbs free energy of formation of $[Al(OH)_4^-]$, diaspore, boehmite and bayerite at 298.15 K and 1 bar, the thermodynamic properties of kaolinite to 800 K and 1 bar, and the heats of solution of several gibbsite samples, *Geochim. Cosmochim. Acta,* 42, 1533, 1978.
44. **Hemingway, B. S.,** Gibbs free energies of formation for bayerite, nordstrandite, $Al(OH)^{2+}$, and $Al(OH)_2^+$, aluminum mobility, and the formation of bauxites and laterites, in *Advances in Physical Geochemistry,* Vol. 2, Saxena, S. K., Ed., Springer-Verlag, New York, 1982, 285.
45. **Hem, J. D. and Roberson, C. E.,** Form and stability of aluminum hydroxide complexes in dilute solution, *U.S. Geol. Surv. Water-Supply Pap.,* 1827-A, 1967.
46. **Hem, J. D., Roberson, C. E., Lind, C. J., and Polzer, W. L.,** Chemical interactions of aluminum with aqueous silica at 25°C, *U.S. Geol. Surv. Water-Supply Pap.,* 1827-E, 1973.
47. **Dyrssen, D.,** The solubility and complex formation of aluminum hydroxide, *Vatten,* 48, 3, 1984.
48. **Moolenaar, R. J., Evans, J. C., and McKeever, L. D.,** The structure of the aluminate ion in solutions at high pH, *J. Phys. Chem.,* 74, 3629, 1970.
49. **Akitt, J. W. and Gessner, W.,** Aluminum-27 nuclear magnetic resonance investigations of highly alkaline aluminate solutions, *J. Chem. Soc. Dalton Trans.,* p. 147, 1984.
50. **Jackson, J. A., Lemons, J. F., and Taube, H.,** Nuclear magnetic resonance (NMR) studies on hydration of cations, *J. Chem. Phys.,* 32, 553, 1960.
51. **Fratiello, A., Lee, R. E., Nishida, V. M., and Schuster, R. E.,** Proton magnetic resonance coordination number study of aluminum (III), beryllium (II), gallium (III), indium (III), and magnesium (II) in water and aqueous solvent mixtures, *J. Chem. Phys.,* 48, 3705, 1968.

52. **Apps, J. A., Neil, J. M., and Jun, C.-H.,** Thermochemical properties of gibbsite, boehmite, diaspore and the aluminate ion between 0 and 350°C, Lawrence Berkeley Laboratory, Report #LBL-21482, in press.
53. **Hemingway, B. S. and Robie, R. A.,** The entropy and Gibbs free energy of formation of the aluminum ion, *Geochim. Cosmochim. Acta,* 41, 1402, 1977.
54. **Rossini, F. D., Wagman, D. D., Evans, W. H., Levine, S., and Jaffe, I.,** Selected values of chemical thermodynamic properties, *Nat. Bur. Stand. U.S. Circ.* 500, 1952.
55. **Wagman, D. D., Evans, W. H., Parker, V. B., Halow, I., Bailey, S. M., and Schuman, P. H.,** Selected values of chemical thermodynamic properties, *Nat. Bur. Stand. U.S. Tech. Note,* 270-3, 1968.
56. **Kittrick, J. A.,** The free energy of formation of gibbsite and $Al(OH)_4^-$ from solubility measurements, *Soil Sci. Soc. Am. Proc.,* 30, 595, 1966.
57. **Singh, S. S.,** The solubility product of gibbsite at 15, 25, and 30°C, *Soil. Sci. Soc. Am. Proc.,* 38, 415, 1974.
58. **Beck, M. T.,** The stability of complexes in solution, in *Reaction Mechanisms in Inorganic Chemistry,* Ser. 1, Vol. 9, Tobe, M. L., Ed., Butterworths, London, 1972, 1.
59. **Stumm, W. and Morgan, J. J.,** *Aquatic Chemistry,* 2nd ed., Wiley-Interscience, New York, 1981.
60. **Bond, A. M. and Hefter, G. T.,** *Critical Survey of Stability Constants and Related Thermodynamic Data of Fluoride Complexes in Aqueous Solution,* IUPAC Chemical Data Ser. No. 27 (Part A), Pergamon Press, Elmsford, NY, 1980.
61. **Baumann, E. W.,** Determination of stability constants of hydrogen and aluminum fluorides with a fluoride-selective electrode, *J. Inorg. Nucl. Chem.,* 31, 3155, 1969.
62. **Agarwal, R. P. and Moreno, E. C.,** Stability constants of aluminum fluoride complexes, *Talanta,* 18, 873, 1971.
63. **Paul, A. D.,** The fluoride complexing of Sc(III), Cu(II), Pb(II), Zn(II), Hg(II), Sn(II), and Ag(I) in aqueous solution, Ph.D. thesis, University of California, Berkeley, 1955.
64. **Brosset, C.,** Ph.D. thesis, University of Lund, Sweden, 1942.
65. **King, E. L. and Gallagher, P. K.,** Thermodynamics of aluminum (III) fluoride complex ion reactions. The graphical elevation of equilibrium quotients from $\bar{n}$ (X), *J. Phys. Chem.,* 63, 1073, 1959.
66. **Hem, J. D.,** Graphical methods for studies of aqueous aluminum hydroxide, fluoride, and sulfate complexes, *U.S. Geol. Surv. Water-Supply Pap.,* 1827-B, 1968.
67. **Akitt, J. W., Fransworth, J. A., and Letellier, P.,** Nuclear magnetic resonance and molar-volume studies of the complex formed between aluminium(III) and the sulphate anion, *J. Chem. Soc. Faraday Trans. 1,* 85, 193, 1985.
68. **Kalidas, C., Knoche, W., and Papadopoulos, D.,** Mechanism of ligand substitution in weak complexes. III, *Ber. Bunsenges. Phys. Chem.,* 75, 106, 1971.
69. **Garrels, R. M. and Christ, C. L.,** *Solutions, Minerals and Equilibria,* Harper & Row, New York, 1965.
70. **Bjerrum, N. and Dahm, C. R.,** The aluminum phosphates. I. Complex formation in acid solutions, *Z. Phys. Chem. Bodenstein-Festband,* p. 627, 1931.
71. **Bohn, H. L. and Peech, M.,** Phosphatoiron (III) and phosphatoaluminum complexes in dilute solutions, *Soil Sci. Soc. Am. Proc.,* 33, 873, 1969.
72. **Langmuir, D.,** Techniques of estimating thermodynamic properties for some aqueous complexes of geochemical interest, in *Chemical Modeling in Aqueous Systems: Speciation, Sorption, Solubility, and Kinetics,* Jenne, E. A., Ed., ACS Symp. Ser. No. 93, American Chemical Society, Washington, D.C., 353, 1979.
73. **Sillén, L. G. and Martell, A. E.,** *Stability Constants of Metal-Ion Complexes,* Suppl. 1, Spec. Publ. No. 25, The Chemical Society, London, 1971.
74. **Smith, R. M. and Martell, A. E.,** *Critical Stability Constants,* Vol. 4, Plenum Press, New York, 1976.
75. **Hogfeldt, E.,** *Stability Constants of Metal Ion Complexes,* Part A, IUPAC Chemical Data Ser. No. 21, Pergamon Press, Elmsford, NY, 1982.
76. **Turner, D. R., Whitfield, M., and Dickson, A. G.,** The equilibrium speciation of dissolved components in fresh water and seawater at 25°C and 1 atm pressure, *Geochim. Cosmochim. Acta,* 45, 855, 1981.
77. **Ohman, L.-O. and Forsling, W.,** Equilibrium and structural studies of silicon (IV) and aluminum (III) in aqueous solution. III. A potentiometric study of aluminum (III) hydrolysis and aluminum (III) hydroxo carbonates in 0.6 *M* NaCl, *Acta Chem. Scand.,* A35, 795, 1981.
78. **Ohman, L.-O. and Sjoberg, S.,** On the insignificance of aluminum borate complexes in aqueous solution, *Mar. Chem.,* 17, 91, 1985.
79. **Ohman, L.-O.,** Equilibrium Studies of Ternary Aluminum (III) Complexes with Ligands Related to Conditions in Natural Waters, Ph.D. thesis, University of Umea, Sweden, 1983.
80. **Chase, M. W., Davies, C. A., Downey, J. R., Frurip, D. J., McDonald, R. A., and Syverud, A. N.,** *JANAF Thermochemical Tables,* 3rd ed., *J. Phys. Chem. Ref. Data,* 14(Suppl. 1), 1985.
81. **Hovey, J. K. and Tremaine, P. R.,** Thermodynamics of aqueous aluminum: standard partial molar heat capacities of Al^{3+} from 10 to 55°C, *Geochim. Cosmochim. Acta,* 50, 453, 1986.

82. **Millero, F. J.,** The molal volumes of electrolytes, *Chem. Rev.*, 71, 147, 1971.
83. **Head, J. H.,** FITIT, a computer program to least squares fit non-linear theories, U.S. Air Force Academy, Tech. Rep. 70-5, 1970.
84. **Nordstrom, D. K. and Munoz, J. L.,** *Geochemical Thermodynamics,* Blackwell Scientific, Palo Alto, CA, 1986.
85. **Ley, H.,** Studium uber die hydrolytische dissoziation der Salzlosungen, *Z. Phys. Chem.*, 30, 193, 1899.
86. **Kahlenberg, L., Davis, D. J., and Fowler, R. E.,** The inversion of sugar by salts, *J. Am. Chem. Soc.*, 21, 1, 1899.
87. **Latimer, W. M. and Jolly, W. L.,** Heats and entropies of successive steps in the formation of AlF_6^{3-}, *J. Am. Chem. Soc.*, 75, 1548, 1953.
88. **Izatt, R. M., Eatough, D., Christensen, J. J., and Bartholomew, C. H.,** Calorimetrically determined log K, $\Delta H°$ and $S°$ values for the interaction of sulphate ion with several bi- and ter-valent metal ions, *J. Chem. Soc. A*, 47, 1969.
89. **Kryzhanovskii, M. M., Volokhov, Y. A., Pavlov, L. N., Eremin, N. I., and Mironov, V. E.,** Aluminum sulfate complexes, *Zh. Prikl. Khim.*, 44, 476, 1971.
90. **Driscoll, C. T., Baker, J. P., Bisogni, J. J., and Schofield, C. L.,** Aluminum speciation and equilibria in dilute acidic surface waters of the Adirondack region of New York State, in *Geological Aspects of Acid Deposition,* Bricker, O. P., Ed., Butterworths, Boston, 1984, 55.
91. **Campbell, P. G. C., Thomassin, D., and Tessier, A.,** Aluminum speciation in running waters on the Canadian pre-cambrian shield: kinetic aspects, *Water Air Soil Pollut.*, 30, 1023, 1986.
92. **Thurman, E. M.,** *Organic Geochemistry of Natural Waters,* Martinus Nijhoff/Dr. W. Junk, Boston, 1985.
93. **Schnitzer, M. and Khan, S. U.,** *Humic Substances in the Environment,* Marcel Dekker, New York, 1972.
94. **Lind, C. J. and Hem, J. D.,** Effects of organic solutes on chemical reactions of aluminum, *U.S. Geol. Surv. Water Supply Pap.*, 1827-G, 1975.
95. **Ohman, L.-O. and Sjoberg, S.,** Equilibrium and structural studies of silicon (IV) and aluminum (III) in aqueous solution. IX. A potentiometric study of mono- and polynuclear aluminum (III) citrates, *J. Chem. Soc. Dalton Trans.*, p. 2513, 1983.
96. **Ohman, L.-O. and Sjoberg, S.,** Equilibrium and structural studies of silicon (IV) and aluminum (III) in aqueous solution. I. The formation of ternary mononuclear and polynuclear complexes in the system Al^{3+} - gallic acid - OH^-. A potentiometric study in 0.6 *M* Na(Cl), *Acta Chem. Scand.*, A35, 201, 1981.
97. **Ohman, L.-O. and Sjoberg, S.,** Equilibrium and structural studies of silicon (IV) and aluminum (III) in aqueous solution. IV. A potentiometric study of polynuclear aluminum (III) hydroxo complexes with gallic acid in hydrolyzed aluminum (III) solutions, *Acta Chem. Scand.*, A36, 47, 1982.
98. **Ohman, L.-O. and Sjoberg, S.,** Equilibrium and structural studies of silicon (IV) and aluminum (III) in aqueous solution. VIII. A potentiometric study of aluminum (III) salicylates and aluminum (III) hydroxysalicylates in 0.6 *M* Na(Cl), *Acta Chem. Scand.*, A37, 875, 1983.
99. **Ohman, L.-O. and Sjoberg, S.,** Equilibrium and structural studies of silicon (IV) and aluminum (III) in aqueous solution. X. A potentiometric study of aluminum (III) pyrocatecholates and aluminum (III) pyrocatecholates in 0.6 *M* Na(Cl), *Polyhedron,* 2, 1329, 1983.
100. **Perrin, D. D.,** *Stability Constants of Metal-Ion Complexes,* Part B, IUPAC Chemical Data Ser. No. 22, Pergamon Press, Elmsford, NY, 1979.
101. **Martell, A. E. and Smith, R. M.,** *Critical Stability Constants,* Vol. 1, Plenum Press, New York, 1974.
102. **Martell, A. E. and Smith, R. M.,** *Critical Stability Constants,* Vol. 3, Plenum Press, New York, 1977.
103. **Smith, R. M. and Martell, A. E.,** *Critical Stability Constants,* Vol. 2, Plenum Press, New York, 1975.

Chapter 3

INORGANIC ALUMINUM BEARING SOLID PHASES

Bruce S. Hemingway and Garrison Sposito

TABLE OF CONTENTS

I. STRUCTURE AND CHEMICAL COMPOSITION

A. The Nature of Ionic Solids

The chemical elements in minerals occur typically as ionic species whose electron configuration is unique and independent of whatever other ions may occur in the mineral structure. The interaction between one ion and another of opposite charge in the structure nonetheless is strong enough to form a chemical bond, termed an ionic bond. Ionic bonds differ from covalent bonds which involve a distortion of the electron configurations of the participating atoms that results in the sharing of electrons.

Ionic and covalent bonds are conceptual idealizations that real chemical bonds only approximate. In general, a chemical bond shows some degree of ionic character and some degree of electron sharing. The Si–O bond, for example, is said to be an even partition between ionic and covalent character, and the Al–O bond is thought to be about 40% covalent and 60% ionic.[1] Aluminum, however, is exceptional in this respect, for almost all of the metal-oxygen bonds that occur in minerals are ionic. Covalence thus plays a minor role in the overall structure of most minerals, aside from the important feature that Si–O bonds impart to them particular stability.[2] In respect to metal-oxygen bonds, covalence figures prominently only in the short-range molecular environment of the metal, and this environment does not change much from one mineral to another. The overall structure and relative stability of a mineral are determined by the longer-range ionic interactions whose behavior can be understood on the basis of classical electrostatics and the Coulomb law.

The electrostatic picture of ionic solids has significant implications for what kinds of atomic structure these solids can have in nature. The structures of most minerals in weathering environments can be rationalized on the physical grounds that the atomic configuration observed is that which tends to minimize the total electrostatic energy of the crystal. This concept has been formulated in a most useful fashion through a set of descriptive statements known as *Pauling's Rules:*[1]

Rule 1. A polyhedron of anions is formed about each cation. The cation-anion distance is determined by the sum of the respective radii and the corrdination number is determined by the radius ratio of cation to anion.

Rule 2. In a stable crystal structure, the sum of the strengths of the bonds that reach an anion from adjacent cations is equal to the absolute value of the anion valence.

Rule 3. The cations maintain as large a separation as possible and have anions interspersed between them so as to screen their charges. In geometric terms, this means that anion polyhedra tend not to share edges or especially faces. If edges are shared, they are shortened.

Rule 4. In a structure comprising different kinds of cations, those of high valence and small coordination number tend not to share polyhedron elements with one another.

Rule 5. The number of essentially different kinds of ion in a crystal structure tends to be as small as possible. Thus the number of types of coordination polyhedra in a close-packed array of anions tend to be a minimum.

Pauling Rule 1 is illustrated in Figure 1. The anion polyhedra mentioned in the rule are shown in the middle of the figure and the bottom row of "ball and stick" drawings shows the cation-anion bonds whose lengths are determined by the ionic radii. The radius of the smallest sphere that can reside in the central void created by packing anions in the four ways shown at the top of the figure can be calculated by Euclidean geometry. It turns out that this radius is always proportional to the radius of the coordinating anion. For example, in the case of tetrahedral coordination, the smallest sphere that can fit inside the four coordinating anions has a radius that is 22.5% of the anion radius. These minimum radius

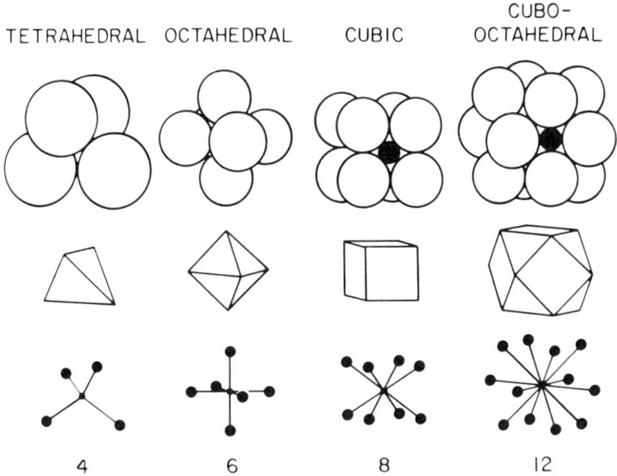

FIGURE 1. Anion configurations for the principal cation coordination numbers found in minerals.

ratios (RR) and coordination numbers (CN) are RR = 1.0, CN = 12; RR = 0.732, CN = 8; RR = 0.414, CN = 6; and RR = 0.225, CN = 4.

Pauling Rule 2 is realted closely to local charge balance in a crystal. The *strength* of an ionic bond from a cation to an anion is defined by the equation:

$$s = Z/CN \tag{1}$$

where Z is the valence of the cation. Rule 2 states that the sum of all s values for the cations bonded to a given anion equals the magnitude of the anion valence. As an example, consider the oxygen ions in α-quartz (SiO_2), which are corrdinated to Si^{4+} ions. The radius of Si^{4+} is 0.026 nm and its coordination number is 4.[3] It follows from Equation 1 that $s = 1.0$ for Si^{4+}. Since the absolute value of the valence of O^{2-} is 2, Pauling Rule 2 permits only two Si^{4+} to bond to an O^{2-} in SiO_2. This means that each O^{2-} in quartz must serve as a corner of no more than two silica tetrahedra like the one pictured on the left in Figure 1. Hypothetical atomic structures for quartz that involve, say, O^{2-} at the corners of an isolated tetrahedron alternating with three tetrahedra linked together are ruled out, even though they would satisfy the chemical formula, SiO_2.[1]

Pauling Rule 3 is based on the Coulomb law applied to cations. The repulsive electrostatic interaction between the cations in a crystal is screened by the negatively charged anions in the respective coordination polyhedra of the cations. If the cations have large valence as does, for example, Si^{4+}, then the polyhedra can do no more than share corners if the cations are to be kept as far apart as possible in a structural arrangement that achieves the lowest possible total electrostatic energy. If the cation valence is somewhat smaller, as it is for Al^{3+}, the sharing of polyhedron edges becomes possible. Edge sharing brings the cations closer together than corner sharing, however, so the task of charge screening by the anions is more difficult. They respond to this difficulty by approaching one another slightly along the shared edge to enhance screening. Doing so, they shorten the edge relative to unshared edges of the polyhedra.[1]

Pauling Rules 4 and 5 continue in the spirit of Rule 3. They reflect the fact that stable ionic crystals containing different kinds of cations cannot tolerate much sharing of the coordination polyhedra nor much variability in the type of coordination environment. These and the other three Pauling Rules serve as useful guides to the atomic structural interpretation of the chemical formulas for aluminum-bearing minerals.

B. Polymorphs

Identical elemental compositions may be arranged in one or more structures. Where more than one structure may exist, the crystals comprising the two or more structures are called polymorphs. The different structures of the polymorphous crystals result in differences in the physical properties and in the appearance of the crystals.

Polymorphism may be illustrated through a discussion of the close packing of like (equal sized) spheres. Close packing represents that arrangement of spheres that permits each sphere to contact the largest number of neighboring spheres. For a layer (sheet), each sphere would be in contact with six nearest neighbors. In a crystal (a three-dimensional solid) there would be six additional neighboring spheres, three above and three below. There are also six voids arround each sphere in the layer. Because there may only be three contacting spheres in the overlying layer that may occupy the six voids, there are two different orientations that may be assumed. If we call the original layer the A layer and the two possible orientations of the overlying layers B and C, then one can see that differences in stacking of A, B, and C may occur. Crystals built from the stacking sequences ABAB and ABCABC would represent polymorphous crystals.

Differences in the stacking of $Al(OH)_6$ octahedra[4] result in polymorphism in the $Al(OH)_3$ phases, *bayerite, doyleite, nordstrandite,* and *gibbsite* (see also Chapter 7). These stacking differences are reflected in differences in molar volume, density, planes of perfect cleavage, and in the thermodynamic properties of the phases. Similarily, *kyanite, andalusite,* and *sillimanite* are aluminosilicate polymorphs with the composition Al_2SiO_5, and *kaolinite, dickite,* and *halloysite* are clay mineral polymorphs.

C. Oxide and Hydroxide Structures

The structures developed by oxide and hydroxide crystals may be most easily visualized in terms of the idealized model presented to describe polymorphism. Oxide and hydroxide structures may be viewed as symmetrically packed sheets of oxygen and/or hydroxyl anions (the like spheres) with metal cations occupying voids between the anions. The geometry of the void in which the metal cation is located is defined by the geometry of the surrounding anions (polyhedron) as shown in Figure 1.

The oxide and hydroxide structures may be classified on the basis of their coordination polyhedra[5] as tetrahedral, octahedral, mixed tetrahedral and octahedral, cubic, and mixed octahedral and cuboctahedral. *Corundum*, Al_2O_3, is an example of an octahedral oxide structure and *gibbsite*, $Al(OH)_3$, an octrahedral hydroxide structure. In each structure only two thirds of the octrahedral voids contain aluminum and the sheet is said to be *dioctahedral.* The corundum structure is built of an AB sequence of layers whereas gibbsite is built from an ABBA sequence in which the AA and BB layer voids are vacant. Bonding between the AA and BB layers in gibbsite results from polarization of the charge on the hydroxyl anions by the proximal aluminum cations in the AB layer voids. The polarized hydroxyl anion exhibits a negative charge and a positive charge (associated with the proton) that are located at the outer A and B surfaces of the AB layer. Hydroxyl bonding occurs between these charged hydroxyl anions on opposing AA and BB layers.

The gibbsite AB layer is a common component of sheet structure aluminosilictes. Isomorphic substitutions of cations for aluminum can alter the net surface charge and the occupancy of the octahedral voids. These effects are discussed in the following section and in Chapter 7.

D. Sheet Structure

The *micas* are built up from two sheets of silica tetrahedra ($Si_2O_5^{2-}$ repeating unit) fused to each planar side of a sheet of metal cation octahedra.[6] The octahedral sheet typically contains Al, Mg, and Fe ions corrdinated to O^{2-} and OH^-. If the metal cation is trivalent, only two of the three possible cationic sites in the octahedral sheet can be filled to achieve charge balance and the sheet is dioctahedral. If the metal cation is bivalent, all three possible sites are filled and the sheet is *trioctahedral*. Isomorphic substitution of Al for Si, Fe(III) for Al, and Fe or Al for Mg occurs typically in the micas along with many trace element substitutions.

Muscovite and *biotite* are the common soil micas, the former being dioctahedral, the latter trioctahedral. In both minerals, Al^{3+} substitutes for Si^{4+}. The resulting charge deficit is balanced by K^+ that corrdinates to 12 oxygen ions in cavities of 2 opposing tetrahedral sheets belonging to a pair of mica layers on top of one another. Thus the K^+ link adjacent mica layers together. The initial weathering reaction of muscovite is characterized by a reduction in the amount of substituted aluminum in tetrahedral coordination in silica sheets. This reduction also occurs with biotite, except that it can be accomplished by the oxidation of ferrous iron as well as by the loss of tetrahedrally corrdinated aluminum.[6]

Clay minerals are aluminosilicates that predominate in the clay fractions of soils at intermediate stages of weathering. These minerals, like the micas, are sandwiches of tetrahedral and octahedral sheet structures.[6] This bonding together of the tetrahedral and octahedral sheets occurs through the apical oxygen ions in the former and always produces a significant distortion of the anion arrangement in the final layer structure formed. The principal distortion is caused by the fact that the apical oxygen ions in the tetrahedral sheet cannot be fit to the corners of the octahedra to form a layer while preserving the ideal hexagonal pattern of the tetrahedra. In order to fuse the two sheets, pairs of adjacent tetrahedra must rotate alternately clockwise and counterclockwise about an axis perpendicular to their basal plane. This rotation, in turn, distorts the symmetry of the hexagonal cavities in the basal plane of the tetrahedral sheet. Besides this distortion, the sharing of edges in the octahedral sheet shortens them (Pauling Rule 3), and isomorphic substitutions of the cations in both sheets tends to make the thickness of the layer structure smaller and its basal surfaces less planar (slightly corrugated) than what would occur ideally. These effects occur both in the micas and the clay minerals.

The clay minerals usually are classified into three *layer types,* distinguished by the number of tetrahedral and octahedral sheets combined, and five *groups,* differentiated by the kinds of isomorphic cation substitutions that occur.[7] The layer types are shown in Figure 2 and the groups are described in Table 1. The 1:1 layer type consists of one tetrahedral and one octahedral sheet. In soil clays, it is represented by the *kaolinite* group, with the chemical formula $[Si_4]Al_4O_{10}(OH)_8$, where the cation enclosed in the square brackets is in tetrahedral corrdination. Normally there is no significant isomorphic substitution for Si or Al in this clay mineral (substitution of Fe[III] for Al up to 3 mol % is observed in tropical weathering environments). As is common with soil clay minerals, the octahedral sheet has two-thirds cation site occupancy (dioctahedral sheet). The 2:1 layer type has two tetrahedral sheets that sandwich an octahedral sheet. The three soil clay mineral groups with this structure are *illite, vermiculite,* and *smectite.* If a, b, and c are the stoichiometric coefficients of Si, octahedral Al, and Fe(III), respectively, in the chemical formulas of these groups, then

$$x = 12 - a - b - c \quad (2)$$

is the *layer charge,* the number of moles of excess electron charge per chemical formula that is produced by isomorphic substitution.

As indicated in Table 1, the three 2:1 groups differ from one another in two principal ways. The layer charge decreases in the order illite > vermiculite > smectite and the vermiculite group is further distinguished from the smectite group by the extent of isomorphic substitution in the tetrahedral sheet. Among the smectites, those in which the substitution of Al for Si exceeds that of Fe(II) or Mg for Al are called *beidellite,* and those in which the reverse is true are called *montmorillonite.* The sample chemical formula in Table 1 for

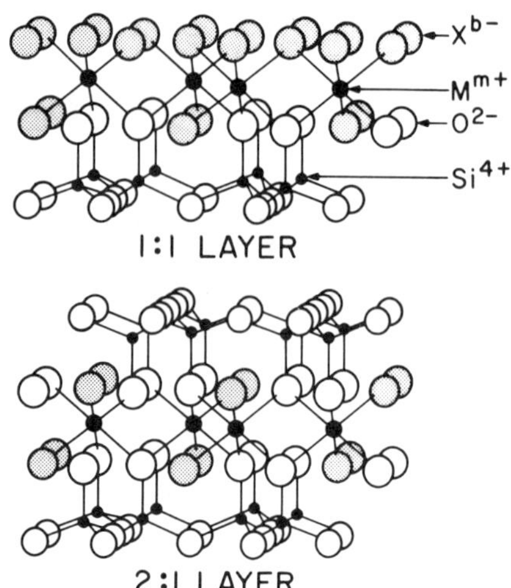

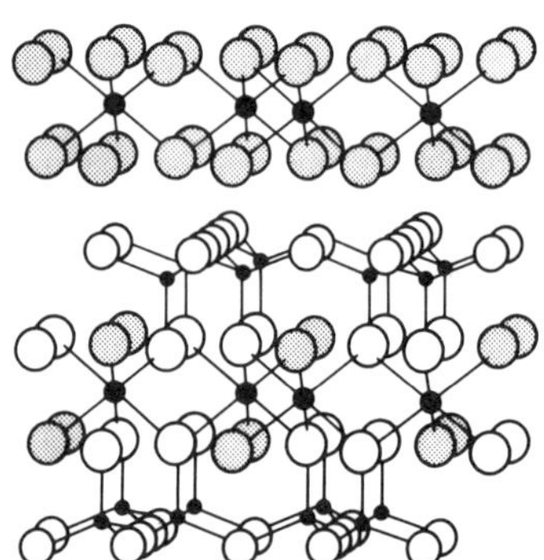

FIGURE 2. The three layer types of clay minerals. (Reprinted with permission from Sposito, G., *The Surface Chemistry of Soils,* Oxford University Press, New York, 1984, chap. 1.)

Table 1
CLAY MINERAL GROUPS

Group	Layer type	Layer charge (x)	Typical chemical formula[a]
Kaolinite	1:1	<0.01	$[Si_4]Al_4O_{10}(OH)_8{\cdot}nH_2O$ (n = 0 or 4)
Illite	2:1	1.4—2.0	$M_x[Si_{6.8}Al_{1.2}]Al_3Fe_{0.25}Mg_{0.75}O_{20}(OH)_4$
Vermiculite	2:1	1.2—1.8	$M_x[Si_7Al]Al_3Fe_{0.5}Mg_{0.5}O_{20}(OH)_4$
Smectite[b]	2:1	0.5—1.2	$M_x[Si_8]Al_{3.2}Fe_{0.2}Mg_{0.6}O_{20}(OH)_4$
Chlorite	2:1 with hydroxide interlayer	variable	$(Al[OH]_{2.55})_4{\cdot}[Si_{6.8}Al_{1.2}]Al_{3.4}Mg_{0.6}O_{20}(OH)_4$

[a] n = 0 is kaolinite and n = 4 is halloysite; M = monovalent interlayer cation.
[b] Principally montmorillonite and beidellite in weathering environment.

smectite represents montmorillonite. In any of these 2:1 minerals, the layer charge is balanced by cations that reside on or in cavities of the basal plane of the oxygen atoms of the tetrahedral sheet. These interlayer cations are represented by M in the chemical formula (Table 1). The 2:1 layer type with hydroxide interlayer is represented by dioctahedral *chlorite* in soil clays. The octahedrally corrdinated cations in chlorite reside in two sheets: one comprising $M(OH)_2O_4^{(m-10)}$ octahedra (with $M^{m+} = Al^{3+}$, Fe^{3+}, or Mg^{2+}) sandwiched in a 2:1 layer, and one comprising principally $Al(OH)_6^{3-}$ octahedra situated on a basal surface of the 2:1 layer. To preserve the electroneutrality of the whole structure, the octahedral occupancy often is larger than the expected value of 8 per chemical formula for the two dioctahedral sheets, such that the excess positive charge balances the negative charge created by isomorphic substitution in the tetrahedral sheets.[6]

Structural disorder in all of the clay minerals listed in Table 1 is created through isomorphic substitutions for their principal cation components. The range of these substitutions in the Periodic Table is very broad. Even more pronounced structural disorder exists in silica and in aluminosilicates freshly precipitated in weathering environments, since these compounds typically are X-ray amorphous. Among the crystalline clay minerals, there is wide variation in molecular order, with disorder in the absence of isomorphic substitution created by dislocations (microcrevices between offset rows of atoms) and irregular stacking of crystalline layers. This kind of disorder exists, for example, in the kaolinite group minerals.[8]

Structurally disordered aluminosilicates, known collectively as *allophane* and *imogolite,* are common in the clay fractions of soils formed on volcanic ash parent material. The atomic structure of allophane is not well understood, but is thought to consist of a 1:1 aluminosilicate layer or possibly an imogolite structure riddled with defects (vacant ion sites) and containing aluminum in both the tetrahedral and octahedral sheets. These many defects promote a curling of the layer into the form of a hollow spherule about 5 nm in diameter whose outer boundary would contain many apertures through which small molecules or ions could enter. As this structural concept would suggest, allophane often is found in association with clay minerals of the kaolinite group, especially the hydrated species, *halloysite* (Table 1). Imogolite, with the general empirical formula, $Si_2Al_4O_{10} \cdot 5H_2O$, exhibits a tubular morphology. The tube unit in the structure contains aluminum only in octahedral coordination and exposes a defective, gibbsite-like surface.

E. Framework Structures

Sheet structure aluminosilicates are characterized by a two-dimensional tetrahedral network of silica tetrahedra in which three oxygens from each tetrahedron are shared with other tetrahedra. Framework structures (also network structures[5]) are characterized by a continuous three-dimensional network of silica tetrahedra in which the four oxygens are shared with other tetrahedra. Differences in the structure of the tetrahedral frame of these groups are reflected in the relative mechanical and chemical stabilities of the groups.

Quartz and *feldspar,* framework structure phases, are found in the clay-sized (<2 μm) fraction of soils. The Si–O bonds of the tetrahedra of silicates and aluminosilicates are stronger than associated octahedral or other cation bonds.[5] Consequently, framework structure minerals can be more resistant to chemical weathering and to mechanical abrasion than aluminum bearing phases with other structures.

Substitution of Al^{3+} or other cations for Si^{4+} in the tetrahedra of the frame weakens the frame and produces a residual charge on the frame that must be balanced by the addition of other cations in the structure (e.g., Si^{4+} is replaced by Na^+ and Al^{3+} in the feldspar low albite). Additional consequences of this type of substitution are (1) the expansion of the frame to accomodate the presence of the additional cation and (2) the replacement of Si^{4+} from a preferred site in the structure yielding an ordered pattern of replacement or a random replacement of Si^{4+} leading to a disordered pattern of Si^{4+} and Al^{3+} tetrahedra in the structure. Both the chemical and physical substitutions effect the relative stability of the aluminum-bearing phases.

F. Chain and Band Structures

Chain and band structure aluminosilicates are characterized by a one-dimensional silica tetrahedral frame[5] in which two oxygens are shared with other tetrahedra forming an infinite chain as in *pyroxene* minerals (or double chain as in the case of *amphibole* minerals which also have periodic sharing of a third oxygen between tetrahedra of adjacent chains). The one-dimensional frame carries a residual charge that must be balanced by the addition of cations to the structure. The charge compensating cations are present in octahedral bands.

The chains of silica tetrahedra are commonly kinked which produces bands of octahedra of different sizes and shapes into which cations of different sizes may fit. Substitution of one or more cations for another (e.g., Fe^{2+} for Mg^{2+} or Na^{+} and Al^{3+} for Mg^{2+}) may result in *solid solutions,* and/or ordered or disordered configurations.

G. Ortho and Ring Structures

Ortho and ring structures are characterized by isolated silica tetrahedra or clusters of tetrahedra and are considered to be zero-dimensional networks.[5] Linkage between isolated tetrahedra or tetrahedral clusters occurs through oxygens with one or more types of cation polyhedra. For example, the three Al_2SiO_5 polymorphs (*kyanite, andalusite,* and *silimanite*) have all silicon in tetrahedral coordination and one aluminum in octahedral coordination, but the other aluminum is coordinated by four oxygens in sillimanite, five in andalusite, and six in kyanite. Chemical substitutions in the cation sites of some of the ortho and ring structure phases produce solid solutions and may produce disordered configurations within the cation sites.

H. Other Structures

Zoltai and Stout[5] provide a good review of the structures of minerals having sulfide, sulfate, carbonate, nitrate, halide, and phosphate as essential anions of their structures. The authors also provide a more comprehensive description of the structures of oxides, hydroxides, silicates, and aluminosilicates than is appropriate here.

II. EXPERIMENTAL DETERMINATION OF THERMODYNAMIC PROPERTIES

A. Introduction

Thermodynamic properties for inorganic aluminum-bearing phases are determined by a number of experimental techniques, e.g., calorimetry, solubility measurements, phase equilibria studies, and electrochemical methods. Often several experimental techniques have been used to derive the same thermodynamic property, e.g., the Gibbs energy of formation, ΔG_f°. Also — more often than one would like — results from the several techniques yield disparate thermodynamic values. For this reason, the evaluation and selection of recommended thermodynamic data requires the simultaneous analysis of data sets that represent more than one experimental technique in order to identify and eliminate potential bias from a particular technique or observation.

Brief descriptions are given of several techniques applied to aluminous solids together with discussions of the strengths and weaknesses of these techniques. This survey includes most, but not all, common techniques.

B. Calorimetry

1. Heat Capacities, Enthalpies, and Entropies

Several calorimetric techniques are used to derive thermodynamic properties. Among the most precise and accurate are the adiabatic calormetric systems[9] used to measure heat capacities at subambient temperatures and from which entropies at 298.15 K (T/K = 273.15 + t/°C) are reported. The U.S. Bureau of Mines (Kelley, King, and co-workers), The University of Michigan (Westrum and co-workers), and the U.S. Geological Survey (Robie, Hemingway, and co-workers) have provided extensive sets of heat-capacity data for aluminum-bearing phases. Heat capacities, C_p, are measured from near 5 K (or about 50 K in older measurements) to about 350 K. Heat-capacity results are extrapolated to 0 K using, in general, a Debye type of calculation,[9] e.g., a plot of C_p/T vs. T^2. The integration of C_p and C_p/T from 0 K to any temperature provides the enthalpy ($H_T - H_0$) and entropy ($S_T - S_0$), respectively, of the phase:

$$H_T - H_0 = \int_0^T C_p \, dT \tag{3}$$

$$S_T - S_0 = \int_0^T (C_p/T) \, dT \tag{4}$$

The relative precision of modern low-temperature adiabatic calorimeters is better than 0.2% in the entropy and the agreement between laboratories is also about 0.2%, despite the use of different practical temperature scales.

The primary advantage of the adiabatic calorimeter over some newer calorimetric techniques [ac-calorimetry (ac) and differential scanning calorimetry (dsc)] has been the precision and accuracy of the technique. The major disadvantage is the sample size requirement (generally about 30 g), although smaller calorimeters have been developed recently at some cost in precision. Differential scanning calorimeters and ac-calorimeters use less than 50 mg of sample and are capable of producing heat capacities with a precision of 0.5 to 1.0%. For example, subambient heat capacities from adiabatic calorimetry and superambient heat capacities form dsc can be introduced into Equation 4 to calculate 68.44 J/(mol · K) for the entropy of gibbsite at 298.15.[11]

Superambient heat capacities have been determined by ac, dsc, and adiabatic calorimetry to temperatures of about 1000 K and with a precision of about 0.5% in the entropy. These methods compare favorably with the older heat-content (enthalpy) calorimetric technique, being both faster and as precise and, in the case of ac and dsc, having the added advantage of requiring a smaller sample. However, the typical construction of a heat-content calorimeter allows the instrument to be used to higher temperatures (near 1800 K). Heat-content and high-temperature adiabatic calorimeters require about 10 g of sample.

Superambient heat capacities and heat contents are subject to larger systematic errors than are heat capacities derived from subambient adiabatic calorimetry. The differences arise largely from errors or differences in the calibration of the practical temperature scales used in different laboratories, which may vary by more than 10 K at temperatures above 1000 K. Entropies derived from superambient heat capacities or hear contents are probably not known to better than 0.5%.

An additional contribution to the entropy of a crystal arises from disorder of the crystal structure. Three main types of disorder[10] may exist in mineral structures, disorder of position, magnetic disorder, and orientational disorder. When disordered crystals are cooled to near 0 K two responses, in principle, may occur: either the crystal atoms will become ordered, with an associated anomalous heat capacity that can be measured, or the crystal retains the disordered state or configuration. In the latter case, the contribution of the disorder to the entropy cannot be measured and the disorder is said to be "frozen-in" at 0 K. The entropy, S_0, then is not zero and must be estimated from a knowledge of the structure of the crystal.

Systematic errors in both super- and subambient heat capacities and heat contents, and in thermodynamic properties derived from these data, can arise from excess heat effects

Table 2
REACTIONS FOR THE DETERMINATION OF THE ENTHALPY OF FORMATION OF GIBBSITE, Al(OH)₃, BY SOLUTION CALORIMETRY[14]

(1)	$Al(c) + [1849HF + 8163H_2O](aq) = [AlF_3 + 1846HF + 8163H_2O](aq) + 3/2H_2(g)$	ΔH_1
(2)	$3H_2O(l) + [AlF_3 + 1846HF + 8163H_2O](aq) = [AlF_3 + 1846HF + 8166H_2O](aq)$	ΔH_2
(3)	$Al(OH)_3(c) + [1849HF + 8163H_2O] = [AlF_3 + 1846HF + 8166H_2O](aq)$	ΔH_3
(4)	$Al(c) + 3H_2O(l) = Al(OH)_3 + 3/2H_2(g)$	ΔH_4
(5)	$3H_2(g) + 3/2O_2(g) = 3H_2O(l)$	ΔH_5
(6)	$Al(c) + 3/2H_2(g) + 3/2O_2 = Al(OH)_3(c)$	ΔH_6

Note: (c) = crystal, (g) = gas, (l) = liquid, (aq) = aqueous solution.

caused by small grain size. The grain size at which the effects become significant varies with chemistry and structure, but samples composed of grains smaller than a few microns must be viewed as suspect.

2. Enthalpy of Formation

Three types of calorimetric procedure have been used to estimate enthalpies of formation for aluminum-bearing phases: aqueous acid solution calorimetry,[9] molten salt calorimetry,[12] and combustion calorimetry.[13] In general, these methods require from ten to hundreds of milligrams of sample and yield enthalpies of formation with a precision of 0.3 to 0.5%.

The enthalpy of formation of an unknown material is determined through a comparison of the heat or enthalpy of solution (combustion) of the unknown with the enthalpies of solution (combustion) of chemical components that, when added together, can represent the stoichiometry of the unknown material. This procedure, called Hess's law, is illustrated in Table 2 for the formation of gibbsite. Reaction (4) represents the enthalpy of formation (ΔH_4) of gibbsite and hydrogen gas from the components, aluminum metal and water: $\Delta H_4 = \Delta H_1 + \Delta H_2 - \Delta H_3$. Often, as in this example, the components that are appropriate for use as enthalpy reference phases are not the most appropriate phases to which to reference the unknown in tabulations of thermochemical data. For this purpose, the compound is usually referenced to its constituent elements ($\Delta H_6 = \Delta H_5 + \Delta H_4$). Following the procedure described above, Hemingway and Robie[14] obtained −1293.13 kJ/mol for the enthalpy of formation of gibbsite from the elements.

Calorimetric procedures developed for the measurement of enthalpies of formation have in common the requirement that the initial states of the reference components (e.g., aluminum metal and water in Table 2) and of the unknown (e.g., gibbsite in Table 2) are known and not significantly different from the states of the ideal reference components and unknown. It is also required that the states of the products of the reactions of the components with the solvent are the same as the states of the products of the reaction of the unknown with the solvent. Failure to meet either of these requirements may result in a significant systematic error in the calculated enthalpy of formation of the unknown. As an example, the enthalpy of solution of MgO prepared by the dehydration of $Mg(OH)_2$ at 400°C is significantly larger[15] than the enthalpy of solution of the same material following sintering for 1 h at 1400°C. For use as a reference component, MgO must be prepared such that the physical properties are equivalent to the MgO produced in the combustion reaction that established the value for the enthalpy of formation of MgO from the elements.

Acid aqueous[9] and molten salt[12] solution calorimeters are more commonly used to determine enthalpies of reaction of minerals at this time. Acid aqueous calorimeters may use acids, bases, or pure water as a solvent and are operated in the temperature range of 20 to 90°C. Molten salt calorimeters typically use mixtures of PbO and B_2O_3 as the solvent and are operated at temperatures near 700°C. Acid aqueous calorimeters typically cannot be used for refractory minerals (e.g., kyanite, Al_2SiO_5) and molten salt calorimeters typically cannot be used for minerals containing volatile components such as OH or H_2O, or for minerals that are not stable up to 700°C (e.g., gibbsite). The typical procedure followed in acid aqueous calorimetry is to dissolve the reference components sequentially in the solvent providing a final solution that is equivalent in chemistry to that resulting from the dissolution of the unknown. The components are typically measured separately in the molten salt procedure. In both procedures the solvent mass is large with respect to the sample and the solutions remain dilute with little or no concentration dependence for the enthalpies of solution.

3. Gibbs Energy of Formation

The Gibbs energy of formation, ΔG°_f, of an aluminum-bearing phase is calculated from the enthalpy (ΔH°_f) and entropy (ΔS°_f) of formation according to the expression:

$$\Delta G^\circ_f = \Delta H^\circ_f - T\Delta S^\circ_f \quad (5)$$

The entropy of formation is the difference between the entropy of the phase and the sum of the entropies of the reference components. Using gibbsite again as an example, the enthalpy of formation and entropy were given above; the entropies[16] of Al(c), H_2(g), and O_2(g) are 28.35, 130.68, and 205.15 J/(mol · K), respectively, at 298.15 K and 1 bar; therefore, the entropy of formation of gibbsite is −463.65 J/(mol · K), and the Gibbs energy of formation[14] at 298.15 K and 1 bar is −1154.89 kJ/mol.

C. Solubility Measurements

A considerable research effort into practical and theoretical aspects of the aqueous chemistry of aluminum in natural waters and soils has been undertaken by the scientific community in the last 30 years. Many of these topics are covered in Chapters 2 and 8. In this section we shall discuss the procedures through which solubility measurements may be used to evaluate thermodynamic properties of aluminum-bearing phases or ionic species.

The mathematical relationships used to extract thermodynamic data from solubility (dissolution) measurements are[17,18]

1. The change in the standard Gibbs energy for a dissolution reaction, ΔG°_r, is related to the *equilibrium constant,* K_r, for the reaction by:

$$\Delta G^\circ_r = -RT \ln K_r \quad (6)$$

2. The enthalpy change for a dissolution reaction may be calculated from the van't Hoff equation:

$$d\ln(K_r)/dT = \Delta H^\circ_r/RT^2 \quad (7)$$

where, for practical reasons, ΔH°_r is assumed to be constant over short temperature intervals, so that Equation 7 becomes:

$$2.303 \log(K_2/K_1) = -(\Delta H^\circ_r/R)(1/T_2 - 1/T_1) \quad (8)$$

A plot of log(K) vs. 1/T will yield a straight line with a slope of $-\Delta H^\circ_r/2.303R$.

3. The enthalpy and entropy change are related to the change in heat capacity as shown in Equations 3 and 4. Thus, if the solubility product constant is known as a function of temperature and the heat capacity change is known or can be estimated, the entropy change for the dissolution reaction can be estimated from Equation 9.[17]

$$\ln(K_2) = (T_1/T_2)\ln(K_1) + (1/RT_2)[(T_2 - T_1)\Delta S^\circ_1 + T_2\int_1^2 \Delta C^\circ_p \, d\ln(T) - \int_1^2 \Delta C^\circ_p \, dT] \quad (9)$$

A review of the procedures and assumptions associated with the calculation of the Gibbs energy of formation of the Al^{3+} ion from the dissolution of gibbsite will acquaint us with the solubility method and use of Equation 6. (Singh[18] has discussed the application of Equations 7 and 9.) The dissolution reaction for gibbsite may be written in the form:

$$Al(OH)_3 = Al^{3+} + 3OH^- \quad (10)$$

for which the equilibrium constant is

$$K_r = [Al^{3+}][OH^-]^3/[Al(OH)_3] \quad (11)$$

where the quantities in brackets represent the thermodynamic activities of the components. The application of Equation 11 in practical experiments requires that two underlying assumptions be met. First, the actual state of the gibbsite sample must be identical with the standard state for gibbsite. Second, it must be shown that the mineral gibbsite actually controls the concentration of Al^{3+} ions in the solution.

Deviations of a mineral from the standard state of that mineral will effect the results of all experimental procedures. Factors such as small grain size (high surface area grains have a large surface energy contribution) or poor crystallinity may produce systematic errors in measured properties. The effect on calorimetric measurements was noted in Section II.B.1. Solubility measurements (and phase equilibria measurements) are sensitive to the surface structure of the mineral grains used in the dissolution experiment. This is somewhat in contrast with the calorimetric procedure, because in calorimetry the total grain is dissolved and the enthalpy of dissolution is averaged over the mass of the material dissolved. Consequently, a very significant surface energy contribution is required to shift the average enthalpy of solution, whereas solubility and phase equilibria measurements represent reactions with the surfaces of the mineral grains.

Equilibrium must be shown to exist between the mineral undergoing dissolution and the dissolution products. The phase undergoing dissolution must control the concentration of the dissolution products and the dissolution reaction must be shown to be reversible; i.e., at equilibrium, addition of dissolution products to the solution must result in precipitation of the mineral in the standard state, and removal of dissolution products from the solution must result in further dissolution of the mineral under study.

Careful sample preparation can eliminate most problems associated with differences in the state of a mineral, but establishing that equilibrium exists in solubility experiments is difficult. Few, if any, aluminosilicate minerals will equilibrate directly with dissolved ions. Most often, the dissolution of an aluminosilicate is accompanied by precipitation of one or more of the dissolution products as an amorphous or poorly crystalline phase (incongruent dissolution). For example, aluminum released by the dissolution of an aluminosilicate will precipitate as amorphous $Al(OH)_3$, but not as one of the crystalline phases, bayerite, doyleite, nordstrandite, or gibbsite that can be properly characterized. Addition of aluminum to such a system will result in the precipitation of additional $Al(OH)_3$ and not in the reversal of the dissolution reaction, i.e., not in the precipitation of the aluminosilicate phase. In other cases, the mineral undergoing dissolution is metastable with respect to another mineral composed of the same elements (e.g., gibbsite is metastable with respect to diaspore). For a metastable phase, the reaction cannot be reversed and approach to "metastable equilibrium" can be reached only from conditions of supersaturation (a higher Gibbs energy state).[19] Once the concentration of solution components falls below the metastable equilibrium value (a natural consequence of near-equilibrium conditions, because some solution components will be lost through precipitation in the more stable phase), the less stable phase begins to dissolve. Also, addition of aluminum ions to the solution to promote equilibration from oversaturation may increase the concentration of aluminum and other components to a value that exceeds the (metastable) equilibrium constant of a less stable phase causing that phase to begin to precipitate, a condition which could be misinterpreted in terms of a reversal of the relative stability of the phases.

Kittrick[20] has reported five values for the solubility product ($K_{SO} = -\log K_r$) of gibbsite based on Equation 10. The average of these values is -34.03 ± 0.12. From Equation 6, the Gibbs energy of the dissolution reaction is 194.3 kJ/mol. The standard Gibbs energy[21,22] of Al^{3+} may be calculated from the Gibbs energy of formation of gibbsite given above and the standard Gibbs energy[16] of OH^- (-157.328 kJ/mol) as -489.4 kJ/mol.[21] In similar treatments[19,23] based upon solubility data for the same sample of gibbsite, the Gibbs energies of Al^{3+} and $Al(OH)_4{}^-$ were found to be -489.8 and -1305 kJ/mol, respectively (see also Chapter 2 for a discussion of ΔG_f° for Al^{3+}).

The interpretation of data from research on the Al-O-H system is not straightforward. Differneces of opinion[19,22,24] exist as to the correct values for the free energies of theAl^{3+} and $Al(OH)_4^-$ (Chapter 2). These differences in interpretation cannot be resolved easily because no single study contains all the information necessary to support the chosen interpretations. The complexity of these problems arises from the fact that none of the $Al(OH)_3$ phases is the stable aluminum hydroxide phase (i.e., diaspore). We have chosen those results that are consistent with the greatest amount of solubility data for the Al-O-H system and with additional phase equilibrium data.[25] However, resolution of the disparate interpretations must await the results of new experiments designed to answer specific questions about the properties of the Al-O-H system.

Thermodynamic data often are used in conjunction with models of water chemistry to develop solutions to societal problems (e.g., to mitigate contamination of ground water by toxic wastes or to increase permeability in deep reservoirs containing hydrocarbons). Such models are based upon a reversal of the calculations and procedures discussed above and, thus, are subject to similar limitations.

D. Phase Equilibria

The third-law method[26] provides a rigorous means of extracting values of the enthalpy of reaction, ΔH_r°, at 298.15 K from equilibrium measurements and simultaneously provides a critical test of the accuracy of the equilibrium data. Knowledge of the entropies and heat capacity functions for all phases involved in the reaction permits the enthalpy of reaction at 298.15 K to be calculated using the relation:

$$-\Delta H_r^\circ = T\Delta[(G_T^\circ - H_{298}^\circ)/T] + \int_1^P \Delta V_T^\circ \, dP + RT \ln f(T,H_2O) \qquad (12)$$

where ΔV_T° is the difference in the molar volumes of the solid phases, $f(T,H_2O)$ is the fugacity of water, and $\Delta[(G_T^\circ - H_{298}^\circ/T]$ is the difference in the Gibbs energy function. The third-law method is based upon heat capacity data — in general the most accurate data used in thermodynamic calculations — and the method is independent of the enthalpies of formation and their associated errors. The Gibbs energy function of a phase is defined by the relation:

$$(G_T^\circ - H_{298}^\circ)/T = (H_T^\circ - H_{298}^\circ)/T - S_T^\circ \qquad (13)$$

The difference in the Gibbs energy function is a continuous, monotonic function of temperature that approaches linearity at higher temperatures allowing relatively accurate extrapolations to temperatures beyond those to which the experimental heat-capacity or heat-content data refer. For the pressure range of most equilibria measurements, the approximation can be made that

$$\int_1^P \Delta V^\circ_T \, dP \approx (P - 1)\Delta V^\circ_{298} \tag{14}$$

Molar volume data are available for many minerals at 298.15 K and for a few phases at higher temperatures.

Six temperature brackets[27] representing the reversal of the reaction:

$$\text{muscovite} + \text{quartz} = \text{sanidine} + \text{corundum} + \text{steam} \tag{15}$$

and four brackets[28] representing the equilibria for the reaction:

$$\text{pyrophyllite} = \text{andalusite} + 3\ \text{quartz} + \text{steam} \tag{16}$$

have been reported. Three temperature brackets representative of each reaction have been used to illustrate the calculation of the enthalpies of reaction for Reactions 15 and 16 using the third-law method. The results are shown in Table 3. The enthalpies of reaction calculated for Reaction 16 represent an internally consistent set of values, whereas those for Reaction 15 show a distinct dependence upon temperature and pressure changes and suggest that a systematic error exists in one or more values of the experimental data used (but not necessarily in the phase equilibria brackets).

Thermodynamic information may also be extracted from phase equilibria data using the relations:

$$(dP/dT)_T = \Delta S^\circ_T/\Delta V^\circ_T \tag{17}$$

and

$$\Delta G^\circ_T = \Delta H^\circ_r + T\Delta[(G^\circ_T - H^\circ_{298})/T] \tag{18}$$

where dP/dT is the slope of the reaction boundary and ΔG°_T is the Gibbs energy of the reaction at temperature T.

III. METHODS OF ESTIMATING THERMODYNAMIC PROPERTIES

A. The Chen Extropolation

Perhaps the most straightforward method of estimating the standard Gibbs energy of formation of aluminum-bearing solid phases is the extrapolation algorithm worked out by Chen.[30] This algorithm is based on the assumption that ΔG°_r for the formation of an inorganic solid from a suite of solid-phase reactants should approach zero as the number of reactants decreases to two and the composition of at least one of them approaches that of the solid in question. Table 4 illustrates this perspective with the formation of low albite ($NaAlSi_3O_8$). The most direct way to express the formation reaction is to use the component oxides as reactants (row 1). The value of ΔG°_r is then −163 kJ/mol, as shown in the third column of the table. The composition of albite is approached more closely, however, if andalusite is

Table 3
CALCULATION OF THE ENTHALPY OF REACTION AT 298.15 K USING THE THIRD-LAW METHOD AND SELECTED EXPERIMENTAL DATA[27,28] FOR REACTIONS 15 AND 16

T K	P bars	$\Delta\left[\frac{(G^\circ_T - H^\circ_{298})}{T}\right]$ J/K	$(P-1)\Delta V^\circ_{298}$ J/K	f_T[a] bars	ΔH°_r J
		Reaction 15			
793	500	−160.2	−0.1924	340.0	88750
833	500	−159.4	−0.1831	369.0	91990
863	2000	−158.8	−0.7081	1015	87980
878	2000	−158.4	−0.6960	1062	88820
963	5000	−156.6	−1.587	3728	86480
978	5000	−156.3	−1.562	3849	87260
		Reaction 16			
668	3500	−165.23	−4.423	694.8	76980
678	3500	−165.16	−4.358	744.5	77660
687	4800	−165.10	−5.898	1252	76740
697	4800	−165.03	−5.814	1329	77400
728	7000	−164.81	−8.118	3203	77040
736	7000	−164.75	−8.030	3323	77540

[a] f_T — values for the fugacity of water were taken from Burnham.[29]

Table 4
TREND IN ΔG°_r FOR THE FORMATION OF LOW ALBITE FROM AN INCREASINGLY COMPLEX SUITE OF REACTANTS

Reactants	$\Sigma_i n_i \Delta G^{\circ}_i$[a] (kJ/mol)	ΔG°_r[b] (kJ/mol)
$0.5Na_2O(c) + 0.5Al_2O_3(c) + 3SiO_2(s)$	−3548.8	−162.9
$0.5Na_2O + 0.5Al_2SiO_5(c) + 2.5SiO_2(c)$	−3550.7	−161.0
$0.5Na_2SiO_3 + 0.5Al_2SiO_5(c) + 2SiO_2(c)$	−3666.05	−45.67
$NaAlSiO_4(c) + 2SiO_2(c)$	−3691.38	−20.34
$NaAlSi_2O_6(c) + SiO_2(c)$	−3708.74	−2.98
$NaAlSi_3O_8(c)$	−3711.7[c]	0.0

[a] Sum of ΔG°_f for a solid reactant times its stoichiometric coefficient (n) in the formation reaction.

[b] $\Delta G^\circ_r = \Delta G^\circ_f[\text{low albite}] - \Sigma_i n_i \Delta G^\circ_i$.

[c] Value from Reference 16.

used instead of corundum (row 2) and Na_2SiO_3 is used instead of Na_2O (row 3). Even better — with less reliance on α-quartz — is to use nepheline (row 4) or jadeite (row 5) as the sodium-containing reactant. For each of these improvements there is an incremental approach of ΔG°_r toward zero evident in the third column of Table 4.

The Chen algorithm consists of developing four or more suites of reactants with which to describe the formation of an inorganic solid, then extrapolating $\Sigma_i n_i \Delta G^\circ_i$ according to a standardized procedure to obtain an estimate of ΔG° for the solid. Chen[30] proposed that the expression

$$\Sigma_i n_i \Delta G_i^\circ = a \exp(-bx) + c \quad (19)$$

be fit to sets of $\Sigma_i n_i \Delta G_i^\circ$ values, where a, b, and c are adjustable parameters and x is the integer index of the suite of reactants (x = 0, 1, 2,...) assigned in order of decreasing $\Sigma_i n_i \Delta G_i^\circ$. For example, the data in Table 4 can be described by Equation 19 with[30] a = 144 kJ/mol, b = 1.13, and c = −3694 kJ/mol if the first five rows are assigned x = 0 to 4. The parameter c is the extrapolated value of $\Sigma_i n_i \Delta G_i^\circ$ as $x \rightarrow \infty$ and is taken as an estimate of ΔG_f° for low albite. In this example, the estimated ΔG_f° is within about 0.5% of measured values.[14] Chen[30] has compiled estimates of ΔG_f° for about 20 aluminosilicates with an average difference of 0.6% from measured values.

The Chen extrapolation has not been applied to aluminum-bearing solids other than aluminosilicates of relatively complex composition. It is evident that the method cannot be used for simple solid phases like Al_2SiO_5 because the number of formation reactions would be too small to fit Equation 19 accurately. Inaccuracy in the estimate provided by the extrapolation procedure also can arise if the index x cannot be assigned unambiguously or if the ΔG° data for the reactant solids are not accurate.

B. The Tardy-Garrels Correlation

In the Chen extrapolation, the integer index x = 0 is assigned to the reaction in which an aluminosilicate is formed from its component oxides. Thus the parameter a in Equation 19 is an estimate of ΔG_r° for this reaction if the parameter c is a good estimate of ΔG° for the product aluminosilicate. Chen[30] noted that a correlated inversely with the ionic potentials of the metal cations in the component oxides, an effect that he ascribed to "internuclear and interionic repulsion". Tardy et al.[31-35] have expanded on this relationship in a correlation technique that can be used to estimate ΔG_f° for aluminum-bearing solids comprising a single anionic component like silicate, phosphate, or hydroxide. The essence of the method is that ΔG_r° for the formation of an aluminum-bearing solid phase should depend sensitively on the properties of its constituent metal cations.

Consider the formation, from oxide solids, of a binary solid containing the metal cation M^{m+} (e.g., Al^{3+}) and the oxyanion $XO_k^{(2k-n)-}$, where X^{n+} is a cation like Si^{4+} or P^{5+}:

$$(a/p)M_pO_q(c) + (b/s)X_sO_t(c) = M_a(XO_k)_b(c) \quad (20)$$

For this reaction,

$$\Delta G_r^\circ = \Delta G_f^\circ[M_a(XO_k)_b] - (b/s)\Delta G_f^\circ[X_sO_t] - (a/p)\Delta G_f^\circ[M_pO_q] \quad (21)$$

It is reasonable to assume that ΔG_r° will be the more negative, the more stable is the metal M in $M_a(XO_k)_b$ as compared to M_pO_q. (No change in the oxidation state of M or X is involved in Equation 20). One measure of the relative stability of M_pO_q is ΔG_r° for the precipitation reaction:

$$pM^{m+}(aq) + qH_2O(l) = M_pO_q(c) + 2qH^+(aq) \quad (22)$$

This standard Gibbs energy change can be expressed:[31]

$$\Delta G^\circ[M_pO_q] - p\Delta G^\circ[M^{m+}] - q\Delta G^\circ[H_2O] \equiv q(\Delta_M - \Delta G^\circ[H_2O]) \quad (23)$$

where

$$\Delta_M = (1/q)(\Delta G_f^\circ[M_pO_q] - p\Delta G_f^\circ[M^{m+}]) \quad (24)$$

Table 5
VALUES OF THE PARAMETER A_X AND B_X IN THE TARDY-GARRELS MODEL[31-35]

X	A_X	B_X (kJ/mol)	X	A_X	B_X (kJ/mol)
H(I)	0.84	−221.8	Si(IV)	1.01	−189.1
C(IV)	1.16	−258.7	P(V)	1.40	−321.7
N(V)	1.30	−259.3	S(VI)	1.33	−385.6

The parameter Δ_M is the change in $\Delta G°$ for the transformation of picamoles of M^{m+} ions from aqueous solution to the solid oxide, divided by the mole number of O^{2-} in the oxide. This parameter is expected to be the more negative, the more stable is M^{m+} in the oxide relative to the fully solvated metal cation. In the case of Al^{3+}, $\Delta_{Al} = -204.1$ kJ/mol.

Since Δ_M is a measure of the stability of M_pO_q, ΔG_r° in Equation 21 should be an increasing function of $-\Delta_M$. Tardy and coworkers[31-35] have shown that ΔG_r° in Equation 21 is in fact correlated linearly with Δ_M:

$$\Delta G_r^\circ = (1/2)A_X\{[(am)(bn)]/[am + bn]\}(\Delta_M - B_X) \tag{25}$$

where A_X and B_X are adjustable parameters dependent only on the nature of the cation X^{n+} in the oxyanion part of $M_a(XO_k)_b$. Table 5 lists available values of these two parameters for hydroxide (H), carbonate (C), nitrate (N), silicate (Si), phosphate (P), and sulfate (S) solids.[31-35]

With the data in Table 5 and the value of Δ_{Al} given above, Equations 21 and 25 can be combined to provide an estimate of ΔG_f° for an aluminum-bearing solid phase. Consider as an example andalusite, Al_2SiO_5. According to the Tardy-Garrels correlation method,

$$\begin{aligned}\Delta G_f^\circ[Al_2SiO_5] &= \Delta G_f^\circ[SiO_2] + \Delta G_f^\circ[Al_2O_3] \\ &\quad - (1/2)(1.01)\{[(6)(4)]/[10]\}(-204.1 + 189.1) \\ &= -856.6 - 1582.3 + 18.2 = -2421 \text{ kJ/mol}\end{aligned}$$

This estimate may be compared to the measured value,[16] −2442.8 kJ/mol. Similar estimates for other binary aluminum-bearing solids are listed in Table 6. It is evident that the inaccuracy of the correlation prediction is about 1%, which makes this approach only of semiquantative value.

Tardy and Garrels[31,32] have interpreted the parameter B_X in Equation 25 as the analog of Δ_M for the cation X^{n+}. If the precipitation reaction for $X^{n+}(aq)$ is written:

$$sX^{n+}(aq) + tH_2O(l) = X_sO_t(c) + 2tH^+(aq) \tag{26}$$

then the corresponding standard Gibbs energy change is:

$$\Delta G_f^\circ[X_sO_t(c)] - s\Delta G_f^\circ[X^{n+}] - t\Delta G_f^\circ[H_2O] \equiv t(\Delta_X - \Delta G_f^\circ[H_2O]) \tag{27}$$

where

$$\Delta_X = (1/t)(\Delta G_f^\circ[X_sO_t] - s\Delta G_f^\circ[X^{n+}]) \tag{28}$$

Tardy and Garrels[31] proposed that $B_X = \Delta_X$ in Equation 25. This interpretation is consistent

Table 6
ESTIMATES OF STANDARD GIBBS ENERGY OF FORMATION FOR ALUMINUM-BEARING SOLID PHASES WITH TARDY-GARRELS CORRELATION METHOD

Solid	$\Delta G_f^\circ(est)^a$ (kJ/mol)	$\Delta G_f^\circ(exp)^b$ (kJ/mol)
$Al(OH)_3$, gibbsite	−1138	−1154.9
Al_2SiO_5, andalusite	−2421	−2442.8
$Al_2(SO_4)_3$, aluminum sulfate	−3249	−3099.9
$AlPO_4$, berlinite	−1690	−1617.9

[a] Estimated values.
[b] Experimental or evaluated values, from Table 7.

with the idea, mentioned above, that increasing stability of X_sO_t (i.e., a more negative Δ_X) produces a more stable $M_a(XO_k)_b$ and a more negative ΔG_r° in Equation 25. Unfortunately, it is not possible to test this concept independently of Equation 25 because ΔG_f° for the species $X^{n+}(aq)$ is not available — only the oxyanion $XO_k^{(2k-n)-}$ is stable in aqueous solution. Moreover, the solid species $X_sO_t(c)$ is not always accessible to direct thermodynamic characterization in the standard state (e.g., $H_2O[c]$ for Δ_H and $CO_2[c]$ for Δ_C). This problem also attends the use of Equations 20 and 31 for hydroxide and carbonate solids,[31-34] in which $\Delta G_f^\circ[H_2O(c)]$ and $\Delta G_f^\circ[CO_2(c)]$ have been estimated indirectly.

C. The Polymer Model

The polymer model[36-38] is a chemically based correlation method that can be used to predict ΔG_f° for layer aluminosilicates (phyllosilicates). The conceptual basis of this model has been discussed by Nriagu[36] and Sposito[38]. Briefly, it is assumed that phyllosilicates like kaolinite or illite can be pictured as condensation copolymers of hydroxide solids. Thus the formation reaction for kaolinite is expressed

$$2Si(OH)_4(c) + 2Al(OH)_3(c) = Al_2Si_2O_5(OH)_4(c) + 5H_2O(l) \qquad (29)$$

and that for a dioctahedral vermiculite would be:

$$0.35Ca(OH)_2(c) + 3.3Si(OH)_4(c) + 2.7Al(OH)_3(c)$$
$$= Ca_{0.35}[Si_{3.3}Al_{0.7}]Al_2O_{10}(OH)_2(c) + 9.65H_2O(l) \qquad (30)$$

where conventional half-unit-cell chemical formulas for the layer silicates have been used. The principal objective of the polymer model is to establish correlation equations for ΔG_r° corresponding to Equations 29 and 30. Since

$$\Delta G_r^\circ = \Delta G_f^\circ[\text{layer silicate}] - \Sigma_i n_i \Delta G_i^\circ + (\Sigma_i n_i Z_i - n_o)\Delta G_f^\circ[H_2O] \qquad (31)$$

where n_i is the stoichiometric coefficient of the ith reactant hydroxide containing a cation of valence Z_i and n_o is the mole number of oxygen in the layer silicate (e.g., 9 for kaolinite), ΔG_f°[layer silicate] can be calculated once ΔG_r° has been determined.

Three correlation expressions have been proposed for ΔG_r° in Equation 31:

$$\Delta G_r^\circ = -1.63\ n_w \qquad \text{(Nriagu}^{36}\text{)} \qquad (32a)$$

$$|\Delta G_r^\circ| = A \exp(ax + bR - cZ) \qquad \text{(Mattigod and Sposito}^{37}\text{)} \qquad (32b)$$

$$|\Delta G_r^\circ| = |\Delta G_{rem}^\circ| + B(xR/Z) \qquad \text{(Sposito}^{38}\text{)} \qquad (32c)$$

The expression proposed by Nriagu[36] develops from the hypothesis that the most important feature of Equations 29 and 30, insofar as ΔG_r° is concerned, is n_w, the number of moles of water liberated (e.g., 5 for 1:1 kaolinite and 9.65 for vermiculite). Typically $n_w = 5$ for 1:1 layer type aluminosilicates and $9 < n_w < 10$ for 2:1 layer type aluminosilicates. In the case of Equation 29, the Nriagu correlation leads to:

$$\Delta G_f^\circ[\text{kaolinite}] = -1.63(5) + 2\Delta G_f^\circ[Si(OH)_4] + 2\Delta G_f^\circ[Al(OH)_3] - 5\Delta G_f^\circ[H_2O] = -3778 \text{ kJ/mol}$$

which can be compared to the experimental value,[14] -3799.4 kJ/mol.

The Mattigod-Sposito correlation applies only to 2:1 layer type aluminosilicates. It recognizes explicitly a major contribution to ΔG_r° for these clay minerals from the change in corrdination as a metal cation of radius R and valence Z transfers from an hydroxide solid to an exchangeable cation site in satisfaction of the layer charge x created by isomorphic substitutions. For smectites, the adjustable parameters in Equation 32b are[37] $A = 34.56$ kJ/mol, $a = 1.9283$, $b = 3.501$/nm, and $c = 0.2819$. Calculations of ΔG_f° for 14 homoionic smectites produced estimates that agreed to within 13 kJ/mol with reported experimental values.[37]

The correlation proposed by Sposito[38] also applies to 2:1 layer type aluminosilicates. The parameter ΔG_{rem}° in Equation 32c is ΔG_r° for the reaction:

$$4Si(OH)_4(c) + n2Al(OH)_3(c) + n3Mg(OH)_2(c) = [Si_4]Al_{n2}Mg_{n3}O_{10}(OH)_2(c) + 10H_2O(l) \qquad (33)$$

where $n2 = 2$, $n3 = 0$ for dioctahedral minerals and $n2 = 0$, $n3 = 3$ for trioctahedral minerals. Thus ΔG_{rem}° is the standard Gibbs energy change for the formation of pyrophyllite or talc from solid hydroxides; these are the "end member" 2:1 layer type aluminosilicates for the clay minerals smectite, vermiculite, and illite. The effect of isomorphic substitutions which produce the layer charges of these clay minerals is represented in the second term on the right side of Equation 32c. For smectites,[38] $B = 921.7$ kJ/(nm·mol) and Equations 31 and 32c lead to estimates of ΔG_f° that agree to within 11 kJ/mol with reported experimental values. As an example, consider the smectite, Wyoming bentonite (montmorillonite): $Mg_{0.195}[Si_{3.81}Al_{1.52}]Al_{1.52}Fe(III)_{0.22}Mg_{0.29})O_{10}(OH)_2$. In respect to Equation 32c and this dioctahedral clay mineral, $\Delta G_{rem}^\circ = -39.4$ kJ/mol, $x = 0.39$, $R = 0.072$ nm, $Z = 2$, and $|\Delta G_r^\circ|$ is estimated as 52.3 kJ/mol. It follows from Equation 31 that ΔG_f° for the smectite is estimated as:

$$\Delta G_f^\circ[\text{smectite}] = -52.3 + 3.81\Delta G_f^\circ[Si(OH)_4] + 1.71\Delta G_f^\circ[Al(OH)_3] + 0.22\Delta G_f^\circ[Fe(OH)_3] + 0.485\Delta G_f^\circ[Mg(OH)_2] - 10\Delta G_f^\circ[H_2O] = -5254 \text{ kJ/mol}$$

which compares well with the reported experimental value,[38] -5262 kJ/mol.

A characteristic feature of 2:1 clay minerals like smectities is the fractional values of the stoichiometric coefficients of the cations in their chemical formulas. Fractional stoichiometric coefficients imply a solid solution phenomenon, with attendant problems of metastability and solubility disequilibrium.[39-41] These difficulties may be enough in some quarters to vitiate any attempt to call clay minerals single phases and impart thermodynamic significance to ΔG° values for them derived in conventional ways from solubility experiments.[39] This caveat must accompany the chemical interpretation of the polymer model as a predictor of thermodynamic properties.

D. Methods for Estimating Entropy and Heat Capacity

Early methods[42,43] of estimating the heat capacity and entropy made use of the Dulong-Petit rule, that the heat capacity of an element is approximately 25.9 J/K at room temperature, the approximation that the heat capacity is about 30.3 J/K at the first transition temperature,[43] and the assumption that the entropy, S_e, of an element[42] at room temperature is

$$S_e = (3/2)\ln(\text{atomic weight}) - 0.94 \quad (34)$$

A heat capacity function can be estimated from the two heat capacity estimates, the temperature of the first transition, and a linear relation of heat capacity with temperature.[43] Summation of the heat capacity or entropy estimates for each element (weighted by the stoichiometric proportion of that element in the chemical formula) provides the estimate of the heat capacity, heat capacity function, or entropy of the phase.

These early empirical methods provide quick and easily developed estimates for the entropy and heat capacity; but, in general, the estimated values for complex chemical phases are not accurate enough for use in model calculations. Keiffer,[44] using the principles of lattice dynamics, has attempted to provide a more theoretical basis for estimating heat capacities by developing a simplified lattice vibrational spectrum that can be used to calculate heat capacities as a function of temperature. This model represents only a simple approximation to real lattice vibrational spectra, but good estimates of the heat capacities of several minerals have been reported.[44] Use of the Keiffer model requires a knowledge of the elastic wave velocities, crystallographic parameters, and spectral frequencies combined with some relatively formidable calculations. Accordingly, the method does not provide an easily applicable procedure.

A procedure utilizing coordination polyhedra (also called "fictive components") can provide good estimates for the heat capacity of aluminosilicate minerals.[45] Heat capacity functions for the fictive components have been calculated from multiple least squares regression analysis of the heat capacities of a large group of silicate minerals. The crystal structure of the phase for which an estimate of the heat capacity is required must be known well enough to specify the coordination polyhedra for the cations. A heat capacity function is developed for the phase through the summation of the heat capacity functions for the constituent cations in the stoichiometric porportions. The procedure was used by Robinson and Hass[45] to estimate the heat capacity of illite. The results of that study show that the estimated values in the temperature range of 250 to 370 K reproduce the experimental data[46] to about 0.9%.

IV. THERMODYNAMIC PROPERTIES FOR SELECTED ALUMINUM-BEARING PHASES

Modern procedures of data evaluation involve the simultaneous analysis of all types of experimental data at the conditions of the observation (i.e., not extrapolated to 298.15 K and 1 bar) and without the approximation shown in Equation 14.[25] These procedures represent the best method for the identification of experimental data containing systematic errors and for the production of computer data bases and tabulations of internally consistent thermodynamic properties for minerals and related materials. Unfortunately, this is a massive task requiring extensive human and computer resource and has not yet been completed for all chemical systems of interest to natural waters.

The thermodynamic data provided in Table 7 represents a combination of levels of evaluation, with those values representing estimates based upon model calculations listed in parentheses. Every effort has been made to present the best values available for the phases listed; however, values given in Table 7 may be expected to change as larger groups of data undergo simultaneous evaluation and correlation.

Table 7
THERMODYNAMIC PROPERTIES FOR SELECTED ALUMINUM-BEARING MINERALS AND RELATED SUBSTANCES

Name and formula	Properties at 298.15 K				Formation from elements		Equation coeff[a]					Valid temp range	Ref.		
	Relative molecular mass	Molar volume (J/bar)	Heat capacity (J/mol·K)	Entropy (J/mol·K)	Enthalpy (J/mol)	Gibbs energy (J/mol)	C_1 $(\times 10^{-2})$	C_2 $(\times 10^{2})$	C_3 $(\times 10^{4})$	C_4 $(\times 10^{-2})$	C_5 $(\times 10^{-6})$	(K)	C_p° / S°	ΔH_f° / ΔG_f°	V
Oxides, Hydroxides, and Aluminum Metal															
Aluminum	26.9815	0.9999	24.31	28.35	0.00	0.00	0.27237	−0.65674	0.14310	—	−0.19916	298	16	16	16
Al			0.05	0.08								933	16	16	
Corundum	101.962	2.5575	79.01	50.92	−1675.70	−1582.23	1.5736	0.071899	—	−9.8804	−1.8969	298	16	16	16
Al_2O_3			0.16	0.11	1.30	1.32						1800	16	16	
γ-Alumina	101.962	—	79.02	59.83	−1653.52	−1562.70	1.5343	0.19681	—	−9.0063	−2.0307	298	16	16	—
Al_2O_3			0.20	0.22	1.26	1.30						1800	16	16	
Boehmite	59.989	1.9535	54.24	37.19	−996.4	−918.4	2.05721	−3.49206	—	−26.3527	1.026656	200	57	57	47
AlO(OH)			0.11	0.11	0.90	0.90						700	57	57	
Diaspore	59.989	1.7760	53.10	35.34	−1000.6	−922.0	1.50556	—	—	−17.3002	0.243069	200	47	16	16
AlO(OH)			0.09	0.09	0.50	0.50						800	47	16	
Bayerite	78.004	3.1143	—	65.05	−1288.25	−1149.00	—	—	—	—	—	—	—	47	4
$AlO(OH)_3$				0.18	2.00	2.00						—	49	19	
Doyleite	78.004	3.1471	—	(66.45)	(−1289.3)	(−1150.5)	—	—	—	—	—	—	—	49	4
$Al(OH)_3$				0.50	2.5	2.5						—	49	49	
Nordstrandite	78.004	3.1837	—	(67.98)	(−1289.9)	−1151.5	—	—	—	—	—	—	—	49	4
$Al(OH)_3$				0.50	2.5	2.0						—	49	19	
Gibbsite	78.004	3.1956	91.73	68.44	−1293.13	−1154.89	2.20851	6.01252	—	−26.6764	0.661704	200	11	19	16
$Al(OH)_3$			0.16	0.14	0.90	0.90						800	11	19	
Al^{3+}(aq)	26.981	—	−119.29	—	—	−489.8	—	—	—	—	—	—	50	—	—
						4.0						—	—	19	
$Al(OH)^{2+}$(aq)	43.988	—	—	—	—	−701.4	—	—	—	—	—	—	—	—	—
						4.0						—	—	19	
$Al(OH)_2^{+}$(aq)	60.996	—	—	—	—	−903.4	—	—	—	—	—	—	—	—	—
						4.0						—	—	19	
$Al(OH)_4^{-}$(aq)	95.010	—	—	—	—	−1305.0	—	—	—	—	—	—	—	—	—
						1.3						—	—	19	
Multiple Oxides															
Chrysoberyl	126.973	3.432	105.4	66.25	−2298.49	−2176.16	3.62701	−8.3527	0.22482	− 40.3369	−6.7976	200	51	51	51
$BeAl_2O_4$			0.3	0.30	3.18	3.18						1800	51	51	

CA[b]	158.041	—	120.6	114.2	−2326.2	−2208.6	1.5062	2.4937	—	—	−3.330	298	52	53	—
$CaO·Al_2O_3$			0.3	0.8	5.0	5.0						1800	52	49	
CA_2[b]	260.003	—	199.6	177.8	−4023.8	−3816.5	2.7652	2.2928	—	—	−7.448	298	52	54	—
$CaO·2Al_2O_3$			0.5	1.3	4.8	5.0						1800	52	49	
CA_6[b]	667.851	—	—	432.0	−10813.0	−10247.0	—	—	—	—	—	—	—	54	—
$CaO·6Al_2O_3$				10.0	20.0	20.0						—	54	49	
C_2A[b]	214.120	—	—	155.2	−2958.0	−2809.6	—	—	—	—	—	—	—	48	—
$2CaO·Al_2O_3$				0.5	5.0	5.0						—	49	49	
C_3A[b]	270.199	8.9261	209.8	205.4	−3587.8	−3411.4	2.6058	1.9163	—	—	−5.025	298	52	53	55
$3CaO·Al_2O_3$			0.6	1.3	10.8	10.8						1800	52	49	
C_4A[b]	326.278	—	—	255.1	−4217.0	−4012.4	—	—	—	—	—	—	—	48	—
$4CaO·Al_2O_3$				0.7	10.0	10.0						—	49	49	
Mayenite	1386.682	51.819	1084.8	1044.7	−19430.6	−18465.6	12.6340	27.405	—	—	−23.138	298	52	53	56
$12CaO·7Al_2O_3$			2.8	6.3	75.0	75.0						1310	52	49	
Hercynite	173.808	4.075	123.02	106.27	−1966.48	−1850.80	8.90366	−60.1774	2.6514	−117.502	6.13417	298	57	16	16
$FeAl_2O_4$			0.30	0.82	8.50	8.50						1000	16	16	
Lithium aluminate	65.921	2.5210	67.81	53.35	−1188.67	−1126.28	0.95382	1.1387	—	−0.75392	−2.3647	298	16	16	16
$LiAlO_2$			0.15	2.10	4.18	4.18						1800	16	16	
Spinel	142.267	3.9710	115.94	80.63	−2299.32	−2174.86	2.2291	0.61267	—	−15.512	−1.6857	298	16	16	16
$MgAl_2O_4$			0.25	0.42	0.75	0.76						1800	16	16	
Sodium aluminate	81.970	2.9764	73.34	70.40	−1135.99	−1072.10	0.80249	2.9874	—	—	−1.406	298	58	14	56
$NaAlO_2$			0.20	0.35	1.26	1.30						740	59	49	

Halides, Carbonates, Sulfates, Phosphates, and Nitrates

Aluminum tribromide	266,708	—	101.7	147.0	−527.2	−494.4	0.78408	7.8073	—	—	—	298	49	48	—
$AlBr_3$			0.4	5.0	1.5	2.5						370	48	49	
Aluminum trichloride	133.341	5.355	91.84	110.67	−704.2	−628.8	0.5691	11.7152	—	—	—	298	48	48	56
$AlCl_3$			0.40	0.65	1.2	1.2						465	48	48	
Chloraluminite	241.433	14.492	296.2	318.0	−2691.6	−2261.1	—	—	—	—	—	—	48	48	56
$AlCl_3·6H_2O$			0.9	2.0	2.5	2.5						—	48	48	
Aluminum trifluoride	83.977	2.6150	75.13	66.48	−1510.4	−1431.3	−1.4615	15.039	—	41.536	−5.6988	298	16	60	16
AlF_3			0.4	0.42	2.0	2.0						728	16	60	
Aluminum triiodide	407.695	—	98.7	159.0	−313.8	−300.8	0.70626	9.4809	—	—	—	298	58	48	—
AlI_3			0.4	0.7	1.6	1.6						464	48	48	
Cryolite	209.942	7.0810	215.72	238.45	−3309.54	−3144.91	1.8277	1354.20	—	—	−196.93	298	16	16	16
Na_3AlF_6			1.65	1.67	4.18	4.30						838	16	16	
Dawsonite	143.996	5.93	142.59	132.00	−1963.97	−1785.99	0.34405	33.4695	—	—	0.74643	260	16	16	16
$NaAlCO_3(OH)_2$			0.40	0.50	2.93	2.95						477	16	16	
Aluminum sulfate	396.182	11.951	259.41	239.3	−3440.84	−3099.94	2.8769	15.871	—	—	−2003.7/T	298	16	16	56
$Al_2(SO_4)_3$			1.16	1.2	1.80	1.88						850	16	16	
Alunite	828.405	29.3600	745.17	656.05	—	—	8.5901	41.235	—	—	−6275.6/T	298	16	—	16
$K_2Al_6(OH)_{12}(SO_4)_4$			3.75	3.77								700	16	—	
K-Al-sulfate	258.207	—	192.97	204.6	−2470.2	−2240.0	—	—	—	—	—	—	48	48	—
$KAl(SO_4)_2$			0.45	0.8	5.0	5.0						—	48	48	

Table 7 (continued)
THERMODYNAMIC PROPERTIES FOR SELECTED ALUMINUM-BEARING MINERALS AND RELATED SUBSTANCES

	Properties at 298.15 K						Equation coeff[a]						Ref.		
					Formation from elements										
Name and formula	Relative molecular mass	Molar volume (J/bar)	Heat capacity (J/mol·K)	Entropy (J/mol·K)	Enthalpy (J/mol)	Gibbs energy (J/mol)	C_1 ($\times 10^{-2}$)	C_2 ($\times 10^{2}$)	C_3 ($\times 10^{4}$)	C_4 ($\times 10^{-2}$)	C_5 ($\times 10^{-6}$)	Valid temp range (K)	C_p° / S°	ΔH_f° / ΔG_f°	V
K-Al-sulfate hydrate	312.253	—	—	314.0	−3381.1	−2974.5	—	—	—	—	—	—	—	48	—
$KAl(SO_4)_2 \cdot 3H_2O$				1.2	5.5	5.5						—	48	48	
Berlinite	121.953	4.6580	93.18	90.79	−1733.8	−1617.9	−35.888	−823.15	19.776	3.1323T	—	298	16	48	16
$AlPO_4$			0.20	0.21	2.1	2.1						830	16	48	
Augelite	199.956	7.376	—	—	—	−2809.6	—	—	—	—	—	—	—	—	61
$Al_2PO_4(OH)_3$						8.0						—	—	61	
Aluminum-nitrate hydrate	321.089	—	433.04	467.8	−2850.48	−2203.39	—	—	—	—	—	—	48	48	—
$Al(NO_3)_3 \cdot 6H_2O$			1.20	4.6	4.20	4.20						—	48	48	
Ortho and Ring Structure															
Kyanite	162.047	4.4090	122.35	84.47	−2594.27	−2444.03	3.36114	−2.5960	—	−35.5746	—	200	47	47	47
Al_2SiO_5			0.15	0.44	0.43	0.39						1600	47	47	
Andalusite	162.047	5.1530	121.99	93.77	−2590.27	−2442.81	5.43227	−20.7090	0.66893	−67.5436	—	200	47	47	47
Al_2SiO_5			0.26	0.73	0.64	0.48						1800	47	47	
Sillimanite	162.047	4.9900	124.53	96.09	−2587.77	−2441.00	3.1347	−1.894162	—	−31.6487	—	200	47	47	47
Al_2SiO_5			0.16	0.55	0.54	0.44						1800	47	47	
Mullite	426.056	13.455	326.1	269.57	−6810.42	−6431.29	7.5455	−2.9426	—	−65.757	−3.4541	298	16	16	16
$3Al_2O_3 \cdot 2SiO_2$			1.2	4.18	2.20	2.22						1800	16	16	
Euclase	145.084	4.686	126.7	89.09	−2532.91	−2370.17	5.3292	−15.0729	0.41223	−67.2630	2.1976	200	51	51	51
$BeAlSiO_4(OH)$			0.3	0.40	3.04	3.04						1800	51	51	
Beryl	537.502	20.327	417.8	346.7	−9006.52	−8500.36	16.25842	−42.5206	1.2032	−201.8094	6.82544	200	51	51	51
$Be_3Al_2Si_6O_{18}$			4.0	4.7	7.10	6.39						1800	51	51	
Lawsonite	314.242	10.132	285.2	237.61	−4879.06	−4525.62	—	—	—	—	—	—	16	16	16
$CaAl_2Si_2O_7(OH)_2 \cdot H_2O$			3.0	2.09	4.60	4.70						—	16	16	
Gehlenite	274.205	9.0240	205.39	209.89	−3981.71	−3782.89	5.88351	−13.43066	0.38909	−62.7433	—	200	47	47	47
$Ca_2Al_2SiO_7$			0.57	0.97	2.46	2.33						1800	47	47	
Grossular	450.455	12.530	330.51	255.97	−6636.34	−6274.72	9.85362	19.3287	0.3353	−107.077	—	200	47	47	47
$Ca_3Al_2Si_3O_{12}$			0.15	2.95	3.22	2.58						1400	47	47	
Cordierite	584.957	23.322	452.30	407.2	−9161.52	−8651.11	8.2134	4.3339	—	−50.003	−8.2112	298	16	16	16
$Mg_2Al_3(AlSi_5O_{18})$			1.60	3.8	5.85	5.90						1700	16	16	

Chain and Band Structure

Ca-Al pyroxene	218.125	6.3500	165.66	141.61	−3298.96	−3122.77	3.22848	—	—	−21.8582	−2.72024	200	47	47	47
$CaAl_2SiO_6$			2.74	2.17	1.91	1.47						1800	47	47	
Spodumene	186.090	5.8370	158.90	129.30	−3053.50	−2880.20	4.2117	−2.4005	—	−47.761	1.9100	298	16	16	16
$LiAlSi_2O_6$			0.50	0.80	2.79	2.80						1200	16	16	
Jadeite	202.140	6.0400	159.95	133.47	−3029.87	−2851.30	3.0113	1.0143	—	−20.551	−2.2393	298	16	62	16
$NaAl(SiO_3)_2$			0.50	1.25	4.20	4.20						1300	16	62	

Framework Structures

Anorthite	278.211	10.079	211.40	199.30	−4229.10	−4003.33	5.1683	−9.2492	0.41883	−45.885	−1.4085	298	16	16	16
$CaAl_2Si_2O_8$			0.40	0.30	3.12	3.15						1800	16	16	
Leonhardite	922.867	40.44	953.95	922.2	−14246.5	−13197.1	—	—	—	—	—	—	16	16	16
$Ca_2Al_4Si_8O_{24}{\cdot}7H_2O$			9.00	10.9	9.6	10.2						—	16	16	
Microcline	278.333	10.872	202.40	214.20	−3967.69	−3742.33	7.5955	−21.711	0.64333	−95.268	4.7642	298	62	16	16
$KAlSi_3O_8$			0.40	0.41	3.37	3.40						1400	16	16	
High sanidine	278.333	10.905	204.50	232.90	−3959.56	−3739.78	6.9337	−17.170	0.49188	−83.054	3.4622	298	62	16	16
$KAlSi_3O_8$			0.40	0.48	3.37	3.40						1400	16	16	
Kaliophillite	158.164	5.989	119.79	133.26	−2121.92	−2005.98	1.8888	5.5187	—	−14.787	—	298	16	16	16
$KAlSiO_4$			0.25	1.25	1.44	1.45						810	16	16	
Leucite	218.248	8.839	164.14	200.20	−3038.65	−2875.89	1.4842	13.425	—	—	−2.1645	298	16	16	16
$KAlSi_2O_6$			0.30	1.70	2.76	2.85						955	16	16	
Low albite	262.225	10.007	205.10	207.40	−3935.12	−3711.72	5.8394	−9.2852	0.22722	−64.242	1.6780	298	62	16	16
$NaAlSi_3O_8$			0.40	0.40	3.42	3.44						1400	16	16	
Analbite	262.225	10.043	204.80	226.40	−3924.24	−3706.51	6.7137	−14.671	0.36586	−79.736	3.1740	298	62	16	16
$NaAlSi_3O_8$			0.40	0.40	3.64	3.66						1400	16	16	
Nepheline	142.055	5.416	115.81	124.35	−2092.11	−1977.50	0.2774	29.54	—	—	—	298	16	16	16
$NaAlSiO_4$			0.30	1.25	2.42	2.45						467	16	16	
Sodalite	969.212	—	806.87	848.07	−13457.7	−12703.6	17.83272	−41.5043	2.9464	−141.833	−5.10604	298	49	63	—
$Na_8Al_6Si_6O_{24}Cl_2$			3.25	4.24	15.8	16.6						1000	63	63	
Marialite[d]	845.115	32.93	671.04	705.48	−12200.7	−11506.8	13.2870	17.90096	−0.0114	−123.525	1.2913	298	49	64	65
$Na_4Al_3Si_9O_{24}Cl$			2.80	2.80	11.0	12.0						1000	64	49	
Meionite[d]	934.719	34.15	705.48	697.91	−13835.4	−13081.2	25.8529	−66.1387	1.3203	−318.177	13.18585	298	49	64	65
$Ca_4Al_4Si_6O_{24}CO_3$			2.80	2.80	13.5	14.0						1000	64	49	

Sheet Structures

Imogolite	198.076	—	—	173.1	−3192.1	−2929.2	—	—	—	—	—	—	—	66	—
$HOSiO_3Al_2(OH)_3$				5.0	2.0	2.0						—	49	66	
Allophane[e]	252.122	—	—	—	—	(−3591.0)	—	—	—	—	—	—	—	—	—
$H_4SiO_4{\cdot}2Al(OH)_3$						6.0						—	—	49	
Endellite	294.191	—	—	—	—	(−4252.6)	—	—	—	—	—	—	—	—	—
$Al_2Si_2O_5(OH)_4{\cdot}2H_2O$						6.0						—	—	49	
Halloysite	258.162	—	245.25	203.33	−4101.03	−3780.37	7.7230	−14.51768	—	−87.2948	1.93671	200	47	47	—

Table 7 (continued)
THERMODYNAMIC PROPERTIES FOR SELECTED ALUMINUM-BEARING MINERALS AND RELATED SUBSTANCES

	Properties at 298.15 K						Equation coeff[a]						Ref.		
					Formation from elements										
Name and formula	Relative molecular mass	Molar volume (J/bar)	Heat capacity (J/mol·K)	Entropy (J/mol·K)	Enthalpy (J/mol)	Gibbs energy (J/mol)	C_1 ($\times 10^{-2}$)	C_2 ($\times 10^{2}$)	C_3 ($\times 10^{4}$)	C_4 ($\times 10^{-2}$)	C_5 ($\times 10^{-6}$)	Valid temp range (K)	C_p° / S°	ΔH_f° / ΔG_f°	V
$Al_2Si_2O_5(OH)_4$			0.76	3.07	1.20	1.51						900	47	47	
Dickite	258.162	9.93	239.79	197.06	−4119.78	−3795.95	9.08360	−21.13260	—	−111.953	3.80445	200	47	47	16
$Al_2Si_2O_5(OH)_4$			0.74	3.07	1.24	1.54						900	47	47	
Kaolinite	258.162	9.952	246.14	204.97	−4119.78	−3799.61	7.49175	−13.54204	—	−82.7864	1.49195	200	47	47	16
$Al_2Si_2O_5(OH)_4$			0.77	1.02	1.06	0.98						1000	47	47	
Pyrophyllite	360.317	12.782	294.35	239.42	−5642.02	−5268.13	14.5451	−79.2186	3.9718	−177.428	—	200	47	47	16
$Al_2Si_4O_{10}(OH)_2$			0.32	0.99	1.15	1.04						1200	47	47	
Muscovite[c]	398.311	14.0716	326.1	306.4	−5971.6	−5595.5	9.177	−8.111	—	−103.48	2.834	298	67	67	16
$KAl_2[AlSi_3O_{10}](OH)_2$			0.6	0.6	5.2	5.2						1000	46	67	
Illite	388.451	—	322.8	276.1	—	−5489.0	—	—	—	—	—	—	46	—	—
(K, Na, Ca)[f]			1.0	1.5								—	46	68	
Phlogopite[d]	417.262	14.991	355.1	315.9	−6246.1	−5860.5	8.7213	−7.725	—	−86.006	0.3575	298	69	69	16
$KMg_3[AlSi_3O_{10}](OH)_2$			1.1	1.0	6.4	6.5						1000	69	69	
Phlogopite[c]	421.244	14.637	342.4	336.3	−6392.9	−6053.1	4.9287	4.9103	—	−15.688	−6.5987	298	16	16	16
$KMg_3[AlSi_3O_{10}]F_2$			0.9	2.1	3.7	3.8						1670	16	16	
Vermiculite	405.484	—	—	—	—	(−5551.8)	—	—	—	—	—	—	—	—	—
$(K_{0.01}Ca_{0.06}Fe(III)_{0.46}$ $Fe(II)_{0.02}Mg_{2.71})(Al_{1.14}$ $Si_{2.91})O_{10}(OH)_2$						10.0						—	—	36	
Paragonite	382.201	13.21	321.5	277.1	−5949.33	−5568.46	6.6844	3.627	—	−58.161	−1.8604	298	69	69	65
$NaAl_2[AlSi_3O_{10}](OH)_2$			1.0	0.9	4.60	4.60						800	69	69	
Beidellite	384.141	—	—	—	—	(−5194.0)	—	—	—	—	—	—	—	—	—
(Mg, Ca, Na, K)[f]						10.0						—	—	38	
Montmorillonite	371.162	—	—	—	—	(−5255.0)	—	—	—	—	—	—	—	—	—
$Mg_{0.195}(Al_{0.19}Si_{3.81})$ $(Al_{1.52}Fe(III)_{0.22}$ $Mg_{0.29}O_{10}(OH)_2$						10.0						—	—	38	
Montmorillonite	375.538	—	—	—	—	(−5112.0)	—	—	—	—	—	—	—	—	—
$Mg_{0.208}(Al_{0.18}Si_{3.82})$ $(Al_{1.29}Fe(III)_{0.335}$ $Mg_{0.445})O_{10}(OH)_2$						10.0						—	—	38	
Margarite	398.187	13.38	323.44	263.64	−6239.61	−5854.84	8.26504	−5.02910	—	−84.2744	—	200	47	47	65
$CaAl_2[Al_2Si_2O_{10}](OH)_2$			0.38	0.66	1.95	1.90						1200	47	47	

Prehnite	412.390	14.0330	331.11	292.74	−6193.63	−5816.43	9.46022	−11.50654	—	−105.605	2.75523	200	47	47	47
$Ca_2Al_2Si_3O_{10}(OH)_2$			0.44	0.72	1.70	1.64						1200	47	47	
Clinochlore	555.797	20.98	531.00	397.6	−8610.2	−8154.0	94.7951	−788.342	36.206	−1359.4264	83.1195	280	70	70	70
$(Mg_5Al)(Si_3Al)O_{10}(OH)_8$			0.50	0.9	10.0	10.0						700	70	49	
Zeolites															
Analcime	220.155	9.749	211.53	226.75	−3296.9	−3077.2	1.09974	29.340	—	—	−1.2514	298	71	71	16
$Na_{0.96}Al_{0.96}Si_{2.04}O_6{\cdot}H_2O$			0.21	0.23	3.3	3.3						600	71	71	
Dehydrated analcime	202.140	—	163.59	171.71	−2970.2	−2803.7	1.68469	10.966	—	—	3.3401	298	71	71	—
$Na_{0.96}Al_{0.96}Si_{2.04}O_6$			0.16	0.17	3.5	3.5						1000	71	71	
Heulandite	708.133	—	781.03	767.18	−10491.0	−9675.6	7.4555	65.133	—	—	−14.10835	298	72	72	—
(Ba, Sr, Ca, K, Na)[f]			0.78	0.77	10.2	10.2						480	72	72	
Natrolite	380.223	16.939	359.23	359.73	−5718.6	−5316.6	3.01957	37.688	—	—	−4.89756	298	73	73	65
$Na_2Al_2Si_3O_{10}{\cdot}2H_2O$			0.72	0.72	5.0	5.0						675	73	73	
Mesolite	392.339	—	371.0	363.0	−5947.1	−5513.2	—	—	—	—	—	—	73	73	—
$Na_{0.68}Ca_{0.66}Al_{1.99}$ $Si_{3.01}O_{10}{\cdot}2.647H_2O$			2.0	2.0	5.4	5.4						—	73	73	
Scolecite	387.784	—	382.81	367.42	−6049.0	−5597.9	1.351876	83.053	—	—	—	298	73	73	—
$CaAl_2Si_3O_{10}{\cdot}3H_2O$			0.77	0.73	5.0	5.0						475	73	73	
Merlinoite	264.313	12.141	292.0	348.0	−3772.4	−3508.8	—	—	—	—	—	—	74	75	74
$KAlSi_{2.2}O_{6.4}{\cdot}1.89H_2O$			0.5	0.4	3.2	3.2						—	74	75	
Merlinoite	261.400	12.122	333.7	359.0	−3896.7	−3604.6	—	—	—	—	—	—	74	75	74
$Na_{0.74}K_{0.26}AlSi_{2.20}O_{6.40}{\cdot}2.39H_2O$			0.6	0.4	3.2	3.2						—	74	75	
Merlinoite	237.277	11.984	247.5	304.3	−3359.3	−3122.7	—	—	—	—	—	—	74	75	74
$KAlSi_{1.81}O_{5.62}{\cdot}1.69H_2O$			0.5	0.3	2.8	2.8						—	74	75	
Merlinoite	233.057	12.291	301.5	319.2	−3487.6	−3224.6	—	—	—	—	—	—	74	75	74
$Na_{0.81}K_{0.19}AlSi_{1.81}O_{5.62}{\cdot}2.18H_2O$			0.6	0.4	2.8	2.8						—	74	75	
Glasses															
$KAlSi_3O_8$	278.337	—	209.4	261.6	−3914.7	−3703.5	6.295	−10.84	0.1928	−72.10	2.496	298	67	14	—
			0.5	2.8	3.3	3.4						1300	14	14	
$NaAlSi_3O_8$	262.224	—	209.3	251.9	−3875.5	−3665.3	9.344	−38.91	1.476	−118.20	5.594	298	67	14	—
			0.5	1.8	3.7	3.7						1200	14	14	
$CaAl_2Si_2O_8$	278.210	—	211.4	237.3	−4171.3	−3956.8	3.752	3.197	—	−24.59	−2.815	298	67	14	—
			0.5	2.5	3.3	3.3						1500	14	14	
$Ca_2Al_2SiO_7$	274.205	—	—	(250.0)	−3931.7	−3744.8	—	—	—	—	—	—	—	48	—
				5.0	4.0	6.0						—	—	—	
$12CaO{\cdot}7Al_2O_3$	1386.682	—	—	(1235.0)	−19087.0	−18178.7	—	—	—	—	—	—	—	48	—
				10.0	25.0	30.0						—	—	—	
$CaO{\cdot}Al_2O_3$	158.041	—	—	(139.0)	−2301.0	−2190.8	—	—	—	—	—	—	—	48	—
				5.0	3.3	4.0						—	—	—	

[a] $C_p^\circ = C_1 + C_2T + C_3T^2 + C_4T^{0.5} + C_5T^{-2}$; exceptions, as noted, exist in C_4 for berlinite and C_5 for aluminum sulfate and alunite.

Table 7 (continued)
THERMODYNAMIC PROPERTIES FOR SELECTED ALUMINUM-BEARING MINERALS AND RELATED SUBSTANCES

[b] C = CaO, A = Al_2O_3.
[c] Calculated for disordered structure.[34]
[d] Calculated for ordered structures.[34]
[e] SRO = Short range order, calculated for $Si:Al_2 = 1:1$ where $0.7:1.0 < Si:Al_2 < 2.0:1.0$ represents the general compositional range.
[f] Selected formulas:
Illite $(K_{0.69}Na_{0.03}Ca_{0.05})(Al_{1.69}Mg_{0.40})(Al_{0.42}Si_{3.58})O_{10}(OH)_2$
Heulandite $(Ba_{0.065}Sr_{0.175}Ca_{0.585}K_{0.132}Na_{0.383})Al_{2.165}Si_{6.835}O_{18}{\cdot}6H_2O$
Beidellite $Mg_{0.135}Ca_{0.01}Na_{0.07}K_{0.095}(Al_{0.45}Si_{3.55})(Al_{1.41}Fe(III)_{0.415}Fe(II)_{0.055}Mg_{0.205})O_{10}(OH)_2$

REFERENCES

1. **Pauling, L.,** *The Nature of the Chemical Bond,* Cornell University Press, Ithaca, NY, 1960, chap. 13.
2. **Eggleston, R. A.,** The relations between crystal structure and silicate weathering rates, in *Rates of Chemical Weathering of Rocks and Minerals,* Colman, S. M. and Dethier, D. P., Eds., Academic Press, Orlando, Florida, 1986, 21.
3. **Shannon, R. D.,** Revised effective ionic radii and systematic studies of interatomic distances in halides and chalcogenides, *Acta Crystallogr.,* A32, 751, 1976.
4. **Chao, G. Y., Baker, J., Sabina, A. P., and Roberts, A. C.,** Doyleite, a new polymorph of $Al(OH)_3$, and its relationship to bayerite, gibbsite and nordstrandite, *Can. Mineral.,* 23, 21, 1985.
5. **Zoltai, T. and Stout, J. H.,** *Mineralogy Concepts and Principles,* 1984, Part II.
6. **Dixon, J. B. and Weed, S. B., Eds.,** *Minerals in Soil Environments,* Soil Science Society of America, Madison, WI, 1977, chaps. 7 to 11, 16.
7. **Sposito, G.,** *The Surface Chemistry of Soils,* Oxford University Press, New York, 1984, chap. 1.
8. **Brindley, G. W. and Brown, G., Eds.,** *Crystal Structures of Clay Minerals and Their X-ray Identification,* Mineralogical Society, London, 1980, chap. 2.
9. **Robie, R. A. and Hemingway, B. S.,** Calorimeters for heat of solution and low-temperature heat capacity measurements, *U.S. Geol. Surv. Prof. Pap.,* 755, 1972.
10. **Parsonage, N. G. and Staveley, L. A. K.,** *Disorder in Crystals,* Clarendon Press, Oxford, 1978, chap. 2.
11. **Hemingway, B. S., Robie, R. A., Fisher, J. R., and Wilson, W. H.,** The heat capacities of gibbsite, $Al(OH)_3$, between 13 and 480 K and magnesite, $MgCO_3$, between 13 and 380 K and their standard entropies at 298.15 K, the heat capacity of calorimetry conference benzoic acid between 12 and 316 K, *U.S. Geol. Surv. J. Res.,* 5, 797, 1977.
12. **Navrotsky, A.,** Recent progress and new directions in high temperature calorimetry, *Phys. Chem. Minerals,* 2, 89, 1977.
13. **Coops, J., Jessup, R. S., and van Nes, K.,** Calibration of calorimeters for reaction in a bomb at constant volume, in *Experimental Thermochemistry,* Rossini, F. D., Ed., Wiley-Interscience, New York, 1956, chap. 3.
14. **Hemingway, B. S. and Robie, R. A.,** Enthalpies of formation of low albite, $NaAlSi_3O_8$, gibbsite, ($Al(OH)_3$), and $NaAlO_2$; revised values for $\Delta H^\circ_{f,298}$ and $\Delta G^\circ_{f,298}$ of some aluminosilicate minerals, *U.S. Geol. Surv. J. Res.,* 5, 413, 1977.
15. **Taylor, K. and Wells, L. S.,** Studies of the heat of solution of calcium and magnesium oxides and hydroxides, *J. Res. Nat. Bur. Stand.,* 21, 133, 1938.
16. **Robie, R. A., Hemingway, B. S., and Fisher, J. R.,** Thermodynamic properties of minerals and related substances at 298.15 K and 1 bar (10^5 Pascals) pressure and at higher temperatures, *U.S. Geol. Surv. Bull.,* 1452, 1979.
17. **Lewis, D.,** Studies of redox equilibria at elevated temperatures. I. The estimation of equilibrium constants and standard potentials for aqueous systems up to 374°C, *Ark. Kemi,* 32, 385, 1971.
18. **Singh, S. S.,** The solubility product of gibbsite at 15, 25 and 35°C. *Proc. Soil Sci. Soc. Am.,* 38, 415, 1974; **Singh, S. S.,** Chemical equilibrium and chemical thermodynamic properties of gibbsite, *Soil Sci.,* 121, 332, 1976.
19. **Hemingway, B. S.,** Gibbs free energies of formation for bayerite, nordstrandite, $Al(OH)^{2+}$, and $Al(OH)_2{}^+$, aluminum mobility, and the formation of bauxites and laterites, in Advances in Physical Geochemistry, Vol. 2, Saxena, S. K., Ed., Springer-Verlag, New York, 1982, chap. 9.
20. **Kittrick, J. A.,** The free energy of formation of gibbsite and $Al(OH)_4^-$ from solubility measurements, *Am. Mineral.,* 51, 1457, 1966.
21. **Hemingway, B. S. and Robie, R. A.,** The entropy and Gibbs free energy of formation of the aluminum ion, *Geochim. Cosmochim. Acta,* 41, 1402, 1977.
22. **May, H. M., Helmke, P. A., and Jackson, M. L.,** Gibbsite solubility and thermodynamic properties of hydroxy-aluminum ions in aqueous solutions at 25°C, *Geochim. Cosmochim. Acta,* 43, 861, 1979.
23. **Hemingway, B. S. and Robie, R. A.,** Revised values for the Gibbs free energy of formation of $[Al(OH)^-_{4,aq}]$, diaspore, boehmite and bayerite at 298.15 K and 1 bar, the thermodynamic properties of kaolinite to 800 K and 1 bar, and the heats of solution of several gibbsite samples, *Geochim. Cosmochim. Acta,* 42, 1533, 1978.
24. **Couturier, Y., Michard, G., and Sarazin, G.,** Constantes de formation des complexes hydroxydés de l'aluminium en solution aqueuse de 20 a 70°C, *Geochim. Cosmochim. Acta,* 48, 649, 1984.
25. **Haas, J. L., Jr., Robinson, G. R., and Hemingway, B. S.,** Thermodynamic tabulations for selected phases in the system CaO-Al_2O_3-SiO_2-H_2O at 101.325 kPa (1 atm) between 273.15 and 1800 K, *J. Phys. Chem. Ref. Data,* 10, 575, 1981.
26. **Lewis, G. N. and Randall, M.,** *Thermodynamics,* 2nd ed., revised by Pitzer, K. S. and Brewer, L., McGraw-Hill, New York, 1961, 377.

27. **Chatterjee, N. D. and Johannes, W.,** Thermal stability and standard thermodynamic properties of synthetic $2M_1$-muscovite, $KAl_2[AlSi_3O_{10}(OH)_2]$, *Contrib. Mineral. Petrol.*, 48, 89, 1974.
28. **Haas, H. and Holdaway, M. J.,** Equilibria in the system Al_2O_3-SiO_2-H_2O involving the stability limits of pyrophyllite, and thermodynamic data of pyrophyllite, *Am. J. Sci.*, 276, 449, 1973.
29. **Burnham, C. W., Holloway, J. R., and Davis, N. F.,** Thermodynamic properties of water to 1000°C and 10,000 bars, *Geol. Soc. Am. Spec. Pap.*, 132, 1969.
30. **Chen, C.-H.,** A method of estimation of standard free energies of formation of silicate minerals at 298.15°K, *Am. J. Sci.*, 275, 801, 1975.
31. **Tardy, Y. and Garrels, R. M.,** Prediction of Gibbs energies of formation. I, *Geochim. Cosmochim. Acta,* 40, 1051, 1976.
32. **Tardy, Y. and Garrels, R. M.,** Prediction of Gibbs energies of formation. II, *Geochim. Cosmochim. Acta,* 41, 87, 1977.
33. **Tardy, Y. and Vieillard, P.,** Relationships among Gibbs free energies and enthalpies of formation of phosphate, oxides and aqueous ions, *Contrib. Mineral. Petrol.*, 63, 89, 1977.
34. **Tardy, Y. and Gartner, L.,** Relationships among Gibbs energies of formation of sulfates, nitrates, carbonates, oxides and aqueous ions, *Contrib. Mineral. Petrol.*, 63, 89, 1977.
35. **Tardy, Y.,** Relationships among Gibbs energies of formation of compounds, *Am. J. Sci.*, 279, 217, 1979.
36. **Nriagu, J. O.,** Thermochemical approximation for clay minerals, *Am. Mineral.*, 60, 834, 1975.
37. **Mattigod, S. V. and Sposito, G.,** Improved method for estimating the standard free energies of formation ($\Delta G^{\circ}_{f,298.15}$) of smectites, *Geochim. Cosmochim. Acta,* 42, 1753, 1978.
38. **Sposito, G.,** The polymer model of thermochemical clay mineral stability, *Clays Clay Miner.*, 34, 198, 1986.
39. **May, H. M., Kinniburgh, D. G., Helmke, P. A., and Jackson, M. L.,** Aqueous dissolution, solubilities and thermodynamic stabilities of common aluminosilicate clay minerals: kaolinite and smectites, *Geochim. Cosmochim. Acta,* 50, 1667, 1986.
40. **Sposito, G.,** *The Thermodynamics of Soil Solutions,* Clarendon Press, Oxford, 1981, 53.
41. **Lippman, F.,** The thermodynamic status of clay minerals, in *Proc. Int. Clay Conf. 1981,* van Olphen, H. and Veniale, F., Eds., Elsevier Amsterdam, 1982, 475.
42. **Latimer, W. M.,** *The Oxidation States of the Elements and Their Potentials in Aqueous Solutions,* Prentice-Hall, New York, 1952, 359.
43. **Kubaschewski, O. and Evans, E. L.,** *Metallurgical Thermochemistry,* Butterworth-Springer, Ltd., London, 1951, chap. 3.
44. **Keiffer, S. W.,** Thermodynamics and lattice vibrations of minerals. III. Lattice dynamics and an approximation for minerals with application to simple substances and framework silicates, *Rev. Geophys. Space Phys.*, 17, 35, 1979.
45. **Robinson, G. R., Jr. and Hass, J. L., Jr.,** Heat capacity, relative enthalpy, and calorimetric entropy of silicate minerals: an empirical method of prediction, *Am. Mineral.*, 68, 541, 1983.
46. **Robie, R. A., Hemingway, B. S., and Wilson, W. H.,** The heat capacity of calorimetry conference copper and of muscovite $KAl_2(AlSi_3)O_{10}(OH)_2$, pyrophyllite $Al_2Si_4O_{10}(OH)_2$, and illite $K_3(Al_7Mg)(Si_{14}Al_2)O_{40}(OH)_8$ between 15 and 375 K and their standard entropies at 298.15 K, *J. Res. U.S. Geol. Surv.*, 4, 631, 1976.
47. **Hemingway, B. S., Haas, J. L., Jr., and Robinson, G. R., Jr.,** Thermodynamic properties of selected minerals in the system Al_2O_3-CaO-SiO_2-H_2O at 298.15 K and 1 bar (10^5 Pascals) pressure and at higher temperatures, *U.S. Geol. Surv. Bull.*, 1544, 1982.
48. **Wagman, D. D., Evans, W. H., Parker, V. B., Schumm, R. H., Halow, I., Bailey, S. M., Churney, K. L., and Nuttall, R. L.,** The NBS tables of chemical thermodynamic properties. Selected values for inorganic and C_1 and C_2 organic substances in SI units, *J. Phys. Chem. Ref. Data,* 11, 1982.
49. Estimates and calculations made in this study.
50. **Hovey, J. K. and Tremaine, P. R.,** Thermodynamics of aqueous aluminum: standard partial molar heat capacities of Al^{3+} from 10 to 55°C, *Geochim. Cosmochim. Acta,* 50, 453, 1985.
51. **Hemingway, B. S., Barton, M. D., Robie, R. A., and Haselton, H. T., Jr.,** Heat capacities and thermodynamic functions for beryl, $Be_3Al_2Si_6O_{18}$, phenakite, Be_2SiO_4, euclase, $BeAlSiO_4(OH)$, bertrandite, $Be_4Si_2O_7(OH)_2$, and chrysoberyl, $BeAl_2O_4$, *Am. Mineral.*, 71, 557, 1986.
52. **Kelley, K. K. and King, E. G.,** Contributions to the data on theoretical metallurgy. XIV. Entropies of the elements and inorganic compounds, *U.S. Bur. Mines Bull.*, 592, 1961.
53. **Coughlin, J. P.,** Heats of formation of crystalline $CaO \cdot Al_2O_3$, $12CaO \cdot 7Al_2O_3$, and $3CaO \cdot Al_2O_3$, *Am. Chem. Soc. J.*, 78, 5479, 1956.
54. **Hemingway, B. S.,** Comments on "Thermodynamic properties of calcium aluminates", *J. Phys. Chem.*, 86, 2802, 1982.
55. **Mondal, P. and Jeffery, J. W.,** The crystal structure of tricalcium aluminate, $Ca_3Al_2O_6$, *Acta Crystallogr.*, B31, 689, 1975.

56. **Morris, M. C., McMurdie, H. F., Evans, E. H., Paretzkin, B., de Groot, J. H., Hubbard, C. R., and Carmel, S. J.,** Standard X-ray diffraction powder patterns. Section 16 - Data for 86 substances, *U.S. Bur. Stand. Monogr.*, 25, 1979 (also earlier monographs in this series).
57. **Hemingway, B. S.,** unpublished data.
58. **Kelley, K. K.,** Contributions to the data on theoretical metallurgy. XIII. High-temperature heat-content, heat-capacity, and entropy data for the elements and inorganic compounds, *U.S. Bur. Mines Bull.*, 584, 1960.
59. **Stull, D. R. and Prophet, H.,** JANAF Thermochemical Tables, 2nd ed., NSRDS — NBS 37, U.S. Government Printing Office, Washington, D.C., 1971.
60. CODATA Task Group on Key Values for Thermodynamics, CODATA recommended key values for thermodynamics 1977, *CODATA Bull.*, 28, 1978.
61. **Wise, W. S. and Loh, S. E.,** Equilibria and origin of minerals in the system Al_2O_3-$AlPO_4$-H_2O, *Am. Mineral.*, 61, 409, 1976.
62. **Hemingway, B. S., Krupka, K. M., and Robie, R. A.,** Heat capacties of the alkali feldspars between 350 and 1000 K from differntial scanning calorimetry, the thermodynamic functions of the alkali feldspars from 298.15 to 1400 K, and the reaction quartz + jadeite = analbite, *Am. Mineral.*, 66, 1202, 1981.
63. **Komada, N., Westrum, E. F., Jr., Hemingway, B. S., Zolotov, M. Yu., Semenov, Yu, V., Khodakovsky, I. L., and Anovitz, L. M.,** Thermodynamic properties of sodalite from 15 to 1000 K, *J. Chem. Thermodyn.*, submitted.
64. **Komada, N., Moecher, D. P., Westrum, E. F., Jr., Hemingway, B. S., Zolotov, M. Yu., Semenov, Yu. V., and Khodakovsky, I. L.,** Thermodynamic properties of scapolites form 10 to 1000 K, *J. Chem. Thermodyn.*, submitted.
65. **Robie, R. A., Bethke, P. M., and Beardsley, K. M.,** Selected X-ray crystallographic data, molar volumes, and densities of minerals and related substances, *U.S. Geol. Surv. Bull.*, 1248, 1967.
66. **Farmer, V. C., Smith, B. F. L., and Tait, J. M.,** The stability, free energy and heat of formation of imogolite, *Clay Miner.*, 14, 103, 1979.
67. **Krupka, K. M., Robie, R. A., and Hemingway, B. S.,** High-temperature heat capacities of corundum, periclase, anorthite, $CaAl_2Si_2O_8$ glass, muscovite, pyrophyllite, $KAlSi_3O_8$ glass, grossular, and $NaAlSi_3O_8$ glass, *Am. Mineral.*, 64, 86, 1979.
68. **Reesman, A. L.,** Aqueous dissolution studies of illite under ambient conditions, *Clays Clay Miner.*, 22, 443, 1974.
69. **Robie, R. A. and Hemingway, B. S.,** Heat capacities and entropies of phlogopite ($KMg_3[AlSi_3O_{10}](OH)_2$) and paragonite ($NaAl_2[AlSi_3O_{10}](OH)_2$) between 5 and 900 K and estimates of the enthalpies and Gibbs free energies of formation, *Am. Mineral.*, 69, 858, 1984.
70. **Henderson, C. E., Essene, E. J., Anovitz, L. M., Westrum, E. F., Jr., Hemingway, B. S., and Bowman, J. R.,** Thermodynamics and phase equilibria of clinochlore, $(Mg_5Al)(Si_3Al)O_{10}(OH)_8$, *EOS*, 64, 466, 1983.
71. **Johnson, G. K., Flotow, H. E., O'Hare, P. A. G., and Wise, W. S.,** Thermodynamic studies of zeolites: analcime and dehydrated analcime, *Am. Mineral.*, 67, 736, 1982.
72. **Johnson, G. K., Flotow, H. E., O'Hare, P. A. G., and Wise, W. S.,** Thermodynamic studies of zeolites: heulandite, *Am. Mineral.*, 70, 1065, 1985.
73. **Johnson, G. K., Flotow, H. E., O'Hare, P. A. G., and Wise, W. S.,** Thermodynamic studies of zeolites: natrolite, mesolite, and scolecite, *Am. Mineral.*, 68, 1134, 1983.
74. **Donahoe, R. J.,** An Experimental Investigation of Some Physiochemical Controls on Zeolite Formation: Implications for Authigensis, Ph.D. thesis, Stanford University, Stanford, CA, 1983.
75. **Hemingway, B. S., Nitkiewicz, A., and Donahoe, R. J.,** unpublished data for the enthalpy of solution of quartz and data from reference 74.
76. **Ulbrich, H. H. and Waldbaum, D. R.,** Structural and other contributions to the third-law entropies of silicates, *Geochim. Cosmochim. Acta*, 40, 1, 1976.

Chapter 4

AQUEOUS POLYNUCLEAR ALUMINUM SPECIES

Paul M. Bertsch

TABLE OF CONTENTS

I. INTRODUCTION

The hydrolytic species of aluminum are believed to be important in mineral phase formation and transitions, in the mobility of aluminum in soils and aquatic systems, and in the toxicity of aluminum to plants and aquatic organisms.[1-7] In addition to their importance in natural systems, aluminum polymers are known to be important in water-treatment applications, as a primary constituent in several pharmaceuticals, and as a principal component in certain catalysts.[8-13] Furthermore, aluminum polymeric species along with aluminum monomers may be directly or indirectly involved in a number of human health disorders including dialysis dementia, senile and presenile dementia of the Alzheimer type, osteodystrophy, and osteomalocia.[14,15]

The purposes of this chapter are to provide a basic introduction to the hydrolytic reactions of aluminum in aqueous solutions, to present the salient features of aluminum polymeric structures, to discuss the mechanisms of their formation, to present methods of speciating aluminum hydrolytic products, and finally, to discuss certain implications of aluminum species in soil-water systems (see also Chapters 2 and 8).

II. HYDROLYSIS AND POLYMERIZATION OF ALUMINUM AQUEOUS SOLUTION

Hydrolysis and polymerization reactions of aqueous aluminum and the resultant hydrolytic products have been studied for more than a century; however, there exist few established principles. Many of the seemingly inconsistent and conflicting results reported derive from utilization of different experimental methods and approaches. Other discrepancies have arisen, not as a result of differences in data collection, but rather as a consequence of data interpretation. These ambiguities have ensued largely from analytical limitations which have hindered the acquisition of direct unequivocal experimental evidence. The failure to describe aluminum hydrolytic reactions adequately in relatively simple aqueous solutions debilitates the extension of chemical principles to aluminum chemistry in heterogeneous aquatic, sediment, and soil systems, where many ill-defined factors affect aluminum transformations (see Chapters 8 and 9).

The major points of departure for the numerous investigations on the hydrolysis and polymerization of aluminum include the following questions: (1) What are the primary hydrolytic products?, (2) What is (are) the major mechanism(s) of hydrolysis and polymerization?, and (3) What are the steady-state conditions and associated thermodynamic constants for a given hydrolytic or polymerization reaction? These items remain areas of active investigation and debate. Significant progress has been made in recent years toward answering the first two questions. The third, however, tends to remain elusive because of an inability to identify precisely all reactants and products, the experimental dependence on the apparent steady-state conditions obtained, and the slow kinetics involved in the transformation of metastable species, coupled with the general lack of information regarding the kinetics of these reactions.

A. Monomeric Hydrolytic Aluminum Species

The hydrolytic reactions of monomeric aluminum in dilute acidic solutions can be described equivocally by a simple three-stage equilibrium reaction,

$$Al(H_2O)_6^{3+} + H_2O \rightleftharpoons Al(OH)(H_2O)_5^{2+} + H_3O^+; \quad K_{1,1} \tag{1}$$

$$Al(OH)(H_2O)_5^{2+} + H_2O \rightleftharpoons Al(OH)_2(H_2O)_4^{+} + H_3O^+; \quad K_{1,2} \tag{2}$$

$$Al(OH)_2(H_2O)_4^+ + H_2O \rightleftharpoons Al(OH)_3(H_2O)_3^\circ + H_3O^+; \quad K_{1,3} \tag{3}$$

where $K_{x,y}$ are the stepwise equilibrium hydrolysis constants for the respective reactions. The over-all formation or hydrolysis constants (β) for these reactions, or a reaction involving the formation of a polynuclear hydroxoaluminum complex of the general form

$$xAl^{3+} + yH_2O \rightleftharpoons Alx(OH)y^{(3x-y)+} + yH^+ \tag{4}$$

can be expressed:

$$\beta_{xy} = \frac{[Al_x(OH)_y^{(3x-y)+}][H^+]^y}{[Al^{3+}]^x} \tag{5}$$

where the water of hydration has been deleted for simplicity. Employing the three-stage monomeric aluminum hydrolysis scheme, several investigators have concluded that the major hydrolytic product in dilute solutions is the monohydroxo monomeric species $Al(OH)(H_2O)_{5^{2+}}$. The over-all equilibrium-formation constant for this species, given by

$$\beta_{1,1} = \frac{[Al(OH)(H_2O)_5^{2+}][H^+]}{[Al(H_2O)_6^{3+}]} \tag{6}$$

has been determined with great unanimity in the literature, with $K_{1,1}$ values ranging from 4.89 to 5.10, perhaps representing one of the few areas of consonance in aluminum chemistry, although spurious $-\log k_{1,1}$ values as low as 4.49 and as high as 5.86 have appeared[16-26] (see Chapter 2 for a comprehensive discussion). This agreement is remarkable considering that the $\beta_{1,1}$ values were generated employing vastly different experimental conditions. Most of the investigators employed conductometric or potentiometric techniques, although some values were derived from solubility measurements. However, aluminum concentrations, indifferent electrolyte concentrations, and reaction temperatures among these investigations varied from 10^{-5} to 10^{-2} *M*, 0 to 1.0 *M*, and 10 to 40°C, respectively. Although the temperature and aluminum and background-electrolyte concentrations seem relatively insignificant with respect to influence on measured $\beta_{1,1}$ values, it will become evident later that these are important parameters with respect to other hydroxo-aluminum species, particularly the polymeric forms.

Mesmer and Baes[28] suggested that the consistent data provided by the several investigators indicate that the first-stage hydrolysis of aluminum involves a relatively rapid reaction. It has been demonstrated by ^1H NMR[29] that the acid-dissociation reaction is indeed very rapid, involving a bimolecular proton transfer directly from a coordinated to an adjacent water molecule according to the reaction

$$(D_2O)_5Al(ODH) + OD_2 \underset{k_{-1}}{\overset{k_1}{\rightleftharpoons}} (D_2O)_5AlOD + HOD_2^+ \tag{7}$$

and not the exchange of inner-sphere waters of hydration. The value of k_1 was reported to be $1.1 \times 10^5\ s^{-1}$ at 30°C, which is consistent with the value ($1.09 \times 10^5\ s^{-1}$) reported by Holmes et al.[30] from dissociation field effect relaxation times.

Many other investigators have been unable to interpret these data based solely on the first-stage hydrolysis reaction and have invoked the second stage and, in fewer instances, the third-stage hydrolytic reactions[22,23,25,26] and, at higher pH values, even fourth-stage hydrolysis,

$$Al^{3+} + 4H_2O \rightleftharpoons Al(OH)_4^- + 4H^+ \tag{8}$$

with corresponding expressions for the over-all formation constants:

$$\beta_{1,2} = \frac{[Al(OH)_2(H_2O)_4^+][H^+]^2}{[Al(H_2O)_6^{3+}]} \tag{9}$$

$$\beta_{1,3} = \frac{[Al(OH)_3(H_2O)_3^\circ][H^+]^3}{[Al(H_2O)_6^{3+}]} \tag{10}$$

$$\beta_{1,4} = \frac{[Al(OH)_4^-][H^+]^4}{[Al(H_2O)_6^{3+}]} \tag{11}$$

Nazarenko and Nevskaya[25] determined the formation constants for mononuclear hydroxo-aluminum species, employing an indirect spectrophotometric technique where arsenazo I and pyrocotechol violet were utilized as competing ligands. They obtained reasonable $\beta_{1,n}$ values for first- and second-stage hydrolysis of 1.04×10^{-5} and 1.25×10^{-10}, respectively, and presented one of the few experimentally determined formation constants for the $Al(OH_3)^\circ$ species of $\beta_{1,3} = 2.51 \times 10^{-16}$. May et al.[26] conducted a thorough investigation of gibbsite solubility, approached from both under- and oversaturation, and measured total mononuclear hydroxo-aluminum species, employing the 8-hydroxyquinoline method. This study provided what is generally considered to be the best value for $\beta_{1,2} = 7.36 \times 10^{-11}$ (see Chapter 2). Under the conditions of their experiment, they found the $Al(OH)_3^\circ$ species contributed insignificantly to gibbsite solubility and implicitly questioned its existence at the very low total aluminum concentrations they employed. ^{27}Al nuclear magnetic resonance (NMR) spectroscopic investigations of serially diluted aluminum standard solutions indicated that the line width at half maximum ($V^1/_2$) of the hexaaquaaluminum (III) peak increased much more than could be accounted for based on the sensitivity of the technique (see Figures 1a and 1b). When these diluted aluminum standard solutions were acidified, however, a very sharp resonance was observed (see Figure 1c).[31] The line-broadening on dilution was attributed to the inclusion of the mono- and/or dihydroxo monomeric species under the resonance peak. The resonance would be expected to be broadened, both because of chemical-exchange processes and as a result of the larger electric-field gradients at the aluminum nucleus when the perfect cubic (O_h) symmetry was destroyed on hydrolysis. Acidification of the dilute aluminum solutions would be expected to shift the equilibria back toward the $Al(H_2O)_6^{3+}$ species, thus eliminating exchange and minimizing the electric-field gradients, consistent with the observed changes in line width. A plot of the line widths at half maximum vs. the amount of hydroxomonomeric aluminum in solution results in a straight line whose slope approaches 1 as the di- and then the trihydroxomonomers are included in the calculation. This result suggests indirectly that the trihydroxo-aluminum monomer contributes to the observed line width, although its contribution was relatively insignificant over the narrow pH range investigated. Thus, the predominant monomeric hydroxo-aluminum species expected in dilute acidic solutions are the mono- and dihydroxo forms, while the existence of the trihydroxo-aluminum complex remains unresolved.

The amphoteric nature of Al^{3+} is expressed at higher pH values with the formation of the aluminate ion, $Al(OH)_4^-$. Brosset and co-workers[32] were unable to conclude definitively whether the soluble aluminum species they observed at high pH values was the mononuclear $Al(OH)_4^-$ or a mixture of this species and a polymeric anionic aluminum complex with an identical hydroxyl-to-aluminum molar ratio. Since that time, many experiments have demonstrated the presence of an anionic aluminum species with a single negative

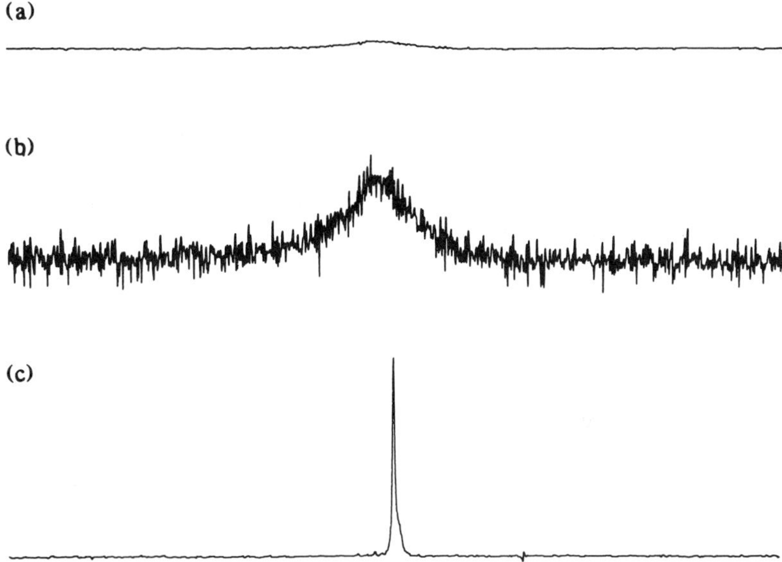

FIGURE 1. ^{27}Al NMR spectra for: (a) 50 mg/l aluminum solution, (b) 50 mg/l aluminum solution with vertical scale 10×, and (c) an acidified 50 mg/l aluminum solution. (From Bertsch, P. M., Barnhisel, R. I., Thomas, G. W., Layton, W. J., and Smith, S. L., *Anal. Chem.*, 58, 2583, 1986. With permission.)

charge.[22,23,33-35] Direct measurements of $Al(OH)_4^-$ in the 0.5 to 6.0 *M* concentration range were made by Moolenaar et al.[36] employing infrared (IR) and Raman (R) spectroscopy. At concentrations below 1.5 *M* Al they observed four vibrational bands, two being IR active at 950 and 725 cm^{-1}, and three R active at 725, 625, and 325 cm^{-1}. They interpreted these data as being consistent with the existence of the tetrahedral $Al(OH)_4^-$ species. They also presented ^{27}Al NMR spectra of these solutions, which displayed a narrow resonance line at ≈80 ppm downfield from the $Al(H_2O)_6^{3+}$ resonance, with both the line width and chemical shift indicative of tetrahedrally coordinated aluminum. At concentrations >1.5 *M* Al they observed new vibrational bands at 900 (IR) 705 (R), and 540 (R) and a broadening of the ^{27}Al NMR resonance with no detectable change in the chemical shift. These observations were interpreted as being indicative of a condensation of the $Al(OH)_4^-$ units to form an anionic polymer of the form $Al_2O(OH)_6^{2-}$. However, the data available to describe the solubility of the aluminum at high OH concentrations clearly indicate that the reaction follows the general form:

$$xAl(OH)_{3_{(s)}} + OH^- \rightleftharpoons Al_x(OH)_{3x+1}^- \qquad (12)$$

where at low solution concentrations x = 1. Several studies have provided $\beta_{1,4}$ constants for $Al(OH)_4^-$ from solubility measurements of aluminum mineral phases. May et al.[26] indicated that the range in equilibrium values for the $Al(OH)_4^-$ species may result from the failure of the various investigations to properly identify the controlling solid phase in alkaline solutions. Mesmer and Baes[28] presented a $\beta_{1,4}$ value of 1×10^{-23} compared to 6.93×10^{-23} reported by May et al.[26] based on gibbsite solubility, although the latter investigators argued that boehmite became the controlling phase in their alkaline suspensions (see Chapter 2).

Baes and Mesmer[37] indicate that of the four formation constants, $k_{1,1}$ and $k_{1,4}$ are probably the most accurately known. Thus, by assuming a linear monotonic decrease in the logarithms of the stepwise formation constants $k_{1,y}/k_{1,y-1}$, they estimated $p\beta_{1,2}$ and $p\beta_{1,3}$ as 10.5 and 16.5, respectively. These values are higher (more negative in $\beta_{1,2}$ and $\beta_{1,3}$) than those determined by Nazarenko and Nevskaya[25]; yet, for the $p\beta_{1,2}$ value, they are closer to that reported by May et al.[26] The change in coordination number anticipated with the formation of the $Al(OH)_4^-$ species, however, would most likely result in a nonlinear change in the ratio of the stepwise formation constants.[38] It is worth noting that studies indicating tetrahedral coordination in the $Al(OH)_4^-$ species were generally conducted at very high base concentrations. In solutions of more moderate base concentrations (1×10^{-6} to 1×10^{-2} *M*), Carreira et al.[39] could not, from their R and IR data, resolve whether the $Al(OH)_4^-$ complex was square planar or octahedral. It is possible that in more dilute aluminum solutions of modest base concentration — similar to those where many $\beta_{1,4}$ values have been determined — $Al(OH)_4^-$ may be a tetragonally distorted octahedral complex representing an intermediate coordination change. This speculation would lend more credence to the $\beta_{1,2}$ and $\beta_{1,3}$ values calculated by Baes and Mesmer,[37] assuming a monotonic decrease from $\beta_{1,4}$ to $\beta_{1,1}$ (see also Chapter 2).

Although the polyvalent $Al(OH)_5^{2-}$ species has been utilized in calculations of aluminum hydrolysis reaction[40] and suggested by Maksimova et al.,[41] there exists no experimental evidence in the literature to substantiate its existence.

B. Polynuclear Hydrolytic Aluminum Species

Since aluminum polymers were proposed as predominant soluble species in certain solutions, there has been continued discussion and debate concerning the appropriateness of utilizing either monomeric or polymeric schemes to describe aluminum hydrolytic reactions in aqueous solutions.[19,21,32,42,43] As mentioned above, the description of aluminum hydrolysis by considering simply monomeric species has been found to be appropriate for systems of low aluminum concentration and OH/Al molar ratios ($\bar{n}$).[19,21,37] Other investigators who have found evidence for polynuclear aluminum complexes often question or criticize these results. The antipodal interpretations can be explained largely by differences in experimental approach employed by the various investigators.

These differences are, barring a few exceptions,[1,28,44] often overlooked or disregarded. Three general types of experimental approach can be identified:

1. Those that study the aluminum hydrolytic reactions concurrent and subsequent to dilution of aluminum salt solutions[16,18-21]
2. Those that study hydrolyzed aluminum species dissolved from a solid phase[22,26]
3. Those that study aluminum hydrolytic reactions concurrent and subsequent to base addition[23,31,32,42,43]

Generally, studies employing approaches 1 and 2 are those for which simple monomeric aluminum hydrolysis can be utilized to describe the data. Studies that report the existence of polymeric aluminum species, on the other hand, have generally employed the third approach. Thus, hydrolysis in solutions containing relatively low aluminum concentrations and low $\bar{n}$ values (i.e., without added base), should be described accurately, employing hydrolysis models which consider only monomeric aluminum species. Hydrolysis in solutions of either higher aluminum concentrations or $\bar{n}$ will generally be better described by including polymeric forms in the calculations. For solutions with aluminum concentrations and $\bar{n}$ values intermediate to these conditions, generalizations are difficult to make, because of discrepancies in the literature and the complexity of the systems. For example, as indicated by Mesmer and Baes,[28] the first-stage hydrolysis of aluminum in a given solution can often be accurately described by considering any species, monomeric or polymeric. This conclusion is accentuated in several investigations where the data could be described adequately considering only simple monomeric hydrolysis of aluminum, even though the authors suspected the presence of polynuclear species.[20,45,46] Even more striking is an extensive compilation[4] of log[Al^{3+}] vs. pH data for acid surface waters, representing a wide range of ionic strength and $SO_4^=$ concentrations, where a consistent break in the slopes of the lines in the pH range 4.5 to 5.0 was observed. This pH range corresponds to the first hydrolysis constant of aluminum, even though some of these acid surface waters undoubtedly contained aluminum polynuclear species. Thus many of the discussions and arguments concerning the appropriateness of either the monomeric or polymeric hydrolysis schemes for describing aluminum hydrolytic reactions in general are irrelevent, since the results are system-specific, being dependent on both the experimental methods and approaches utilized by a given investigator and on the model chosen to describe the results derived from a given experiment (see also Chapter 2).

The hydrolysis and polymerization of aluminum can generally be described by

$$xAl^{3+} + yH_2O \rightleftharpoons Al_x(OH)_y^{(3x-y)+} + yH^+ \tag{13}$$

with the corresponding formation constant

$$K_{xy} = \frac{[Al_x(OH)_y^{(3x-y)+}][H^+]y}{[Al^{3+}]} \cdot \frac{f_{xy}f_{H^+}^y}{f_{Al^{3+}}^x a_{H_2O}^y} \tag{14}$$

where f_i is an activity coefficient. It is difficult to identify unequivocally the species that are produced by aluminum hydrolysis because of the apparent diversity and number of coexisting species. As a result, there is often disagreement about specific aluminum hydrolytic products and, not surprisingly, widely differing values of K_{xy}, even for a given hydrolytic species.

1. Proposed Structures

Jander and Winkel[47] were among the first to suggest the existence of polynuclear aluminum species based on diffusion coefficients measured in solutions of basic aluminum salts. Brosset[48] found, 20 years later, that the simple monomeric hydrolysis scheme was inappropriate for interpreting his potentiometric titrations of aqueous aluminum solutions. Rather, he interpreted his data by assuming the existence of an indefinite series of polynuclear complexes with the general form $[Al_2(OH)_3]_n^{3+}$. These same data were reinterpreted 2 years later,[32] this time with the conclusion that the $[Al_6(OH)_{15}]^{3+}$ polynuclear species was predominant. This was an undemonstrative conclusion, however, since it was also suggested that an indefinite series of polymers of the general form $Al[Al_2(OH)_5]_n^{(3+n)+}$ may be the major products. These species are consistent with the "core-links theory" proposed by Sillén.[49,50] Still later, Biedermann[51] proposed a mixture of the $[Al_{13}(OH)_{34}]^{5+}$ and $[Al_7(OH)_{17}]^{4+}$ species. Matijevic and Tezak[52] first interpreted their colloid-coagulation data by assuming the dimeric aluminum species, but later suggested the octamer $Al_8(OH)_{20}^{4+}$ as the principal hydrolyzed aluminum species in aqueous solutions. Hayden and Rubin[53] supported the existence of the octameric aluminum species, dismissing the possibility of any other polynuclear complex.

Hsu and Bates[54] employed various chemical analyses of solid phases formed in their partially neutralized aluminum solutions and ion-exchange resins to remove soluble polynuclear species, and proposed a continuous series of polynuclear complexes with the basic unit of the form $[Al_6(OH)_{12}(H_2O)_{12}]^{6+}$. Although they had convincing evidence to suggest that OH associated with polynuclear aluminum species was predominately in bridging positions, they had no other direct or indirect structural evidence to support the proposed hexameric ring or "gibbsite fragment" model of aluminum polymerization. The choice of the hexameric aluminum species as a basic structural unit was justified based on previous unsubstantiated conclusions by Brosset[48] and Hsu and Rich.[55] This "hexameric ring scheme" is similar to the "core-links" model of Sillén[49,50] and has been favored by many investigators.[23,35,55-58] However, employing ultracentrifugation, Aveston[59] derived average degrees of polymerization (Nw) as a function of ñ and tested his results against several hydrolysis schemes, concluding that the $[Al_6(OH)_{15}]^{3+}$ species did not fit his data and that the octameric $[Al_8(OH)_{20}]^{4+}$ species was unlikely. Rather, he interpreted his results as consistent with the existence of the dimer $Al_2(OH)_2^{4+}$ and a larger polycation, the $[Al_{13}(OH)_{32}]^{7+}$ species. The $Al_2(OH)_2^{4+}$ structure had been determined by X-ray analysis of crystals rapidly precipitated from partially neutralized aluminum solutions by the addition of $SO_4^=$ or $SeO_4^=$,[60] as had the $[AlO_4Al_{12}(OH)_{24}(H_2O)_{12}]^{7+}$ (Al_{13}) species.[61] The interpretation of potentiomeric data does not allow the discrimination of one O^{2-} or $2OH^-$ ligands, with the result that the $[AlO_4Al_{12}(OH)_{24}(H_2O)_{12}]^{7+}$ species described by Johansson[61] is often reported as $[Al_{13}(OH)_{32}]^{7+}$ by many investigators, particularly in earlier work. It is also clearly distinguishable from the triple-ring $[Al_{13}(OH)_{30}(H_2O)_{18}]^{9+}$ polynuclear species of the "gibbsite fragment" model. Aveston[59] also cited the small-angle X-ray work of Rausch and Bale,[62] who suggested a polymer with a radius of gyration of 4.3Å, assumed to be the Al_{13} polymer, predominated in the solutions they investigated.

Mesmer and Baes[28] interpreted their potentiometric data as indicative of the existence of the dimer $Al_2(OH)_2^{4+}$, trimer $Al_3(OH)_4^{5+}$, and another large polymer, probably with the structure $Al_{14}(OH)_{34}^{8+}$. They were, however, very noncommittal and suggested the possibility of the "Al_{13}" polymer, or a pentadecamer with the formula either $Al_{15}(OH)_{36}^{9+}$ or $Al_{15}(OH)_{37}^{8+}$. Baes and Mesmer[37] later reinterpreted their data, indicating an acceptable fit with the Al_{13} polymer in concurrence with the conclusions of Aveston.[59] Akitt and coworkers[42,63-67] also proposed the Al_{13} cation from their ^{27}Al NMR spectroscopic work. Akitt et al.[63] initially proposed the existence of the $Al(H_2O)_6^{3+}$, $Al_2(OH)_2(H_2O)_8^{4+}$, $Al_{13}O_4(OH)_{24}(H_2O)_{12}^{7+}$, and the $Al_8(OH)_{20}(H_2O)_x^{4+}$ species in hydrolyzed aluminum solutions, but later[42] suggested the predominance of the Al_{13} cation along with small amounts of the dimer and monomeric hexaaqua aluminum in similar solutions. Other hydroxo polynuclear species proposed in the literature include the $[Al_4(OH)_{10}]^{2+}$, $[Al_4(OH)_8]^{4+}$, $(Al_9(OH)_n)^{(27-n)+}$, and polymers of the $Al[(OH)_8Al_3]_n^{(3+n)+}$ series.[68,69] In addition, there is ^{27}Al, ^{31}P, ^{19}F, and ^{13}C NMR evidence supporting a number of mixed monomeric and polynuclear phosphato-aluminum, fluoro-aluminum, phosphato-fluoro aluminum, and citrato-aluminum-fluoro complexes.[70-72]

Of all the polynuclear species proposed, those having the most convincing experimental support, albeit usually indirect, are the $Al_2(OH)_2^{4+}$, $Al_8(OH)_{20}(H_2O)_x^{4+}$, those of the "gibbsite fragment" model, $Al_6(OH)_{12}(H_2O)_{12}^{6+}$ through $Al_{54}(OH)_{144}(H_2O)_{36}^{18+}$, and the $Al_{13}O_4(OH)_{24}(H_2O)_{12}^{7+}$ species. It will become evident below that these species most likely do not coexist in a given solution, and several may not exist at all.

2. *Formation Mechanisms*

Of the various polynuclear aluminum structures proposed, many have been suggested without consideration of structure, and few have direct evidence unequivocally proving their existence. The most popular hydrolysis-polymerization model is the "core-links" or hexameric ring scheme, which can accommodate an opulence of possible species.[48,50,54,55,58,73,74] The basic unit of this model for aluminum polynuclear formation is either the $[Al_6(OH)_{12}(H_2O)_{12}]^{6+}$ (single ring)[23] or the $Al_{10}(OH)_{22}(H_2O)_{16}^{8+}$ (double ring)[1] species (see Figure 2). These polynuclear units are believed to coalesce with aging via deprotonation of edge-group water molecules, with the subsequent formation of double hydroxide bridges between units (see Figure 3). Thus, it is envisaged that the polymerization process involves continued bidimensional growth of the hexameric ring units, with the resulting polymers

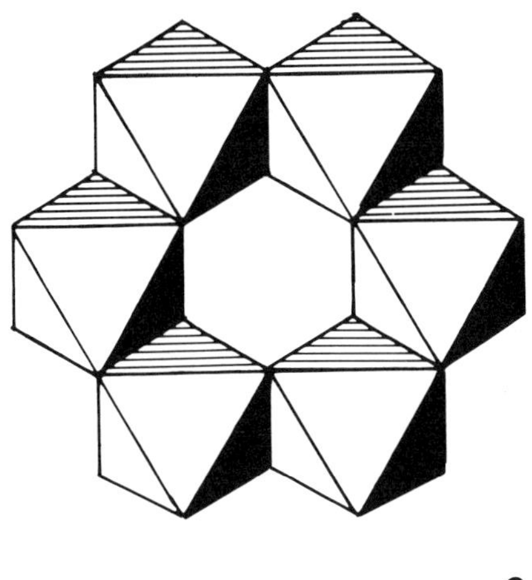

$[Al_6(OH)_{12}(H_2O)_{12}]^{6+}$

FIGURE 2. Representation of the ring structure formed by the aluminum hexamer $Al_6(OH)_{12}^{6+}$. (Adapted from Schutz, A., Stone, W. E. E., Poncelet, G., and Fripiat, J. J., *Clays Clay Miner.*, 35, 251, 1987.)

prefiguring the crystal structure of the aluminum trihydroxide polymorphs, gibbsite, bayerite, and nordstrandite. In essence, the hexameric ring or "core-links" model of aluminum polymerization was originally proposed as a logical extension of the solid-state, crystalline aluminum trihydroxide structure.[54,56,75] An obvious result of increased polymerization is that the OH/Al ratio increases and the average charge per structural aluminum decreases. The observation that hydrogen ion activities of partially neutralized solutions generally increase on aging has been taken in support of this polymerization process; yet, as suggested above, there is no direct or convincing indirect evidence supporting all of these structures.

There has been some disagreement among the proponents of the hexameric ring model as to the smallest polynuclear species present in partially neutralized aluminum solutions. Hem and co-workers[74-76] have proposed the aluminum dimer $[Al_2(OH)_2(H_2O)_8]^{4+}$ with a double-hydroxide bridged structure as the polymer formed initially in partially neutralized solutions. Stol et al.[56] concurred with this conclusion and proposed a trimeric aluminum species of the form $[Al_3(OH)_4(H_2O)_{10}]^{5+}$, which essentially takes a structure resembling half of a hexameric ring unit. Hsu,[1] conversely, questioned the existence of the $[Al_2(OH)_2(H_2O)_8]^{4+}$ species, suggesting that electrostatic repulsive forces between aluminum atoms would be too great for this structure to be stable. Johansson[60] determined the crystal structures of a basic aluminum sulfate and an isomorphous selenate and found that the structure consisted of discrete aluminum-oxygen complexes of the form Al_2O_{10}, which are composed of two AlO_6 octahedral units with a common edge. Based on the shortened 0-0 distances in this structure, it was concluded that the 0-0 contacts are joined by hydrogen bonds consistent with the structural formula $[Al_2(OH)_2(H_2O)_8](SO_4)_2(H_2O)_2$. Therefore, there is good evidence to suggest that $[Al_2(OH)_2(H_2O)_8]^{4+}$ could be a stable polynuclear species, with many investigators inferring its existence from indirect potentiometric studies[20,28,43] and more recently from direct NMR spectroscopic investigations.[63,66,77] Data on the differential reactivity of polymers to added acid or complexation agents would suggest that linear polynuclear species such as the tetramer $[Al_4(OH)_{10}(H_2O)_8]^{2+}$ proposed by Kentamaa,[78] do not exist, since only six of the hydroxyls would occupy bridging positions.[76] Even linear polynuclear species of the general form $[Al_n(OH)_{2n-2}]^{n+2}$ appear unlikely, since at $\bar{n} = 2.0$, the degree of polymerization would be expected to approach infinity — clearly inconsistent with experimental

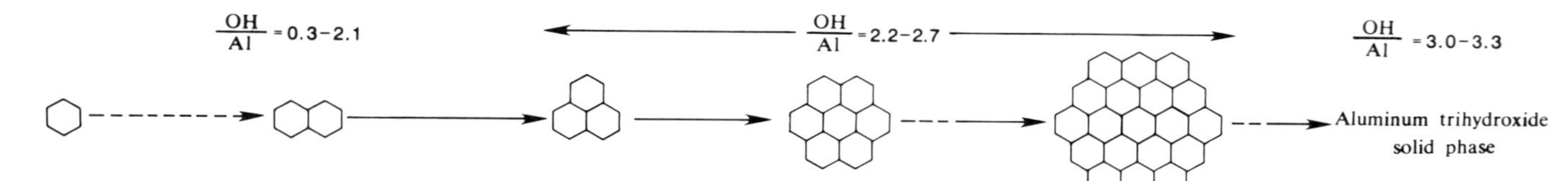

FIGURE 3. Aluminum polymerization via coalescence of the hexamer units according to the "gibbsite-fragment" model. (Adapted from Hsu, P. H., in *Minerals in Soil Environments,* 2nd. ed., Dixon, I. B. and Weed, S. B., Eds., Soil Science Socity of America, Madison, WI, 1988; and Smith, R. W. and Hem, J. D., *U.S. Geol. Surv. Water-Supply Pap.,* 1827-D, 1972.)

observation. Furthermore, aluminum in the interior positions of these linear polynuclear species would most likely require an octahedral to tetrahedral coordination change and, therefore, might only be expected in solutions of very high pH values and aluminum concentration.[36,39] These types of structural limitations led Hem[76] to conclude that polymerization of aluminum is restricted by the requirement that aluminum can share only one doubly bridged OH group in a given plane, this also being a structural feature of the trihydroxide aluminum mineral phases.

A more recently proposed model of aluminum polymerization includes the Al_{13} polynuclear species $[AlO_4Al_{12}(OH)_{24}(H_2O)_{12}]^{7+}$ (although the charge on the Al_{13} polymer has been reported to range from 3+ to 7+[43]), usually in combination with a minimum of other species, such as the dimer $[Al_2(OH)_2(H_2O)_8]^{4+}$ and/or the trimer $[Al_3(OH)_4(H_2O)_{10}]^{5+}$.[37,43,64-67] The Al_{13} polynuclear species was proposed originally by a number of investigators primarily because its existence had been confirmed in crystals of basic aluminum sulfates precipitated from partially neutralized aluminum solutions.[61] The Al_{13} structure elucidated from solid-state analysis and the chemically indistinguishable $Al_{13}(OH)_{32}^{7+}$ were also selected by several investigators to explain small-angle X-ray diffraction,[62] ultracentrifugation,[59] and potentiometric data.[37] The structural analysis of this basic aluminum sulfate indicates that it consists of a highly symmetrical, tetrahedrally coordinated aluminum centralized in a cage-like structure composed of 12 octahedrally coordinated aluminum atoms (see Figure 4). These Al-O octahedra are grouped about each apex and are interconnected by both their common edges and vertices. Thus, it is believed that on rapid addition of sulfate, the Al_{13} polymeric units already existing in solution precipitate rapidly.

Several investigators have provided direct NMR spectroscopic evidence for the existence of the Al_{13} polymer in partially neutralized solutions ranging widely in aluminum concentration, ñ, and in how they were synthesized.[43,63-67,77] The resonance peak assigned to the Al_{13} polynuclear species is ≈63 ppm (±0.5) downfield from the hexaaquaaluminum (III) cation, clearly in the region characteristic of tetrahedrally coordinated aluminum (see Figure 5).[77] The other 12 octahedrally coordinated aluminum atoms in the structure display a very broad resonance as a result of their rather asymmetric environment, which produces a large electric-field gradient at their positions. The two major lines of evidence supporting the assignment of the 63-ppm peak to tetrahedrally coordinated aluminum within the Al_{13} structure are first, that as described earlier, the chemical shift lies in the region characteristic of tetrahedrally coordinated aluminum and, second, the very sharp resonance line suggests both a very symmetrical environment with a correspondingly weak electric-field gradient, and lack of exchange with other aluminum species. The nonlabile nature of this complex may be a requirement for its metastability, since the tetrahedral aluminum component would not be expected to be stable at low pH values. It has been demonstrated that, for a given aluminum concentration, the amount of Al_{13} in partially neutralized aluminum solutions increases linearly with ñ until high ñ values, where the Al_{13} concentration is observed to decrease.[43,77] The exact ñ where this polynuclear species first appears, or when its concentration decreases, is highly dependent on factors such as aluminum concentration, mode of base addition, base injection rate, and mixing conditions during synthesis (see Table 1). The interaction between ñ and base injection rate was found to be extremely sensitive, perhaps explaining many of the discrepancies existent in the literature (see Figure 6).

In addition, the observed influence of preparation parameters on the relative amounts of monomeric aluminum, the Al_{13} polymer, or larger, more polymerized aluminum species formed during synthesis, suggests that the Al_{13} polymer may be an artifact of synthesis procedures requiring inhomogeneous pH conditions at the point of base introduction, or at the solid solution interface of dissolving carbonate phases, or aluminum metal undergoing oxidation.[66,79,80] Thus it is thought that $Al(OH)_4^-$ is a required precursor to Al_{13} polymer formation; i.e., some aluminum already exists in tetrahedral coordination. The Al_{13} polymer

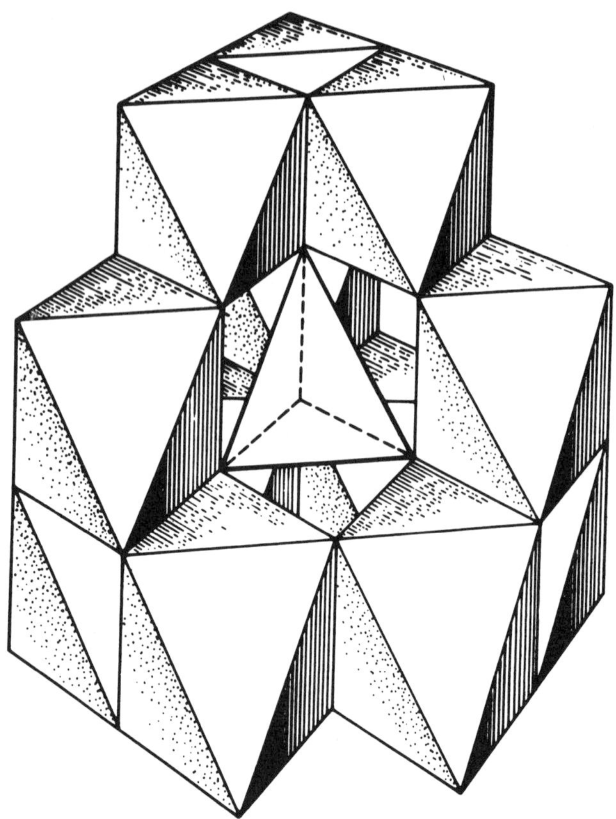

FIGURE 4. Representation of the $[AlO_4Al_{12}(OH)_{24}(H_2O)_{12}]^{7+}$ polynuclear species, demonstrating the tetrahedrally coordinated aluminum at the center of the cage-like structure comprising 12 octahedrally coordinated aluminum atoms joined by common edges. (From Baes, C. F. and Mesmer, R. E., *The Hydrolysis of Cations,* John Wiley & Sons, New York, 1976. With permission.)

must, therefore, form rapidly and irreversibly when $Al(OH)_4^-$ interacts with 12 octahedrally coordinated aluminum ions and not form on aging via polymeric intermediates as suggested by Baes and Mesmer.[37] Several investigators have indicated that the Al_{13} polynuclear species forms only during synthesis, thus providing support for this contention.[43,77,80,81] Akitt and Farthing[66] proposed a mechanism of Al_{13} polynuclear formation whereby six dimeric species nucleate around an $Al(OH)_4^-$. They argued that the Al_{13} structure could be viewed as being comprised of six oxygen-sharing dimeric units which bridge the edges of the central tetrahedron. This mechanism may be reasonable, considering that the 0-0 distance (0.310 nm) in the AlO_4 unit is similar to the separation between a pair of apical O atoms (0.290 nm) in the position normal to the plane of the Al_{13} structure.[48,61] Regardless of the exact mechanism of Al_{13} polymer formation, this species has been demonstrated to be the primary polynuclear species in aluminum solutions of high ñ,[43,63-67,77] the principal component of aluminum chlorohydrate, a highly soluble antiperspirant,[10,79,82,83] a very efficient scavenger of anions in physicochemical water-treatment processes,[8,9] and, as originally hypothesized by Vaughan and Lussier[84] and Pinnavia,[12] the interlayer aluminum polynuclear component of the pillared clays utilized in catalytic processes.[13] Further research to better define the exact conditions required for the formation and metastability of the Al_{13} polymer is necessary before all of the implications regarding its existence can be realized.

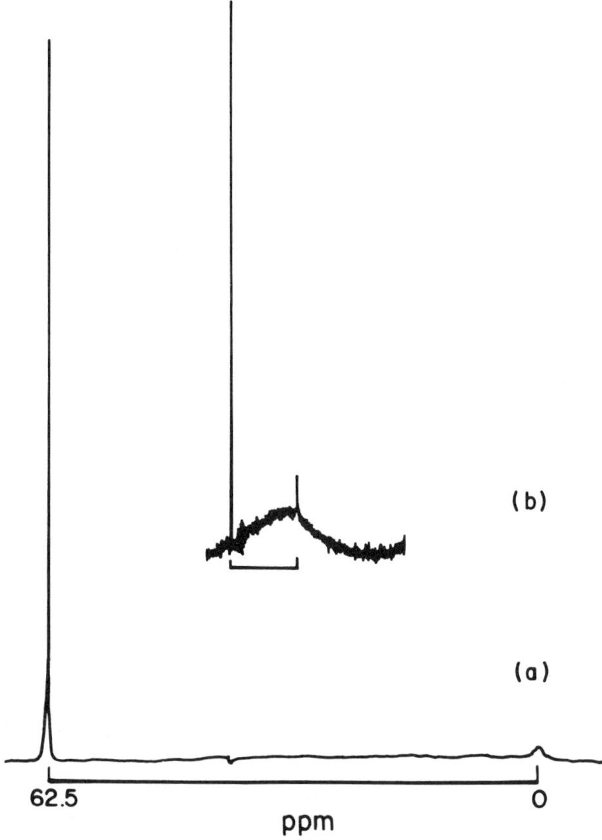

FIGURE 5. ^{27}Al NMR spectrum of a partially neutralized 0.1 mol/l $AlCl_3$ solution at $\bar{n} = 2.5$, demonstrating the monomeric aluminum species at 0 ppm and the tetrahedrally coordinated aluminum of the Al_{13} polymer at ≈63 ppm downfield from the monomeric aluminum peak. In the condensed spectrum, (b), the broad resonance resulting from the 12 octahedrally coordinated aluminum atoms in the Al_{13} structure is apparent. (From Bertsch, P. M., Thomas, G. W., and Barnhisel, R. I., *Soil Sci. Soc. Am. J.*, 50, 825, 1986. With permission.)

A potentially important mechanism for the formation of polynuclear aluminum species involves polymerization at the mineral-water interface. Although the consequences of these reactions with respect to aluminum interlayer formation have received considerable attention,[85] little work has been conducted to define aluminum polymerization reactions at surfaces of clay minerals and their potential importance to the formation of polynuclear aluminum, particularly in moderately acid environments where solution aluminum concentrations are low (see Chapter 6). For example, Bloom et al.[86] suggested that aluminum polynuclear species could be disregarded in their Ca-Al exchange studies because of the low solution-aluminum concentrations utilized. Very conservative estimates of surface-aluminum concentrations in the exchange studies of Bloom et al.[86] can be made with the Boltzmann distribution equation from diffuse double-layer theory,[87]

$$Ni = Nio\,\frac{(1 - \tanh(-Zie\Psi_0/4KT)\exp^{-dx})^2}{(1 + \tanh(-Zie\Psi_0/4KT)\exp^{-dx})^2} \tag{15}$$

where Ni = the concentration of ions i at some distance from a surface with a potential

Table 1
THE FORMATION OF THE Al_{13} POLYMER IN VARIOUS SOLUTIONS AS INFLUENCED BY ALUMINUM CONCENTRATION, ñ, AND SYNTHESIS CONDITIONS

Aluminum conc. (mol/l)	ñ	Base injection rate (cm^3/min)	Al_{13+}[a]	Comments	Ref.
1×10^{-1}	>0.50	3.33	+		43
	>0.5, <2.2	3.33	+	Continual linear increase in Al_{13} concentration	43
	>2.2	3.33	+	Concentration of Al_{13} decreased with ñ	43
3.34×10^{-2}	0.25	1.2	−		77
	0.25	86.0	−		80
	0.40	1.2	+		80
	0.50	1.2	+		77
	2.5	0.6	+	Continual linear increase in Al_{13} concentration from ñ = 2.25	77, 80
	2.5	1.2	+	Concentration of Al_{13} decreased compared to ñ = 2.25	77, 80
3.34×10^{-3}	1.0	1.2	−		80
	>1.5	1.2	+		80
3.34×10^{-4}	2.5	1.2	−		80
	2.5	100.0	+	62.5 ppm peak visible following 30,000 scans; concentration of Al_{13} estimated at $\sim 2.0 \times 10^{-4}$ mol/l	80

[a] +, − indicate the presence or absence respectively, of detectable 62.5 resonance peak indicative of the Al_{13} polymer.

From Bertsch, P. M., *Soil Sci. Soc. Am. J.*, 51, 825, 1987. With permission.

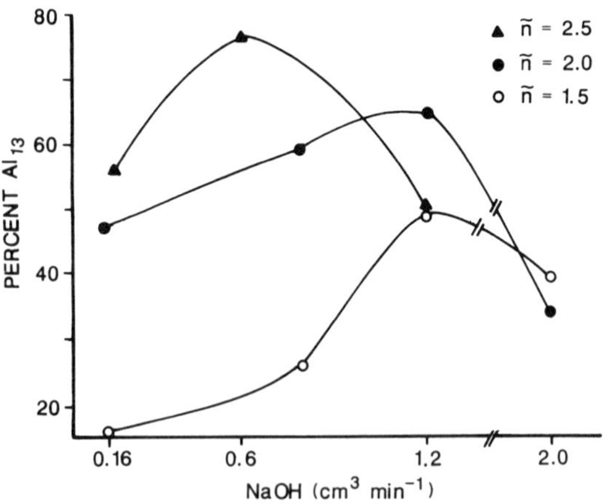

FIGURE 6. The fraction of Al_{13} (of Al_T) formed in 3.34×10^{-2} mol aluminum/l solutions (ñ = 1.5, 2.0, and 2.5) as a function of base injection rate. (From Bertsch, P. M., *Soil Sci. Soc. Am. J.*, 51, 825, 1987. With permission.)

Table 2
ALUMINUM CONCENTRATIONS AT VARIOUS DISTANCES FROM A CLAY SURFACE, CALCULATED FROM DOUBLE-LAYER THEORY EMPLOYING THE Al^{3+}–Ca^{2+} EXCHANGE DATA OF BLOOM ET AL.[86]

Bulk solution conc. of aluminum (mol/l)	Distance from surface (Å)	Conc. some distance x from surface (mol/l)
2.8×10^{-6}	5	1.25×10^{-1}
	50	1.57×10^{-3}
	100	3.96×10^{-4}
77.3×10^{-6}	5	1.53×10^{-1}
	50	1.65×10^{-3}
	100	4.52×10^{-4}
482.0×10^{-6}	5	1.57×10^{-1}
	50	1.94×10^{-3}
	100	7.65×10^{-4}

Ψ_o, Nio = the concentration of ion i in the bulk solution (ions/cm^3), Zi = the valence of ion i, e = the charge on the proton (4.8×10^{-10} esu), Ψ_o is the potential at the interface (mv), K is the Boltzmann constant (1.38×10^{-16} erg deg^{-1}), T = the absolute temperature, x is the distance from the surface or between surfaces, and d = the thickness of the double layer (cm) given by

$$d = \sqrt{\frac{8\pi NZ^2e^2}{\epsilon KT}} \tag{16}$$

where N, Z, e, K, and T have the meaning given above, and ϵ is the dielectric constant of the aqueous medium. Assuming the smectite used in this study has a Ψ_o typical of montmorillonite (200 mV), then the concentration of aluminum at the surface of the clay for the various bulk-solution aluminum concentrations reported can be conservatively estimated (see Table 2). Clearly, even when aluminum concentrations in bulk solutions are low, the surface aluminum concentrations are several orders of magnitude greater, presumably making polymer formation a very favorable process.

The tendency for aluminum to hydrolyze at clay mineral surfaces was demonstrated in many other studies.[85,88] There is strong evidence indicating that aluminum polymers can form at clay mineral surfaces simply by washing a partially aluminum-saturated clay suspension with water, with some indications that the Al_{13} polymer may form in this manner.[89,90] Thus, high concentrations of aluminum at mineral surfaces, coupled with the influences of the surface charge on hydrolytic reactions, may easily result in aluminum polymer formation, even in systems having relatively low aluminum solution concentrations. There is some evidence that hydroxy-interlayered 2:1 clays may contain polymers having both the gibbsite fragment and the Al_{13} polymeric structures.[85] For hydroxy-interlayered clays displaying a stable 1.4 nm d spacing, it has been suggested that aluminum polymers with the gibbsite-fragment structure occupy the interlayer space, while those 2:1 clay minerals with a stable 1.8 nm d spacing, the Al_{13} polymer has been proposed as the interlayering agent with recently advanced, direct magic angle spinning NMR evidence substantiating the latter.[85]

Little is known, however, concerning the type of polynuclear forms associated with clay minerals at moderate aluminum surface coverage, although several studies have demonstrated a preference of surfaces for small polynuclear species.[91,92] Another important finding is that some aluminum polynuclear species are exchangeable to other cations,[85] making clay minerals a potential source of aluminum polynuclear complexes in soil solution and aquatic systems (see Chapter 6 for a comprehensive discussion).

III. IDENTIFICATION AND SPECIATION OF HYDROXO-MONOMERIC AND POLYNUCLEAR ALUMINUM

The differentiation of monomeric and polynuclear aluminum species in solution is problematic. Several indirect methods of investigation have been used, including conductometric, potentiometric, light scattering, ultracentrifugation, ion-exchange or dialysis techniques, and various kinetic methods based on the interaction of aluminum with a complexation agent (see Chapter 1 for a comprehensive discussion). As a result, several ambiguities and little definitive information exist. More recently, direct ^{27}Al NMR spectroscopic investigations have provided some unique information regarding the distribution of monomeric and polynuclear aluminum species coexisting in solution, and on the nature of some of the polymeric species, but even these investigations have been unable to provide unequivocal evidence for all of the polynuclear species present.

Okura and co-workers[93] suggested the 8-quinolinolate method for aluminum analysis, where the aluminum interacting instantaneously was assumed to be mononuclear species. They suggested that, by using this method, monomeric and polynuclear aluminum species could be differentiated. Turner[94] developed this method further and found that three aluminum forms could be differentiated. The first was the aluminum fraction reacting "instantaneously" with 8-quinoline, which, in concordance with Okura et al.,[93] he assumed were monomeric aluminum species (Al_a), including $Al(H_2O)_6^{3+}$, $Al(OH)(H_2O)_5^{2+}$, $Al(OH)_2(H_2O)_4^{+}$, and, in solutions of higher pH values, $Al(OH)_4^-$. He conceded that the initial reaction could include small polynuclear species, such as the dimer. In addition, he identified a slower reacting aluminum species which he assumed to be polynuclear complexes (Al_b), and, finally, a nonreactive aluminum fraction assumed to be one or more forms of a colloidal solid phase (Al_c). Subsequent investigations demonstrated that the method was consistent with predictions based on the nature of the specific partially neutralized aluminum solutions, i.e., aluminum concentration and ñ, where the Al_b and Al_c fractions generally increased with these parameters.[95,96]

The evidence presented, indicating that the two slowly reacting phases were distinct, included: (1) filtration through a 0.01 μm filter, where it was observed that the Al_b fraction remained in solution while the Al_c fraction was removed; (2) the observation that Al_b followed a first-order reaction when reacted with excess 8-hydroxyquinoline, whereas the Al_c fraction followed a two-thirds order process; (3) the calculated activation energies of the reaction of the Al_b and Al_c fractions with 8-hydroxyquinoline were found to be 46.0 and 83.7 kJ/mol, respectively; and (4) the Al_b fraction was removed by clay minerals but was found to be exchangeable, whereas the Al_c fraction, although still removed, was nonexchangeable to other cations. In a subsequent investigation, Turner[95] provided evidence that the Al_c fraction was composed of two phases, one where the integrity of the polynuclear species was maintained within an amorphous precipitate, and one which developed into gibbsite within a few days of aging. The former was produced via rapid neutralization of solutions containing polynuclear species, with the latter being formed by rapid addition of base to a solution containing only monomeric aluminum species. And in yet other studies, Turner[97,98] proposed a second polynuclear aluminum species, similar to the previously defined Al_b fraction with respect to $ñ \approx 2.5$, and presumably size, however, possessing a much slower reaction rate with 8-hydroxyquinoline. It was suggested that appreciable amounts of this second polynuclear species would only be observable if the partially neutralized solutions were prepared in such a manner as to prevent the formation of any solid-phase components. By studying the aging properties of these solutions, Turner[98] concluded that the Al_b polynuclear fraction would slowly convert to the more slowly reacting polymer according to a pseudo-first-order process with experimental rate constants ranging from 0.57×10^{-3} to 3.5×10^{-3}/d and a corresponding activation energy of 87.0 kJ/mol. A significant finding in these aging studies was that the rate constant of the Al_b polynuclear fraction interacting with 8-hydroxyquinoline was not influenced by the $\bar{n}$ or mode of synthesis of the original solution, and that the rate constant remained unchanged with aging even though the amount of Al_b was observed to decrease, i.e., as Al_b converted to the more slowly reacting polynuclear species; clearly suggesting that the Al_b polynuclear cations were not increasing in size or complexity as required by the hexameric ring or gibbsite-fragment model of aluminum polymerization.

Smith[73] and Smith and Hem[74] also identified three forms of aluminum in partially neutralized solutions employing differential kinetic reactions with another complexing agent, ferron (8-hydroxy-7-iodo-5 quinoline-sulfonic acid). Corresponding to the results of Turner, who used a different complexation agent, they suggested that the aluminum reacting instantaneously with ferron was mononuclear, while intermediate polynuclear aluminum species displayed pseudo-first-order reaction kinetics with ferron, and finally that the aluminum fraction having a virtually imperceptible reaction rate with ferron was very large polynuclear species or initial solid phases. Smith and Hem[74] and Hem and Roberson,[75] found that aluminum-hydroxide particles retained on a 0.1 μm filter, some too small to be identified by X-ray diffraction, were demonstrated by electron microscopy to have a hexagonal crystal pattern indicative of gibbsite. Based on these observations it was concluded that Al_c was composed of microcrystalline gibbsite particles and that the polynuclear species not retained on filtration through 0.1 μm filter material contained a similar structure, i.e., the polynuclear species were structurally similar to gibbsite. Results from ferron analyses on a number of partially neutralized aluminum solutions varying in $\bar{n}$ from 0.55 to 3.01 were also interpreted as consistent with the hexameric ring or gibbsite-fragment model of aluminum polymerization.[78] This conclusion was based primarily on the tendency for the Al_b-ferron pseudo-first-order rate constants to decrease with both increasing $\bar{n}$ and aging. Examination of the data, however, indicates that the only clearly discernible decrease in the pseudo-first-order rate constant as a function of $\bar{n}$ is perhaps between $\bar{n} = 0.55$ and 0.94, and, with respect to aging, *only* for $\bar{n} = 0.55$. The data presented seem to conform better to the explanation that the Al_b fraction comprises largely a single polynuclear species. Another puzzling aspect of invoking the "gibbsite-fragment" polymerization mechanism to explain kinetic data of aluminum-complexation agent interactions with partially neutralized aluminum solutions is the distinct boundary observed between Al_b and Al_c. This observation is seemingly inconsistent with the model, making it difficult to reconcile the conclusion of Smith and Hem[74] that Al_b polynuclear species become the less reactive Al_c particles after attaining some "critical size". More recent investigations employing ferron kinetics with partially neutralized aluminum solutions of various ages also support the conclusions of Turner[98] and Turner and Ross[99] that polynuclear aluminum initially formed in solution may be composed largely of a single species that transforms slowly on aging to a polynuclear component having a lower reaction rate with ferron.[81,100,101]

Bersillon et al.[58] developed a speciation scheme to differentiate mononuclear and various polynuclear aluminum forms in partially neutralized aluminum-chloride solutions. They employed (1) a modification of the ferron procedure to estimate monomeric aluminum (2) a kinetically controlled, selective-adsorption procedure for the estimation of "low" polynuclear aluminum species, and (3) a precipitation step employing Na_2SO_4 to estimate "medium" and "high" hydroxo-aluminum polynuclear species. These results were interpreted as consistent with the hexameric ring scheme or "gibbsite-fragment" model of aluminum

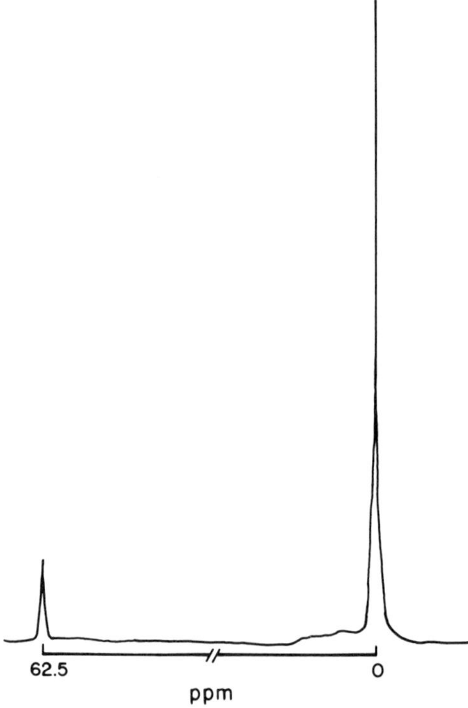

FIGURE 7. The ^{27}Al NMR spectrum of a 0.5 mol/l $AlCl_3$ solution ($\bar{n}$ = 2.0), demonstrating the 0- and 63-ppm peaks indicative of aluminum monomers and Al_b polynuclear species, respectively, and the broad resonance, just downfield of the monomeric aluminum peak, assigned to the bridged dimer. (From Bertsch, P. M., Thomas, G. W., and Barnhisel, R. I., *Soil Sci. Soc. Am. J.*, 50, 825, 1986. With permission.)

polymerization. Recently, direct ^{27}Al NMR investigations have cast doubt on this indirect method, since a single polynuclear species was found to display differential precipitation kinetics with added sulfate and to be adsorbed differentially by a cation-exchange resin.[81] More recent studies, where partially neutralized aluminum solutions were introduced into a spectrophotometer immediately after mixing, the absorbance was monitored continually, and the data were analyzed utilizing simultaneous first-order reaction functions, have clearly differentiated monomeric aluminum from a polynuclear form having a pseudo-first-order rate constant considerably higher than that previously reported.[102] This species, which may be interpreted as a small polymer having a rapid reaction rate with ferron, has probably been overlooked by previous investigators who utilize an arbitrary initial reaction time, usually 15 to 30 s.

Another method that has provided direct evidence to differentiate monomeric and some polynuclear aluminum species is ^{27}Al NMR spectroscopy. Akitt and co-workers[63-67] utilized this technique on more highly concentrated hydrolyzed aluminum solutions and were able to measure directly the coexisting monomeric aluminum and $AlO_4Al_{12}(OH)_{24}(H_2O)_{12}^{7+}$ species. At higher temperatures (~80°C) a broad resonance was observed downfield from the monomeric aluminum peak and was assigned to the dimeric aluminum species. A higher-field instrument equipped with a superconducting magnet allowed the direct observation of the broad resonance at 25°C[77] (see Figure 7). Subsequent studies on an even more sensitive instrument have demonstrated the existence of this resonance at 25°C for solutions 6.0 ×

10^{-2} *M* with respect to aluminum and at ñ = 2.25.[103] This resonance has been assigned to the dimer by other investigators, both because this species has been directly determined by X-ray diffraction analysis and because, given the quadrupolar relaxation of the ^{27}Al nucleus, it is unimaginable that octahedrally coordinated aluminum polynuclear species of greater size would be observable by NMR spectroscopy. When partially neutralized aluminum solutions were prepared by hydrolysis of aluminum metal, other tetrahedrally coordinated aluminum polynuclear species were observed.[67] Although inconclusive, it was proposed that these peaks were associated with species consisting of partial Al_{13} units with octahedra disposed as flexible chains or forming crosslinks. Further work will be required to better elucidate formation conditions and the structure of these other aluminum polynuclear species.

Employing a quantitative ^{27}Al NMR technique, the distribution of monomeric and Al_{13} polynuclear species as a function of solution concentration, ñ, and mode of synthesis has been investigated.[77] The results demonstrate that the Al_{13} polynuclear species forms in detectable quantities in solutions 3.4×10^{-2} *M* in aluminum and at ñ > 0.25 and that its concentration increases linearly as a function of ñ. Correspondingly, the aluminum monomeric species decreased monotonically as a function of ñ. The distribution of these aluminum species and the apparent steady-state conditions obtained were found to be highly dependent on synthesis conditions, with strong interactions between aluminum concentration, ñ, and base-injection methods and rates.

Other recent investigations have coupled direct ^{27}Al NMR spectroscopy with timed spectrophotometric and potentiometric methods of analyses.[8,9,43,81,104] Bottero et al.[43] cast doubt on interpretations of potentiometric data without direct spectroscopic evidence, since good correlations were observed between several polymerization models and the potentiometric data, yet the Al_{13} polymer was directly determined to be the predominant polynuclear species in the solutions they studied. Ferron and ^{27}Al NMR methods have also been combined to study partially neutralized aluminum solutions. It was demonstrated that the utilization of arbitrary reaction times in the ferron assay tends to overestimate the monomeric aluminum fraction, presumably as a result of smaller polynuclear species, such as the dimer, interacting rapidly with ferron.[81] Another important observation in these investigations is that the Al_b fraction measured by ferron and 8-hydroxyquinoline reaction kinetics corresponds closely to the amount of $[AlO_4Al_{12}(OH)_{24}(H_2O)_{12}]^{7+}$ polynuclear species present.[8-10] Also, as indicated above, the SO_4^{2-} precipitation and resin-adsorption methods for differentiating various polymers of the "gibbsite-fragment" structure proposed by Bersillon et al.[58] were found to be erroneous by direct ^{27}Al NMR spectroscopic investigations.[81] In addition to those species directly observable by ^{27}Al NMR spectroscopy, another polynuclear species was inferred from spin lattice relaxation times (T_1) of monomeric aluminum in various diluted and partially neutralized aluminum chloride solutions. The dilution of a 3.34×10^{-2} *M* aluminum solution by an order of magnitude resulted in a decreased T_1 of the monomeric aluminum peaks from 0.220 to 0.140 s. The addition of base to the 3.34×10^{-2} *M* aluminum solution to ñ = 0.25 and 2.00 decreased the measured T_1 of the monomeric peak to 0.014 and 0.012, respectively.[81] The decrease in T_1 on dilution of the aluminum solution was explained as resulting from the rapid chemical exchange between the $Al(H_2O)_6^{3+}$ and the $Al(OH)(H_2O)_5^{2+}$ species, which, as described previously, involves a simple proton-exchange mechanism. The large decrease in the T_1 values following partial neutralization cannot be explained by this mechanism, and indicates that the monomeric aluminum species are in rapid exchange with some polynuclear aluminum complex. The very rapid exchange, as evidenced by the inclusion of both the monomers and polymer as one species on the NMR time scale, supports the hydrated cyclic dimeric structure proposed by Grunwald and Fong[24] (see Figure 8). It is unlikely that the bridged dimeric structure, which displays a unique chemical shift, could be involved in such rapid exchange, since the formation of this species involves the breaking and formation of two bonds with the subsequent exclusion of two

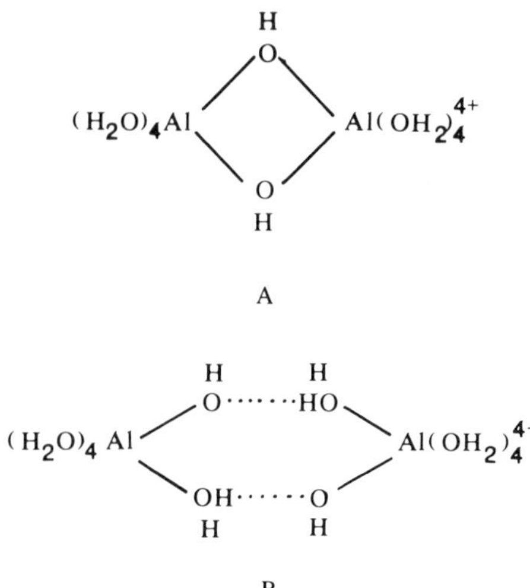

FIGURE 8. Representation of (A) the bridged dimer structure, which displays a ^{27}Al NMR resonance and is in relatively slow exchange with aluminum monomers, and (B) the hydrated cyclic dimer, which is in rapid exchange with hydroxo aluminum monomers and influences the spin lattice relaxation rates of the monomeric aluminum.

water molecules (see Figure 8). ^{27}Al NMR spectroscopy, although not definitive, usually can account for >80% of the aluminum species in solution and has provided significant information on the types of polynuclear species and on the distribution between monomeric and polynuclear species in certain hydrolyzed solutions. Furthermore, the method has provided information on the applicability, limitations, and boundary conditions of certain indirect methods of investigations.

Several general conclusions regarding polynuclear aluminum species can be made, based on the best available data. There is substantial direct and indirect evidence to support the existence of the $Al_2(OH)_2^{4+}$ and $Al_{13}O_4(OH)_{24+n}^{(7-n)+}$ polynuclear complexes. In solutions of higher aluminum concentration and ñ these polynuclear forms clearly predominate. There is strong indirect evidence, however, to indicate that other species are present in the same solutions. At least one group of these polynuclear species reacts more slowly with ferron and 8-hydroxyquinoline than the Al_{13} species, and there is convincing indirect evidence to suggest that Al_{13} transforms to this complex on aging. It is proposed that these polynuclear species are larger polymers of the "gibbsite-fragment" model (see Figure 3) and may represent "clusters" or "nuclei" from which gibbsite crystallites eventually form. There is little convincing evidence to support the large number of possible polynuclear species implicit in the "gibbsite-fragment" model. Thus, if the smaller and intermediate polymeric species depicted in this model are formed at all, they must be transformed rapidly to clusters, nuclei, or perhaps even crystallites and can be viewed as transient species. This view is consistent with the observed data and tends to reconcile the two most convincing, yet disparate, models of aluminum polynuclear formation. There are still many uncertainties concerning the polynuclear species predominating in less concentrated hydrolyzed aluminum solutions ($<1.0 \times 10^{-3}M$ aluminum). Future investigations will undoubtedly provide the information required to better elucidate aluminum polynuclear-formation mechanisms and the transformation of these species to crystalline solid phases.

IV. FLOCCULATION AND PRECIPITATION OF ALUMINUM POLYNUCLEAR SPECIES

Nucleation of monomeric and polymeric aluminum species into colloidal or distinct solid phases is an important phenomenon utilized in the physicochemical treatment of water and wastewater.[8,9,105-109] Particulate phases are clearly important aluminum components in natural aquatic systems and may represent the most important labile phase with respect to aluminum release in acidic surface waters (see also Chapter 7). Many of the large polynuclear or colloidal solid phases initially formed are composed of very fine particulates that pass easily through membrane filters, potentially causing major errors in the determination of "dissolved" aluminum concentrations and thus problems with identifying controlling solid phases or instances of authentic supersaturation.[110,111] As a result, several investigators have devised methods of aluminum analysis that estimate only monomeric aluminum forms or at least attempt to separate "dissolved" from colloidal solid-phase aluminum.[93,112,113] The amount of colloidal solid-phase aluminum produced during synthesis and the nature of this material is highly dependent on solution composition and the experimental conditions employed. The composition of a given solution, including the total aluminum concentration and the type and amount of indifferent or complexing anions present, and specific experimental conditions including neutralization procedures, mixing conditions, and temperature, clearly influence the observed flocculation and precipitation behavior of aluminum polynuclear species.

The influence of anions on the precipitation of aluminum from solutions and on the crystalline solid phase ultimately formed has been studied extensively.[1,107,114,115] Beginning with the classic studies of Marion and Thomas,[114] it was demonstrated that strongly complexing ligands, such as F^- or oxalate, generally increase the pH value of maximum precipitation, while polyvalant anions, such as $SO_4^=$ or $SeO_4^=$, usually decrease the pH value of maximum precipitation as compared to Cl^-, NO_3^-, or ClO_4^-. Consistent with other aspects of aluminum chemistry, there exists confusion and inconsistency in the literature concerning the mechanisms of anion interaction with polynuclear species to form colloidal solid phases, much of which can be attributed to differences in experimental methods or approaches.

The presence of anions has been demonstrated to cause flocculation of polymeric aluminum species and influence crystallization rates with the relative effectiveness following the general order: phosphate > silicate > sulfate > nitrate > perchlorate.[116] Several explanations for this observation have been proposed, including anion inclusion in the resulting solid phase,[117] supersaturation and precipitation of solid phases other than the trihydroxide phases,[117] and physical flocculation resulting from reduction in electrostatic repulsion between aluminum polynuclear centers.[107,118] Clearly, all mechanisms are operative depending on the type of anion considered, or even for a given anion, on the experimental conditions employed. Nail et al.[119,120] proposed that the initial solid aluminum phase which precipitates from neutralized aluminum solutions is composed of polymeric species with the "gibbsite-fragment" structure bound together by anions, which also were thought to inhibit further crystallization by restricting growth along edges. Hem[76] provided electron microscopic evidence for aluminum hydroxide particles, 0.05 μm in size, having a well-defined hexagonal crystal pattern resembling gibbsite. Flocculation of aluminum polynuclear species by added salt was also explained by DeHek et al.,[107] based on the "gibbsite-fragment" polymerization model. Thus, the nucleation-growth phenomenon enhanced by anions was explained as a lowering of the energy barrier by effectively reducing the repulsive interactions between polynuclear centers catalyzing further growth of the gibbsite fragments along the *a* and *b* crystallographic axes. DeHek et al.[107] suggest that NO_3^- and $SO_4^=$ were both effective in the flocculation process via this mechanism but that, as anticipated, $SO_4^=$ would be more effective at lower solution concentrations. The same type of mechanism has been suggested for anion interactions with Al_{13} polynuclear species.[82,84]

Another important aspect that has often been overlooked is the sequence of anion introduction into the aluminum solution. In many studies[121-125] the anions have been introduced prior to neutralization, whereas in others anions are interacted with a partially neutralized aluminum solution.[8,9,52,53,115] Studies utilizing the former approach have generally demonstrated the inhibitory influence of complexing anions on the hydrolysis-precipitation reactions of aluminum, whereas studies employing the latter approach demonstrated that even the addition of F^- to partially neutralized aluminum solutions caused rapid precipitation of aluminum polynuclear species.[8-9] Even when organic compounds are added to aluminum solutions and then hydrolyzed, nucleation behavior is found to vary, depending on the relative concentration of ligand. For example, Tentorio et al.[126] found that adding 3-mordant blue to an $Al_2(SO_4)_3$ solution and then hydrothermally hydrolyzing the solutions resulted in the formation of alumunim-hydroxide solid phases whose properties varied as a function of the mordant blue/aluminum molar ratios. With increasing mordant blue, the average particle size of the resulting aluminum-hydroxide solid phase decreased until high molar ratios (>1.25), where precipiation was inhibited totally. Also, at the lower chelate/aluminum ratios the average particle size of the aluminum-hydroxide precipitate was greater than that of particles formed in a solution without the chelating agent, suggesting that nucleation was rapid under these conditions. These observations may have important implications regarding the nature of organic-aluminum complexes in natural systems containing a variety of lower relative molecular mass organic acids and dissolved or colloidal humic materials[126] (see also Chapter 5).

The influence of dissolved organic constituents on aluminum nucleation reactions has been studied extensively.[121-129] Most of these investigations have focused on the solid phases formed in hydrolyzed aluminum solutions containing dissolved organic solutes rather than on the nature of the soluble aluminum polynuclear species. These reactions are, however, very important to describing the transformations of aluminum in aquatic and soil systems, in which dissolved organic constituents are ubiquitous.[124] Dissolved organic substances that form stable complexes with aluminum have been found to compete with OH^- ligands, effectively perturbing aluminum hydrolytic reactions presumably through a disruption of the hydroxyl bridging mechanism[124] (see Chapter 5). The inhibition of aluminum solid-phase formation by organic acids relates directly to the stability of the soluble organic-aluminum complex, with the experimentally observed sequence: citric acid > malic acid > tannic acid > aspartic acid > *p*-hydroxybenzoic acid.[125] Studies employing both organic and fulvic acids have demonstrated that the presence of dissolved organic compounds greatly inhibits the crystallization of aluminum-hydroxide precipitates.[121-125,127] Many experiments have indicated that dissolved organic constituents and high concentrations of inorganic anions result in the precipitation of noncrystalline aluminum products or poorly crystalline pseudoboehmite,[122-125,128] with several investigators indicating that pseudoboehmite actually may be stabilized by these perturbing ligands.[124] In an extensive investigation of the influence of inorganic and organic ligands on aluminum solid-phase formation, Violante and Huang[128] demonstrated that various ligands promote and stabilize the formation of pseudoboehmite over the crystalline $Al(OH)_3$ polymorphs according to the following sequence: chloride < sulfate < phthalate ≃ succinate < glutamate < silicate < aspartate < phosphate < salicylate ≃ malate < tannate < citrate < tartrate.

Lind and Hem[129] suggested that the organic flavone, queracetin, not only inhibited crystallization of the aluminum solid phases once formed, but also inhibited the polymerization process. This conclusion was based on the lower amount of structural OH associated with the polynuclear compounds: 1.4 to 1.9 for solutions containing queracetin as compared to 1.8 to 2.76 for comparable solutions not containing queracetin. However, the amount of polynuclear aluminum (Al_b), as determined by the ferron assay, tended to be greater in the solutions containing queracetin. These observations suggest that greater amounts of smaller polynuclear species are formed in the presence of the organic compounds, although the authors attributed the greater amounts of Al_b in these solutions to possible differences in synthesis methodology and provided no specific evidence to substantiate this conclusion. More recently, Jardine and Zelazny[130] indicated that citrate and formate did not affect significantly the interaction of ferron with aluminum species present in partially neutralized solutions. Both the amount of Al_a and the rate constant for the Al_b-ferron interaction were similar for partially neutralized solutions with and without the added organic compounds. This suggests that, under the conditions of their experiments, either the amount and type of polynuclear compounds formed were not influenced by the citric or malonic acid or that, if organic-aluminum polymers do form, their interaction with ferron is similar to the aluminum polynuclear species formed in the absence of organic compounds. As indicated earlier, the sequence of anion introduction and the organic/aluminum molar ratio are important variables influencing organic aluminum-polynuclear interactions that need to be considered when comparing data.

Other experimental variables leading to differences in observed nucleation-precipitation reactions of aluminum polynuclear species include the rate and method of base addition and the effect of temperature during synthesis or aging of solutions. Vermeulen et al.[44] suggested that high localized hydroxyl ion concentrations are common in most procedures of neutralization, resulting in solutions tending to be far from equilibrium and often in the formation of some colloidal precipitate. In several studies where a colloidal precipitate was formed during rapid neutralization, it was observed that, upon dissolution, polynuclear species of the Al_b form would result.[95] Direct ^{27}Al NMR studies have demonstrated that heating cloudy solutions resulted in the appearance of Al_{13} polynuclear species,[42] suggesting that the solid phase consisted of discrete Al_{13} units stabilized by a bridging mechanism. Turner[98] suggested that some of the initial solid phases formed on rapid neutralization of aluminum solutions redissolved on aging, yielding increases in the Al_b fraction as estimated by the 8-hydroxyquinoline assay. These data may also be interpreted as indicating the existence of Al_{13} polynuclear units within a colloidal precipitate matrix. It has been demonstrated by ^{27}Al NMR spectroscopy that some aluminum retained by ultrafiltration (13-nm pore size) was composed of Al_{13} polynuclear species,[8,9] providing further support for the ability of this complex to nucleate in discrete units.

Temperature has also been shown to influence the nucleation of aluminum polynuclear species. Some investigators have chosen to synthesize their partially neutralized aluminum solutions at elevated temperatures, citing more favorable reaction kinetics,[28,32] since other studies conducted at 25°C indicate that solutions can require one to several years to reach steady state.[54,75] It has been demonstrated that at higher temperatures, aluminum solutions hydrolyze to a greater extent, and precipitation may occur at lower $\tilde{n}$ values.[28,68] Hydrothermal hydrolysis of Al^{3+} solutions has been utilized in a number of studies to produce boehmite and alumina,[126,131,132] although more recently it has been demonstrated that gibbsite, boehmite, alumina, and alunite were formed by spontaneous hydrolysis of a 0.5 mol Al/l solution at 25°C when prepared from $Al_2(SO_4)_3$.[132] Light-scattering measurements confirmed the presence of colloidal precipitates following 8 months of aging, although sufficient yields for filtration and characterization were not obtained until approximately 42 months of aging. These results are significant, in that they suggest that boehmite and alumina may form without the requirement of thermal conditions or high inorganic anion to aluminum ratios as previously thought. It is not clear what role the accompanying anion may play in these spontaneous hydrolytic reactions. No similar results have been reported for $Al–Cl^-$, $–NO_3^-$, or $–ClO_4^-$ solutions, however, suggesting that specifically interacting anions, such as SO_4^{2-}, may play a unique role in the nucleation processes under these conditions.

V. SUMMARY AND CONCLUSIONS

It has been recognized for many years that, under the proper conditions, aluminum monomeric hydrolytic products combine to form polynuclear species in solution and at clay mineral surfaces. These polynuclear aluminum species represent important metastable dissolved constituents that may remain in solution for many years.

There is good evidence to indicate that the aluminum dimer is perhaps the first polynuclear species formed in solution, with both the hydrated cyclic $[Al_2(OH)_2(H_2O)_{10}]^{4+}$ and bridged structures $[Al_2(OH)_2(H_2O)_8]^{4+}$ being likely. The hydrated cyclic species is apparently in rapid exchange with the mono-hydroxo monomers, whereas the bridged species can be considered more stable, being directly observable by NMR spectroscopy. Presumably, the formation of the bridged dimer requires sufficient energy to overcome the electrostatic repulsive forces between aluminum atoms, with the resulting loss of water molecules from their coordination spheres. Thus, as might be anticipated, elevated temperatures have been demonstrated to increase the exchange between aluminum monomers and bridged dimers.

There is conclusive, direct evidence indicating that the $[AlO_4Al_{12}(OH)_{24}(H_2O)_{12}]^{7+}$ species is an important polymer in water-treatment processes, pharmaceuticals, and certain catalysts. It forms readily in partially neutralized aluminum solutions, although its formation may require specific conditions that may not be typical of many natural water systems. There exists little convincing evidence for the existence of the $[Al_6(OH)_{12}(H_2O)_{12}]^{6+}$ polymer, or the smaller species of the "gibbsite-fragment" polymerization model, although many questions remain, especially for aluminum solutions less concentrated than 10^{-4} *M*. In addition, there has not been a convincing mechanism proposed that would impart kinetic stability to a series of polymers prefiguring the highly thermodynamically favored trihydroxide mineral phases, nor one that would account for the distinct boundary observed between the Al_b and Al_c fractions measured via timed spectrophotometric assays. The Al_{13} polymer, conversely, should possess kinetic stability resulting from the requirement for structural rearrangement prior to its precipitating as a trihydroxide mineral phase and, likewise, would be structurally distinct from the Al_c components which may be composed essentially of microcrystalline trihydroxide mineral phases.

A yet-unresolved polynuclear species is the more slowly reacting component of the Al_b fraction which is measurable via timed spectrophotometric assays and inferred by difference in speciation methods employing ^{27}Al NMR methods. Several studies have indicated that the Al_{13} polymer transforms very slowly on aging to this presumably more stable polynuclear form, suggesting that it represents the product of an Al_{13} rearrangement into a polymer having the gibbsite structure: i.e., larger polymers of the "gibbsite-fragment" model. Based on the evidence available in the literature and on results currently being generated in the laboratory of the author, one may propose several different nucleation pathways whereby solid phases form from aluminum polynuclear species. In solutions where little or no Al_{13} polymer is formed, it is believed that aluminum dimers predominate and combine to form nuclei and eventually Al_c materials having the aluminum-trihydroxide mineral structure. On a relative scale, it is proposed that these systems reach "equilibrium" relatively rapidly. Solutions containing significant quantities of the Al_{13} polynuclear species have multiple nucleation pathways that are determined by the history of a solution. Partially neutralized aluminum solutions synthesized such that no solid phases are formed are stable for relatively long periods of time. Conversely, if solid phases are formed on synthesis, then the rate of Al_{13} nucleation can vary significantly depending on the amounts and nature of the precipitate formed. As described earlier, in the presence of small amounts of Al_c composed primarily of what can be considered microcrystalline gibbsite, the Al_{13} polymer is observed to gradually convert to a polynuclear form presumably having a structure prefiguring the trihydroxide mineral phases. This conversion is more rapid than the indefinitely stable solutions, yet relative to the reactions described subsequently they are relatively slow, occurring over a number of years. The third major nucleation pathway is the precipitation of a noncrystalline aluminum phase composed of linked Al_{13} units. This precipitate can, under certain conditions, remain stable for some time, converting slowly to one of the aluminum trihydroxide polymorphs. However, under other conditions this noncrystalline precipitate can redissolve, releasing Al_{13} polymers to solution. The boundary conditions defining these reactions are not well understood, but they may have significance with respect to determining the specific trihydroxide polymorph formed on aging. These reactions and their implications for the type of solid phase formed will undoubtedly be better elucidated in future studies.

Other areas requiring further research include the mixed-ligand monomeric and polynuclear aluminum species which are now known to include phosphato-fluoro-aluminum, citrato-fluro-aluminum, and citrato-phosphato-aluminum complexes. The conditions under which these complexes form and their relative stabilities need to be defined, since these species are not utilized currently in thermodynamic databases for geochemical models, yet citrate and fluoride are known to be important ligands in aquatic and soil systems. Future studies utilizing direct spectroscopic methods may be required to define some of the parameters associated with the formation and stability of these complexes.

ACKNOWLEDGMENTS

Gratitude is expressed to Ms. Jan Hinton and Ms. Denise Brock for typing this manuscript, and to Jean Coleman for drafting many of the figures. The author was supported during preparation of this manuscript by a grant from Southern California Edison (SCE) and partially by contract DE-AC09-76SROO819 between the University of Georgia and the U.S. Department of Energy. Appreciation is extended to Dr. Carl Fox, SCE project manager, for his support and cooperation.

REFERENCES

1. **Hsu, P. H.,** Aluminum hydroxides and oxyhydroxides, in *Minerals in Soil Environments*, 2nd Ed., Dixon, J. B. and Weed, S. B., Eds., Soil Science Society of America, Madison, WI, 1988.
2. **Johnson, N. M., Driscoll, C. T., Eaton, J. S., Likens, G. E., and McDowell, W. H.,** 'Acid rain' dissolved aluminum and chemical weathering at the Hubbard Brook Experimental Forest, New Hampshire, *Geochim. Cosmochim. Acta,* 45, 1421, 1981.
3. **Lazerte, B. D.,** Forms of aqueous aluminum in acidified catchments of central Ontario: a methodological analysis, *Can. J. Fish Aquat. Sci.,* 41, 766, 1984.
4. **Nordstrom, D. K. and Ball, J. W.,** The geochemical behavior of aluminum in acidified surface waters, *Science,* 232, 54, 1985.
5. **Foy, C. D., Chaney, R. L., and White, M. C.,** The physiology of metal toxicity in plants, *Annu. Rev. Plant Physiol.,* 29, 511, 1978.
6. **Bennet, R. J., Breen, C. M., and Fey, M. V.,** The primary site of aluminum injury in the root of *Zea mays, S. Afr. J. Plant Soil,* 2, 8, 1985.
7. **Baker, J. P. and Schofield, C. L.,** Aluminum toxicity to fish in acidic waters, *Water Air Soil Pollut.,* 18, 289, 1982.
8. **Buffle, J., Parthasarathy, N., and Haerdi, W.,** Importance of speciation methods in analytical control of water treatment processes with application to fluoride removal from waste waters, *Water Res.,* 19, 7, 1985.
9. **Parthasarthy, N. and Buffle, J.,** Study of polymeric aluminum (III) hydroxide solutions for application in waste water treatment. Properties of the polymer and optimal conditions of preparation, *Water Res.,* 19, 25, 1985.
10. **Teagarden, D. L., Kozlowski, J. F., White, J. L., and Hem, S. L.,** Aluminum chlorohydrate. I. Structure studies, *J. Pharm. Sci.,* 70, 758, 1981.

11. **Shah, D. N., White, J. L., and Hem, S. L.,** Mechanism of interaction between Polyols and aluminum hydroxide gel, *J. Pharm. Sci.*, 70, 1101, 1981.
12. **Pinnavia, T. J.,** Intercalated clay catalysts, *Science,* 220, 365 and 371, 1983.
13. **Plee, D., Borg, F., Gatineau, L., and Fripiatt, J. J.,** High resolution solid state ^{27}Al and ^{29}Si nuclear magnetic resonance study of pillared clays, *J. Am. Chem. Soc.*, 107, 2362, 1985.
14. **Platts, M. M., Owen, G., and Smith, S.,** Water purification and the incidence of fractures in patients receiving home haemodialysis supervised by a single centre: evidence for "safe" upper limit of aluminum in water, *Br. Med. J.*, 288, 969, 1984.
15. **Laussac, J. P. and Commenges, G.,** ^{1}H, ^{13}C, ^{31}P and ^{27}Al NMR study of aluminum-ATP complexes. A possible relation with its biological application, *Nouv. J. Chim.*, 7, 579, 1983.
16. **Bronsted, J. N. and Volquartz, K.,** Acid dissociation of aquo ions, *Z. Phys. Chem.*, 134, 97, 1928.
17. **Hartford, W. H.,** Chromic acid anodic baths. Interpretation of glass electrode measurements, *Ind. Eng. Chem.*, 34, 920, 1942.
18. **Ito, T. and Yui, N.,** Hydrolysis constant of the aluminum ion in chloride solutions, *Chem. Abstr.*, 48, 56136, 1954.
19. **Schofield, R. K. and Taylor, A. W.,** The hydrolysis of aluminum salt solutions, *J. Chem. Soc.*, 1954, 4445, 1954.
20. **Kubota, H.,** The hydrolysis of aluminum in dilute solutions, *Diss. Abstr.*, 16, 864, 1956.
21. **Frink, C. R. and Peech, M.,** Hydrolysis of the aluminum ion in dilute aqueous solutions, *Inorg. Chem.*, 2, 473, 1963.
22. **Raupach, M.,** Solubility of simple aluminum compounds expected in soils. I. Hydroxides and oxyhydroxides, *Aust. J. Soil Res.*, 1, 28, 1963.
23. **Hem, J. D. and Roberson, C. E.,** Form and stability of aluminum hydroxide complexes in dilute solutions, *U.S. Geol. Surv. Water-Supply Pap.*, 1827-A, 1967.
24. **Grunwald, E. and Fong, D. W.,** Acidity and association of Al ions in dilute aqueous acid, *J. Phys. Chem.*, 73, 650, 1969.
25. **Nazarenko, V. A. and Nevskaya, E. M.,** Spectrophotometric determination of the constants of mononuclear hydrolysis of aluminum ions, *Russ. J. Inorg. Chem.*, 14, 1696, 1969.
26. **May, H. M., Helmke, P. A., and Jackson, M. L.,** Gibbsite solubility and thermodynamic properties of hydroxy-aluminum ions in aqueous solution at 25°C, *Geochim. Cosmochim. Acta,* 43, 861, 1979.
27. **Couturier, Y., Michard, G., and Sarazin, G.,** Constantes de formation des complexes hydroxydes de l'aluminum en solution aqueuse de 20 a 70°C, *Geochim. Cosmochim. Acta,* 48, 649, 1984.
28. **Mesmer, R. E. and Baes, C. F.,** Acidity measurements at elevated temperatures. I. Aluminum ion hydrolysis, *Inorg. Chem.*, 10, 2290, 1971.
29. **Fong, D. W. and Grunwald, E.,** Kinetic study of proton exchange between the $Al(OH_2)_6^{3+}$ ion and water in dilute acid, *J. Am. Chem. Soc.*, 91, 2413, 1969.
30. **Holmes, L. P., Cole, D. L., and Eyring, E. M.,** Kinetics of aluminum hydrolysis in dilute solutions, *J. Phys. Chem.*, 72, 301, 1968.
31. **Bertsch, P. M., Barnhisel, R. I., Thomas, G. W., Layton, W. J., and Smith, S. L.,** Quantitative determination of aluminum-27 by high resolution nuclear magnetic resonance spectrometry, *Anal. Chem.*, 58, 2583, 1986.
32. **Brosset, C., Biedermann, G., and Sillén, L. G.,** Studies on the hydrolysis of metal ions. XI. The aluminum ion, Al^{3+}, *Acta Chem. Scand.*, 8, 1917, 1954.
33. **Gayer, K. H., Thompson, L. C., and Zajicek, O. T.,** The solubility of aluminum hydroxide in acidic and basic media at 25°C, *Can. J. Chem.*, 36, 1268, 1958.
34. **Reesman, A. L., Pickett, E. E., and Keller, W. D.,** Aluminum ions in aqueous solution, *Am. J. Sci.*, 267, 99, 1969.
35. **Smith, R. W. and Hem. J. D.,** Effect of aging on aluminum hydroxide complexes in dilute solutions, *U.S. Geol. Surv. Water-Supply Pap.*, 1827-D, 1972.
36. **Moolenaar, R. J., Evans, J. C., and McKeener, L. D.,** The structure of the aluminate ion in solution at higher pH, *J. Phys. Chem.*, 74, 3629, 1970.
37. **Baes, C. F. and Mesmer, R. E.,** *The Hydrolysis of Cations,* John Wiley & Sons, New York, 1976.
38. **Cotton, F. A. and Wilkinson, G.,** *Advanced Inorganic Chemistry,* 4th ed., John Wiley & Sons, New York, 1980.
39. **Carreira, L. A., Maroni, V. A., Swaine, J. W., and Plumb, R. C.,** Raman and infrared spectra and structures of the aluminate ions, *J. Chem. Phys.*, 45, 2216, 1966.
40. **Lindsay, W. L.,** *Chemical Equilibria Soils,* John Wiley & Sons, New York, 1979.
41. **Maksimova, I., Mashovets, U. P., and Yushkevich, V.,** Electric conductivity and structure of sodium aluminate in aqueous solutions, *J. Appl. Chem. (USSR)*, 40, 2594, 1967.
42. **Akitt, J. W. and Farthing, A.,** New ^{27}Al NMR studies of the hydrolysis of the aluminum(III) cation, *J. Magn. Reson.*, 32, 345, 1978.

43. **Bottero, J. Y., Cases, J. M., Fiessinger, F., and Poirier, J. E.,** Studies of hydrolyzed aluminum chloride solutions. I. Nature of aluminum species and composition of aqueous solutions, *J. Phys. Chem.*, 84, 2933, 1980.
44. **Vermeulen, A. C., Geus, J. W., Stol, R. J., and DeBruyn, P. L.,** Hydrolysis-precipitation studies of aluminum(III) solutions. I. Titration of acidified aluminum nitrate solutions, *J. Colloid Interface Sci.*, 51, 449, 1975.
45. **Frink, C. R. and Sawhney, B. L.,** Neutralization of dilute aqueous aluminum salt solutions, *Soil Sci.*, 103, 144, 1967.
46. **Dezelic, W., Bilinski, H., and Wolf, R. H. H.,** Precipitation and hydrolysis of metalic ions. IV. Studies on the solubility of aluminum hydroxide in aqueous solution, *J. Inorg. Nucl. Chem.*, 33, 791, 1971.
47. **Jander, G. and Winkel, A.,** Diffusion coefficients of basic aluminum solutions, *Z. Anorg. Chem.*, 200, 257, 1931.
48. **Brosset, C.,** On the reactions of the aluminum ion with water, *Acta Chem. Scand.*, 6, 910, 1952.
49. **Sillén, L. G.,** Quantitative studies of hydrolytic equilibria, *Q. Rev.*, 13, 146, 1959.
50. **Sillén, L. G.,** On equilibria in systems with polynuclear complex formation, *Chem. Scand.*, 15, 1981, 1961.
51. **Biedermann, G.,** *Sven. Kem. Tidskr.*, 76, 362, 1964.
52. **Matijevic, E. and Tezak, B.,** Detection of polynuclear complex aluminum ions by means of coagulation measurements, *J. Phys. Chem.*, 57, 951, 1953.
53. **Hayden, P. L. and Rubin, A. J.,** Systematic investigation of the hydrolysis and precipitation of aluminum(III), in *Aqueous-Environmental Chemistry of Metals*, Rubin, A. J., Ed., Ann Arbor Science, Ann Arbor, MI, 1976.
54. **Hsu, P. H. and Bates, T. F.,** Formation of X-ray amorphous and crystalline aluminum hydroxides, *Mineral. Mag.*, 33, 749, 1964.
55. **Hsu, P. H. and Rich, C. I.,** Aluminum fixation in a synthetic cation exchanger, *Soil Sci. Soc. Am. Proc.*, 24, 21, 1960.
56. **Stol, R. J., Van Helden, A. K., and DeBruyn, P. L.,** Hydrolysis-precipitation studies of aluminum(III) solutions. II. A kinetic study and model, *J. Colloid Interface Sci.*, 57, 115, 1976.
57. **Richburg, J. S. and Adams, F.,** Solubility and hydrolysis of aluminum in soil solutions and saturated-paste extracts, *Soil Sci. Soc. Am. Proc.*, 34, 728, 1970.
58. **Bersillon, J. L., Hsu, P. H., and Fiessinger, F.,** Characterization of hydroxy-aluminum solutions, *Soil Sci. Soc. Am. J.*, 44, 630, 1980.
59. **Aveston, J.,** Hydrolysis of aluminum ion: ultracentrifugation and acidity measurements, *J. Chem. Soc.*, 1965, 4438, 1965.
60. **Johansson, G.,** The crystal structures of $[Al_2(OH)_2H_2O)_8](SO_4)_2 \cdot 2H_2O$ and $[Al_2(OH)_2(H_2O)_8]\ (SeO_4)_2 \cdot 2H_2O$, *Acta Chem. Scand.*, 16, 403, 1962.
61. **Johansson, G.,** On the crystal structures of some basic aluminum salts, *Acta Chem. Scand.*, 14, 771, 1960.
62. **Rausch, M. V. and Bale, H. D.,** Small-angle x-ray scattering from hydrolyzed Al nitrate solutions, *J. Chem. Phys.*, 40, 3391, 1964.
63. **Akitt, J. W., Greenwood, N. N., Khandelwal, B. L., and Lester, G. D.,** ^{27}Al nuclear magnetic resonance studies of the hydrolysis and polymerization of the hexa-aqua-aluminum(III) cation, *J. Chem. Soc. Dalton Trans.*, 1972, 604, 1972.
64. **Akitt, J. W. and Farthing, A.,** Aluminum-27 nuclear magnetic resonance studies of the hydrolysis of aluminum(III). II. Gel-permeation chromatography, *J. Chem. Soc. Dalton Trans.*, 1981, 1606, 1981.
65. **Akitt, J. W. and Farthing, A.,** Aluminum-27 nuclear magnetic resonance studies of the hydrolysis of aluminum(III). III. Stopped-flow kinetic studies, *J. Chem. Soc. Dalton Trans.*, 1981, 1609, 1981.
66. **Akitt, J. W. and Farthing, A.,** Aluminum-27 nuclear magnetic resonance studies of the hydrolysis of aluminum(III). IV. Hydrolysis using sodium carbonate, *J. Chem. Soc. Dalton Trans.*, 1981, 1617, 1981.
67. **Akitt, J. W. and Farthing, A.,** Aluminum-27 nuclear magnetic resonance studies of the hydrolysis of aluminum(III). V. Slow hydrolysis using aluminum metal, *J. Chem. Soc. Dalton Trans.*, 1981, 1626, 1981.
68. **Patterson, J. H. and Tyree, S. Y.,** A light scattering study of the hydrolytic polymerization of aluminum, *J. Colloid Interface Sci.*, 43, 384, 1973.
69. **Fripiat, J. J., Van Cauwelaert, F., and Bosmans, H.,** Structures of aluminum cations in aqueous solutions, *J. Phys. Chem.*, 69, 2458, 1965.
70. **Akitt, J. W., Greenwood, N. N., and Lester, G. D.,** Nuclear magnetic resonance and Raman studies of the aluminum complexes formed in aqueous solutions of aluminum salts containing phosphoric acid and fluoride ions, *J. Chem. Soc.*, 1971, 2450, 1971.
71. **White, R. E., Tiffen, L. O., and Taylor, A. W.,** The existence of polymeric complexes in dilute solutions of aluminum and orthophosphate, *Plant Soil*, 45, 521, 1976.
72. **Bertsch, P. M. and Layton, W. J.,** unpublished, 1987.
73. **Smith, R. W.,** Reactions among equilibrium and non-equilibrium aqueous species of aluminum hydroxy complexes, *Adv. Chem. Ser.*, 106, 250, 1971.

74. **Smith, R. W. and Hem. J. D.,** Effect of aging on aluminum hydroxide complexes in dilute solutions, *U.S. Geol. Surv. Water-Supply Pap.,* 1827-D, 1972.
75. **Hem, J. D. and Roberson, C. E.,** Form and stability of aluminum hydroxide complexes in dilute solutions, *U.S. Geol. Surv. Water-Supply Pap.,* 1827-A, 1967.
76. **Hem, J. D.,** Aluminum species in water, *Adv. Chem. Ser.,* 106, 98, 1971.
77. **Bertsch, P. M., Thomas, G. W., and Barnhisel, R. I.,** Characterization of hydroxy aluminum solutions by aluminum-27 nuclear magnetic resonance spectroscopy, *Soil Sci. Soc. Am. J.,* 50, 825, 1986.
78. **Kentamaa, J.,** The hydrolysis of aluminum chloride, *Acad. Sci. Fenn. Ann. Ser. A.,* 67, 1955.
79. **Teagarden, D. L., Hem. S. L., and White, J. L.,** Conversion of aluminum chlorohydrate to aluminum hydroxide, *J. Soc. Cosmet. Chem.,* 33, 281, 1982.
80. **Bertsch, P. M.,** Conditions for Al_{13} polymer formation in partially neutralized aluminum solutions, *Soil Sci. Soc. Am. J.,* 51, 825, 1987.
81. **Bertsch, P. M., Layton, W. J., and Barnhisel, R. I.,** Speciation of hydroxy-aluminum solutions by wet chemical and aluminum-27 NMR methods, *Soil Sci. Soc. Am. J.,* 50, 1449, 1986.
82. **Teagarden, D. L., Radavich, J. F., White, J. L., and Hem, S. L.,** Aluminum chlorohydrate. II. Physicochemical studies, *J. Pharm. Sci.,* 70, 762, 1981.
83. **Teagarden, D. L., White, J. L., and Hem, S. L.,** Aluminum chlorohydrate. III. Conversion to aluminum hydroxide, *J. Pharm. Sci.,* 70, 808, 1981.
84. **Vaughan, D. E. W. and Lussier, R.,** Preparation of molecular sieves based on pillared interlayered clays (PILC), in *Proc. 5th Int. Conf. Zeolites,* Rees, L. V., Ed., Heyden, London, 1980.
85. **Bertsch, P. M. and Barnhisel, R. I.,** Chlorites and hydroxy-interlayered vermiculite and smectite, in *Minerals in Soil Environments,* 2nd ed., Dixon, J. B. and Weed, S. B., Eds., Soil Science Socity of America, Madison, WI, 1988.
86. **Bloom, P. R., McBride, M. B., and Chadbourne,** Adsorption of aluminum by a smectite. I. Surface hydrolysis during Ca^{2+}–Al^{3+} exchange, *Soil Sci. Soc. Am. J.,* 41, 1068, 1977.
87. **Van Olphen, H.,** *Clay Colloid Chemistry,* 2nd ed., John Wiley & Sons, New York, 1977.
88. **Barnhisel, R. I. and Bertsch, P. M.,** Aluminum in *Methods of Soil Analysis, Part 2,* 2nd ed., Page, A. L., Ed., American Association Agronomy, Madison, WI, 1982, 275.
89. **Perrott, K. W.,** The nature of cationic aluminum species on the cation exchange surface of mica, *J. Colloid Interface Sci.,* 82, 136, 1981.
90. **White, J. L.,** Personal communication.
91. **Hodges, S. C. and Zelazny, L. W.,** Influences of OH/Al ratios and loading rates in aluminum-kaolinite interactions, *Soil Sci. Soc. Am. J.,* 47, 221, 1983.
92. **Jardine, P. M., Zelazny, L. W., and Parker, J. C.,** Mechanisms of aluminum adsorption on clay minerals and peat, *Soil Sci. Soc. Am. J.,* 49, 862, 1985.
93. **Okura, T., Goto, K., and Votuyanagi, T.,** Forms of aluminum determined by an 8-quinolinolate extraction method, *Anal. Chem.,* 34, 581, 1962.
94. **Turner, R. C.,** Three forms of aluminum in aqueous systems determined by 8-quinolinolate extraction methods, *Can. J. Chem.,* 47, 2521, 1969.
95. **Turner, R. C.,** Kinetics of reactions of 8-quinolinol and acetate with hydroxyaluminum species in aqueous solutions. II. Initial solid phases, *Can. J. Chem.,* 49, 1688, 1971.
96. **Turner, R. C. and Sulaiman, W.,** Kinetics of reactions of 8-quinolinol and acetate with hydroxyaluminum species in aqueous solutions. I. Polynuclear hydroxyaluminum cations, *Can. J. Chem.,* 49, 1683, 1971.
97. **Turner, R. C.,** Effect of aging on properties of polynuclear hydroxyaluminum cations, *Can. J. Chem.,* 54, 1528, 1976.
98. **Turner, R. C.,** A second species of polynuclear hydroxyaluminum cation, its formation and some of its properties, *Can. J. Chem.,* 54, 1910, 1976.
99. **Turner, R. C. and Ross, G. J.,** Conditions in solution during the formation of gibbsite in dilute Al salt solutions. IV. Effect of Cl concentration and temperature and a proposed mechanism for gibbsite formation, *Can. J. Chem.,* 48, 723, 1970.
100. **Tsai, P. P. and Hsu, P. H.,** Studies of aged OH-Al solutions using kinetics of Al-ferron reactions and sulfate precipitation, *Soil Sci. Soc. Am. J.,* 48, 59, 1984.
101. **Tsai, P. P. and Hsu, P. H.,** Aging of partially neutralized aluminum solutions of sodium hydroxide/aluminum molar ration = 2.2, *Soil Sci. Soc. Am. J.,* 49, 1060, 1985.
102. **Jardine, P. M. and Zelazny, L. W.,** Mononuclear and polynuclear aluminum speciation through differential kinetic reactions with ferron, *Soil Sci. Soc. Am. J.,* 50, 895, 1986.
103. **Bertsch, P. M. and Layton, W. J.,** unpublished, 1987.
104. **Denney, D. Z. and Hsu, P. H.,** ^{27}Al nuclear magnetic resonance and ferron kinetic studies of partially neutralized $AlCl_3$ solutions, *Clays Clay Miner.,* 34, 604, 1986.
105. **Birkner, F. B. and Morgan, J. J.,** Polymer flocculation kinetics of dilute colloidal suspensions, *J. Am. Water Works Assoc.,* 60, 175, 1968.

106. **Snodgrass, W. J., Clark, M. M., and O'Melia, C. R.,** Particle formation and growth in dilute aluminum (III) solutions, *Water Res.,* 18, 479, 1984.
107. **DeHek, H., Stol, R. J., and DeBruyn, P. L.,** Hydrolysis-precipitation studies of aluminum(III) solutions. III. The role of the sulfate ion, *J. Colloid Interface Sci.,* 64, 72, 1978.
108. **Letterman, R. D. and Iyer, D. R.,** Modeling the effects of hydrolyzed aluminum and solution chemistry on flocculation kinetics, *Environ. Sci. Technol.,* 19, 673, 1985.
109. **Tambo, N. and Watnabe, Y.,** Physical characterestics of flocs. I. The floc density function and aluminum floc, *Water Res.,* 13, 409, 1979.
110. **Kennedy, V. C., Zellweger, G. W., and Jones, B. F.,** Filter pore-size effects on the analysis of Al, Fe, Mn, and Ti in water, *Water Resour. Res.,* 10, 785, 1974.
111. **Jones, B. F., Kennedy, V. C., and Zellweger, G. W.,** Comparison of observed and calculated concentrations of dissolved Al and Fe in stream water, *Water Resourc. Res.,* 10, 791, 1974.
112. **Barnes, R. B.,** The determination of specific forms of aluminum in natural water, *Chem. Geol.,* 15, 177, 1975.
113. **May, H. M., Helmke, P. A., and Jackson, M. L.,** Determination of mononuclear dissolved aluminum in near-neutral waters, *Chem. Geol.,* 24, 259, 1979.
114. **Marion, S. P. and Thomas, A. W.,** Effects of diverse anions on the pH of maximum precipitation of aluminum hydroxide, *J. Colloid Sci.,* 1, 221, and 234, 1946, and references therein.
115. **Ross, G. J. and Turner, R. C.,** Effect of different anions on the crystalization of aluminum hydroxide in partially neutralized aqueous aluminum salt systems, *Soil Sci. Soc. Am. Proc.,* 35, 389, 1971.
116. **Hsu, P. H.,** Effect of phosphate and silicate on the crystallization of gibbsite from OH-Al solutions, *Soil Sci.,* 127, 219, 1979.
117. **Nordstrom, D. K.,** The effect of sulfate on aluminum concentrations in natural waters: some stability relations in the system Al_2O_3-SO_3-H_2O at 298K, *Geochim. Cosmochim. Acta,* 46, 681, 1982.
118. **Serna, C. J., White, J. L., and Hem, S. L.,** Anion-aluminum hydroxide gel interactions, *Soil Sci. Soc. Am. J.,* 41, 1009, 1977.
119. **Nail, S. L., White, J. L., and Hem, S. L.,** Structure of aluminum hydroxide gel. I. Initial precipitate, *J. Pharm. Sci.,* 65, 1188, 1976.
120. **Nail, S. L., White, J. L., and Hem, S. L.,** Structure of aluminum hydroxide gel. II. Aging mechanism, *J. Pharm. Sci.,* 65, 192, 1976.
121. **Wang, M. K., White, J. L., and Hem, S. L.,** Influence of acetate, oxalate, and citrate anions on precipitation of aluminum hydroxide, *Clays Clay Miner.,* 31, 65, 1983.
122. **Kwong, N. K. and Huang, P. M.,** Influence of citric acid on the crystalization of aluminum hydroxides, *Clays Clay Miner.,* 23, 164, 1977.
123. **Kwong, N. K. and Huang, P. M.,** Influence of citric acid on the hydrolytic reactions of aluminum, *Soil Sci. Soc. Am. J.,* 41, 692, 1977.
124. **Huang, D. M. and Violante, A.,** Influence of organic acids on crystallization and surface properties of precipitation products of aluminum, in *Interactions of Soil Minerals with Natural Organics and Microbes,* Huang, P. M. and Schnitzer, M., Eds., Science Society of America, Madison, WI, 1986.
125. **Kwong, N. K. and Huang, P. M.,** The relative influence of low-molecular-weight complexing organic acids on the hydrolysis and precipitation of aluminum, *Soil Sci.,* 128, 337, 1979.
126. **Tentorio, A., Matijevic, E., and Kratohuil, J. P.,** Preparation and optical properties of spherical colloidal aluminum hydroxide particles containing a dye, *J. Colloid Interface Sci.,* 77, 418, 1980.
127. **Kodama, H. and Schnitzer, M.,** Effect of fulvic acid on the crystallization of aluminum hydroxides, *Geoderma,* 24, 195, 1980.
128. **Violante, A. and Huang, P. M.,** Influence of inorganic and organic ligands on the formation of aluminum hydroxides and oxyhydroxides, *Clays Clay Miner.,* 33, 181, 1985.
129. **Lind, C. J. and Hem, J. D.,** Effects of organic solutes on chemical reactions of aluminum, *U.S. Geol. Surv. Water-Supply Pap.,* 1827-G, 1975.
130. **Jardine, P. M. and Zelazny, L. W.,** Influence of organic anions on the speciation of mononuclear and polynuclear aluminum by ferron, *Soil Sci. Am. J.,* 51, 885, 1987.
131. **MacDonald, D. D., Butler, P., and Owen, D.,** Hydrothermal hydrolysis of Al^{3+} and the precipitation of boehmite from aqueous solution, *J. Phys. Chem.,* 77, 2474, 1973.
132. **Singh, S. S.,** The formation and coexistence of gibbsite, boehmite, alumina and alunite at room temperature, *Can. J. Soil. Sci.,* 62, 327, 1982.
133. **Schutz, A., Stone, W. E. E., Poncelet, G., and Fripiat, J. J.,** Preparation and characterization of bidimensional zeolitic structures obtained from synthetic beidellite and hydroxy-aluminum solutions, *Clays Clay Miner,* 35, 251, 1987.

Chapter 5

NATURALLY OCCURRING ALUMINUM-ORGANIC COMPLEXES

F. J. Stevenson and G. F. Vance

TABLE OF CONTENTS

I. INTRODUCTION

The binding of aluminum by naturally occurring organic substances is a subject of considerable importance in several scientific disciplines, including crop science, forestry, geochemistry, limnology, soil science, and environmental chemistry. Organic substances are involved in the weathering and neogenesis of aluminum-bearing minerals, and they serve as agents for the transport of aluminum in leached terrestrial soils, notably those classified as Spodosols[1] or Podzols.[2,3] Aluminum can act as a bridging cation for the binding of humic substances to clay minerals, thereby affecting the physical properties of soils.

The interaction of aluminum with organic substances is of considerable importance in controlling soil solution levels of the highly toxic Al^{3+} ion in acid soils and natural waters. Research on acid deposition has implicated organic substances in the solubilization and transport of aluminum from terrestrial environments to natural waters, a subject covered in Chapters 9 and 10. A knowledge of the nature of organic ligands that form complexes with aluminum, and of the properties of the complexes thus formed, will lead to a better understanding of the behavior of aluminum in terrestrial and aquatic systems.

Emphasis will be given herein to the significance of aluminum-organic matter interactions in soils, the distribution of naturally occurring organic ligands that bind aluminum, and the nature and stability of aluminum-organic matter complexes. Although primary consideration will be given to terrestrial systems, many of the concepts will apply to aquatic systems.

II. SIGNIFICANCE OF ALUMINUM-ORGANIC MATTER INTERACTIONS IN THE PEDOSPHERE

Essentially every aspect of the chemistry of aluminum in soils, sediments, and aquatic systems is influenced by reactions involving organic substances. The properties of many modern soils, notably those of humid and semihumid environments, are directly related to the effect of organic substances on aluminum transformations during pedogenesis. In most soils, organic matter occurs in intimate association with mineral matter, and aluminum plays a key role in the association. The availability of native and fertilizer phosphorus in acid soils is regulated to some extent by phosphate reactions involving aluminum-organic matter complexes. Toxicities of Al^{3+} in acid soils can be reduced or eliminated by practices that promote complexation of aluminum with organic substances.

A. Pedogenic and Environmental Aspects

Plants and microorganisms have long been regarded as vital factors in weathering processes and soil genesis by their influence on the following:

1. The decay of organic matter by microorganisms leads to the formation of CO_2, which serves as a weathering agent because of its tendency to form carbonic acid with water ($H_2O + CO_2 \rightarrow H_2CO_3$).
2. Water-soluble organic substances produced in the biosphere lead to the complexation and mobilization of metal ions (including aluminum).
3. The uptake of elements (including aluminum) by plant roots results in their translocation from the deeper soil horizons to the surface layer, where they are solubilized and transported to lakes and streams as complexes with organic substances.

Several processes are involved in the mobilization and transport of aluminum in the pedosphere, including the weathering of parent rock materials, eluviation of aluminum to lower soil horizons during pedogenesis, uptake by plant roots and translocation into leaf tissue, incorporation into the humus layer of the soil, and transport to natural waters as soluble aluminum-organic matter complexes. Although aluminum-organic interactions will be emphasized, it should be noted that a variety of other processes also are involved in aluminum transformations, as discussed in other chapters of this book.

1. Weathering and Neogenesis of Minerals

There is little doubt that organic compounds produced in the biosphere, together with humic and fulvic acids, enhance the weathering of aluminum-bearing rocks and minerals, and that they serve as transporting agents for aluminum during pedogenesis. The initial stage is characterized by colonization of rock surfaces by algae, lichens, and fungi, all of which produce chelating agents that solubilize aluminum and other cations from minerals such as silicates, sulfides, and oxides. With time, organic chelates of plant origin, or synthesized by microorganisms, further contribute to the weathering process.

Mineral weathering by organic acids and other chelating organic biochemicals occurs through two distinctive modes of action, namely, lowering of pH value because of ionization of acidic functional groups (e.g., $COOH \rightleftharpoons COO^- + H^+$) and formation of chelate complexes. For any given pH value, weathering is normally enhanced in the presence of a complexing agent. Organic acids that form weak complexes with aluminum, such as acetic acid, attack minerals primarily through the protons they produce. These substances are generally less effective in mineral weathering than those organic compounds that form highly stable complexes with aluminum, such as citric and oxalic acids. By forming aluminum-organic complexes, organic acids and other complexing agents not only accelerate the decomposition of aluminum-containing minerals but facilitate the movement of aluminum to lower horizons in the soil profile.

An extensive literature has developed on the dissolution of rocks and minerals through the action of organic substances, and the reader is referred to several recent reviews for additional information.[4-7] The nature and distribution of biochemicals produced in the pedosphere are discussed in Section III.

In addition to their effects on mineral weathering, organic substances can retard or enhance the formation of aluminum minerals.[6] It is known that a wide variety of organic acids (e.g., citric, maleic, tannic, and 4-hydroxybenzoic, among others) can hinder the precipitation of solid-phase products of aluminum. The mechanism that has been postulated for this effect is that the ligand, through occupation of coordination sites of aluminum hydroxides, imposes a constraint on subsequent hydrolysis reactions, as illustrated with citrate in Structure 1.

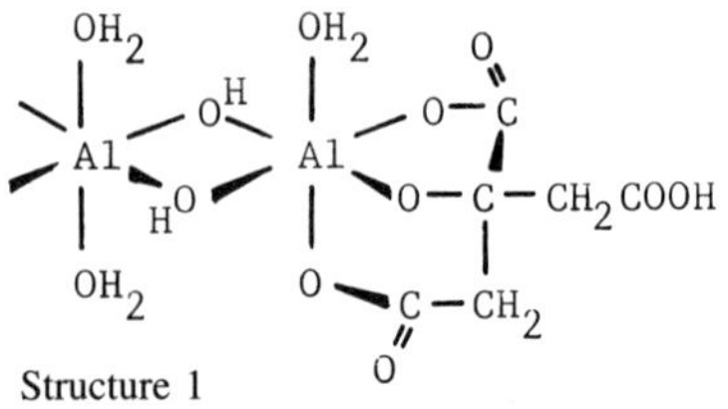

Structure 1

Organic chelating agents also have the ability to distort the arrangement of the unit sheets normally found in crystalline aluminum hydroxides, leading to the formation of short-range ordered aluminum precipitation products. The action of citric acid in perturbing aluminum-hydroxy interlayering in montmorillonite was postulated by Kwong and Huang[8] to result from the formation of linkages of the type depicted in Structure 1. The formation of crystalline aluminum hydroxides in soils and sediments can be hampered in environments where organic acids tend to accumulate; acidic environments are especially noted for this. Organic acids can also influence the types of aluminum hydroxide polymorphs thus formed.[6]

2. Eluviation and Transport of Aluminum in Soils

The mobilization and transport of aluminum and other polyvalent cations in leached soils are facilitated through the formation of complexes with organic substances. At the pH values found in many soils, hydrolysis reactions occur that lead to the formation of insoluble aluminum hydroxides; when complexed by organic ligands, aluminum can be maintained in solution and transported into lower horizons of the soil profile or to lakes and streams.

The formation of Spodosols is often cited as a prime example of the role of organic matter in the eluviation of aluminum during pedogenesis. These soils develop under climatic and biologic conditions that result in the translocation of considerable quantities of sesquioxides and organic matter into the subsoil. An organic-rich mineral layer of the soil (O and A horizons) consisting of decomposition products of the forest litter is underlain by a light-colored eluvial horizon (E), which has lost substantially more aluminum and iron than silicon. This horizon is followed by a dark-colored illuvial horizon (B) in which the major accumulation products are aluminum, iron, and organic matter.[1-3] Other soils with an E horizon also show evidence for transport of aluminum in association with organic matter.

It is well known that aluminum, iron, and organic matter accumulate in appreciable quantities in spodic horizons. What is not known is the mechanism(s) responsible for the translocation of sesquioxides. One popular theory is that aluminum and iron are complexed in the eluvial horizons (A and E) and translocated downward as soluble complexes with organic substances, with precipitation occurring in the lower mineral horizons because of metal saturation, polymerization of organic substances, or changes in ionic strength or pH values.[9] A more recent theory is that aluminum and iron are transported as hydroxy-aluminum silicate sols (imogolite and allophane); soluble organic colloids migrating downward then are precipitated on the previously deposited imogolite and allophane.[10] In all likelihood both processes occur, the relative importance of each being dependent on environmental conditions.

Soils forming in volcanic ash (Andepts) lead to the retention and accumulation of organic matter in the surface layer because of reactions with aluminum on oxide surfaces. In this case, little, if any, eluviation of aluminum or organic matter occurs.

3. Effect of Acidic Inputs (Acid Rain) on Mobilization of Organic Ligands and Transport of Aluminum

Acid deposition occurs over broad areas of northeastern U.S., eastern Canada, and northern Europe. The most noticeable effect of acid deposition has been a lowering of pH values in numerous lakes in eastern North America and in Scandinavia. Accompanying the decrease in pH values has been an increase in dissolved trivalent monomeric aluminum, which is toxic to aquatic organisms. As is the case with terrestrial soils, the concentration of the highly toxic Al^{3+} species can be reduced through complexation with organic substances (see Chapters 9 and 10).

An additional concern is the acidification of soils, with subsequent solubilization and transport of aluminum to lakes and streams in surface and ground waters, thereby adversely affecting aquatic life. Since organic constituents form highly stable complexes with aluminum, their role in the transport of aluminum and in aquatic acidification processes must be defined. One of the dominant forms of aluminum in the throughfall and leachates of the O, E, and B horizons of Spodosols (Typic Haplorthods) in the Adirondack mountains of New York is organically complexed aluminum.[11] A positive correlation has been observed between organic carbon levels and soluble aluminum plus iron in natural waters.[12] Existing information on external and internal sources of acidification of terrestrial and aquatic systems and the role of dissolved organic carbon and aluminum is reviewed in Chapter 9.

B. Effects of Aluminum on Plant Growth

Much public interest has recently been focused on aluminum toxicities believed to arise from the movement of aluminum from forest ecosystems into lakes and streams as a result of acid rain. Of equal importance is the toxicity of aluminum to plants grown on acidic soils. Organically complexed forms of aluminum in soil solutions and natural waters are much less toxic to plants and aquatic life than Al^{3+} or its hydrated monomers ($Al[OH]^{2+}$, $Al[OH]_2^+$) (see Chapter 9).

1. Organically Bound Forms of Aluminum in the Soil Solution

The speciation of aluminum in the soil solution depends primarily on pH values, the content of dissolved organic matter, and the types of competing inorganic ligands. Several recent studies indicate that organic ligands play a dynamic and important role in defining the speciation of aluminum in the aqueous phase of forest soils.[11,13]

Methods for determining organically bound aluminum in the soil solution include: (1) reaction of inorganic aluminum with a specific chelating agent, (2) removal of charged inorganic species by sorption on a cation-exchange resin, (3) reduction in F^- through complexation with Al^{3+} (as determined with a F^- ion-selective electrode), and (4) separation of the aluminum-organic complex through dialysis.[14] In the cation-exchange resin method, neutral or anionic inorganic forms of aluminum are not adsorbed but are calculated from known thermodynamic data, thereby yielding organically bound aluminum by difference. The quantification of aqueous aluminum is covered in Chapter 1; thermodynamic data for monomeric aluminum species are given in Chapter 2.

Acidification of the O horizons of forest soils may lead to changes in aluminum concentration and speciation through increased solubilization of aluminum and/or organic matter. Additional consequences are changes in the pH-dependent characteristics of the solubilized humic matter, namely, a decrease in charge density as a result of decreased ionization of COOH groups and configurational changes in humic macromolecules.[15]

Since dissolved organic matter in soils and natural waters varies both in content and composition, the percentage of aluminum in organically bound forms would be expected to be highly variable. Research on dissolved aluminum in forest soils indicates that aluminum-organic forms dominate in the forest floor, but decline in importance with increasing soil depth.[16]

2. Influence of Organic Matter in Ameliorating Aluminum Toxicities

Aluminum is a constituent of all plants, but the amount varies greatly depending on soil and plant factors. The aluminum content of most plants is of the order 200 μg/g; aluminum-accumulating species may contain more than 0.1% aluminum on a dry-weight basis.[17] The physiological function of aluminum in plants, if any, is unknown.

Aluminum toxicity is a major problem in acid soils (pH < 5.5). Toxicities because of the Al^{3+} species have been noted in several regions of eastern U.S., Canada, and the tropics, where acid soils are found.[18-20] However, acid soils that are rich in native organic matter, or amended with large quantities of organic residues, give low Al^{3+} concentrations in the soil solution and permit good growth of crops under conditions where toxicities would otherwise occur. Results of studies conducted at various soil pH values, aluminum concentrations, and amounts of organic amendments have indicated that lower exchangeable aluminum contents, decreased aluminum toxicity, and better plant growth are achieved with an increase in the amount of organic matter added.[19]

Liming a soil reduces aluminum toxicity by decreasing aluminum solubility and replacing aluminum on exchange sites with Ca. Ahmad and Tan[21] discuss several disadvantages of liming and conclude that organic matter is as effective in reducing aluminum toxicities on a short-term basis. Addition of organic matter and lime together gave the best results.

C. Role in Formation of Stable Aggregates

The physical properties of soil (i.e., texture, structure, and bulk density) influence plant growth through their effects on aeration, water penetration and retention, and mechanical impedance to roots. Organic matter is considered to be indispensable for promoting good structure in a wide range of soils, particularly those representative of Mollisols, Alfisols, Ultisols, and Inceptisols. A factor of some importance is the cohesion of clay and/or clay-humus complexes into larger particles through bridging with polyvalent cations (e.g., Al^{3+}, Fe^{3+}, Ca^{2+}).

The stability of soil aggregates depends primarily on the interactions between micro- and macro-units, which in large part are held together by cementing agents, such as gums or gels produced by microorganisms.[22,23] However, linkages between organic matter, polyvalent cations, and clay minerals are also involved. High relative molecular mass mixed complexes of the type shown below (see Structure 2) may act as cementing agents between clay particles, thereby promoting soil aggregation.

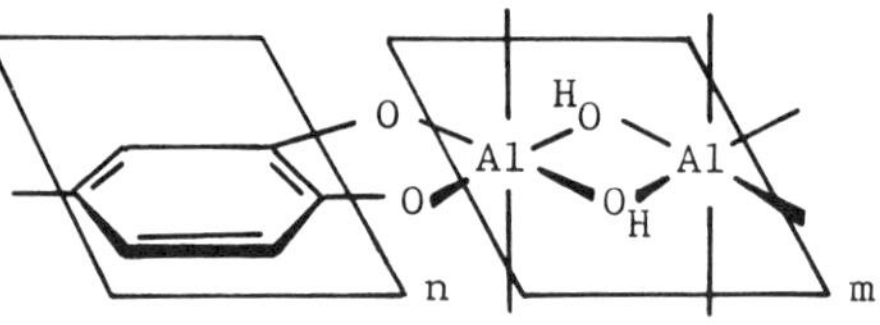

Structure 2

According to Edwards and Bremner,[24] in soils well supplied with exchangeable cations, a microaggregate can be represented as:

$$([\text{Clay-P-OM}]_x)_y$$

where P is a polyvalent cation (Al^{3+}, Fe^{3+}, Ca^{2+}, etc.), OM is the humified organic matter, and Clay–P–OM is the primary particle; x and y are finite whole numbers dictated by the size of the primary particles. The size and stability of the microaggregates are partially determined by the absolute and relative amounts of clay and humified organic matter, and by the nature of the polyvalent cation. Binding through aluminum and iron would be of greatest importance in acid soils, such as Alfisols and Ultisols.

A modified mechanism of aggregation, believed to be more consistent with conditions existing in the natural soil, was proposed by Mortland,[25] who concluded that polyvalent cations and organic matter are linked via H_2O bridges (Clay–P–H_2O–OM). According to

Reid et al.,[26] aggregate stability may be affected adversely by rhizosphere organic compounds that form highly stable complexes with the bridging polyvalent cation.

D. Sorption-Desorption of Phosphate

The plant availability of native and fertilizer phosphorus in soil is affected by a variety of independent, but not necessarily exclusive, reactions, including: (1) chelation of aluminum, iron, and calcium by naturally occurring organic ligands, (2) competition between humates and phosphate ions for adsorbing surfaces, (3) formation of protective coatings over colloidal sesquioxides, and (4) formation of phospho-humic complexes through bridging with aluminum.[22]

The action of organic acids and related compounds in solubilizing mineral phosphates has been attributed, in part, to the formation of soluble complexes with Al^{3+} and Fe^{3+} in acid soils and with Ca^{2+} in calcareous soils. Similar reactions are undoubtedly involved in preventing the adsorption or precipitation of phosphate added as fertilizer, or formed *in situ* by the weathering of minerals or decay of organic matter. Research reviewed elsewhere[4-6,27-29] has shown that the most effective compounds in releasing phosphate from insoluble aluminum and iron phosphates, or in reducing precipitation, are those that form highly stable chelate complexes with metal ions.

The competition of humates for adsorbing surfaces and the formation of protective coatings are of special importance in allophanic soils, in that the precipitation of phosphate is reduced. These soils are unique in that they adsorb appreciable amounts of phosphate from reactions at aluminum-hydroxide surfaces. Through the interactions of humates with surface aluminum, the adsorption of phosphate is reduced, thereby increasing its availability to plants.

Little work has been done on the role of phospho-humic complexes in affecting phosphate bioavailability.[30-32] The suggestion has been made that aluminum serves as a link between the phosphate anion and the negatively charged humic molecule, such as in acid peats (see Structure 3) and in allophanic soils (see Structure 4). Complexes also can be formed with soluble aluminum fulvates, and it appears that some of the phosphate in the soil solution and leachate waters of forest soils occurs as soluble FA-Al-phosphate complexes.

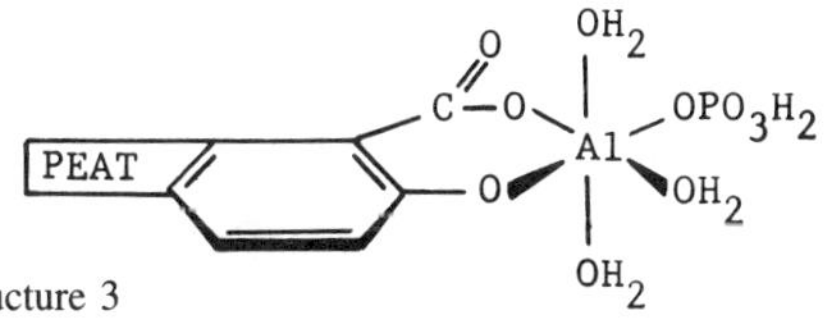

Structure 3

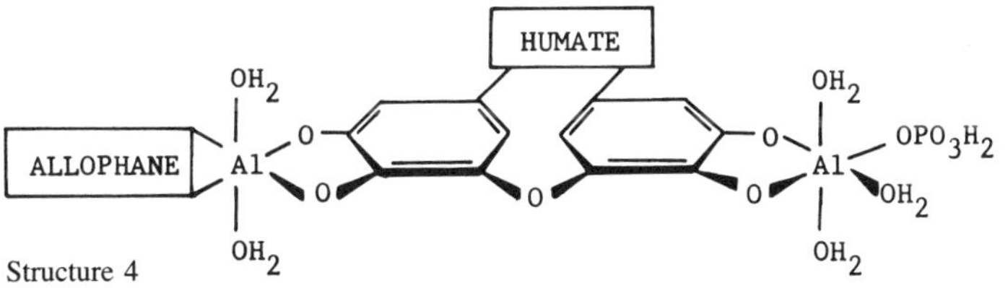

Structure 4

The sorption of phosphate by acid organic soils has been attributed to the formation of organic aluminum-phosphate complexes.[31]. A major part of the total phosphorus in some allophanic soils occurs as humus-phosphate complexes.[32]

III. NATURE AND DISTRIBUTION OF WELL-DEFINED ORGANIC LIGANDS

The soil organic compounds that form stable complexes with aluminum and other polyvalent cations are of two kinds: (1) well-defined biochemical compounds synthesized by living organisms, such as simple aliphatic acids, phenols and phenolic acids, hydroxamate siderophores, sugar acids, and complex polymeric phenols, and (2) a series of acidic, yellow- to black-colored substances formed by secondary synthesis reactions and referred to as humic and fulvic acids.[22,29,33-35] The relative importance of the two types of compound is difficult to evaluate and will vary with soil and environmental conditions. Biochemical compounds are ubiquitous in soils and natural waters, and these constituents would be expected to play a dominant role where microbial activity is intense. On the other hand, humic substances may be of greater importance in: (1) modifying the reactivity of aluminum on oxide surfaces, (2) reducing toxicities of aluminum in acid soils, and (3) binding aluminum in most natural waters. Complexes of aluminum with humic and fulvic acids will be discussed in Section IV.

It should be noted that many of the chelating effects attributed to biochemical compounds have been extrapolated from laboratory studies. Some investigators have questioned the importance of these effects under conditions existing in nature because biological compounds are subject to rapid destruction through microbial metabolism.[36] A wide array of biochemicals are produced periodically in soil through the activities of microorganisms; others are found in excretions from plant roots and in leachates of plant residues, including leaf litter of the forest floor.[6,7,27,29] Organic chelating substances are also found in the stemflow of forest trees, as well as in canopy drippings. Biochemical chelating agents do normally have only a transitory existence in the pedosphere, and the amounts found at any one time represent a balance between synthesis and destruction by microorganisms. Measurable quantities usually can be found in zones favorable for the proliferation of microorganisms, such as in the rhizosphere and in the organic layers of Alfisols and Spodosols. An (incomplete) list of biochemical chelating compounds, along with a description of the environments where they occur, is presented in Table 1.

The amounts of potential chelate formers in the aqueous phase at any one time is normally low and variable in most agricultural soils. Approximate concentrations of individual biochemical species in the soil solution are[29,34]

Simple organic acids	1×10^{-3} to 4×10^{-3} *M*
Amino acids	8×10^{-5} to 6×10^{-4} *M*
Phenolic acids	5×10^{-5} to 3×10^{-4} *M*
Hydroxamate siderophores	1×10^{-8} to 1×10^{-7} *M*

Chelate structures of Al^{3+} with known biochemical compounds can be illustrated as follows:

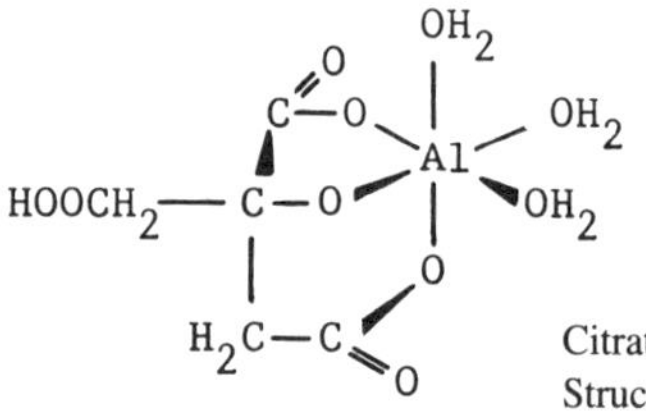

Citrate
Structure 5

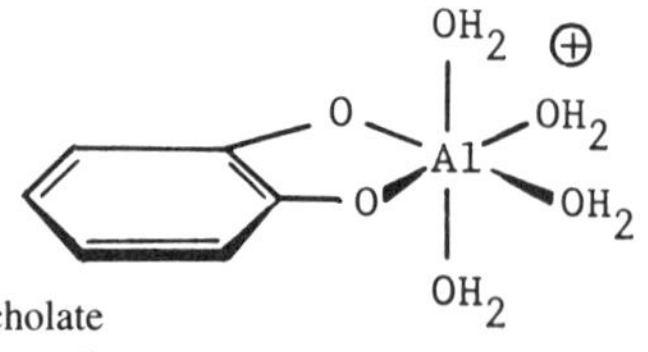

Catecholate
Structure 6

Table 1
OCCURRENCE OF IMPORTANT BIOCHEMICAL COMPOUNDS THAT COMPLEX WITH ALUMINUM AND OTHER CATIONS

Compound	Occurrence
Citric, tartaric, lactic, malic acids	Produced by bacteria in the rhizosphere and during decay of plant remains, identified in root exudates, aqueous extracts of forest litter and canopy drippings
Oxalic acid	Produced by fungi in forest soils, including mycorrhizal fungi, particularly abundant in acid soils
2-Ketogluconic acid	Synthesized by bacteria living on rock surfaces and in the rhizosphere, particularly abundant in habitats rich in decaying organic matter
Hydroxamate siderophores	Produced in the rhizosphere and by ectomycorrhizal fungi, greater amounts may be produced when organisms are under Fe stress
Phenolic acids	Formed through decay of plant residues (lignin), abundant in canopy drippings and leachates of forest litter, involved in the mobilization and transport of aluminum in acid soils
Polymeric phenols	Present in high amounts in leachates of forest litter, produced by lichens growing on rock surfaces

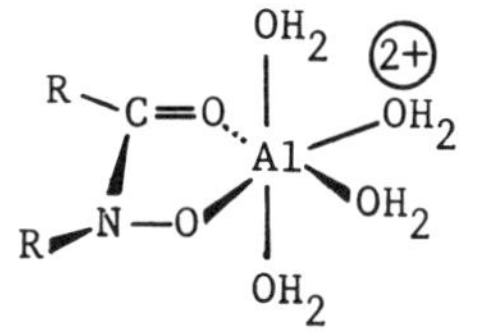

Hydroxamate
Structure 7

The relative importance of any given biochemical compound or ligand type in aluminum binding will vary with time and environmental conditions, including the nature of the organic ligand being synthesized and the kind and amount of competing cations. In most ecosystems, binding of aluminum occurs by a relatively large number of ligands present in small amounts rather than by a few dominant species present at high concentrations. Aluminum, being trivalent as a cation, would be expected to compete favorably with divalent cations (Ca^{2+}, Mg^{2+}, Zn^{2+}, etc.) for available organic ligands. On the other hand, many natural ligands form more stable complexes with Fe^{3+} than with Al^{3+}, as shown in Table 2.

Aluminum complexation occurs predominantly with organic groups containing oxygen; those containing nitrogen generally form weak interactions. This fact indicates that the natural compounds of interest in aluminum binding are those that contain COOH, phenolic-, enolic-, and aliphatic-OH groups, and possibly ketonic C=O and ester functional groups; nitrogenous substances (amino acids and porphyrins) are of less importance even though they are present in most soils.

A. Low Relative Molecular Mass Organic Acids

Low relative molecular mass organic acids are of particular interest because they are ubiquitous in nature and form stable chelate complexes with Al^{3+} and other polyvalent cations, as noted above. Studies on the isolation and identification of organic acids in relation to pedogenesis and general soil conditions are numerous. The review of Stevenson[27] reveals that the following low relative molecular mass aliphatic acids have been reported in soil: formic, HCOOH; acetic, CH_3COOH; propionic, CH_3CH_2COOH; butyric, $CH_3CH_2CH_2COOH$; α-crotonic, $CH_3CH{=}CHOOH$; lactic, $CH_3CHOHCOOH$; oxalic, $(COOH)_2$; succinic, $(CH_2COOH)_2$; fumaric, $(CHCOOH)_2$; tartaric, $(CHOHCOOH)_2$; and citric, $COH(CH_2$

Table 2
STABILITY CONSTANTS OF AL^{3+}, FE^{3+}, AND CU^{2+} COMPLEXES OF SOME REPRESENTATIVE SOIL ORGANIC LIGANDS[a]

	Ligands per metal ion	Al^{3+}	Fe^{3+}	Cu^{2+}
Formic Acid	1	1.36 (1.0)[b]	3.10 (1.0)	1.38 (1.0)
Acetic Acid	1	1.51 (1.0)	3.38 (0.1)	1.33 (0.1)
	2	3.76 (0.3)[c]	6.50 (0.1)	3.09 (0.1)
Propionic acid	1	1.69 (1.0)	3.40 (0.1)	1.66 (1.0)
Oxalic Acid	1	6.10 (1.0)	7.53 (0.1)	4.84 (0.1)
	2	11.09 (1.0)	13.64 (0.1)	9.21 (0.1)
	3	15.12 (1.0)	18.49 (0.1)	
Citric Acid	1	8.32 (0.25)[d]	11.20 (0.1)	5.90 (0.1)
	2			9.04 (0.1)
Catechol	1	16.30 (0.1)	20.00 (0.1)	13.90 (0.1)
	2	29.30 (0.1)	34.70 (0.1)	24.90 (0.1)
	3	37.60 (0.1)	43.80 (0.1)	

[a] Data taken from Martell and Smith[37] unless otherwise noted; see Chapter 2 for a complete list.
[b] Numbers in parentheses represent ionic strength (in mol/l) at which the stability constant was determined.
[c] Young and Bache.[38]
[d] Sillén and Martell.[39]

$COOH)_2COOH$, among others. Hydroxy acids, such as citric, form stronger complexes (see Structure 5) than those acids containing a single COOH group (see Table 2 for representative stability constants and Chapter 2 for a comprehensive discussion).

Although the concentration of organic acids in the soil solution is normally low (1×10^{-3} to 4×10^{-3} *M*), substantially higher amounts can be found in the rhizosphere of crop plants. Aliphatic organic acids often accumulate in water-logged soils, such as rice paddy fields, where volatile acids tend to persist, such as acetic, butyric, formic, fumaric, propionic, valeric ($CH_3CH_2CH_2COOH$), succinic, and lactic acids.[27] High relative molecular mass aliphatic acids are also known to accumulate in poorly drained, organic-rich environments. Various aspects of the dynamics of organic acids in soil have been discussed by Wang et al.[40]

The leaves of many plants contain high concentrations of nonvolatile organic acids, mostly citric and malic ($COOHCH_2CH[OH]COOH$) and to a lesser extent succinic, fumaric, and oxalic.[27] Aqueous extract of forest litter have been found to contain a variety of low relative molecular mass organic acids, many of which are capable of complexing aluminum and other metal ions.[6,7,27,29,36] Leachates from the forest canopy may also contribute to the aliphatic acid content of the soil solution, as well as natural waters.

The rhizosphere appears to be a favorable habitat for organic acid-producing microorganisms.[41,42] Aliphatic organic acids synthesized by these organisms include: formic, acetic, propionic, butyric, oxalic, fumaric, glycolic ($CH_2[OH]COOH$), succinic, tartaric, citric and 2-ketogluconic acids.

Considerable emphasis recently has been given to the importance of oxalic acid as a chelator of aluminum and iron in acid forest soils.[43-45] Many fungi are prolific producers of oxalic acid, including the vesicular-arbuscular mycorrhizal fungi, where calcium oxalate crystals can form at the soil-hyphae interface.[45] Bacteria capable of growing under strongly acid conditions, such as species of *Pseudomonas*, are also capable of producing copious amounts of oxalic acid (see review of McKeague et al.[7]).

B. Hydroxamate Siderophores

Hydroxamate siderophores produced by soil microorganisms are believed to play an important role in the iron nutrition of plants.[46-48] They consist of a group of microbially produced iron-transport moieties containing the anionic reactive group (R–CO–NO^-). Complexes are also formed with Al^{3+} (see Structure 7), although of lower stabilities than for Fe^{3+}.

Biologically significant levels of hydroxamate siderophores have been observed in soil (10^{-8} to 10^{-7} *M*). The amounts contained in the rhizosphere of plants appear to be 10 to 50% higher than in bulk soil.[46] Hydroxamate siderophores have also been shown to be produced by soil fungi, including the ectomycorrhizal fungi, which live in intimate association with plant roots.[48]

C. Sugar Acids

Sugar acids may also be important natural chelators of aluminum in soils.[7] Gluconic (Structure 8), glucuronic (Structure 9), and galacturonic (Structure 10) acids are all common metabolites of microorganisms. Habitats rich in organic matter have been found to contain large numbers of microorganisms that synthesize 2-ketogluconic acid (see Structure 11).[49] This compound was shown to make up more than 25% of the organic acids in the rhizosphere.[50] A high proportion of the bacteria in soil, as well as those living on rock surfaces, produce 2-ketoglutonic acid.[51]

Structure 8	Structure 9	Structure 10	Structure 11
COOH	CHO	CHO	COOH
HCOH	HCOH	HCOH	C=O
HOCH	HOCH	HOCH	HOCH
HCOH	HCOH	HOCH	HCOH
HCOH	HCOH	HCOH	HCOH
CH_2OH	COOH	COOH	CH_2OH

D. Phenols and Phenolic Acids

Phenolic compounds are widely distributed in soils. They are synthesized by a variety of microorganisms and have been found in root exudates, in forest canopy leachates, and in decomposing plant and animal remains. The concentration of phenolic acids in the soil solution has been estimated at 5×10^{-5} to 3×10^{-4} *M*.[29,34]

Phenolic compounds released during the decay of lignin include: 4-hydroxybenzoic, protocatechuic (3,4-dihydroxybenzoic), vanillic (4-hydroxy-3-methoxybenzoic), syringic (3,5-dimethoxy-4-hydroxybenzoic), 4-hydroxycinnamic, gallic (3,4,5-trihydroxybenzoic), and ferulic (4-hydroxy-3-methoxycinnamic) acids and their aldehydes. Whitehead et al.[52] found that 4-hydroxybenzoic, vanillic, 4-hydroxycinnamic, and ferulic acids were widely distributed in agricultural soils. Aldehyde and methoxy derivatives (i.e., 4-hydroxybenzaldehyde, vanillin [4-hydroxy-3-methoxybenzaldehyde], and syringaldehyde [3,5-dimethoxy-4-hydroxybenzaldehyde] would be expected to be less effective in binding aluminum than carboxylic and hydroxy derivatives.[53] Of the phenolic acids, those containing adjacent OH groups would be the most effective in binding Al^{3+}, such as protocatechuic (Structure 12), gallic (Structure 13), and caffeic (Structure 14) acids.[53,54]

Structure 12 Structure 13 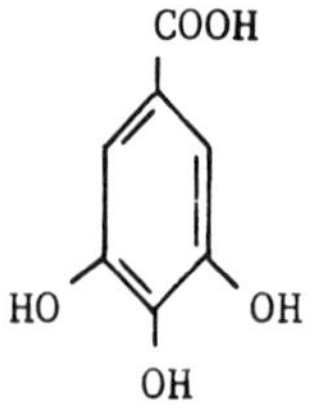 Structure 14

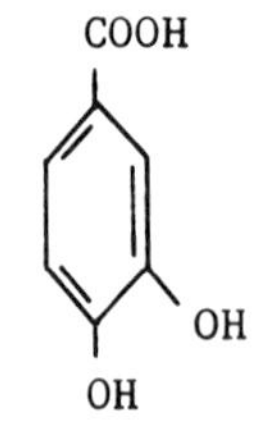

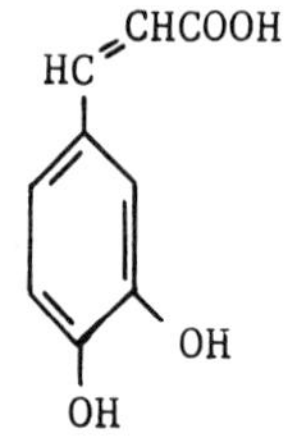

A wide variety of phenols and phenolic acids have been observed in the products synthesized by the microscopic fungi *Stachybotrys atra, S. chartarum,* and *Epicoccum nigrum* when grown on nonlignin carbon sources.[55] Together with products of lignin origin, they serve as precursors for the formation of humic substances.

Root exudates contain phenolic acids in addition to the aliphatic acids previously discussed. Decomposition products of plant roots also contain various phenolic acids, including: phenylacetic ($C_6H_5CH_2COOH$), cinnamic ($C_6H_5CH{=}CHCOOH$), 4-hydroxycinnamic, 4-hydroxyphenylpropionic (4-OH[C_6H_4]CH_2CH_2COOH), and 3,4-dihydroxyphenylpropionic (3,4-OH[C_6H_3]CH_2CH_2COOH).[27]

Phenols and phenolic acids are believed to be of considerable importance in the complexation and translocation of aluminum and iron in forest soils.[53,56] Among the compounds detected in the humus layers, in canopy leachates, and in leaf-litter leachates are vanillic, 4-hydroxybenzoic, and 4-hydroxycinnamic acids (as well as citric, malic, oxalic, and succinic acids).

E. Polymeric Phenols

As used herein, the term "polymeric phenols" refers to those substances containing more than one aromatic ring and the phenolic OH group. They include the flavanoids of plant origin and comprise one of the largest and most widespread group of secondary plant products. Structures for quercetin (15), catechin (16) and gallocatechin (17) are shown below.

Quercitin
Structure 15

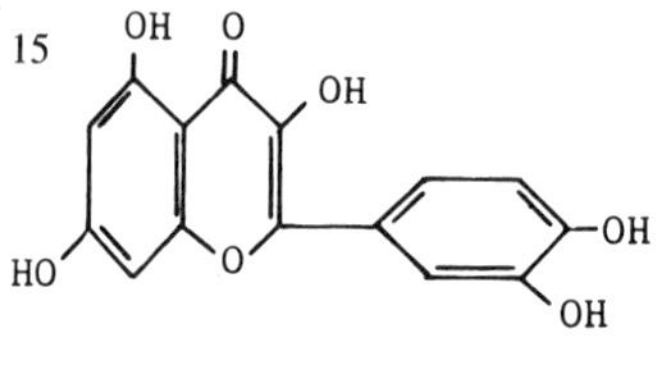

Catechin
Structure 16

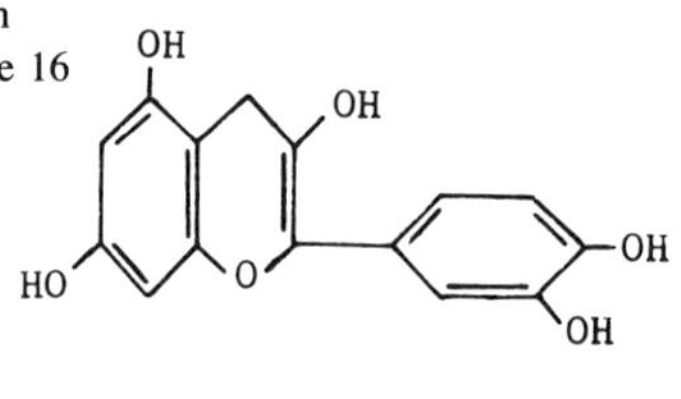

Gallocatechin
Structure 17

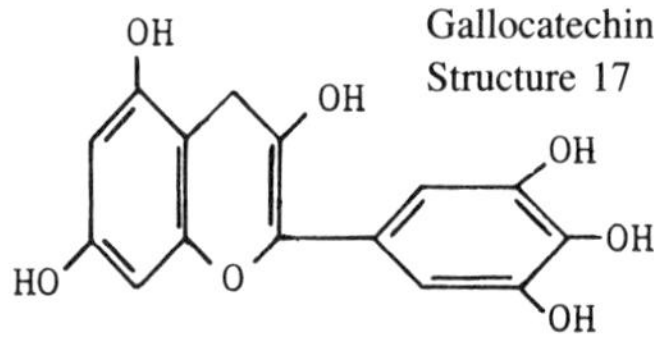

Flavanoids have been identified in aqueous extracts of the leaves and needles of several plants,[57] including the catechin isomers (D- and epicatechin), quercetin (see Structure 15), and the two isomers of gallocatechin (see Structure 17.)[58] Coulson et al.[59] concluded that catechin isomers were of considerable importance in soil formation processes due to their ability to form highly stable complexes with aluminum and iron.

The ability of lichens to dissolve mineral substances during the weathering of rocks and minerals is well known. These organisms synthesize a variety of complex phenolic compounds that form highly stable complexes with metal ions. The basic building units of the lichen acids are based on orsellic (Structure 18) and b-orsellic (Structure 19) acids, with orsellic acid commonly found as dimers (Structures 20 and 21).

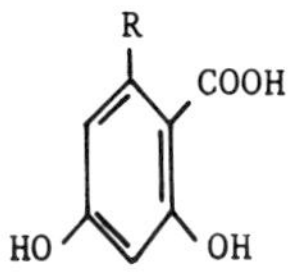

Structure 18

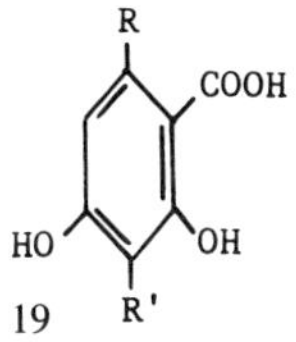

Structure 19

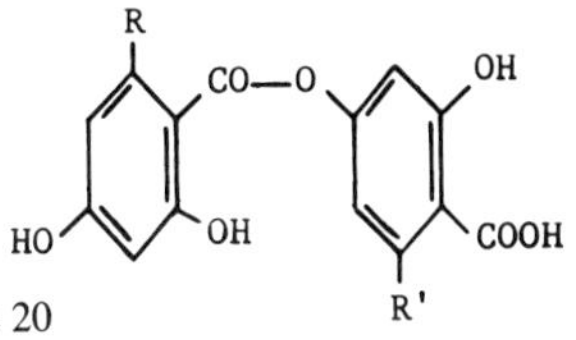

Structure 20

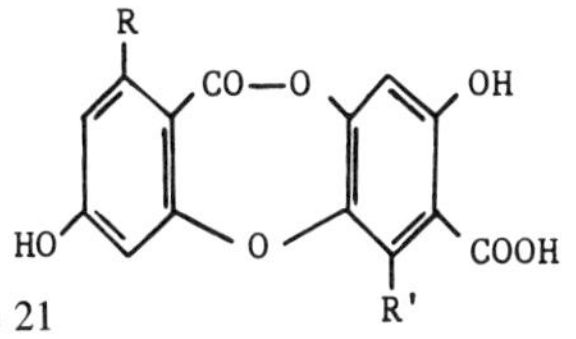

Structure 21

Geographically, lichens are widely distributed in nature. They are often the initial colonizers of virgin landscapes. Iskandar and Syers[60] demonstrated the complexing ability of six lichen acids — salazinic, strictic, evernic, lecanoric, roccellic, and atranorin; silicate minerals, biotite, granite, and basalt were all degraded, with release of aluminum and other cations.

The biological stability of low relative molecular mass phenolic acids is enhanced through their polymerization and condensation into more complex products, thereby extending the time over which they can interact with aluminum. Enzymatic and autoxidation polymerization reactions are common occurrences in soil and water systems.[35] It can be visualized that low relative molecular mass phenolic compounds are first polymerized to complex polyphenolic constituents and then further to humic and fulvic acids.

The tannins are a rather ill-defined group of substances with relative molecular masses from 500 to 3000 and that contain at least 1 or 2 phenolic OH groups per each 100 relative molecular mass. They are of two main types: condensed and hydrolyzable forms. Condensed tannins consist of derivatives of a flavanoid nature, for which catechin (flavan-3-ols) and flavan-3,4-diols are primary constituents. Hydrolyzable tannins contain gallic and hexahydroxydiphenic acids (digallic acid), which are bound to a sugar moiety through a glycosidic linkage. Tannins contain numerous phenolic OH groups and thus are potential chelators of aluminum and metal ions in soils.[6]

F. Miscellaneous Compounds

Proteins and carbohydrates are also capable of forming complexes with aluminum, but their importance in soil is unknown. As much as 30% of the soil organic matter occurs as saccharides, but only a small portion has been accounted for as polysaccharides. Mucilaginous coatings of microbial origin adhering to mineral particles have been found to contain aluminum (see McKeague et al.[7]). Polysaccharides extracted from soil usually contain complexed aluminum, iron, and silicon. Evidence for complexing of aluminum by soil polysaccharides has been obtained by Saini.[61]

The bonding of polysaccharides to clay minerals through an aluminum linkage has been suggested as a mechanism for the formation of stable soil aggregates.[23] Also, reduction in the biodegration of polysaccharides because of aluminum binding may be an important means in maintaining organic matter levels in soil.[62] Montmorillonite saturated with aluminum has been shown to have a strong affinity for polygalacturonic acid.[63]

Amino acids, peptides, and proteins are capable of forming complexes with metal ions, but they are believed to play a role subservient to other naturally occurring organic ligands.[4-7] These substances have a strong affinity for binding to silicate minerals. Amino acids have been shown to perturb the hydrolytic reactions of aluminum.[6]

IV. COMPLEXATION OF ALUMINUM BY HUMIC AND FULVIC ACIDS

A. Chemical Properties of Humic Substances

Humic and fulvic acids are complex organic substances that have defied complete characterization despite extensive study. They are best described as a series of acidic, yellow- to black-colored polyelectrolytes with highly variable relative molecular masses.[22,23] As usually defined, humic acid is the material extracted from soil with an alkaline solution (usually 0.1 to 0.5 *M* NaOH) and precipitated upon acidification; fulvic acid is the fraction that remains in solution. The organic matter that is not solubilized by alkali is referred to as the ''humin'' fraction and usually makes up about 20% of soil organic matter. The nature of this material is unknown, but is believed to consist, at least in part, of humic and fulvic acids so intimately bound to clay minerals (such as though bonds with aluminum or iron) that they cannot be solubilized by the procedure used.

The ''fulvic acid fraction'' normally contains appreciable amounts of low relative molecular mass biochemical compounds and polysaccharides. Separation of ''true'' fulvic acid can be accomplished by sorption-desorption on a nonionic macroreticular resin (e.g., XAD-8). The term ''fulvic acid'' should be reserved as a generic name for the yellow- to brown-colored constituents in the acid filtrate.[22]

Fulvic acids are lower in relative molecular mass than humic acids, and, for this reason, they are the predominant form of humic matter in the soil solution. According to Malcolm,[64] every natural water sample thus far tested has been shown to contain humic substances, of which approximately 90% is fulvic acid.

The amounts of humic matter vary considerably from soil to soil, and are closely related to total organic matter content, which ranges from 1.5% or less in coarse-textured soils (sands) to 6.5% in prairie grassland soils (e.g., Mollisols). Poorly drained soils and wetlands often have organic matter contents approaching 20%; Histosols (organic soils) have very high organic matter contents that generally exceed 30%. The humus of Histosols and grassland soils (i.e., Mollisols) is characterized by a high content of humic acid; that of forest soils (Alfisols, Spodosols, and Ultisols) contains high amounts of fulvic acid. The ratio of humic acid to fulvic acid usually decreases with depth in the soil profile.

Isolated humic and fulvic acids from mineral soils invariably contain appreciable amounts of metal ions, including aluminum.[65-67] Attempts to obtain ash-free preparations by further extraction and fractionation have generally been unsuccessful. Griffith and Schnitzer[65] found that aluminum constituted the highest percentage of bound metals in partially purified metal-humic complexes from tropical soils. Alberts and Dickson[68] found that aluminum was associated more with the fulvic acid fraction of a pond sediment than with the humic acid fraction.

B. Nature of Reactive Sites

The abilities of humic substances to form stable complexes with aluminum and other polyvalent cations can be accounted for by their high content of oxygen-containing functional groups, which include COOH, phenolic-, enolic-, and alcoholic-OH, and C=O. The distribution of functional groups in humic and fulvic acids, as recorded in the recent literature, is summarized in Table 3. Total acidities of fulvic acids (640 to 1420 cmol$[-]$kg^{-1}) are substantially higher than for humic acids (560 to 890 cmol$[-]$kg^{-1}). Both COOH and acidic OH groups (generally presumed to be phenolic OH) contribute to the acidic nature of these substances, with COOH being the most important, especially in fulvic acids.

Fulvic acids are highly variable in composition. Those of lakes and streams appear to be more ''aliphatic'' in nature than those of terrestrial soils. Irrespective of source, humic acids have somewhat similar chemical characteristics. The basic structure of humic acid is believed to be an aromatic ring of the di- or trihydroxy-phenyl type bridged together by –O–, $-CH_2-$,

Table 3
OXYGEN-CONTAINING FUNCTION GROUPS IN HUMIC AND FULVIC ACIDS[22]

	Total acidity	COOH	Acidic OH[a]	Weakly acidic plus alcoholic OH	C=O
Humic acids	560—890	150—570	210—570	20—490	10—560
Fulvic acids	640—1420	520—1120	30—570	260—950	120—420

[a] Usually reported as "phenolic OH".

–NH–, –N=, –S–, and other groups and containing OH groups and the quinone linkage. Typical structural elements of humic acids are shown in Structures 22 to 25.

Structure 22

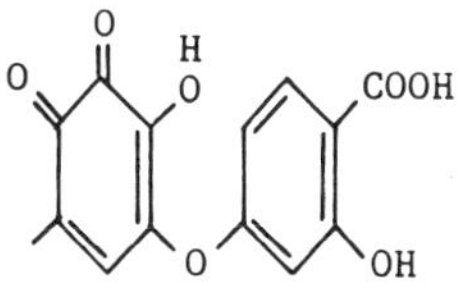

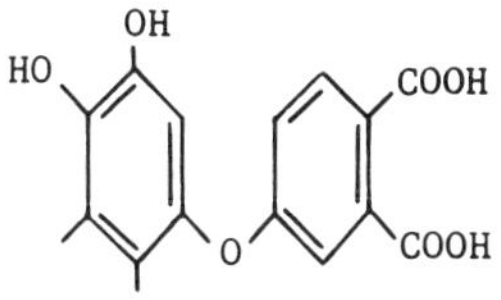

Structure 23

Structure 24

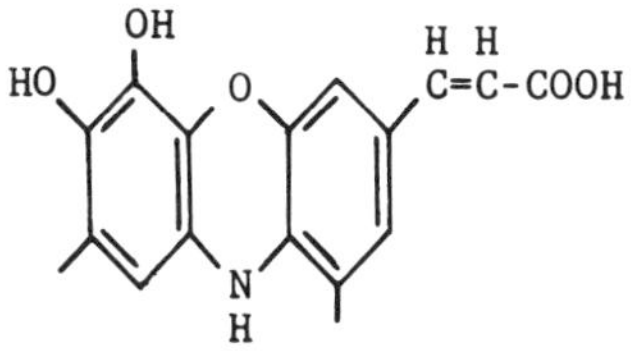

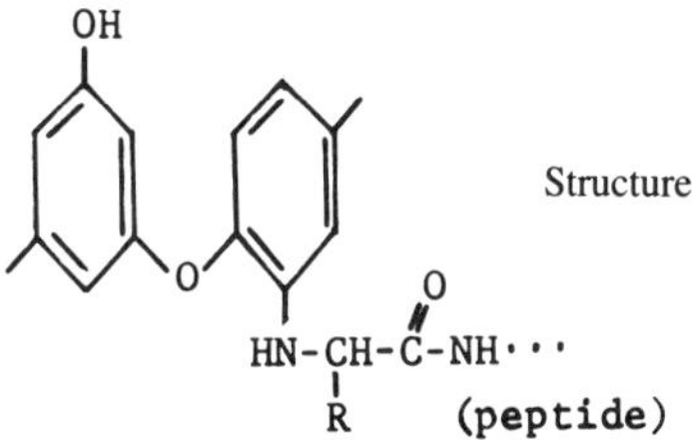

Structure 25

HN-CH-C-NH··· (peptide)

These structures show that COOH groups (as well as OH) are neither identical with respect to positions occupied by other groups nor uniformly spaced on the molecule. Most COOH groups are believed to be attached directly to the aromatic ring. In the natural state, humic and fulvic acids contain attached proteinaceous (see Structure 25) and carbohydrate residues, which are also capable of forming complexes with aluminum.

Binding sites that have been reported from time to time in humic and fulvic acids are shown in Structures 26 to 33.

Structure 26

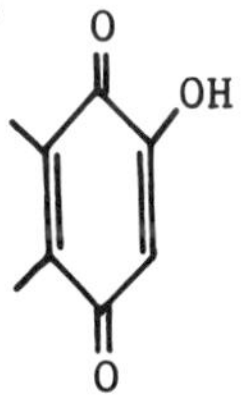

Structure 27

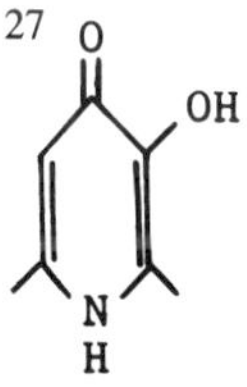

Structure 28

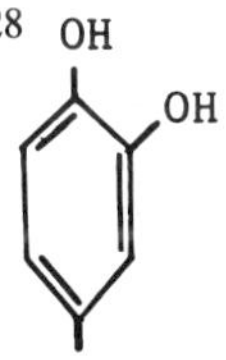

Structure 29

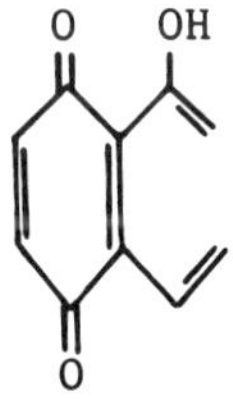

Structure 30

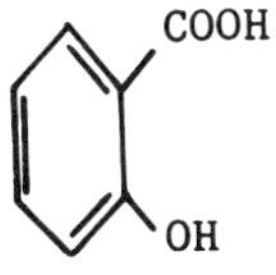

Structure 31

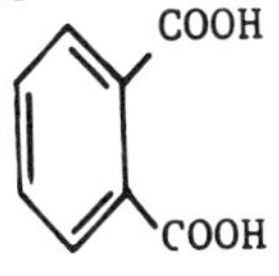

Structure 32

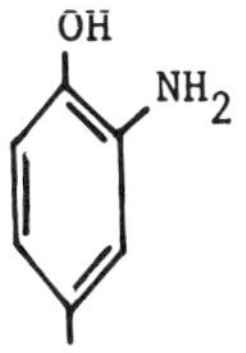

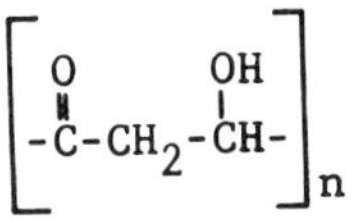

Structure 33

According to Schnitzer and Khan,[33] the main reaction for the binding of metal ions by humic substances is at a COOH-phenolic OH site (Structure 30); a reaction of less importance is with adjacent COOH groups (Structure 31). Phenolic compounds containing ortho-phenolic OH groups (Structure 29) have been credited for the complexation, mobilization, and translocation of aluminum in processes leading to the formation of spodic horizons.[53]

The binding of aluminum by humic substances can occur through: water bridges (Structure 34), electrostatic (coulombic) attraction (Structure 35), formation of a coordinate linkage with a single donor group (Structure 36), and formation of a chelate (ring) complex (Structure 37).

Structure 34

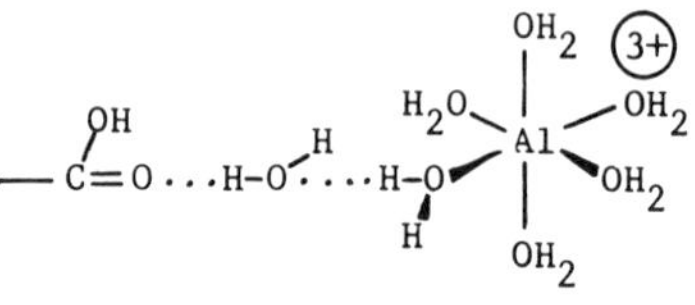

Structure 35

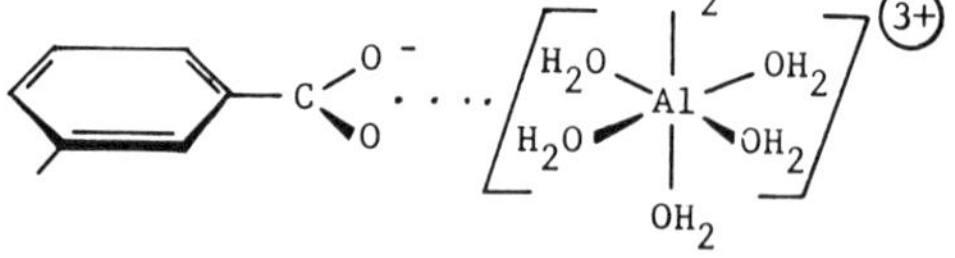

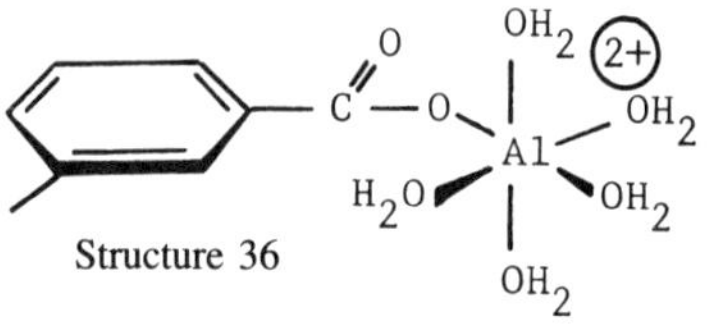

Structure 36

Structure 37

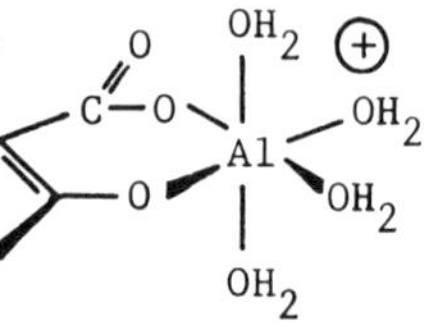

Sites forming the strongest complexes (see Structures 36 and 37) are expected to react first and to represent the predominant form of complexed aluminum in environments where humic substances are present in excess; binding at the weaker sites (Structures 34 and 35) would become increasingly important as binding sites become saturated with aluminum (and Fe^{3+}, which forms complexes similar to that of Al^{3+}). In addition to the above, aluminum can serve as a linkage between two or more humic molecules, as illustrated below:

Structure 38

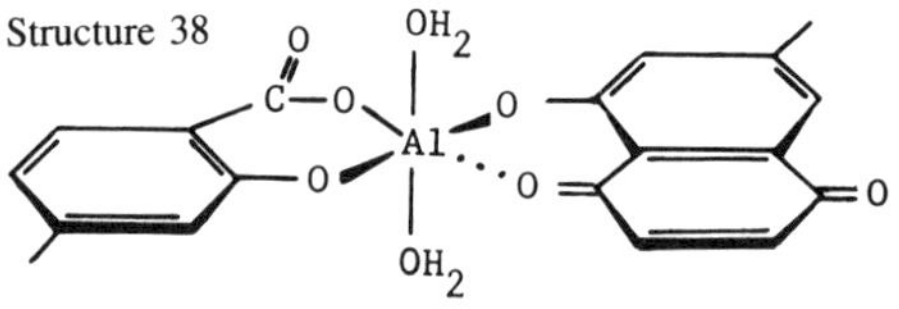

C. Solubility Characteristics of Aluminum-Humate Complexes

Humic substances form both soluble and insoluble complexes with aluminum, depending on pH value, presence of salt or electrolyte (i.e., ionic strength effect), and degree of saturation of binding sites. Other factors affecting coagulation include the concentration and source of the humic substance. The higher relative molecular mass humic acids are more susceptible to precipitation than fulvic acids.

In mineral soils, most of the organic matter is bound to mineral surfaces through linkages with aluminum and other polyvalent cations. The binding of humic and fulvic acids at oxide surfaces is also believed to involve linkages with aluminum (see Structure 3).

Mechanisms that influence the solubility of humic substances in the presence of aluminum can be summarized as follows:

1. Precipitation because of protonation and subsequent reduction of charges on the humic polymer (i.e., molecule becomes more hydrophobic)
2. Formation of chain-like structures through aluminum bridges (see Structure 38)
3. Attachment to clay particles and oxide surfaces, such as through an aluminum linkage[69]

Structure 39

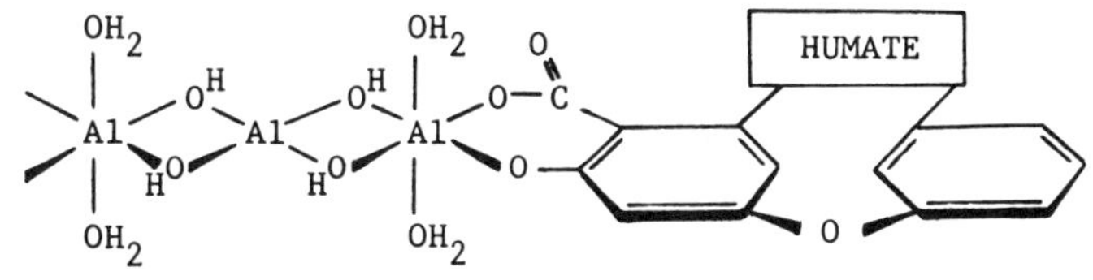

4. Formation of hydroxy-aluminum complexes at high pH values:

Structure 40 Structure 41

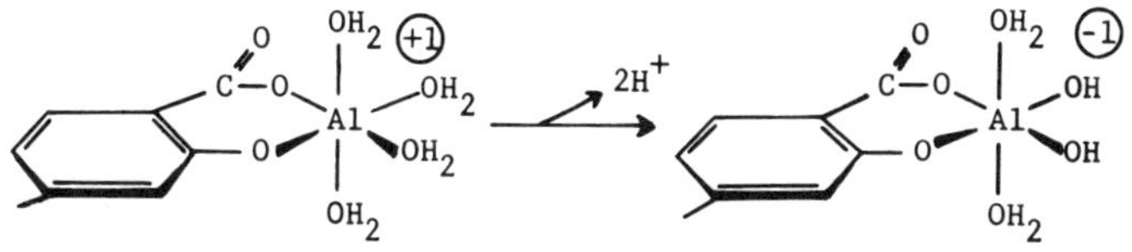

Charge reduction of soluble humic substances occurs though reaction of a COOH group, or more precisely the carboxylate ion (COO^-), with aluminum, which in turn causes the "stretched" configuration of the humic molecule to collapse, thereby reducing its solubility. As chain-like structures are formed, and as oxygen-containing functional groups become neutralized, precipitation increases. At high pH values the complexes can again be solubilized because of formation of aluminum-hydroxy complexes (see Structure 41). Results of studies on the coagulation of humic substances with aluminum suggest that coagulation will occur over a narrow pH range (4 to 7) in natural systems.[69]

The cross-linking of humic substances to clay particles and oxide surfaces through aluminum, iron, and other cations is believed to be of importance in the formation of stable aggregates in soil (see Section II.C). Spodosols have an illuvial spodic horizon composed of aluminum- and iron-fulvate complexes; these complexes, in turn, generally coat sand particles.[9] The high organic matter contents of allophanic tropical soils has been attributed to stabilization of humic matter through linkages with aluminum.

D. Binding Capacities

Both direct and indirect methods have been used to determine the binding capacities of humic substances for aluminum. Direct methods involve the determination of binding sites occupied by aluminum, for which UV and fluorescence spectroscopic techniques have been used. Because of its greater sensitivity, fluorescence spectroscopy provides data for binding under conditions that are more typical of the soil solution and natural waters (i.e., low organic matter concentrations). Fluorescence of organic ligands is quenched when complexation with aluminum occurs; thus, differentiation of free from bound sites can be made. An important consequence of the fluorescence method is that, by making a distinction between intensity and scattering, solution-phase complexation can be separated from solid-phase adsorption.[70]

Indirect methods include the separation of organic free aluminum species (i.e., Al^{3+} and aluminum-hydroxide cations) from the organically bound aluminum, such as by extraction with a known chelating agent. The free species are then determined by atomic absorption spectroscopy (flame and flameless). Colormetric and fluorimetric methods have also been used.[71] Another indirect approach has been through use of cation-exchange resin to separate free and bound aluminum species[72] (see Chapter 1 for a comprehensive discussion). This latter procedure will be discussed in Section V as a way to determine stability constants of aluminum-humate complexes. Binding capacities of humic substances for aluminum are influenced by such factors as pH value, ionic strength of the aqueous solution, and nature of the competing cation.

1. pH Effect

The effect of pH on aluminum complexation results from: (1) hydrolysis reactions involving Al^{3+} (formation of monomeric species and polymers) and (2) changes in charge characteristics of the humic matter (i.e., degree of ionization of COOH groups). The effect of pH on the formation of monomeric and polymeric species of aluminum is discussed in Chapters 5 and 6; in brief, an increase in pH value above about 4.5 leads to hydrolysis of aluminum with formation of aluminum hydroxides.

Humic and fulvic acids act as weak-acid polyelectrolytes in which the ionization of COOH groups is strongly influenced by pH value, thereby affecting their ability to bind aluminum. Aluminum binding can also be affected in other ways, such as through configurational changes in the macromolecule.

The addition of aluminum to solutions of humic acids leads to a release of protons, with a corresponding drop in pH value.[73] As compared to other heavy metals, the magnitude of the drop in pH value is exceeded only by Fe^{3+}, which may be because of stronger binding of Fe^{3+} by humic acids.

2. Influence of Salts and Competing Cations

The binding of aluminum by humic substances is influenced by the presence of electrolytes (ionic strength effect) in at least two ways. First, activity coefficients of charged inorganic species are dependent on the ionic composition of the solution. At the same ionic strength, the activities of trivalent cations are reduced more so than divalent cations, which, in turn, are reduced to a greater extent than monovalent cations. For solutions with ionic strengths between 0.001 and 0.1, the physical size of the ion must also be taken into consideration.

A second effect is because of competition of cations for binding sites on the ligand. For macromolecules, there exists the potential for a variety of configurational arrangements based on the type and nature of the interacting cation. Research conducted by Ghosh and Schnitzer[74] suggests that humic substances behave like rigid "spherocolloids" at high humic concentrations, at low pH values, or at high concentrations of a neutral electrolyte. However, they behave like "flexible linear" colloids at low humic concentrations, neutral pH values, or at low ionic strength.

The following effects must be taken into account when considering competing cations.[75]

1. Modifications in the configuration of the macromolecule with changes in pH value or ionic strength
2. Condensation of counterions in the diffuse layer of the macromolecule
3. Changes in activity coefficients of inorganic ions in the reaction mixture

E. Methods of Study

Several methods are presently available for examining ways in which aluminum interacts with humic and fulvic acids. The three most commonly used techniques, potentiometry, spectroscopy, and thermogravimetry, will be discussed first, following which some less frequently used or recently developed methods will be noted.

1. Potentiometric Titrations

The formation of complexes between aluminum and humic and fluvic acids leads to the release of protons from COOH and other acidic groups that participate in complexation. With COOH, the reaction occurs through the COO^- ion, then protons are released during reestablishment of the ionization equilibrium:

$$RCOOH \rightleftharpoons RCOO^- + H^+$$

$$Al^{3+} \qquad RCOOAl^{2+}$$

If more than one donor atom is involved (i.e., formation of a chelate structure — see Structure 37), additional protons will be released per aluminum atom bound. A further complication is the release of protons from the complex through the formation of aluminum hydroxides at pH values >4.5 (see Structures 40 and 41). Side reactions involving inorganic aluminum species must also be taken into account (see Chapter 4)

Two approaches have been used in potentiometric studies of aluminum-humate complexes, namely, base titrations of humic or fulvic acid in the presence of a known amount of aluminum and the determination of the magnitude of the pH value drop after addition of aluminum. The former method was applied to soil humic acids by Shah et al.,[76] who found that both aluminum-organic complexes and aluminum hydroxides were formed. The latter was evident by an inflection in the titration curve at pH 4.5, corresponding to the formation of aluminum hydroxides. However, the appearance of a distinct inflection depends upon the extent to which reactive functional groups are saturated with aluminum.[73] Only when aluminum is present in excess does an inflection become evident. At high saturation of binding sites, precipitation of humic acid occurs, but as the pH value increases above about 7, the precipitate slowly dissolves. On the basis of the number of protons released from an aluminum-fulvate complex per mole of base added, Schnitzer and Skinner[77] concluded that the complexes began to dissociate at pH >8.

2. Spectroscopic Methods

Spectroscopic approaches show considerable promise for examining complexes of aluminum with humic substances. They include UV, IR, and ^{13}C nuclear magnetic resonance (NMR) spectroscopy. These methods are direct in that they apply to functional groups involved in binding.

Only limited use has been made of UV spectroscopy in studies of aluminum, undoubtedly because of the fact that the spectra of humic and fulvic acids are broad and rather featureless. The approach was used by Alberts and Dickson[78] to study aluminum binding by a humic acid, tannic acid, and lignosulfonic acid. For the aluminum-humic acid complex, no prominent shifts were noted, but a small plateau was observed near 270 nm.

IR spectroscopy has provided information on sites that are involved in the binding of aluminum by humic substances. Spectra for an aluminum-humic acid complex showed increased COO– absorption at 1600 and 1400 cm^{-1} as compared to the humic acid spectrum at the same pH value, indicating participation of COOH groups.[79] A new band was observed at 1080 cm^{-1}, and assigned to aluminum-hydroxide formation. Similar results were reported by Shinagawa et al.,[80] in which case spectra of aluminum humates with increasing aluminum contents showed evidence for sharpening of an OH band at 3400 cm^{-1} and introduction of a broad band near 1000 cm^{-1}. The presence of aluminum-hydroxide linkages may account for these peaks.

The technique of ^{13}C NMR spectroscopy has provided valuable information on structural components of humic substances, and this approach may ultimately prove useful for examination of their interactions with aluminum. Another potential technique is aluminum NMR. Because the aluminum nucleus has a spin quantum number different from zero (e.g. $^{27}_{13}Al$, spin quantum number 5/2), it is amenable to examination by NMR.

3. Thermogravimetric Analysis

In thermogravimetry (TG) analysis, the mass of the substance being analyzed is monitored as the temperature is increased. Two approaches are used, namely, differential thermogravimetry (DTG), in which the rate of weight loss is determined, and differential thermal analysis (DTA), where the difference in temperature between a sample and a reference is measured. Both approaches have been applied to aluminum complexes of humic substances.[33,79,81]

For soil fulvic acids, the main decomposition reaction determined by DTG occurs at 420°C.[33] When molar 1:1 aluminum-fulvic acid complexes (number of average relative molecular mass of the fulvic acid was 670) were analyzed, DTG curves were similar to the original fulvic acid. However, upon increasing the molar ratio to 3:1 and 6:1, rather broad peaks appeared at 350 to 450°C. A weaker band at 50 to 100°C increased in size when the amount of aluminum in the complexes increased. Since the pH value was maintained at 4 to minimize the formation of aluminum hydroxide, this last peak may be because of increased water retention. In a study using DTA, Tan[79] found that the thermal stability of humic acid was increased through the binding of aluminum. Contrary to this, trivalent cations were reported to cause lower thermal stability of humic and fulvic acids because of an increased strain within the molecule when complexed with aluminum.[33]

4. Other Approaches

The characterization of aluminum-humic acid complexes based on relative molecular mass distribution has provided information on the major binding components contained in a humic acid sample. By Sephadex® gel chromatography, Kribek et al.[82] observed that a reduction in the low relative molecular mass components was accompanied by an increase in high relative molecular mass components upon aluminum complexation. Davis and Gloor[83] used gel chromatography to separate the humic substances of a Swiss lake into three relative molecular mass classes. Humic material in the intermediate range (1000 to 3000 relative molecular mass) was the most effective in reacting with suspended aluminum particles. In the study of Kribek et al.,[82] acid hydrolysis of the high relative molecular mass aluminum complexes did not produce low relative molecular mass material. A possible explanation of these results is that the low relative molecular mass substances participated in aluminum complexation to a greater extent than the high relative molecular mass substances, and that the aluminum complexes were stable toward acid degradation.

Analysis of soil humic acid and its aluminum complex by differential pulse polarography (DPP) showed that although the humic acid produced a reduction peak at pH 3.4 or less, the complex was electro-inactive.[84] Stable complexes were formed at pH 4.0 with free aluminum in equilibrium with the complex being rather low ($<0.1\%$ of the total aluminum). These results confirm that humic acids have a high affinity for aluminum, and that they serve as carriers of aluminum in the soil solution as well as natural waters.

V. CONDITIONAL STABILITY CONSTANTS OF AL^{3+} COMPLEXES WITH HUMIC SUBSTANCES

For a complete understanding of aluminum interactions with complex macromolecules, information is required on: (1) thermodynamic stability constant of the complexes, (2) binding capacity as a function of pH value and ionic strength, (3) dissociation rate or lability, and (4) kinetics of complex formation.[85] For humic and fulvic acids, none of the above has been adequately investigated.

The discussion that follows will be limited to item 1, namely, stability constants of aluminum complexes with humic substances. This information is required for use in computer models (e.g., GEOCHEM, MINEQL, MICROQL, etc.) designed for predicting the speciation of metal ions in the soil solution and natural waters using analytical data for total cations, anions, and organic ligands. A survey of mathematical models for determining stability constants of metal complexes with soil humic substances has been presented elsewhere.[34]

A. General Considerations

The reaction between a metal ion, M, and an organic ligand, L, is given by:

$$jM + iL \rightleftharpoons M_jL_i \tag{1}$$

where j is the number of moles of metal ion combined per complex molecule and i is the number of moles of organic ligand molecules. The terms M, L, and M_jL_i represent molar concentrations of free M, free L, and the complex, respectively. The overall formation (equilibrium) constant is given by:

$$\beta_i = \frac{(M_jL_i)}{(M)^j(L)^i} \tag{2}$$

When the metal ion is the central group, which is usually the case with small molecules, a series of species of the type ML_i are obtained and equilibria are described by a series of successive constants (b_1, b_2 . . . b_i). In recent studies with humic substances, the assumption has been made that the macromolecule is the central group to which several metal ions are bound (formation of M_jL complexes).

Humic substances are extremely heterogeneous and most values are reported as apparent or conditional stability constants that are functions of pH value, ionic strength, and concentrations of aluminum and humic material.[86] In practice, polynuclear complexes are probably formed, a subject that deserves greater attention in the future (see Chapter 4).

Calculations of stability constants require that accurate values be obtained for the concentration of free ligand, L, and/or the free metal ion, M (see Chapter 2). In many cases, allowance has not been made for side reactions involving the metal ion (for aluminum formation of hydroxide species) or the ligand (protonation of reactive sites, $L^- + H^+ \rightarrow HL$). Only rarely has the concentration of humic matter been expressed in molar units; most often, this parameter has been expressed in terms of the reactive site concentration, nL_t. Approaches for estimating nL_t have been described by Stevenson and Fitch.[34]

Methods used for the determination of stability constants of aluminum complexes with humic substances are (1) potentiometric titration,[38,87,88] (2) ion exchange,[89-91] (3) resin separation of charged aluminum species,[72] and (4) continuous variation, or Job's method.[89] Each method has advantages and disadvantages, and results obtained by the three approaches are not readily comparable. Only methods 1 and 2 have been given serious attention. Method 3 has been found to be useful for the separation of aluminum-organic complexes and free aluminum species from natural waters (see Chapter 1).

A résumé of conditional stability constants reported for aluminum complexes of humic substances is given in Table 4. Most values are for assumed 1:1 complexes.

B. Potentiometric Titration Method

In this method, the metal ion is the assumed central group (j = 1). As applied to humic substances, complex formation is regarded as a competitive reaction between the metal ion

Table 4
CONDITIONAL STABILITY CONSTANT OF ALUMINUM COMPLEXES OF HUMIC AND FULVIC ACIDS

Sample[a]	Method[b]	# Ligand bound aluminum	I (*M*)	pH	Log β[c]	Ref.
SHA	IEE	0.83	0.2	4.0	4.50	90
SHA	IEE	0.87	0.2	6.0	5.98	90
SHA	IEE	0.90	0.2	4.0	3.15	91
CHA	CRS	1.0	—	3—5	6.8	72
Peat	PT	1.0	0.1	2.75—4.50	1.78—3.90	88
SFA	IEE	1.0	0.1	2.35	3.7	89
SFA	CV	1.0	0.1	2.35	3.7	89

[a] SHA = soil humic acid; SFA = soil fulvic acid; CHA = commercial humic acid.
[b] Methods include IEE, ion-exchange equilibrium; PT, potentiometric titration; CV, continuous variation; CRS, cation-exchange resin separation.
[c] Most values are for assumed 1:1 complexes. See text for additional constants.

and H^+ for the reactive site. The approach has been referred to as the "constant capacitance" method. For aluminum, the reactions are

$$Al^{3+} + HL \overset{b_1^*}{\rightleftharpoons} AlL^{2+} + H^+ \tag{3}$$

$$AlL^{2+} + HL \overset{b_2^*}{\rightleftharpoons} AlL_2^+ + H^+ \tag{4}$$

$$AlL_2^+ + HL \overset{b_3^*}{\rightleftharpoons} AlL_3 + H^+ \tag{5}$$

where HL represents the associated form of the binding site (e.g., COOH or phenol-OH). The overall stability constant, β_3^*, is given by:

$$\beta_3^* = b_1^* b_2^* b_3^* = \frac{(AlL_3)(H^+)^3}{(HL)^3(Al^{3+})} \tag{6}$$

The concentration of the free ligand or binding site (HL) is determined for each point of the titration curve, from which a formation function $\bar{n}$ is calculated:

$$\bar{n} = \frac{AlL^{2+} + 2AlL_2^+ + 3AlL_3}{Al^{3+} + AlL^{2+} + AlL_2^+ + AlL_3} = \frac{L_t\text{-}HL\text{-}L^-}{Al_t} \tag{7}$$

where (L_t) and (Al_t) are the total concentrations of ligand and aluminum ion, respectively. In practice, HL is calculated from the titration data, and L^- is obtained from the relationship:

$$L^- = \frac{K_i(HL)}{(H^+)} \tag{8}$$

where K_i is the ionization constant of the acidic functional group.

The value of $\bar{n}$ indicates the number of ligand binding sites bound per metal ion. The general form of $\bar{n}$ is

$$\bar{n} = \frac{\sum_{i=1}^{N} i\beta_i^* (HL/H^+)^i}{1 + \sum_{i=1}^{N} \beta_i^* (HL/H^+)^i} \quad (9)$$

where β_i^* is the product of the stepwise formation constants.

A functional relationship exists between constants obtained in this manner and those obtained for the reaction of the metal ion with the dissociated form of the reactive site ($Al^{3+} + L^- \rightleftharpoons AlL^{2+}$; $b_1 = [AlL^{2+}]/[Al^{3+}][L^-]$). The relationship is given by:

$$b_i^* = K_i b_i \quad (10)$$

To obtain values for the successive constants, $\bar{n}$ is plotted against $p(HL/H^+)$ and the successive constants are obtained at half-integer values of $\bar{n}$ (β_1^*, β_2^*, β_3^* at $\bar{n} = 0.5$, 1.5, and 2.5 respectively). Plots obtained by Arai and Kumada[86] for aluminum complexes of four humic acids showed that successive formation constants were not well-separated. Accordingly, values at $\bar{n} = 0.5$ (β_1^*) were taken as an index of binding affinity. Log β_1^* values for the humic acids ranged from 0.15 to 0.37.

A more sophisticated potentiometric titration approach was applied to aluminum-organic matter complexes by Young and Bache.[38] To avoid significant hydrolysis of aluminum, the calculations were made from data obtained at pH < 4.0. The Bjerrum $\bar{n}$ function ranged from about 0.5 to 2.5. Best fit of the data was achieved by considering only the complexing species AlL_2^+ and AlL_3. Values for β_2^* and β_3^* were obtained by regression analysis of the following equation (obtained by expansion of Equation 9 and elimination of the β_1^* term):

$$\frac{\bar{n}}{(2 - \bar{n})\{[LH][H^+]^{-1}\}^2} = \beta_2^* + \frac{(3 - \bar{n})\{[LH][H^+]^{-1}\}}{(2 - \bar{n})} \beta_3^* \quad (11)$$

Log β_i^* values were then converted to log β_i values from the relationship:

$$\log \beta_i^* = \log \beta_i - pK_i' - i\log \gamma_H \quad (12)$$

where γ_H is the H^+ ion activity coefficient and pK_i' is the intrinsic acidity constant.

In terms of the reaction of aluminum with the dissociated form of the reactive site (see Equation 10), log β_2 values for a peat and Spodosol fulvic acid at the pH value of half-neutralization of acidic functional groups and an ionic strength of 0.3 *M* were 5.78 and 6.08, respectively; corresponding log β_3 values were 8.40 and 8.37, respectively.[38]

C. Ion-Exchange Equilibrium Method

Early work on the complexation of aluminum by humic substances was done by competition of the metal ion with a cation-exchange resin, using the Schubert ion-exchange equilibrium method. The assumptions are made that the metal ion is the central group, and that the ligand is present in large excess (e.g., a 1:1 complex is formed). In the presence of a cation-exchange resin, the concentration of metal ion is distributed among the resin, M_R, the metal complex, M_C, and the aqueous phase as a free cation, M_f. The distribution coefficient, λ, is given by the equation:

$$\lambda = M_R/(M_f + M_c) \quad (13)$$

In the absence of a complexing agent, the distribution coefficient is

$$\lambda_o = M_R/M_F \tag{14}$$

By appropriate substitutions into Equation 13 and from the assumptions noted earlier, the following simplified equation is obtained:

$$\log[(\lambda_o/\lambda) - 1] = \log \beta + i\log (L) \tag{15}$$

From an experimental standpoint, λ is measured for several values of L. A plot of the left side of Equation 15 vs. log L yields a straight line with i as the slop and log β as the intercept.

Limitations of the Schubert approach as applied to humic substances have been discussed elsewhere.[92] Schnitzer and Hansen[89] applied the method to an aluminum-fulvate complex, for which a log β value of 3.7 was obtained.

A modified version of the ion-exchange equilibrium method was developed by Ardakani and Stevenson,[28] and subsequently used by Adhikari and Ray[90] to determine stability constants of aluminum complexes with humic acids. The basis for the method as applied to aluminum complexation is as follows.

From Equation 1, substituting Al for M, the concentration of aluminum in the complex is

$$Al_c = j(Al_jL_i) \tag{16}$$

or

$$Al_c = j\beta(Al_f)^j(L)^i \tag{17}$$

By taking the common logarithm of both sides, the following is obtained:

$$\log (A1_c) = \log j + \log \beta + j\log (Al_f) + i\log (L) \tag{18}$$

Within a small range of free-aluminum concentration, J and β can be assumed to be constant and Equation 18 becomes

$$\log (Al_c) = j\log (Al_f) + i\log (L) + C \tag{19}$$

Partial differentiation of log (Al_c) with respect to log (Al_f) with (L) constant and with respect to log (L) with (Al_f) constant, gives

$$\left.\frac{\partial[\log (Al_c)]}{\partial[\log (Al_f)]}\right|_L = j \tag{20}$$

and

$$\left.\frac{\partial[\log (Al_c)]}{\partial[\log (L)]}\right|_{AL_f} = i \tag{21}$$

Equation 20 indicates that if (L) is kept constant, a plot of log (Al_c) vs. log (Al_f) will be a straight line of slope j and an intercept I_a defined by

$$I_a = \log \beta + \log j + i\log (L) \tag{22}$$

Similarly, Equation 21 indicates that if (Al_f) is kept constant, a plot of log (Al_c) vs. log (L) yields a straight line of slope i and an intercept, I_b, given by

$$I_b = \log \beta + \log j + j\log (Al_f) \tag{23}$$

Using this approach, Adhikari and Ray[90] obtained β values of 4.50 and 5.98 for the aluminum complexes of two soil humic acids.

D. Other Approaches

Resin separation of charged aluminum species — A resin-separation method was used by Potts et al.[72] for determining stability constants of aluminum complexes with a commercial humic acid (Aldrich) and the dissolved organic matter of a water sample. Cationic aluminum species in the solutions were separated from organic complexed and uncharged aluminum-hydroxy species by passage through a cation-exchange resin. The thermodynamic equilibrium computer model GEOCHEM was then used to correct for uncharged aluminum-hydroxy species. Stability constants (log β) of the aluminum-humate complex, as obtained by regression analysis of the binding data (e.g., Langmuir plot), were of the order of 6.8, with little variation between pH 3 to 5. The conclusion was reached that aluminum was bound through a single type of binding site, most probably COOH.

Method of continuous variation — The method of continuous variation, or Job's method, is based on absorbance measurements for a series of solutions containing variable ratios of ligand and aluminum while maintaining a constant reactant concentration. The point at which absorbance is maximum corresponds to the maximum number of ligands bound. Once this information is obtained, successive formation constants of the various complexes can be obtained.[33] By this approach, a log β value of 3.7 was obtained for an aluminum-fulvate complex.[89] In a critical examination of the method, MacCarthy and Mark[93] concluded that results obtained with humic substances are affected by light scattering (i.e., because of precipitation of complexes), thereby leading to erroneous log β values.

VI. SUMMARY

Naturally occurring organic substances of various types play a prominent role in the binding of aluminum in soils and sediments. Both soluble and insoluble complexes are formed. Biochemical compounds produced by microorganisms (e.g., aliphatic organic acids, phenolic constituents, and hydroxamate siderophores) are particularly important in environments where microbial activity is intense, such as in the rhizosphere and the organic rich layers of forest soils, notably Alfisols and Spodosols. Humic substances are of greater importance in modifying the reaction of aluminum on clay and oxide surfaces, in reducing toxicities of aluminum in acid soils, in regulating the sorption-desorption of phosphate in allophanic soils, and in transporting aluminum in natural waters.

ACKNOWLEDGMENTS

Gratitude is expressed to the U.S.-Israel Agricultural Research and Development Fund (BARD I-500-82) for support of research on cation-anion interactions with humic substances.

REFERENCES

1. Soil Survey Staff, Soil Taxonomy: A Basic System of Soil Classification for Making and Interpreting Soil Surveys, Agricultural Handbook No. 436, U.S. Government Printing Office, Washington, D.C., 1975.
2. Canada Soil Survey Committee, The Canadian System of Soil Classification, Canada Department of Agriculture, Publ. 1455, Information Canada, Ottawa, 1978.
3. Food and Agriculture Organization, Soil Map of the World, Vol. 1, UNESCO, Paris, 1974.
4. **Robert, M. and Berthelin, J.,** Role of biological and biochemical factors in soil mineral weathering, in *Interaction of Soil Minerals with Natural Organics and Microbes,* Spec. Publ. No. 17, Huang, P. M. and Schnitzer, M., Eds., Soil Science Society of America, Madison WI, 1986, 453.
5. **Tan, K. H.,** Degradation of soil minerals by organic acids, in *Interaction of Soil Minerals with Natural Organics and Microbes,* Spec. Publ. No. 17, Huang P. M. and Schnitzer, M., Eds., Soil Science Society of America, Madison, WI, 1986, 1.
6. **Huang, P. M. and Violante, A.,** Influence of organic acids on crystallization and surface properties of precipitation products of aluminum, in *Interactions of Soil Minerals with Natural Organics and Microbes,* Spec. Publ. No. 17, Huang, P. M. and Schnitzer, M., Eds., Soil Science Society of America, Madison, WI, 1986, 159.
7. **McKeague, J. A., Cheshire, M. V., Andreux, F., and Berthelin, J.,** Organo-mineral complexes in relation to pedogenesis, in *Interactions of Soil Minerals with Natural Organics and Microbes,* Spec. Publ. No. 17, Huang, P. M. and Schnitzer, M., Eds., Soil Science Society of America, Madison, WI, 1986, 549.
8. **Kwong, N. K. K. F. and Huang, P. M.,** The relative influence of low-molecular-weight complexing organic acids and the hydrolysis and precipitation of aluminum, *Soil Sci.,* 128, 337, 1979.
9. **De Coninck, F.,** Major mechanisms in formation of spodic horizons, *Geoderma,* 24, 101, 1980.
10. **Anderson, H. A., Berrow, M. L., Farmer, V. C., Hepburn, A., Russell, J. D., and Walter, A. D.,** A reassessment of podzol formation processes, *J. Soil Sci.,* 33, 125, 1982.
11. **David, M. and Driscoll, C. T.,** Aluminum speciation and equilibria in soil solutions of a Haplorthod in the Adirondack mountains (New York, U.S.A.), *Geoderma,* 33, 297, 1984.
12. **Perdue, E. M., Beck, K. C., and Reuter, J. H.,** Organic complexes of iron and aluminum in natural waters, *Nature,* 260, 418, 1976.
13. **Driscoll, C. T., van Breeman, N., and Mulder, J.,** Aluminum chemistry in a forested Spodosol, *Soil Sci. Soc. Am. J.,* 49, 437, 1985.
14. **Hodges, S. C.,** Aluminum speciation: a comparison of five methods, *Soil Sci. Soc. Am. J.,* 51, 57, 1987.
15. **James, B. R. and Riha, S. J.,** Soluble aluminum in acidified organic horizons of forest soils, *Can. J. Soil Sci.,* 64, 637, 1984.
16. **Nilsson, S. I. and Bergkvist, B.,** Aluminum chemistry and acidification processes in a shallow Podzol on the Swedish westcoast, *Water Air Soil Pollut.,* 20, 311, 1983.
17. **Kabata-Pendias, A. and Pendias, H.,** *Trace Elements in Soils and Plants,* CRC Press, Boca Raton, FL, 1984.
18. **Foy, C. D.,** Effects of aluminum on plant growth, in *The Plant Root and its Environment,* Carson, E. W., Ed., University Press of Virginia, Charlottesville, 1974, 601.
19. **Hoyt, P. B. and Turner, R. C.,** Effects of organic materials added to very acid soils on pH, aluminum, exchangeable NH_4^+, and crop yields, *Soil Sci.,* 119, 227, 1975.
20. **Hargrove, W. L. and Thomas, G. W.,** Effect of organic matter on exchangeable aluminum and plant growth in acid soils, in *Chemistry of the Soil Environment,* Spec. Publ. No. 40, Dowdy, R. H., et al., Eds., American Society of Agronomy, Madison, WI, 1981, 151.
21. **Ahmad, F. and Tan, K. H.,** Effect of lime and organic matter on soybean seedlings grown in aluminum-toxic soil, *Soil Sci. Soc. Am. J.,* 50, 656, 1986.
22. **Stevenson, F. J.,** *Humus Chemistry: Genesis, Composition, Reactions,* Wiley-Interscience, New York, 1982.
23. **Emerson, W. W., Foster, R. C., and Oades, J. M.,** Organo-mineral complexes in relation to soil aggregation and structure, in *Interactions of Soil Minerals with Natural Organics and Microbes,* Spec. Publ. No. 17, Huang, P. M. and Schnitzer, M., Eds., Soil Science Society of America, Madison, WI, 1986, 521.
24. **Edwards, A. P. and Bremner, J. M.,** Microaggregates in soils, *J. Soil Sci,* 18, 64, 1967.
25. **Mortland, M.,** Clay-organic complexes and interactions, *Adv. Agron.,* 22, 75, 1970.
26. **Reid, J. B., Goss, M. J., and Robertson, P. D.,** Relationship between the decreases in soil stability effected by the growth of maize roots and changes in organically bound iron and aluminum, *J. Soil Sci.,* 33, 397, 1982.
27. **Stevenson, F. J.,** Organic acids in soil, in *Soil Biochemistry,* Vol. 1, McLaren, A. D. and Peterson, G. H., Eds., Marcel Dekker, New York, 1967, 119.

28. **Ardakani, M. S. and Stevenson, F. J.,** A modified ion-exchange technique for determination of stability constants of metal-soil organic matter complexes, *Soil Sci. Soc. Am. J.*, 36, 884, 1972.
29. **Stevenson, F. J. and Ardakani, M. S.,** Organic matter reactions involving micronutrients, in *Micronutrients in Agriculture,* Mortvedt, J. J., et al., Eds., American Society of Agronomy, Madison, WI, 1972, 79.
30. **White, R. E. and Thomas, G. W.,** Hydrolysis of aluminum in weakly acidic organic exchangers: implications for phosphate adsorption, *Fert. Res.*, 2, 159, 1981.
31. **Bloom, P. R.,** Phosphorus adsorption by an aluminum-peat complex, *Soil Sci. Soc. Am. J.*, 45, 267, 1981.
32. **Borie, F. and Zunino, J.,** Organic matter-phosphorus associations as a sink in P-fixation processes in allophanic soils of Chile, *Soil Biol. Biochem.*, 15, 599, 1983.
33. **Schnitzer, M. and Khan, S. U.,** *Humic Substances in the Environment,* Marcel Dekker, New York, 1972, 327.
34. **Stevenson, F. J. and Fitch, A.,** Chemistry of complexation of metal ions with soil solution organics, in *Interactions of Soil Minerals with Natural Organics and Microbes,* Spec. Publ. No. 17, Huang, P. M. and Schnitzer, M., Eds., Soil Science Society of America, Madison, WI, 1986, 29.
35. **Wang, T. S. C., Huang, P. M., Chou, C.-H., and Chen, J.-H.,** The role of soil minerals in the abiotic polymerization of phenolic compounds and formation of humic substances, in *Interactions of Soil Minerals with Natural Organics and Microbes,* Spec. Publ. No. 17, Huang, P. M. and Schnitzer, M., Eds., Soil Science Society of America, Madison, WI, 1986, 251.
36. **Stevenson, F. J.,** *Cycles of Soil: Carbon, Nitrogen, Phosphorus, Sulfur, Micronutrients,* John Wiley & Sons, New York, 1986.
37. **Martell, A. E. and Smith, R. M.,** *Critical Stability Constants; Other Organic Ligands, Vol. 3,* 1977; and *First Supplement, Vol. 5,* 1982, Plenum Press, New York.
38. **Young, S. D. and Bache, B. W.,** Aluminum-organic complexation: formation constants and a speciation model for the soil solution, *J. Soil. Sci.*, 36, 261, 1985.
39. **Sillén, L. G. and Martell, A. E.,** *Stability Constants of Metal-Ion Complexes, Spec. Publ. No. 25,* The Chemical Society, London, 1971.
40. **Wang, T. S. C., Cheng, S.-Y., and Tung, H.,** Dynamics of soil organic acids, *Soil Sci.*, 104, 138, 1967.
41. **Rovira, A. D. and McDougall, B. M.,** Microbiological and biochemical aspects of the rhizosphere, in *Soil Biochemistry,* Vol. 1, McLaren, A. D. and Peterson, G. H., Eds., Marcel Dekker, New York, 1967, 417.
42. **Rouatt, J. W. and Katznelson, H.,** A study of the bacteria on the root surface and the rhizosphere soil of cropped plants, *J. Appl. Bacteriol.*, 24, 164, 1961.
43. **Graustein, W. C., Cromack, K., Jr., and Sollins, P.,** Calcium oxalate: occurrence in soils and effect on nutrient and geochemical cycles, *Science,* 198, 1252, 1977.
44. **Cromack, K., Sollins, P., Graustein, W. C., Speidel, K., Todd, A. W., Spycher, G., Li, C. Y., and Todd, R. L.,** Calcium oxalate accumulation and soil weathering in mats of hypogenous fungi *Hysterangum crtoslsum, Soil Biol. Biochem.*, 11, 463, 1979.
45. **Jurinak, J. J., Dudley, L. M., Allen, M. F., and Knight, W. G.,** The role of calcium oxalate in the availability of phosphorus in soils of semiarid regions: a thermodynamic study, *Soil Sci.*, 142, 255, 1986.
46. **Powell, P. E., Szaniszlo, P. J., Cline, G. R., and Reid, C. P. P.,** Hydroxamate siderophores in the iron nutrition of plants, *J. Plant Nutr.*, 5, 653, 1982.
47. **Akers, H. A.,** Multiple hyroxamic acid microbial iron chelators (siderophores) in soils, *Soil Sci.*, 135, 156, 1983.
48. **Szaniszlo, P. J., Powell, P. E., Reid, C. P. P., and Cline, G. R.,** Production of hydroxamate siderophore iron chelators by ectomychorrhizal fungi, *Mycologia,* 73, 1158, 1981.
49. **Webley, D. M. and Duff, R. B.,** The incidence, in soils and other habitats, of microorganisms producing 2-ketogluconic acid, *Plant Soil,* 22, 307, 1965.
50. **Moghimi, A., Tate, M. E., and Oades, J. M.,** Characteristics of rhizosphere products, especially 2-ketogluconic acid, *Soil Biol. Biochem.*, 10, 283, 1978.
51. **Webley, D. M., Henderson, M. E. K., and Taylor, I. F.,** The microbiology of rocks and weathered stones, *J. Soil Sci.*, 14, 102, 1963.
52. **Whitehead, D. C., Dibbs, H., and Hartley, R. D.,** Phenolic compounds in soil as influenced by the growth of different plant species, *J. Appl. Ecol.*, 19, 579, 1982.
53. **Vance, G. F., Mokma, D. L., and Boyd, S. A.,** Phenolic compounds in soils of hydrosequences and developmental sequences of Spodosols, *Soil Sci. Soc. Am. J.*, 50, 992, 1986.
54. **Sikora, F. J. and McBride, M. B.,** Complexation of Aluminum by Protocatechuic and Caffeic Acids as Determined by Ultraviolet Spectrophotometry, unpublished manuscript, 1987.
55. **Martin, J. P. and Haider, K.,** Microbial activity in relation to soil humus formation, *Soil Sci.*, 111, 54, 1971.
56. **Bruckert, S.,** Influence des compos's organiques hydrosolubles sur la p'dogen'se en mileau acide, *Ann. Agron.*, 21, 421 and 725, 1970.

57. **King, H. G. C. and Bloomfield, C.,** The reaction between water-soluble tree leaf constituents and ferric oxide in relation to podzolisation, *J. Sci. Food Agric.*, 17, 39, 1966.
58. **Malcolm, R. L. and McCracken, R. J.,** Canopy drip: a source of mobile soil organic matter for mobilization of iron and aluminum, *Soil Sci. Soc. Am. Proc.*, 32, 834, 1968.
59. **Coulson, C. B., Davis, R. I., and Lewis, D. A.,** Polyphenols in plant, humus and soil. II. Reduction and transport by polyphenols of iron in model soil columns, *J. Soil Sci.*, 30, 20, 1960.
60. **Iskandar, I. K. and Syers, J. K.,** Metal-complex formation by lichen compounds, *J. Soil Sci.*, 23, 255, 1972.
61. **Saini, G. R.,** Sequestering of iron and aluminum by soil polysaccharides, *Curr. Sci.*, 10, 259, 1966.
62. **Martin, J. P., Ervin, J. P., and Richards, S. J.,** Decomposition and binding action in soil of some mannose-containing microbial polysaccharides and their Fe, Al, Zn and Cu complexes, *Soil Sci.*, 113, 322, 1972.
63. **Parfitt, R. L. and Greenland, D. J.,** Adsorption of polysaccharides by montmorillonite, *Soil Sci. Soc. Am. Proc.*, 34, 862, 1970.
64. **Malcolm, R. L.,** Geochemistry of stream fulvic and humic substances, in *Humic Substances in Soil, Sediment, and Water,* Aiken, G. R. et al., Eds., Wiley-Interscience, New York, 1985, 181.
65. **Griffith, S. M. and Schnitzer, M.,** The isolation and characterization of stable metal-organic matter complexes from tropical volcanic soils, *Soil Sci.* 120, 126, 1975.
66. **Nissenbaum, A. and Swaine, D. J.,** Organic matter-metal interactions in recent sediments: the role of humic substances, *Geochim. Cosmochim. Acta,* 40, 809, 1976.
67. **Cheshire, M. V., Berrow, M. L., Goodman, B. A., and Mundie, C. M.,** Metal distribution and nature of some Cu, Mn, and V complexes in humic and fulvic fractions of soil organic matter, *Geochim. Cosmochim. Acta,* 41, 1131, 1977.
68. **Alberts, J. J. and Dickson, T. J.,** Organic carbon and cation associations in humic material from pond water and sediment, *Org. Geochem.*, 8, 55, 1985.
69. **Dempsey, B. A., Ganho, R. M., and O'Melia, C. R.,** The coagulation of humic substances by means of aluminum salts, *J. Am. Water Works Assoc.*, 76, 141, 1984.
70. **Saar, R. A. and Weber, J. H.,** Fulvic acid: modifier of metal-ion chemistry, *Envron. Sci. Technol.*, 16, 510A, 1982.
71. **Bloom, P. R., Weaver, R. M., and McBride, M. B.,** The spectrophotometric and fluorometric determination of aluminum with 8-hydroxyquinoline and butyl acetate extraction, *Soil Sci. Soc. Am. J.*, 42, 713, 1978.
72. **Potts, D. B., Alberts, J. J., and Elzerman, A. W.,** The influence of pH on the binding capacity and conditional stability constants of aluminum and naturally-occurring organic matter, *Chem. Geol.*, 48, 293, 1985.
73. **Khan, S. U.,** Interaction between the humic acid fraction of soils and certain metallic cations, *Soil Sci. Soc. Am. Proc.*, 33, 851, 1969.
74. **Ghosh, K. and Schnitzer, M.,** Macromolecular structures of humic substances, *Soil Sci.*, 129, 266, 1980.
75. **Cleven, R. F. M. J.,** Heavy Metal/Polyacid Interaction; an Electrochemical Study of the Binding of Cd(II), Pb(II), and Zn(II) to Polycarboxylic and Humic Acids, Thesis, Agricultural University, Wageningen, The Netherlands, 1984, 225.
76. **Shah, R. K., Chokshi, M. R., and Soni, K. P.,** Interaction between certain cations and the humic acids extracted from different Gujarat soils, *J. Indian Chem. Soc.*, 54, 912, 1977.
77. **Schnitzer, M. and Skinner, S. I. M.,** Organo-metallic interactions in soils. I. Reaction between a number of metal ions and the organic matter of a Podzol B horizon, *Soil Sci.*, 96, 86, 1963.
78. **Alberts, J. J. and Dickson, T. J.,** Organic carbon and cation associations in humic material from pond water and sediment, *Org. Geochem.*, 8, 55, 1985.
79. **Tan, K. H.,** Formation of metal-humic acid complexes by titration and their characterization by differential thermal analysis and infrared spectroscopy, *Soil Biol. Biochem.*, 10, 123, 1978.
80. **Shinagawa, A., Miyauchi, N., and Higashi, T.,** Preparation of Al-humates and their aluminum content and cation-exchange capacity, *Soil Sci. Plant Nutr. (Tokyo),* 28, 1, 1982.
81. **Dupuis, T.,** Characterization par analyse thermique differentielle des complexes de l'aluminum avec les acides fulviques et humiques, *J. Therm. Anal.*, 3, 281, 1971.
82. **Kribek, B., Kaigl, J., and Oruzinsky, V.,** Characteristics of di- and trivalent metal-humic acid complexes on the basis of their molecular weight distribution, *Chem. Geol.*, 19, 73, 1977.
83. **Davis, J. A. and Gloor, R.,** Adsorption of dissolved organics in lake water by aluminium oxide: effect of molecular weight, *Environ. Sci. Technol.*, 15, 1223, 1981.
84. **Ritchie, G. S. P., Posner, A. M., and Ritchie, I. M.,** The polarographic study of the equilibrium between humic acid and aluminum in solution., *J. Soil Sci.*, 33, 671, 1982.
85. **Tuschall, J. R. and Brezonik, P. L.,** Complexation of heavy metals by aquatic humus: a comparative study of five analytical methods, in *Aquatic and Terrestrial Humic Materials*, Christman, R. F. and Gjessing, E. T., Eds., Ann Arbor Science, Ann Arbor, MI, 1983, 275.

86. **Arai, S. and Kumada, K.,** Factors controlling stability of Al-humate, *Geoderma,* 26, 1, 1981.
87. **Adhikari, M., Roy, J., and Hazra, G.,** Potentiometric study of humic acid metal interactions, *J. Indian Chem. Soc.,* 55, 332, 1978.
88. **Hargrove, W. L. and Thomas, G. W.,** Conditional formation constants for aluminum-organic matter complexes, *Can. J. Soil Sci.,* 62, 571, 1982.
89. **Schnitzer, M. and Hansen, E. H.,** Organo-metallic interactions in soils. VIII. An evaluation of methods for the determination of stability constants of metal-fulvic acid complexes, *Soil Sci.,* 109, 333, 1970.
90. **Adhikari, M. and Ray, J. N.,** Stability constants of humic acids and their phosphorylated derivatives with Cu^{2+} and Al^{3+}, *J. Indian Chem. Soc.,* 53, 238, 1976.
91. **Adhikari, M., Chakrabarti, G., and Hazra, G.,** Measurement of stability constant of humic acid metal complexes, *Agrochimica,* 21, 134, 1977.
92. **Clark, J. S. and Turner, R. C.,** An examination of the resin exchange method for the determination of stability constants of metal-soil organic matter complexes, *Soil Sci.,* 107, 8, 1969.
93. **MacCarthy, P. and Mark, H. B., Jr.,** An evaluation of Job's method of continuous variations as applied to soil organic matter-metal ion interactions, *Soil Sci. Soc. Am. J.,* 40, 267, 1976.

Chapter 6

SURFACE REACTIONS OF AQUEOUS ALUMINUM SPECIES

Lucian W. Zelazny and Philip M. Jardine

TABLE OF CONTENTS

I. INTRODUCTION

In nature, aluminum has only a single stable oxidation state (+3) and can be found as a tetrahedrally coordinated cation (ionic radius = 0.046 nm) or as an octahedrally coordinated cation (ionic radius = 0.061 nm) in coordination with oxygen.[1] Tetrahedral aluminum is present as an essential constituent in feldspars, some clay minerals, some high-temperature silicates and oxides, and as the aluminate ion in alkaline solutions. Octahedral aluminum is present as a solution cation in acid and neutral conditions and in many clay minerals and many oxides, hydroxides, and oxyhydroxides (see Chapter 2 and 3).

Solution aluminum is highly hydrated, although the primary water shell can be displaced by very electronegative ligands such as F^-. The most striking feature of aluminum in solution is its tendency to hydrolyze, and the literature addressing this subject is voluminous, as discussed in Chapter 4. The mononuclear hydrolysis products, $[AlOH{\cdot}5H_2O]^{2+}$, $[Al(OH)_2{\cdot}4H_2O]^+$, $[Al(OH)_3{\cdot}3H_2O]^0$, and $Al(OH)_4^-$ have all been proposed to result from the successive deprotonation of coordinated H_2O.[2,3] Hydrolysis will proceed spontaneously when an aluminum salt is dissolved in water,[4] with the resulting drop in pH value being determined by the total aluminum concentration and the satisfaction of equilibrium relations. Unfortunately, estimates of the formation constants for each product have varied from study to study (see Chapter 2). The difficulty in arriving at reliable thermodynamic stability constants is caused in part by their proximity (as compared to a typical multiprotic acid), but, more importantly, by the dominance of polynuclear species upon the addition of hydroxide to the solution.[2,4,5]

The nature and extent of the polynuclear aluminum hydroxide species has been the subject of innumerable studies (see Chapter 4). A host of different formulas have been proposed by various authors.[3,5,6] Whether this lack of agreement should be attributed to difficulties in determining structure or to a real variability in the products formed is still unclear. We believe that the polynuclear species are transient, metastable intermediates in the precipitation of crystalline $Al(OH)_3$. If a partially neutralized solution is at true equilibrium, it should contain only the mononuclear species and the stable solid phase.[2]. However, opinion is not uniform on this point,[3] and the achievement of equilibrium is hampered by slow kinetics, especially at low values of n, the OH/Al molar ratio.[2] Moreover, definitional difficulties are involved. As polymerization proceeds, the aluminum units go from true solutes to macromolecules, to colloids, and to visible crystals, and delineating these phases is problematic.[4]

Variability in the reported polynuclear species may be caused, in part, by the fact that the nature and distribution of these species can be a function of ionic strength, the total aluminum concentration, the total OH added, the pH value, the temperature, the anions present, the time, and the method of preparation.[5] It is noteworthy, however, that the fraction of mononuclear aluminum present appears to be relatively independent of aging or the rate of base addition for a given value of n.[5,7,8] The distribution of polynuclear to solid-phase aluminum changes, with the latter forming at the expense of the former over time.[5,9-11] The presence of additional reactive ligands, such as PO_4^{-3}, SO_4^{-2}, and F^-, in combination with added base, introduces further complications. Consequently, it is often necessary to utilize empirical techniques to quantitate mono- and polynuclear forms of aluminum (see Chapters 1 and 4 for a complete discussion).

With these uncertainties and recognized difficulties with the understanding of aluminum hydrolysis and polymerization in solution, it is our objective to expound further on these reactions in the presence of charged surfaces. We shall then review the influence of inorganic and organic anions on the surface reactions of adsorbed aluminum species, and conclude with a discussion of the changes in chemical and physical properties of clays and soils produced by the adsorption of these species. Extensive reviews[4,12-19] exist in this area of study, and these sources also should be consulted.

II. INTERACTION OF AQUEOUS ALUMINUM SPECIES WITH CLAY MINERAL AND SOIL SURFACES

Aluminum adsorbed on clays or soils may be exchanged by other cations, may be strongly held or "fixed" (i.e., nonexchangeable), or may precipitate as various solid phases of aluminum. The natural occurrence of the various solid phases of aluminum, their characteristics, surface chemistry, and solubility are examined in detail in Chapters 3, 5, 7, and 8 of this book.

Many early experiments were conducted using aluminum-saturated soils or clays. Experiments by Rich,[20,21] Kaddah and Coleman,[22,23] and Kissel et al.[24] indicated that the removal of excess saturating salts by repeated washing may result in considerable aluminum hydrolysis. This could occur when any solution of pH > 3.0 is used to remove excess salts, since above this value pH aluminum may begin to hydrolyze.[3] Veith and Sposito[25] concluded that in the presence of a sulfonic resin, hydrolysis could even occur at pH > 2.7. The failure to anticipate the hydrolysis of adsorbed aluminum has greatly affected the conclusions of many earlier studies, since the initial charge of the adsorbed species were unknown. Nevertheless, the general trends and observations of these reports laid the foundation for our current knowledge of aluminum chemistry and are worthy of consideration in proper perspective.

A. Exchangeable Aluminum

Efforts to characterize the nature of soil acidity has prompted studies to measure the exchange of Al^{3+} from soils by an unbuffered salt solution.[26-28] Since the removal of aluminum is highly influenced by the technique, type and concentration of extracting agent, and pH value, it is sometimes termed "extractable" instead of exchangeable.[29,30] Although many reagents have been used to extract aluminum from soils, unbuffered salt solutions are the most common. Unbuffered salt solutions are believed to remove exchangeable aluminum ions bonded electrostatically to permanent charged sites. Whereas buffered salt solutions will replace exchangeable aluminum as well as some fraction of nonexchangeable and precipitated forms of aluminum.[27] Skeen and Sumner[31] showed that the quantity of aluminum exchanged or extracted by NH_4^+, alkali metal, and alkaline earth cations is inversely related to their hydrated radius. Also, the affinity of a cation for a site is related directly to its effectiveness as a salt for aluminum extraction.[24] The extracting solution should contain a high concentration of ions to displace aluminum, and should have a low pH value to maintain aluminum in a soluble form.[33] McLean[33] maintained that KCl-exchangeable aluminum forms soluble $AlCl_3$, which will partially hydrolyze to the corresponding acid, keeping the pH value of the leaching front sufficiently acid to prevent readsorption or precipitation of aluminum.

Earlier, McLean et al.[29] concluded that 1 *M* NH_4OAc at pH 4.8 was the superior extractant to use because, at this pH value, it extracted enough aluminum from most soils for accurate measurement, yet did little damage to the clay crystal. However, Pratt and Bair[26] found that this reagent extracted less aluminum from soils that had a pH value lower than 4.8 and more aluminum from soils with a higher pH value. In other words, the buffered 1*M* NH_4OAc solution controlled the pH value of the extraction instead of the soil, and the use of unbuffered 1*M* KCl was recommended. Lin and Coleman[34] showed that unbuffered KCl was more effective as a displacing agent than $CaCl_2$ or NaCl. On the other hand, Yuan and Fiskell[30] found that, for some soils, the salts of divalent cations, particularly $BaCl_2$, extracted more aluminum than KCl or NaCl. Skeen[35] indicated that Cl^- was a suitable anion for the removal of exchangeable aluminum because of its low degree of penetration into the structure of clays and oxides.

1. Inorganic Surfaces

Initial studies reported that only the Al^{+3} species was exchangeable from soil constituents.[20,34,36] This was also reported by Thomas[37] in a study of $Ca(OH)_2$-treated Madison soil, aluminum clay, and Dowex®-50 cation-exchange resin. He used conductometric titrations and a colorimetric procedure to determine the number of millimoles of change and extracted aluminum respectively, and concluded that no hydroxy-aluminum species were extracted. The only sample of unlimed soil examined, however, yielded a computed charge <3 for exchangeable aluminum, but this result was ignored. Hsu[38] studied the interaction of hydroxy-aluminum solutions of varying basicity with montmorillonite and showed that, when a large amount of clay was treated with a small amount of hydroxy-aluminum solution, the polymers were held in positions of strong affinity, being difficult to extract and unavailable for conversion to gibbsite upon aging.

Several studies[39,40] since have shown that the number of moles of exchangeable cation change removed from soil by an unbuffered solution were greater than the total cation-exchange capacity (CEC), including the exchange of partially hydrolyzed aluminum ions. Other reports have shown that both monomeric and polymeric aluminum species are involved in the exchange from soils and clay minerals.[28,41-44] Bache and Sharp[28] reported that the amount of polymeric aluminum extracted from two British soils increased with increasing concentration and valence of the extracting cation.

Hodges and Zelazny[43] found that hydrolyzed aluminum species could be exchanged from kaolinite, montmorillonite, peat, and a sulfonic-exchange resin when the aluminum concentration in solution was less than equivalent or equivalent to the CEC of the exchangers. The amount of hydrolyzed species that were exchanged decreased with increasing equilibration time. In a second study, these authors[44] noted that hydroxy-aluminum species were exchanged from a poorly crystalline Georgia kaolinite at loading rates $>100\%$ of the CEC. The amount of polynuclear aluminum exchanged, as determined by reaction with ferron, increased with increasing basicity; up to 70% of the displaced aluminum was in the form of polymers when the equilibrating solution had an OH/Al molar ratio of 2.4.

Successive extractions or leaching of soils have been used many times in an attempt to characterize the exchanger composition of soils. Lin and Coleman[34] leached soils with acidified and unacidified 0.1 and 1.0 *M* KCl to see if they could identify and determine exchangeable aluminum. They found that 0.1 *M* KCl was not nearly as effective as 1.0 *M* KCl, and that neutral salt solutions displaced only "exchangeable aluminum," whereas acidified KCl solutions extracted aluminum in excess of the CEC. Shen and Rich[36] performed successive extractions of an aluminum-saturated montmorillonite with 1.0 *M* NaCl in an attempt to study fixed aluminum. They also found that the amount of aluminum removed decreased with each extraction and that the initial extractions removed all of the "exchangeable aluminum."

Skeen and Sumner[31,45,46] studied the effect of cation and anion species, pH value and solution concentration on the exchange and dissolution of aluminum from several highly weathered, acidic Natal soils. Plots of cumulative aluminum removed vs. the number of extractions were initially curvilinear (thought to represent both exchangeable and nonexchangeable aluminum), followed by a linear portion (thought to represent removal of only nonexchangeable aluminum). They reasoned that, if the linear portions of the curves were extrapolated to the ordinate, this would be a measure of "exchangeable aluminum." Their results suggested that decreasing pH value had minimal effect on exchangeable aluminum, but increasing concentration of the extracting salt increased the apparent amount of exchangeable aluminum. This was also reported by Sivasubramaniam and Talibudeen,[47] who indicated that increased extractable aluminum with decreasing pH values was the result of dissolution of nonexchangeable forms, since exchangeable aluminum should be independent of the pH value of the extracting solution. Skeen and Sumner[46] found that the amount of aluminum released was influenced by the cation in the order $NH_4 > K > Ca > Mg > Na$. The anions most effective at minimizing the release of nonexchangeable aluminum were Cl^- and NO_3^-.

Amedee and Peech[48] extracted soils with 1.0 and 0.1 *M* KCl in a similar manner, and agreed with Skeen and Sumner[46] that 1.0 *M* KCl extracted a greater quanity of aluminum than 0.1 *M* KCl. They also found that the value of (pH-1/3pAl) in the KCl extract was nearly constant at 2.8, corresponding approximately to the solubility of gibbsite (see Chapter 3). They further noted that aluminum could be extracted by 1.0 *M* KCl from freshly precipitated $Al(OH)_3$, but not from gibbsite, and suggested that aluminum also may be extracted from amorphous $Al(OH)_3$ in soils.

Since past research had indicated that more aluminum could be exchanged from montmorillonite than from vermiculite,[20,34,49] it was postulated that extracted aluminum underwent hydrolysis and polymerization, followed by irreversible adsorption to a greater extent in the higher-charged vermiculite. To study the nature of the aluminum ions displaced from vermiculite, Kaddah and Coleman[22,23] leached aluminum-saturated trioctahedral vermiculites with solutions of NaCl, KCl, and $CaCl_2$. By potentiometric titration of the leachates, they were able to quantify the amount of "exchangeable acidity" ($Al^{3+} + H^+$) and aluminum. They found that KCl extracted the smallest amount of exchangeable acidity and had the lowest ratio of aluminum to exchangeable acidity. It was also noted that a more dilute extracting solution increased the amount of hydrolysis during the exchange.

Sequential extraction of soils with 1 *M* KCl was used by Kissel et al.[24] to study the hydrolysis of nonexchangeable hydroxy-aluminum. Using conductimetric titrations, no exchangeable H^+ was detected on the soils, but only in KCl extracts. They noted that, with continued extraction, a decreasing but continuous amount of acidity was released. It required 9.0 l of 1 *M* KCl for the hydrolysis of the soil to cease. They believed this was the point where the hydroxy-aluminum had completely hydrolyzed to $Al(OH)_3$.

2. *Organic Surfaces*

Evans and Kamprath[50] observed less exchangeable aluminum in organic soils than mineral soils, even when the pH value of the organic soil was lower. Thomas[51] sampled a Maury soil from six depths which varied in organic matter content from 5.1 to 0.8%. He found that, at any pH value, the aluminum exchanged by KCl was lower as organic matter increased and concluded that small increases in organic matter may result in substantial reductions in exchangeable aluminum. Hargrove and Thomas[52] equilibrated 10 g of a Zanesville subsoil and 25 ml of 0.1 *M* KCl with various amounts of H-peat and 0.1 *M* KOH. Decreasing amounts of aluminum solution were observed with increasing peat additions at any given pH value.

Bloom et al.[32] used the Skeen and Sumner method[46] to extract two soils and an aluminum-peat with 1 *M* KCl and 0.33 *M* $LaCl_3$. They demonstrated that $LaCl_3$ extracted significantly more aluminum than KCl, and that the quantity of aluminum extracted, expressed as a fraction of $BaCl_2$-TEA acidity, was greater for the peat than the two mineral soils. Exchange equilibrium was obtained more quickly in the peat and a significantly greater amount of the aluminum on organic matter sites was exchangeable with neutral salts. Oates and Kamprath[53] reported that unbuffered chloride salts of lanthanum and potassium removed similar amounts of aluminum from mineral soils, but that lanthanum removed more aluminum from organic soils. Juo and Kamprath[54] suggested the use of $CuCl_2$ solution as an extractant for estimating the potentially reactive aluminum pool in soils. The relative effectiveness for displacing aluminum from organic matter was $Cu^{2+} > La^{3+} > K^+$.

B. Nonexchangeable Aluminum

1. *Hydrolysis*

Early research on the speciation of adsorbed aluminum focused on the effect of an exchanger on the hydrolysis of aluminum in solution. Ragland and Coleman[55] leached an aluminum-saturated montmorillonite with 10^{-4} *M* $AlCl_3$ solution and observed that the aluminum content and pH value of the leachate decreased. They concluded that the hydrolysis of aluminum was enhanced in the presence of clays when compared with hydrolysis in an aqueous solution. The driving force for the enhanced hydrolysis was suggested to be the preferential adsorption of the hydrolyzed products as compared with Al^{3+}. This induced further hydrolysis in order to maintain equilibrium conditions in solution. Ragland and Coleman[55] noted very little decrease in the CEC of the montmorillonite, even though considerable amounts of aluminum were sorbed. They concluded that the fixed aluminum was condensed as $Al(OH)_3$. They further supported this hypothesis by noting that removal of adsorbed aluminum with acid required three H^+ per aluminum released. Frink and Peech[56] agreed that $Al(OH)_3$ was the form of fixed aluminum, but argued that, in the presence of montmorillonite, the hydrolysis of aluminum in solution decreased as a result of the lower observed solution pH value. Unfortunately, the methodology they used prevents the resolution of this paradox on the basis of the experiments that were reported. Some of the difficulties are exemplified by the reported values of +3.1 to +3.8 for the charge on each aluminum ion exchanged from montmorillonite,[56] with even higher values reported for hectorite. Furthermore, possible instabilities of the montmorillonites used were not considered by these researchers.[55,56] It has been demonstrated repeatedly[18] that the nonexchangeable aluminum forms may have OH/Al ratios of <3.0. The fixation of hydroxy-aluminum with a composition of approximately $Al(OH)_2^+$ (i.e., $n = 2.0$) by a Dowex® resin was reported by Hsu and Rich.[57] Rich[20] described a decrease in CEC from 134 to 1 meq/100 g for a Lincoln, Montana vermiculite which had been aged 8 d in partially neutralized $AlCl_3$. He assumed this decrease was caused by the adsorption of aluminum polymers with an average n value of 2.0. Shen and Rich[36] added partially neutralized $AlCl_3$ solutions to calcium- and aluminum-saturated montmorillonite and found that 80% of the added aluminum was fixed with $n = 2.0$ after 6 months of aging. Their conclusions were based on acid dissolution of the newly formed interlayer material. They pointed out, however, that small amounts of residual interlayer aluminum and gibbsite remained. The inclusion of these materials in calculating the average OH/Al ratio could have increased the ratio substantially above 2.0.

Similar studies were conducted by Clark,[58] Turner and Brydon,[59,60] and Turner.[61,62] Turner[62] pointed out that experimental techniques could have significant effects on the speciation of aluminum exchanged and retained by exchangers. Unfortunately, all of the above studies were conducted using "Al^{3+}-saturated" exchangers prepared by leaching with an aluminum salt followed by washings with H_2O to remove the excess salt. Thus, the OH/Al ratio of the initially adsorbed aluminum, assumed to be zero, was actually unknown. Substantial gains in controlling experimental conditions were achieved by using sodium- or calcium-saturated samples. These results are presented below in a discussion of the influence of surface properties on adsorbed species (see Section III).

2. Polymerization and Precipitation

Hu[4] proposed a scheme for aluminum polymerization involving two hydroxyl bridges between each aluminum atom with the remaining coordination occupied by water molecules. His studies of aluminum-resin[57] and aluminum vermiculite systems[49] indicated a +1 charge per aluminum atom, corresponding to the composition $[Al_6[OH]_{12}]^{6+}$ analogous to the ring structure of gibbsite. Examining the neutralization of $AlCl_3$ or $Al_2(SO_4)_3$ in the absence of resin or vermiculite, Hsu and Bates[63] noted a charge of +0.8 per aluminum atom, corresponding to the $[Al_{10}[OH]_{22}]^{8+}$ double-ring structure. Subsequently, Hsu[4] proposed a polymerization scheme (see Figure 1) consisting of single or double gibbsite-like rings at $n \leq 2.1$. Further increases in pH ($n = 2.2 - 2.7$) transformed these cyclic derivatives into larger polymers with a reduced net positive charge per aluminum atom, with the ionic charge

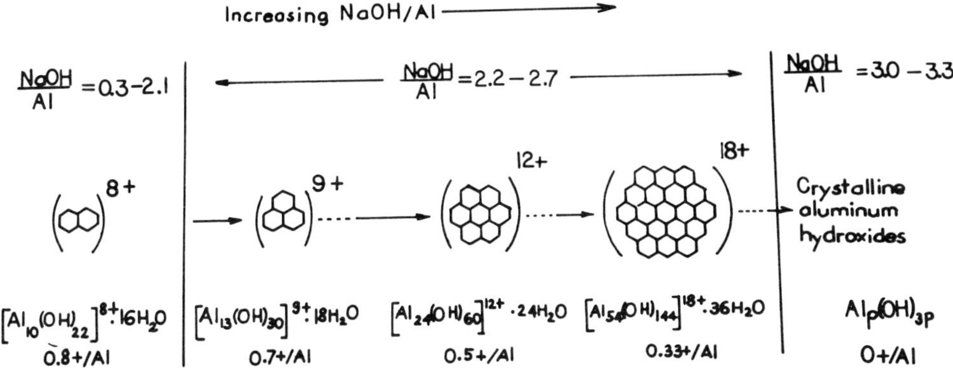

FIGURE 1. Proposed development of aluminum hydroxide.[4] (Revised from Hsu, D. H. and Bates, T. F., *Soil Sci. Soc. Am. Proc.*, 28, 763, 1964.)

remaining positive until n = 3. This residual positive charge must be balanced by counter anions in solution or by the negative charge of clay minerals. Polynuclear species other than hexamers and their cyclic derivatives may be present,[64] but their contribution toward the total species population was perceived as unimportant.[65] Because of their large size and high positive charge, these reaction products remain nonexchangeable in interlayer positions and are essentially unreactive except for their ability to release or accept protons in response to changes in pH value or cation content of the vicinal aqueous solution.

The mechanism of aluminum sorption in interlayer positions of clays is still not completely understood. Grim and Johns[66] proposed that hydroxy-aluminum interlayers existed as "islands" which were randomly distributed in the interlayer spaces of charged-minerals. Chakravarti and Talibudeen[67] and Dixon and Jackson[68] believed that the hydroxy-aluminum species were concentrated near clay mineral edges. This was supported by Frink,[69] who suggested that aluminum interlayers exist as "atolls" where the concentrated portion of hydroxy-aluminum polymers occurred at the clay mineral edges. Barshad[70] suggested further that polynuclear aluminum species were preferentially adsorbed by 2:1 clay minerals relative to monomeric species.

The formation of an "atoll" structure[69] in contrast to a uniform distribution of hydroxy-aluminum polymers is illustrated in Figure 2. Harsh and Doner[71] observed a random distribution of hydroxy-aluminum "islands" in the interlayers of Wyoming montmorillonite by transmission-election microscopy of freeze-fracture replicas of the clay surface. The concept that hydroxy-aluminum "islands" occur randomly in the interlayers of montmorillonite and as "atolls" in the interlayers of vermiculite is supported by the relationship of surface area and CEC presented by Barnhisel[72] and Kozak and Huang,[73] as well as the work of Sawhney[74] and Novak et al.[75] The "atoll" structure, concentrated near clay mineral edges, can result in a steric blocking of exchange sites occurring within the center of particles where an insufficient number of "props" occurs. Thus, hydroxy-aluminum "props" at the mineral edge cannot support or maintain the crystal structure at a 14 Å d-spacing when potassium-saturated and heated, and the structure will collapse toward 10 Å.

At n ≥ 3, aluminum is completely neutralized and crystalline $Al(OH)_3$ can form rapidly in a matter of hours. These undercharged species are thought to be expelled from the interlayers of clay minerals because of the existence of a strong, negatively charged environment that must be satisfied by cations. In addition to growth in the x and y direction, the polymers in solution may also grow in the z direction through the orientation of platelets via H-bonding and van der Waals forces. Severe constraints are further placed on the system during this ordering process, and more water must be removed from the structure.

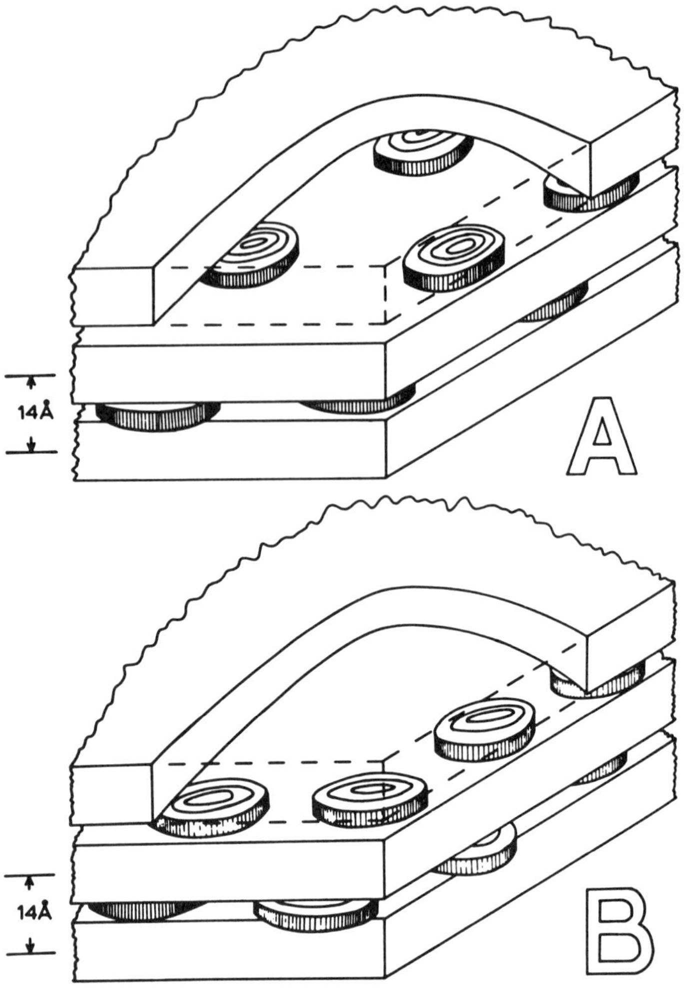

FIGURE 2. Illustration of the distribution of hydroxy-aluminum polymers in the interlayer space of 2:1 clay minerals. (A) A uniform distribution; (B) an "atoll" arrangement.[60] (Modified from Dixon, J. B. and Jackson, M. L., *Soil Sci. Soc. Am. Proc.*, 26, 358, 1962.)

Aluminum hydroxide may crystallize as bayerite, gibbsite, or nordstrandite, with similar a- and b-crystallographic lengths and differences only in the stacking of sheets along the c-crystallographic axis. Rapid precipitation appears to favor the bayerite structure, whereas slow crystal growth favors the gibbsite structure.[76] Furthermore, pH value may play a significant role in governing the formation of these three polymorphic structures.[77,78] An alkaline medium favors the bayerite structure, whereas an acidic medium promotes the gibbsite structure.[77] A mixture of nordstrandite with bayerite and/or gibbsite was obtained in an intermediate pH value range. It has also been reported that gibbsite can be prepared at pH > 12 on an industrial scale, whereas bayerite forms between pH 9 and 12 when a supersaturated sodium-aluminate solution is allowed to cool.[4] Furthermore, it is known that nordstrandite and gibbsite, not bayerite, are more common in alkaline soils or in bauxite deposits.[79-81] Inorganic anions may also play an important role in governing the formation of the three polymorphic structures.[63,77,82] A mixture of bayerite and nordstrandite was formed when $Al_2(SO_4)_3$ was neutralized with NaOH, whereas pure bayerite was formed when $AlCl_3$

was neutralized.[63] In Jamaica, nordstrandite is predominant in high-silica bauxite deposits, whereas gibbsite occurs in low-silica bauxite.[82] The mechanisms governing the development of these polymorphic structures still remain obscure (see also Chapter 3).

III. INFLUENCE OF SURFACE PROPERTIES ON ADSORBED SPECIES

The early history of both the theoretical and applied chemistry of soil acidity has been reviewed by Jenny.[12] His title, ''Reflections on the Soil Acidity Merry-Go-Round'', previews the soil acidity carousel, beginning at the turn of the century with early aluminum theories, through the H-clay theory, to newer aluminum theories. The final rejection of the H-clay theory, which occurred in the early to mid 1950s, prompted countless investigations on the interaction of aluminum with various constituents of soil and the implications for soil acidity as a whole. Coleman and Harward[83] and Low[84] provided rather convincing laboratory evidence that aluminum was the adsorbed species responsible for soil acidity, not H_3O^+. This was verified for natural systems by Rich and Obenshain,[85] who found evidence of nonexchangeable aluminum in the interlayers of vermiculite in an acid soil. Furthermore, it was also observed that H-saturated clays prepared by mineral acid or H^+-exchange-resin treatments were not stable, but spontaneously changed into Al^{3+} or Al^{3+}-Mg^{2+} clays.[70,86,87] This explains why appreciable amounts of neutral salt-exchangeable H^+ are not present in mineral soils. In an effort to understand the behavior of acid soils, numerous studies were conducted on the interaction of aluminum with various soil constituents to model actual reactions in heterogeneous soil systems.

A. Inorganic Surfaces

1. Vermiculite and Smectite

The interaction of aluminum with vermiculite and montmorillonite has been an attractive topic of study, because these two phyllosilicates occur in abundance in many soils (see Chapter 3). Hsu and Bates[49] treated vermiculite with aluminum solutions of varying basicities and concluded that the clay interlayer became interstratified with aluminum polymers. Analysis with X-ray diffraction after potassium-saturation verified their contention, since the complete collapse of the vermiculite interlayer to 10 Å was not evident (see Figure 3). When relatively few polymers were adsorbed on the clay, a strong, assymmetric, 10-Å peak broadened on its low-angle side, indicating incomplete collapse. With increasing amounts of aluminum polymers in the interlayer, a broad hump and then a broad, 12-Å peak gradually developed. A strong, 14-Å peak did not develop until about 70% of the CEC of the vermiculite was occupied by aluminum-hydroxy polymers. With high concentrations of aluminum in the interlayer, Hsu and Bates[49] suggested that further sorption may be impeded either physically (because of the size of the polymer) or chemically (because of unfavorable charge distribution) so that polymerization is limited. However, they found that aluminum hydrolysis was enhanced in the presence of the vermiculitic clay. Rich[20] also suggested enhanced hydrolysis of adsorbed aluminum on vermiculite. During this period, Jackson[64] proposed the concept of ''steric pinching'', which suggested that ions located in a high-charge density interlayer, such as interlayers of vermiculite, have a greater opportunity for hydrolysis and olation relative to ions located in low-charge density interlayers, such as those of montmorillonite. Closer association between the Al^{3+} ion and its coordinated water molecules was thought to account for its enhanced polymerization and hydrolysis on vermiculite. This idea soon met with disfavor through the experimental efforts of several researchers.

Artificial aluminum interlayers were synthesized in montmorillonite and vermiculite by Carstea.[88] He found that the formation of aluminum interlayers in both phyllosilicates increased with increasing temperature, and that the greater degree of interlayering in montmorillonite as compared to vermiculite of a similar size fraction was related to the difference

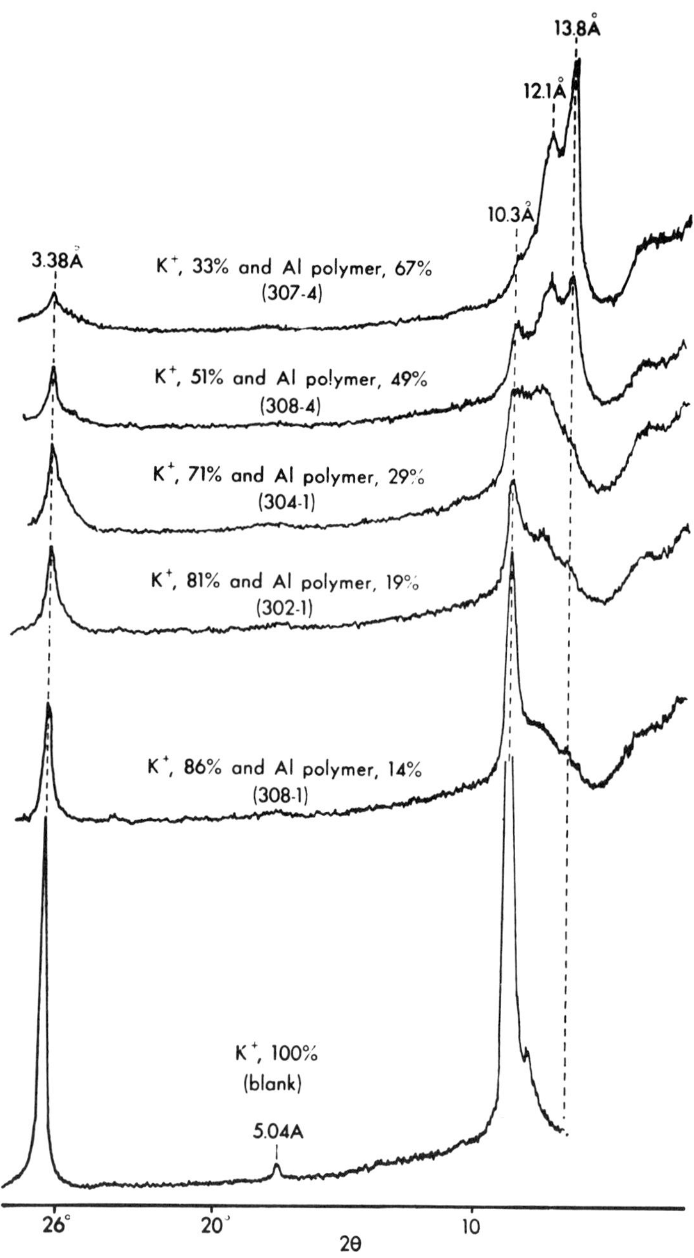

FIGURE 3. X-ray diffraction patterns of vermiculite pretreated with OH-Al solution of NaOH/Al = 2.55, followed by K^+ saturation. (From Sawhney, B. L., *Clays Clay Miner.*, 16, 157, 1968. With permission.)

in the amount and location of the charge (see Table 1). Since vermiculite has a higher-charge density, the aluminum polymer should be held more tightly, limiting the degree of hydrolysis. Sawhney[74] also proposed that the expanded interlayer space in montmorillonite provides an ideal locale for the enhanced polymerization of aluminum, whereas the interlayer space of vermiculite restricts polymerization. The increased number of negative charges per unit area on vermiculite may delay polymerization and precipitation of aluminum, as do anions such as Cl^- and SO_4^{2-}.[77,89-91] The formation of hydroxy-aluminum species in beidellite

Table 1
PHYSICAL AND CHEMICAL PROPERTIES OF HYDROXY-ALUMINUM-INTERLAYERED MONTMORILLONITE AND VERMICULITE AFTER 10 D OF EQUILIBRATION

Particle size (μm)	pH		CEC			mg Al[a]/me of blocked CEC	Basal spacings (A) X-ray pretreatments				
	1 min	10 d	cmol/kg	Percent reduction of CEC	Percent of aluminum adsorbed		Ca 54%RH	Ca Solv.	K 105°C	K 300°C	K 550°C
Montmorillonite (3°C)											
<0.2	5.6	5.3	18.8	85.2	70.5	94	17.2	17.9	14.5	14.4	13.6
2—0.2	5.5	5.1	25.6	78.9	56.2	84	16.7	17.3	14.2	14.0	13.2
Vermiculite (3°C)											
2—0.2	5.6	5.3	95.9	35.8	35.1	96	14.7	14.8	11.8	11.0	10.4
50—2	5.3	5.1	134.2	24.5	18.4	62	14.7	14.2	14.0[b], 11.9	10.3, 11.1	10.1
Montmorillonite (21°C)											
<0.2	5.6	4.9	11.6	90.0	92.5	114	14.7	14.2	14.8	14.7	14.0
2—0.2	5.1	4.6	15.2	87.4	83.5	114	15.0	15.0	14.7	14.6	13.7
Vermiculite (21°C)											
2—0.2	6.1	5.3	83.7	43.8	46.2	102	14.8	14.7	11.8	11.2	10.4
50—2	5.7	4.8	100.3	43.4	53.5	98	15.0	14.0	11.8	11.0	10.5

[a] Initially, 72 mg of aluminum per 0.5 g of mineral were added.
[b] Where two peaks were observed, the more prominent one is underlined.

From Carstea, D., *Clays Clay Miner.*, 16, 231, 1968. With permission.

also suggest restricted alluminum polymerization in high-charge density minerals.[92] Surface area determination by argon adsorption implied that the interstratified aluminum-hydroxy material was highly porous. Possible dehydroxylation of the hydroxy-aluminum resulted in the formation of stable anhydrous alumina ''pillars'', which moved the beidellite interlayers apart and occupied only a small internal area. Smectites pillared with aluminum-hydroxy polymers has also been suggested by Plee et al.[93]

Strong evidence suggests that precipitates of amorphous or crystalline $Al(OH)_3$ can form in the interlayers of montmorillonite, but not in those of vermiculite.[41,60,74,94] Multilayer sorption of hydroxy-aluminum polymers was noted in montmorillonite interlayers by Sawhney.[74] With aging, the sorbed aluminum species further polymerized to form gibbsite. This phenomenon was not observed in vermiculite and was therefore attributed to a favorable condition in montmorillonite interlayers for the organization of hydroxy-aluminum ions into a gibbsite structure. Turner and Brydon[60] also demonstrated gibbsite formation in aged, partially neutralized aluminum-montmorillonite samples. By monitoring the solubility product of $Al(OH)_3$ with time, they showed that montmorillonite served as a template for the growth of crystalline gibbsite. With excess amorphous aluminum in the system, the solubility product of crystalline gibbsite,[95] $10^{-34.0}$, was approached rapidly. Once gibbsite had formed in the interlayer space, the clay-aluminum complex was no longer stable and gibbsite was eventually released from the interlayer.

Singh and Brydon[96] also formed artificial interlayers in montmorillonite using the salt $Al_2(SO_4)_3$. The material that initially formed disappeared from the interlayer space and a crystalline basic aluminum sulfate formed in solution. The precipitated aluminum interlayer material was not stable with sulfate present at the low OH/Al molar ratio of 2.2. Veith[94] suggested that the formation of gibbsite on clays requires extensive aging and may occur at an n value as low as 0.6, provided ''enough'' solution is added to increase the basicity of adsorbed aluminum to the point beyond which gibbsite is formed. Hodges and Zelazny[43] found that montmorillonite fixed aluminum with basicities ranging from 2.1 to 2.9, suggesting that precipitates may have formed (see Table 2). The formation of gibbsite was later verified using differential scanning calorimetry (DSC).

Studies on the interaction of aluminum with soil clays generally indicate that polynuclear aluminum species are preferred related to aluminum monomers by interlayered clays.[70] Veith[41,94] has shown that montmorillonite adsorbs rather large polynuclear aluminum species from solution, whereas vermiculite adsorbs smaller aluminum-hydroxide complexes (see Figure 4). He suggested that this is related to the relative nonexpandability of the vermiculite clay. Hodges and Zelazny[43] have also shown that the amount of aluminum fixed by montmorillonite decreased with decreasing concentration of aluminum in solution, suggesting the preferential adsorption of hydrolyzed polynuclear species. The more dilute solutions contain a larger percentage of mononuclear aluminum. Investigating the interaction of partially neutralized aluminum with montmorillonite using miscible displacement, Jardine et al.[97] found that the clay initially preferred monomeric aluminum, followed by polymer preference at longer times (see Figure 5a). Plotting the fraction of monomer or polymer adsorbed on the montmorillonite (reduced concentration) vs. time, Jardine et al.[97] showed that the decrease in monomer adsorption was more rapid than the decrease in polymer adsorption. This result reflects the ability of montmorillonite to accumulate aluminum polymers in its interlayers. The selectivity reversal for monomers and polymers of aluminum on montmorillonite was also evident from the time dependence of effluent basicities see (Figure 5b). Initially high effluent basicities suggest that low-basicity aluminum species (most likely aluminum monomers) were removed from solution and adsorbed on montmorillonite. As leaching continued, effluent basicities monotonically decreased below the influent basicity, reflecting the accumulation of aluminum polymers in the interlayers of montmorillonite.

Table 2
CHEMICAL CHARACTERISTICS OF $BaCl_2$-EXCHANGEABLE ALUMINUM AND AVERAGE BASICITIES FOR ADSORBED, EXCHANGEABLE, AND NONEXCHANGEABLE ALUMINUM ON KAOLINITE, MONTMORILLONITE, AND PEAT

Sample	BaCl			Av OH/Al	Av OH/Al (total ads. Al)	Av OH/Al (nonexchangeable Al)
	$BaCl_2$ ext. Al / Total ads. Al	Monomer Al / $BaCl_2$ ext. Al	Polymer Al / $BaCl_2$ ext. Al			
Kaolinite	0.353	0.140	0.860	2.10	2.71	2.86
Montmorillonite	0.290	0.299	0.701	1.82	2.41	2.85
Peat	0.359	0.381	0.619	1.39	2.08	0.88

From Jardine, P. M., Zelazny, L. W., and Parker, J. C., *Soil Sci. Soc. Am. J.*, 49, 862, 1985. With permission.

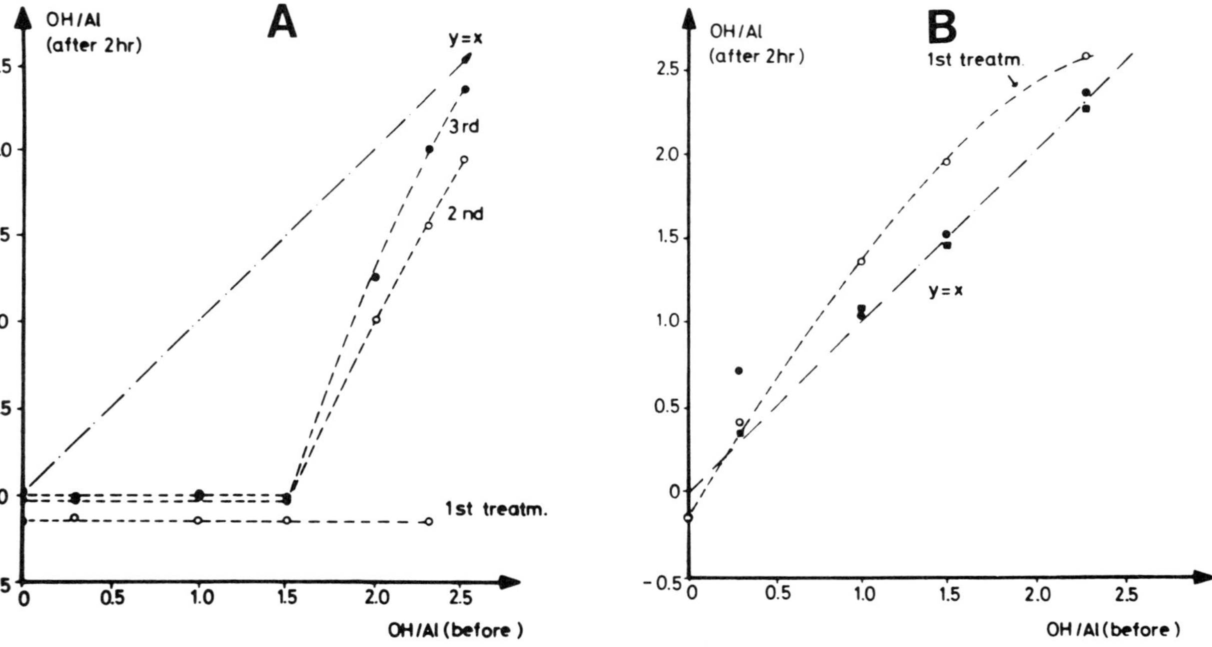

FIGURE 4. (A) Average OH/Al ratios of aluminum in solution before and after the first, second, and third 2-h treatment of sodium-montmorillonite (Wyoming) with equivalent $Al(OH)_BCl_{3-B}$. (B) Average OH/Al ratios of aluminum in solution before and after the first, second, and third 2-h treatment of sodium-vermiculite (S.A.) with equivalent $Al(OH)_BCl_{3-B}$ (○ = first, ● = second, and ■ = third treatment). (From Veith, J. A., *Proc. Soil Sci. Soc. Am. J.*, 41, 865, 1977. With permission.)

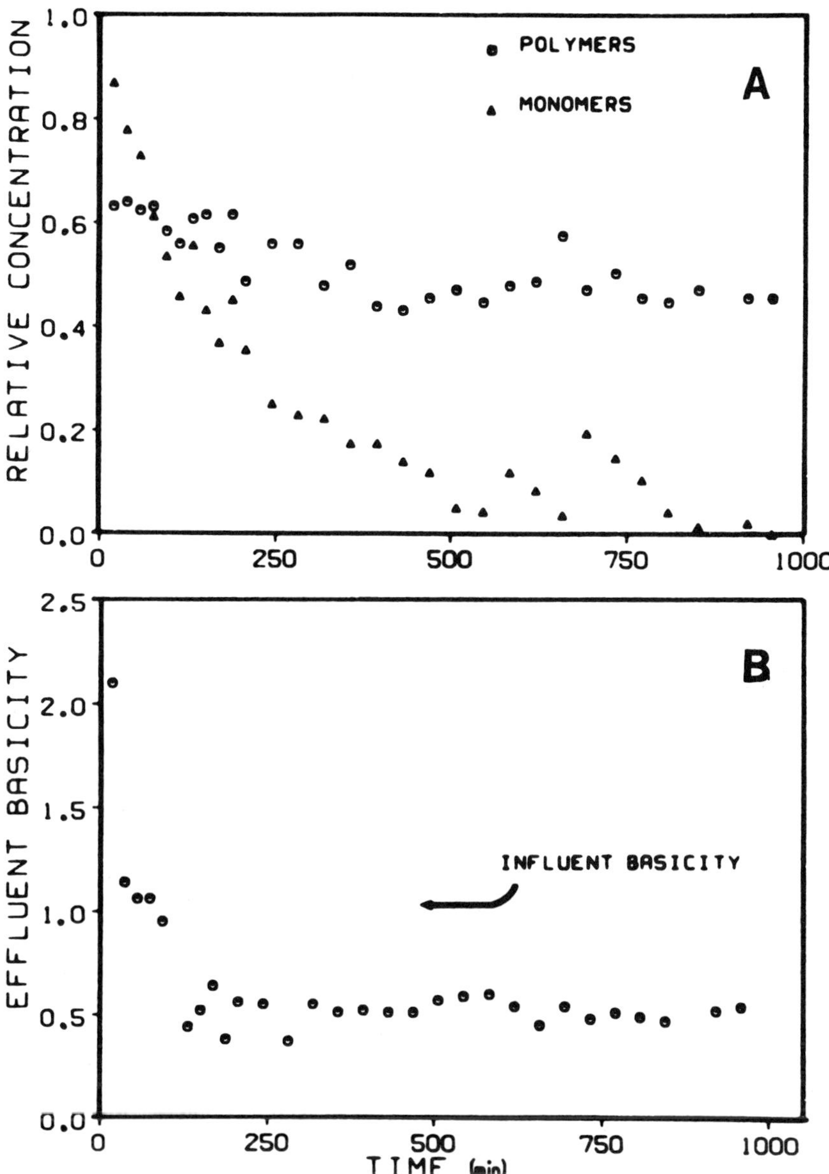

FIGURE 5. (A) Relative concentration of total polymeric and monomeric aluminum adsorbed as a function of time on montmorillonite. (B) Effluent aluminum basicities as a function of time on montmorillonite. (From Jardine, P. M., Zelazny, L. W., and Parker, J. C., *Soil Sci. Soc. Am. J.*, 49, 862, 1985. With permission.)

2. *Kaolinite*

Numerous studies in recent years have addressed the interactions of aluminum with kaolinite, a major constituent of acidic soils in the tropics and the southeastern U.S. Turner and Brydon[60] noted that high solution aluminum concentrations produced rapidly formed precipitates on kaolinite which were soon released because of an apparent instability of the sorbed precipitate. When smaller amounts of aluminum were added to the clay, however, aluminum precipitates formed which were held permanently on the surface of the kaolinite. It was assumed that the precipitates occurred on the clay mineral edges where the primary

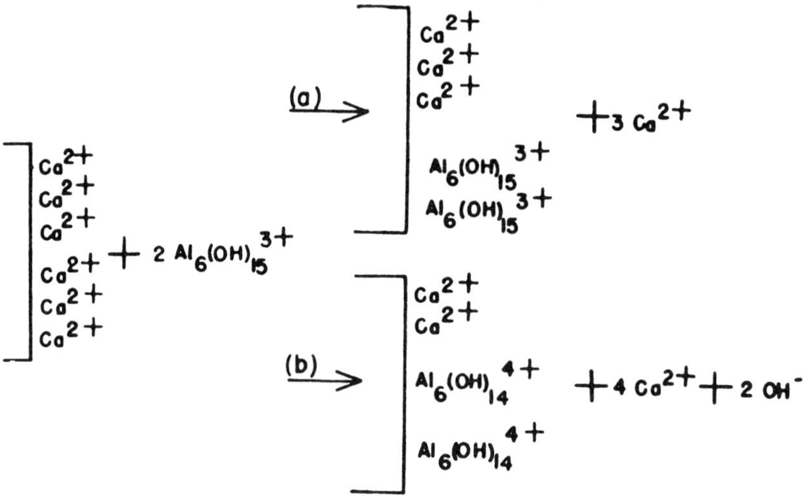

$$\left]\begin{matrix}Ca^{2+}\\Ca^{2+}\\Ca^{2+}\\Ca^{2+}\\Ca^{2+}\\Ca^{2+}\end{matrix}\right. + 2\,Al_6(OH)_{15}^{3+} \xrightarrow{(a)} \left]\begin{matrix}Ca^{2+}\\Ca^{2+}\\Ca^{2+}\\Al_6(OH)_{15}^{3+}\\Al_6(OH)_{15}^{3+}\end{matrix}\right. + 3\,Ca^{2+}$$

$$\xrightarrow{(b)} \left]\begin{matrix}Ca^{2+}\\Ca^{2+}\\Al_6(OH)_{14}^{4+}\\Al_6(OH)_{14}^{4+}\end{matrix}\right. + 4\,Ca^{2+} + 2\,OH^-$$

FIGURE 6. A hypothetical mechanism for the adsorption of hydroxy-aluminum. (a) With no change in polymer structure; (b) with dehydroxylation of adsorbed polymer. (From Hodges, S. C. and Zelazny, L. W., *Soil Sci. Soc. Am. J.*, 47, 221, 1983. With permission.)

seat of charge is located, not on the planar surfaces (see Chapter 7). Brown and Hem,[98] however, found that the adsorption of polymeric aluminum species on clay surfaces was more closely related to surface area than to CEC. They also suggested that polymeric aluminum adsorbed on mineral surfaces was not readily transformed to gibbsite. This idea was supported further by El-Swaify and Emerson,[99] who investigated the effect of aluminum and iron oxides on the aggregate stability of kaolinite. They suggested that the presence of kaolinite prevented aluminum hydroxides from crystallizing as gibbsite. In the absence of the clay, however, identical hydroxy-aluminum solutions were able to precipitate gibbsite. This apparent ''anti-gibbsite'' effect, first proposed by Jackson,[100] was also noted by Hodges and Zelazny.[43,44] Investigating the interaction of dilute, partially neutralized aluminum solutions with kaolinite, these authors[44] found that the average basicity of adsorbed aluminum decreased with increasing age. This implied that kaolinite was capable of dehydroxylating hydrolyzed aluminum (see Figure 6). The fixation of aluminum on the kaolinite resulted in basicity values ranging from 2.1 to 2.4. Even so, no gibbsite was detected using DSC.

Hodges and Zelazny[44] determined that the addition of amounts of aluminum less than or equivalent to the exchange capacity of kaolinite resulted in (1) an *increase* in pH value relative to solutions with no kaolinite, as compared to reported *decreases* when excess aluminum was added, and (2) depolymerization of aluminum added ($n = 2.0$ and 2.4) by kaolinite when aluminum equivalent to 40 and 60% of the exchange capacity was added. When excess aluminum was added to kaolinite, the pH value decreased compared to the pH value of the solutions with no kaolinite, the amount of exchangeable and nonexchangeable aluminum increased, and the basicity of nonexchangeable aluminum increased. Other effects of aluminum additions of less than the CEC included decreased hydrolysis of aluminum adsorbed from solutions of low initial n value and decreased basicity of the exchangeable aluminum as compared with previously reported data. Also, nonexchangeable aluminum often was observed to have increased basicities as compared to data reported previously.

The kinetics and mechanism of aluminum adsorption on calcium-saturated kaolinite were studied by Jardine et al.[97,101] using miscible displacement methods. Observed breakthrough curves for continuous injection of unneutralized and partially neutralized aluminum solutions through kaolinite suggested the existence of at least two mechanisms for aluminum adsorption (see Figure 7). It was shown that a relatively rapid reaction, with solution aluminum in

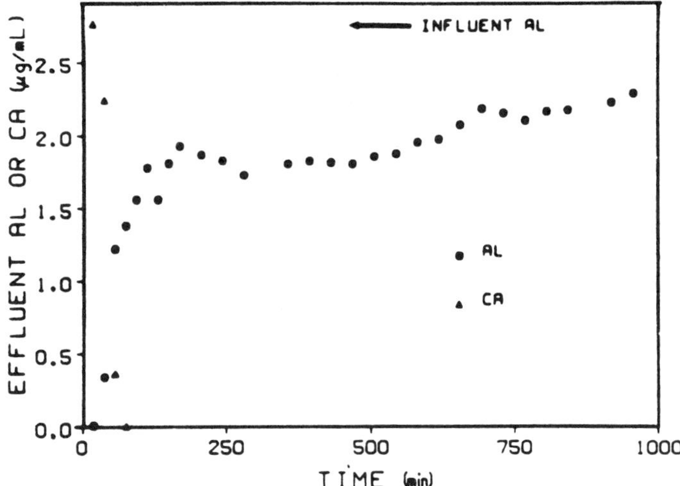

FIGURE 7. Effluent aluminum and calcium as a function of time for kaolinite. (From Jardine, D. M., Zelazny, L. W., and Parker, J. C., *Soil Sci. Soc. Am. J.*, 49, 862, 1985. With permission.)

equilibrium with a solid phase, involved calcium-aluminum exchange, and a slower reaction involved aluminum polymerization on the adsorbent. This scenario was supported by modeling observed breakthrough curves with nonequilibrium transport theory.[101] These authors[97] found that the polymerization of aluminum was more extensive on kaolinite than on montmorillonite and peat (see Table 2). The average OH/Al molar ratios for adsorbed, exchangeable, and nonexchangeable aluminum, using influent aluminum solutions with OH/Al = 1.0, were much higher for kaolinite relative to the other exchangers, suggesting extensive aluminum polymerization on kaolinite.

3. Illite and Mica

A meager amount of research has been conducted on the interactions of aluminum with illite and the primary minerals biotite and muscovite. Huang and Kozak,[102] investigating the adsorption of hydroxy-aluminum polymers by muscovite and biotite, suggested that the increase in adsorption with a decrease in particle size was attributable to an increase in external surface area. This seems reasonable, since potassium-saturated mica particles should have interlayer space of limited accessibility. The depletion of some K^+ from the interlayers of micas resulted in a larger quantity of adsorbed aluminum on biotite, but had a negligible effect on the adsorption of aluminum on muscovite (see Table 3).[73] This appeared to be caused by charge effects and steric hindrance of the aluminum polymers in the interlayers of potassium-depleted muscovite.

El-Swaify and Emerson,[99] investigating the effect of precipitated aluminum and iron hydroxides on the swelling and aggregate stability of illite, suggested that this clay is interstratified by hydroxy-aluminum species, which reduces its diffuse double layer and, hence, swelling tendencies. Columbera et al.,[103] however, found that illite adsorbed hydroxy-aluminum polymers (see Figure 8), and suggested that the adsorption process was not one of exchange, but instead involved the formation of H bonds between the OH groups of the polymer and siloxane groups on the illite surface. Investigating the interaction of aluminum with illite and montmorillonite, Edwards et al.[104] observed positive Cl^- adsorption by the aluminum clays and suggested that it was associated with the tendency of this cation to undergo hydrolysis. The authors[104] noted that aluminum hydrolysis was less on illite relative

Table 3
ADSORPTION OF HYDROXY-ALUMINUM BY VARIOUS MINERALS AFTER AGING IN HYDROXY-ALUMINUM POLYMER SOLUTIONS

	mg Al adsorbed/g clay	
Mineral[a]	OH/Al = 2.0	OH/Al = 2.5
Na-montmorillonite	25.8	27.0
	58.7[b]	57.8[b]
Na-vermiculite	21.6	21.0
	27.6[b]	24.6[b]
K-vermiculite	8.6	9.9
Biotite	5.0	5.5
48% K-depleted biotite	23.2	14.9
63% K-depleted biotite	27.0	17.5
	30.6[b]	22.5[b]
70% K-depleted biotite	26.9	14.8
82% K-depleted biotite	27.0	15.1
Muscovite	4.5	4.4
44% K-depleted muscovite	4.8	4.3
49% K-depleted muscovite	6.0	4.6
57% K-depleted muscovite	7.0	6.6

[a] All samples were aged in 50 ml of polymer solution per g of clay unless otherwise stated.
[b] 250 ml of polymer solution per g of clay.

From Kozak, L. M. and Huang, P. M., *Clays Clay Miner.*, 19, 95, 1971. With permission.

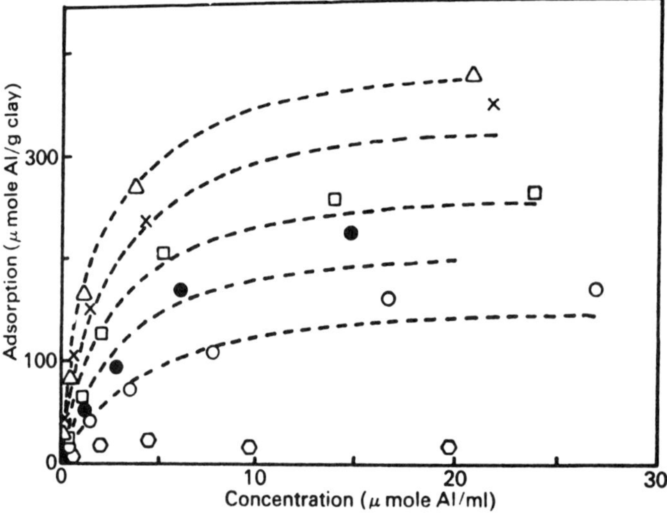

FIGURE 8. The adsorption of aluminum onto Fithian illite from hydroxy-aluminum perchlorate solutions. △ Adsorption from OH/Al = 2.0, 0.5 *M* $NaClO_4$ solutions; × Adsorption from OH/Al = 1.5, 0.6 *M* $NaClO_4$ solutions; □ Adsorption from OH/Al = 1.5, 0.5 *M* $NaClO_4$ solutions; ● Adsorption from OH/Al = 1.0, 0.5 *M* $NaClO_4$ solutions; ○ Adsorption from OH/Al = 0.5, 0.5 *M* $NaClO_4$ solutions; and ⬡ Adsorption from acidified (pH 1.5—2) aluminum perchlorate, 0.5 *M* $NaClO_4$ solutions. (From Columbera, P. M., Posner, A. M., and Quirk, J. P., *J. Soil Sci.*, 22, 118, 1971. With permission.)

to montmorillonite, and suggested that it was the result of the higher-charge density of the former clay that protected the bound aluminum from extensive hydrolysis.

B. Organic Surfaces

1. Resins

Exchange resins often are used to simulate soil minerals under conditions where instability of the minerals would seriously affect the results. To avoid reactions with structural aluminum in clays, Hsu and Rich[57] used a synthetic cation-exchange resin to study aluminum fixation. They found that aluminum-hydroxy ions, with an OH/Al ratio of 2, became nonexchangeable on the resin and reduced its CEC. Veith and Sposito[25] treated a calcium-saturated resin with partially neutralized aluminum and noted that the basicity of adsorbed aluminum was always higher than the basicity of the initial aluminum solution. They concluded that aluminum hydrolysis could occur at pH values >2.7 in the presence of a sulfonic resin. Cotten[105] also found that the extent of aluminum hydrolysis increased as the degree of cross-linkage and charge density of the sulfonic resin decreased. The enhanced hydrolysis and polymerization of aluminum in the presence of cationic-exchange resins was also shown by Hodges and Zelazny,[43] who detected gibbsite in their samples after several months of aging.

2. Natural Organic Compounds

Another important constituent of heterogeneous soil systems is organic matter. Its complex polymeric structure has been discussed by Theng[106] and Schnitzer.[107] Aluminum interactions on peat, humic acid, and fulvic acid have been studied by many researchers to model aluminum reactions with organic matter (see Chapter 5 for a complete discussion). Pionke and Corey[108] suggested that aluminum-organic matter complexes were quite strong, since the amount of nonexchangeable aluminum was highly correlated with organic matter content. They assumed that nonexchangeable aluminum was the difference between 1 *M* NH_4OAc at pH 4.8 extracted aluminum and KCl extracted aluminum. Spyridakis et al.[109] observed the formation of kaolinite from the alteration of biotite to hydroxy-aluminum interlayered vermiculite in sand cultures in which coniferous seedlings were grown. Where organic matter was added in the culture medium, kaolinite did not form, and it was suggested that the organic matter may have complexed the aluminum released by weathering, which was essential for kaolinite formation.

Most of the pH buffering in acid soils containing organic matter is because of aluminum hydrolysis on organic matter exchange sites and is not necessarily associated with organically derived H ions.[14,19] This agrees with the observations of Bloom et al.,[110] who concluded that the hydrolysis of organically bound aluminum is a major source of buffering in the pH range of 4 to 5. White and Thomas[111] suggested that weakly acidic carboxyl groups act as a sink for H_3O^+ released during the hydrolysis and polymerization of aluminum on organic matter. They found a close correlation between the apparent pK_a of the acid groups of organic matter and the basicity of the adsorbed aluminum, suggesting that aluminum was more extensively hydrolyzed when the average Brønsted acid strength of the carboxyl groups of the exchanger was weaker.

Using muck samples and a carboxylic resin to serve as a model for soil organic matter, Hargrove and Thomas[112] suggested that it would be unlikely that aluminum could occupy all of the reactive sites of organic matter, since the former is always hydrolyzed to some degree and H_3O^+ ions will be adsorbed on weakly ionized acidic groups. The average basicities of adsorbed aluminum on the mucks ranged from 0.88 to 2.07. The upper limit of this range agreed well with the results of White and Thomas,[111] who used acid peat and humic acid to simulate soil organic matter. Bloom et al.[110] and Bloom and McBride[113] proposed that hydroxy-aluminum ions were preferred more by organic matter than Al^{+3}. They suggested that hydroxy-aluminum ions have a lower charge and hydration energy, allowing them to approach more closely a binding site and possibly maximize the binding energy. Bloom[114] further hypothesized that the hexameric polymers postulated by Hsu and Rich[57] are not likely to be sorbed by peat, since the geometry of the peat carboxyl sites would limit, if not prevent, association with the six charges of the hexameric ring. Hargrove and Thomas,[112] however, suggested that three, or even two, simultaneously dissociating carboxyl groups are not necessary for complexing Al^{+3}. The stability of aluminum-humic acid complexes were shown by Arai and Kumada[115] to increase with decreasing distance between two adjacent carboxylic groups on the humic acid, or with increasing basicity of sorbed aluminum. Young and Bache[116] and Jardine and Zelazny[117] have also shown that the chemical affinity of fulvic acid for aluminum is no greater than that of a weak complexant, such as adipic or acetic acid. Hodges and Zelazny[43] investigated the interaction of dilute, partially neutralized aluminum solutions with peat. They noted that the exchanger fixed aluminum with basicities very near that of the equilibrium solution used, and that no gibbsite could be detected on the solid using DSC. The lack of aluminum precipitation on the peat was indirectly associated with the high-charge density of the exchanger, as is the case with vermiculite. Investigating the kinetics and mechanisms of mononuclear and polynuclear aluminum adsorption on peat, Jardine et al.[97] also found very limited hydrolysis and polymerization of these species on this exchanger. Further discussion of natural organic complexes is given in Chapter 5.

IV. INFLUENCE OF ANIONS ON THE SURFACE REACTIONS OF ADSORBED SPECIES

A. Inorganic Anions

The adsorption of polynuclear and mononuclear aluminum onto soil constituents is not affected only by the surface properties of the latter, but also depends on associated complexing ligands that compete for aluminum. Some inorganic anions present in the soil solution compete strongly with mineral and organic surfaces for reactive sites with the aluminum ion. Most studies investigating the influence of anions on the adsorption of aluminum have used Cl^-.[36,59,61,118,119] Hsu[76] showed that partially neutralized $AlCl_3$ solutions with OH/Al molar ratios in the range 1.8 to 2.7 were metastable for about 1 year, and then began to precipitate slowly as gibbsite. Partially neutralized $AlCl_3$ solutions (OH/Al = 2.7), which were dialyzed to remove excess chlorine, formed a precipitate within 2 weeks. This suggested that high chlorine concentrations inhibited the spontaneous hydrolysis and polymerization of aluminum as the solutions aged.

Hsu[76] assumed that polynuclear aluminum coordinated with Cl^- ions and that further hydrolysis and polymerization must proceed via dissociation of these ions. The dissociation of Cl^- from the polymer is inversely related to the Cl^- ion concentration in the bulk solution. Thus the Cl^- ion may inhibit the formation of gibbsite, both in pure solution and in the presence of a soild-phase adsorbent.[77,118,119] The presence of a negatively charged solid adsorbent may minimize the effect of Cl^-, since double-layer repulsion will reduce the concentration of Cl^- in the vicinity of surface-bound aluminum. If the concentration of chlorine is high enough, repulsive forces cannot completely eliminate the association of chlorine with the aluminum polymer. Barnhisel and Rich[77] found kaolinite to be a suitable template for gibbsite formation; however, poor crystallinity of the gibbsite may have been caused by the inclusion of Cl^- ions in the gibbsite lattice, thus preventing the growth of larger crystals. In clay-free systems, Violante and Jackson[120] formed stable pseudoboehmite in neutralized aluminum solutions at pH 8 to 10. When montmorillonite was present in these solutions, the crystallization of pseudoboehmite was inhibited and nucleation of gibbsite and/or nordstrandite occurred on the clay surface. Only in the presence of very high concentrations of electrolyte (2 to 3 *M* NaCl at pH 9) did pseudoboehmite crystallize as a separate phase with montmorillonite present.

The rate of $Al(OH)_3$ crystallization is generally inversely related to the tendency of anion inclusion in aluminum-hydroxy precipitates, and this tendency decreases with an increase in size and structural complexity of monovalent inorganic anions. Ross and Turner[89] found that the ability of anions to become included and reduce the rate of gibbsite crystallization followed the order $Cl^- > NO_3^- > ClO_4^-$, where ClO_4^- was shown to have an insignificant effect on the rate of aluminum crystallization. The interaction of Cl^- and Na_3^- are outside the coordination sphere of the aluminum ion and therefore mainly electrostatic. Anions such as SO_4^{2-}, CO_3^{2-}, $H_2PO_4^-$, and F^- show evidence of inner- or outer-sphere coordination with aluminum[90,121-123] and this will undoubtedly affect the formation and adsorptive properties of mononuclear and polynuclear aluminum (see Chapter 4 for a comprehensive discussion). Investigating the effect of SO_4^{2-} on the adsorption of partially neutralized aluminum by montmorillonite, Singh and Brydon[96] reported that the interlayer material which formed initially disappeared from the interlayer space and a crystalline basic aluminum sulfate formed. The ion activity product, $(Al^{3+})\ (OH^-)^3$, was $10^{-33.6}$, slightly greater than that for gibbsite ($10^{-34.0}$). In chlorine systems, supersaturation can exist with respect to gibbsite, but, once gibbsite is formed, the interlayer material, which is approaching OH/Al = 3, is not stable. Singh and Brydon[96] found that interlayer material in the presence of SO_4^{2-} was not stable at the low OH/Al molar ratio of 2.2.

The anions SO_4^{2-}, CO_3^{2-}, $H_2PO_4^-$, and F^- are much more effective than Cl^- or NO_3^- in preventing the spontaneous formation of $Al(OH)_3$ in supersaturated solutions.[89,96,124] Jardine and Zelazny[125] have shown that SO_4^{2-}, $H_2PO_4^-$, and F^- form strong bonds with mononuclear and polynuclear aluminum, which did not readily dissociate until high solution pH values were obtained. The products that form at any given pH value are stable and the anions associated with aluminum will reduce the spontaneous formation of $Al(OH)_3$ which occurs upon aging. The coordination of SO_4^{2-}, CO_3^{2-}, $H_2PO_4^-$, and F^- with mononuclear and polynuclear aluminum will undoubtedly affect the formation and adsorptive properties of these aluminum species on soil constituents. Extensive research is required to understand more fully the interaction of inorganic anion-aluminum complexes with solid-phase adsorbents (see also Chapter 4).

B. Organic Anions

The association of organic anions with aluminum and the interaction of these species with soil constituents has not been extensively studied. As is the case with many inorganic anions, solution-phase organic ligands may coordinate with aluminum and inhibit the hydrolytic reactions of the cation, thus retarding the crystallization of aluminum hydroxides.[126-130] As the affinity of organic ligands for aluminum increases, the effectiveness of the anion to hinder the hydrolysis and polymerization of aluminum also increases. Kwong and Hwang[129] found that the ability of several organic ligands to perturb the precipitation of solid-phase aluminum was in the order citrate > malate > tannate > aspartate > *p*-hydroxybenzoate. Increased concentrations of these organic ligands increased the degree of hindrance of the hydrolytic reactions of aluminum in solution. The mechanism by which organic ligands impede aluminum hydrolysis and polymerization is believed to involve disruption of aluminum hydroxyl bridges, which are essential for the formation of crystalline aluminum hydroxides.[127-131]

The effect of organic ligands on the formation of hydroxy-aluminum clay complexes depends greatly on the initial organic ligand/aluminum molar ratio. Goh and Huang[132] reacted aluminum-citrate solutions (OH/Al = 0.0 and 2.5) with montmorillonite for 30 d (see Table 4). A low citrate/aluminum ratio of 0.1 was found to not significantly affect the CEC or the structure of the hydroxy-aluminum-montmorillonite complexes. An intermediate citrate/aluminum ratio of 0.5 resulted in the precipitation of aluminum-citrate complexes in the interlayer of montmorillonite. Structural distortions of the clay interlamellar space resulted,

Table 4
THE pH VALUE AND CEC OF MONTMORILLONITE AFTER EQUILIBRATION WITH HYDROXY-ALUMINUM AND CITRIC ACID

Sample no.	NaOH/Al		Citrate/Al	pH Initial	pH Final	CEC[a] before extraction by 0.2 *M* HCl (cmol$_c$/kg)	CEC[a] after extraction by 0.2 *M* HCl (cmol$_c$/kg)
1		In water		7.4	7.4	96	95
2		In $AlCl_3$		4.0	3.9	78	83
3	0[b]		0.1	3.5	3.6	92	88
4	0		0.5	3.0	3.0	84	86
5	0		1.0	2.8	3.0	89	85
6	2.5		0	4.5	4.1	31	43
7	2.5		0.1	4.6	4.7	34	47
8	2.5		0.5	4.3	4.6	61	76

[a] Mean error of duplicates = ±3 cmol$_c$/kg.
[b] Al NaOH/Al = 0, no NaOH was added to the montmorillonite suspension in the $AlCl_3$ solution.

From Goh, T. B. and Huang, P. M., *Can. J. Soil Sci.*, 64, 411, 1984. With permission.

perturbing the crystallization of orderly aluminum-hydroxy interlayers. At high citrate/aluminum ratios (citrate/aluminum = 1.0), most of the aluminum remained in solution. Suspensions aged in the presence of unneutralized and partially neutralized aluminum with no added citrate resulted in a decrease in pH value because of the hydrolysis of surface-bound aluminum with aging. Samples that were aged in the presence of citric acid exhibited no change in pH value or a slight increase in pH value, since citrate was most likely inhibiting aluminum hydrolysis and polymerization (see Table 4).

The influence of various concentrations of citrate on the CEC of the hydroxy-aluminum montmorillonite is clearly shown in the data for samples 6 to 9, which were aged at an initial OH/Al ratio of 2.5 (see Table 4). The adsorption of hydroxy-aluminum by montmorillonite decreased with increasing citrate/aluminum ratio. Thus, less charge blocking of the montmorillonite was occurring at the higher ratio. Although the presence of negative charge because of the dissociation of citrate is possible, the net effect of increasing the citrate/aluminum ratio with montmorillonite was an increase in CEC (see Table 4). The stability of hydroxy-aluminum-montmorillonite complexes under alkaline conditions (pH 7 to 9) increases in the presence of electrolytes. Violante and Jackson[120] found that the effectiveness of anions to stabilize interlayering was of the order citrate $\gg SO_4^- > Cl^-$. These results contrast with those of Goh and Huang,[132] who performed their studies under acidic conditions (pH 3 to 5).

A contrasting, but most interesting approach to the formation of aluminum-hydroxy interstratified minerals in the presence of organic anions was undertaken by Vincente et al.[133] Instead of adding hydrolized aluminum to mineral suspensions, these authors allowed the aluminum to originate from the mineral itself via reactions with added organic acids. The acidyifing conditions created by the organic acids resulted in a spontaneous release of structural aluminum, which then was readsorbed on the mineral surface. The interaction of oxalate and citrate with trioctahedral micas resulted in the destruction of the mineral because of rapid complexation of the released aluminum by the organic acids. Lack of aluminum readsorption on the solid phase caused mineral destruction. Tartrate and salicylate, which do not bind aluminum as well as oxalate and citrate, caused the autotransformation of trioctahedral mica to interstratified vermiculite-mica. Organic acids which complex aluminum rather weakly (e.g., succinic, aspartic, and acetic acids) react with mica to produce hydroxy-aluminum vermiculite. Vincente et al.[133] further suggest that galacturonic acid interactions with mica form a high-charge density smectite. Appreciable research is necessary to evaluate more fully the interaction of organic anion-aluminum complexes with soil constituents.

V. CHANGES IN CHEMICAL AND PHYSICAL PROPERTIES OF CLAYS AND SOILS BY ADSORBED SPECIES

A. Chemical

1. CEC Reduction

The cation-exchange capacity of clay minerals can be drastically reduced[58,69,134] by the occupancy of exchange sites by positively charged, nonexchangeable hydroxy-cation groups.[36,135] The CEC of many soil clays has been increased by the extraction of interlayer aluminum.[58,68,69,85,134,136-141] Similarly, many researchers have reported reductions in the CEC of clays following adsorption of hydroxy-aluminum polymers.[38,42,61,62,73,142-150] However, others have reported little if any such reduction.[55,56] Hydroxy-aluminum interlayers may affect CEC[18] by (1) the occupancy of exchange sites, (2) the physical blocking of the exchange reaction, not allowing the saturating cation to come into contact with the exchange sites, and (3) the influence of hydroxy-material on pH values and vice versa. Consequently, many conditions must be clarified to determine the influence of hydroxy-aluminum interlayers on resulting CEC.

Keren[148] evaluated the influence of final pH value and rate of aluminum neutralization on the reduction in CEC of a montmorillonite containing hydroxy-aluminum interlayers. A constant addition of 5.33 mols aluminum/kg montmorillonite was selected to provide sufficient coverage of the clay surface under the assumption that (1) the specific surface area of the clay was 76 ha/kg, (2) the clay platelets were mutually parallel, and (3) the hydroxy-aluminum precipitate had a structure similar to gibbsite. Neutralization of this aluminum-montmorillonite at a constant rate of 0.1 mmol OH/min to a final pH of 5, 6, 7.5, 9, and 10 provided reductions in CEC of 88, 85, 66, 41, and 0%, respectively. Neutralization of this aluminum-montmorillonite to a final pH of 7.5 at rates of 0.025, 0.1, 0.4, and 0.8 meq OH/min provided reductions in CEC of 71, 66, 58, and 51%, respectively. Thus, low equilibrium pH values and slow rates of aluminum neutralization lead to greater reductions in CEC.

The CEC of sodium-montmorillonite-hydroxy-aluminum mixtures was measured at 495, 483, and 468 mmol/kg for systems containing aluminum coverage at 10, 25, and 50% of the CEC, respectively.[149] These results indicate that the hydroxy-aluminum polymers effectively reduced the CEC of sodium-montmorillonite by about 46, 47, and 49%, regardless of coverage. This reduction was found for a wide range of aluminum contents: from 0.1[149] to 17[150] times the CEC; however, the degree of reduction in CEC depended to some degree on the aluminum content. For the same clay, equilibrium pH value, and neutralization rate, the CEC reduction was 71% at an aluminum content of 5.33 mol/kg[148] to 46% at an aluminum content of 0.03 mol Al/kg[149] (see Table 5). Despite large differences in aluminum content, only a relatively small change was noted in the CEC, indicating that only a small amount of aluminum-hydroxy polymer was responsible for the CEC reduction in montmorillonite. This strong effect appears to have been caused by the precipitation of hydroxy-aluminum polymers on the planar surfaces of single platelets. These aluminum polymers may have adsorbed strongly to the clay planar surfaces, interacting with other platelets free of hydroxy-aluminum polymers to form a quasicrystal in which a significant part of the adsorbed sodium could be trapped and become nonexchangeable.

Hydroxy-aluminum polymers were less effective in reducing the CEC of calcium-mont-

Table 5
SODIUM SATURATION AND ^{22}Na ADSORPTION (COUNTS/MIN) ON MONTMORILLONITE-HYDROXY-ALUMINUM MIXTURES AT pH 7.5 FOLLOWING SUCCESSIVE SATURATION AND EXTRACTION TREATMENTS

	Percent of adsorbed ion[a]			A[b]			B	C
System	Na	Ca	Al	^{22}Na (cpm/sample)	Adsorbed Na (mmol/kg)	CEC reduction (mmol/kg)	^{22}Na (cpm/sample)	Adsorbed Na (mmol/kg)
Na-clay				2530 + 212	914 + 76		6	—
Ca-clay				2412 + 121	871 + 44		70	—
Al-clay[c]				389 + 3	872 + 11		0	—
Clay-hydroxy-Al	90	0	10	1371 + 56	495 + 21	46	0	442 + 45
Clay-hydroxy-Al	75	0	25	1337 + 167	483 + 60	47	0	462 + 45
Clay-hydroxy-Al	50	0	50	1296 + 118	468 + 43	49	2	435 + 36
Clay-hydroxy-Al	0	90	10	2198 + 56	791 + 20	9	17	781 + 12
Clay-hydroxy-Al	0	75	25	2175 + 21	786 + 7	10	0	801 + 15
Clay-hydroxy-Al	0	50	50	1791 + 27	647 + 10	26	2	675 + 10

[a] The percent of adsorbed ions before titration with OH^- took place.
[b] A = five washes with ^{22}Na-labeled 0.1 NaCl plus five ethanol washes; B = five washes with 0.1 *M* KCl solution; C = five washes with ^{22}Na-labeled 0.1 *M* NaCl plus five ethanol washes.
[c] Five washes with ^{22}Na-labeled 1 *M* NaCl at pH 3.5 plus five ethanol washes.

From Keren, R., *Clays Clay Miner.*, 34, 534, 1986. With permission.

morillonite systems (<26%) than sodium-montmorillonite (>46%).[149] This difference may be ascribed to differences in the arrangement of the clay platelets. Because of limited mobility of hydroxyl ions between the platelets of calcium-montmorillonite, the amount of hydroxy-aluminum precipitation may be less in the calcium systems and thus the CEC reduction would be correspondingly smaller.

Keren[149] noted that adsorbed ^{22}Na in the montmorillonite-hydroxy-aluminum mixtures (which remain after CEC determination; see column A, Table 5) was removed by washing with 0.1 *M* KCl (see column B, Table 5), showing that sodium adsorbed on exposed surfaces was exchangeable. Moreover, after these washings, most of the blockage of the exchange sites remained (see column C, Table 5), indicating that the interaction between the hydroxy-aluminum precipitate and the clay platelets was relatively stable.

Pure aluminum hydroxides have been reported to adsorb divalent cations (M^{2+}) through some form of $OH–M^{2+}$ interaction.[151] Kwong and Huang[130,152] measured calcium adsorption on aluminum-precipitated products of 50 mmol/kg at pH 5 to 130 mmol/kg at pH 9. The adsorption of calcium on the aluminum precipitates was enhanced measurably by the initial presence in solution of 1.0×10^{-4} *M* tannate > citrate > malate > aspartate > *p*-hydroxybenzoate (see Figure 9). At an initial concentration of 1.0×10^{-6} *M*, only tannic acid and citric acid enhanced calcium adsorption on the aluminum-precipitated products.

2. *Cation Fixation and Selectivity*

Since hydroxy interlayers result in the reduction of the effective CEC of clay minerals, the presence of these interlayer species should counter the effect of high-charge density and decrease the potassium fixation capacity of minerals.[15] The presence of hydroxy-aluminum islands can also act as a "prop" between silicate layers and inhibit the collapse of the layers, subsequently reducing potassium fixation in vermiculite and in soil containing vermiculite.[153] Fixation may also be reduced by the presence of "exchangeable aluminum" which is not easily displaced by low concentrations of KCl.[85,153-158]

The presence of hydroxy interlayers can also affect the selectivity for various ions[159] by permitting the entry of potassium-sized ions into partially opened mica layers while preventing collapse at the mineral edges. These effects of hydroxy-interlayer species on the selectivity of some clay minerals for potassium,[73,141,159-163] cesium,[164-166] and rubidium[167] have been demonstrated experimentally.

Rich and Black[160] noted that the introduction of hydroxy-aluminum groups into vermiculite increased the Gapon selectivity coefficient (k_G) for a K/Ca system from 5.7 to 11.1 (see Table 6). However, potassium fixation was reduced markedly. When vermiculite saturated with the mixed calcium-potassium solution was dried, 2.8 cmol/kg potassium was removed by 1 *M* NH_4OAc, but in the case of the aluminum-treated sample 23.2 cmol/kg potassium was removed. Prior to drying, 17.8 and 28.9 cmol/kg potassium had been removed, respectively. The k_G for a number of soil clays and specimen clay minerals are presented in Table 6. Apparently, potassium ions are more mobile in "propped" interlayers than in interlayers where the silicate layers tend to collapse about the potassium ions (see Figure 10). Thus, in the first instance, potassium can reach the wedge zones and can be adsorbed preferentially.

Hydroxy aluminum interlayer material apparently also has a chemical as well as a physical effect on K/Ca selectivity.[168] Zelazny and Rich[169] found that the K/Ca ratio for adsorbed ions increased as temperature increased in systems containing hydroxy-aluminum in expandable layer silicates and in a synthetic high-charge, cation-exchange resin. In the absence of hydroxy-aluminum, but in the presence of many other cations, temperature had little effect on the K/Ca ratio. Similar effects of hydroxy-aluminum interlayer material on the temperature dependence of cesium selectivity have also been reported.[170]

The position of the wedge zone appears to influence the extent of the "catalytic" effect

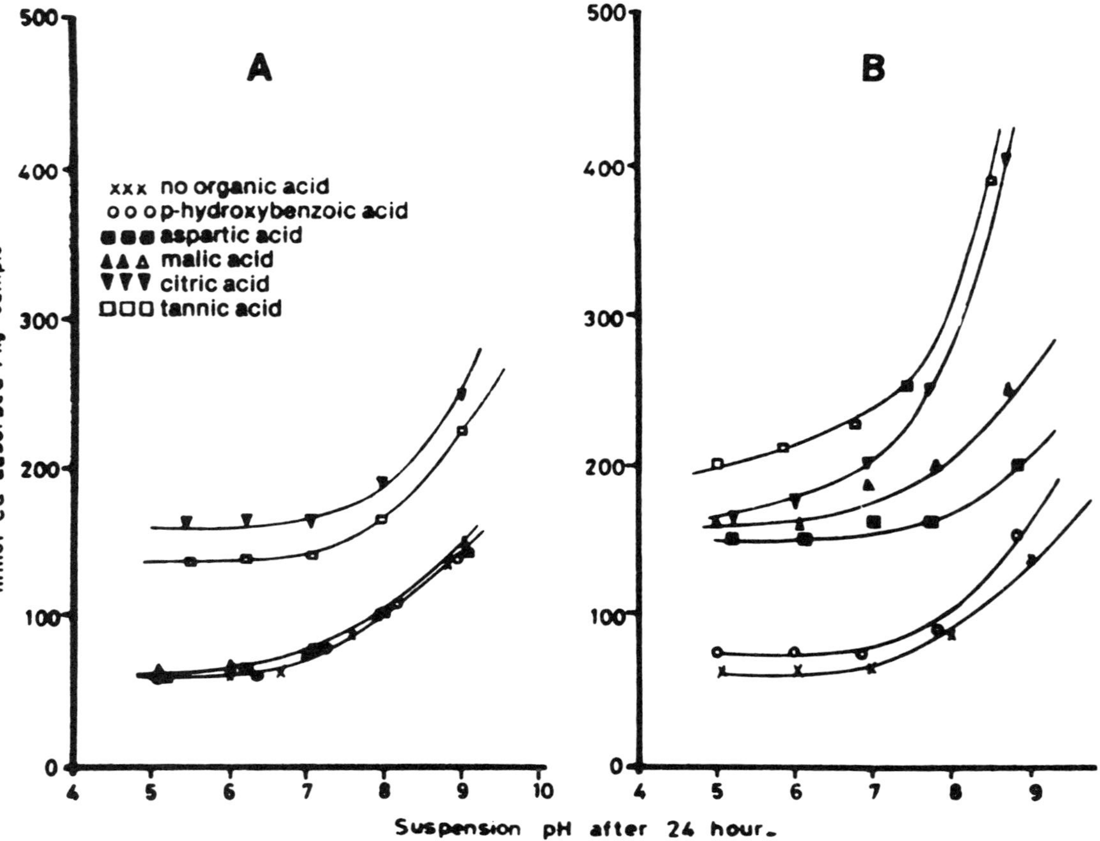

FIGURE 9. Comparison of tannic acid and selected low relative molecular mass organic acids on the retention of calcium as a function of pH by precipitation products of aluminum formed after the 1-d aging at room temperature in systems at the initial aluminum concentration of 1.1×10^{-3} *M* and OH/Al molar ratio of 3.0. The initial concentrations of organic acids present during the precipitation of aluminum are (A) 1.0×10^{-6} *M* and (B) 1.0×10^{-4} *M*. (From Kwong, K. F. and Huang, P. M., *Soil Sci.*, 128, 337, 1979, and *Geoderma*, 26, 179, 1981. With permission.)

Table 6
EXCHANGEABLE CALCIUM AND POTASSIUM AND THE GAPON-EXCHANGE COEFFICIENT FOR CLAYS SATURATED WITH A MIXED $CaCl_2$-KCl (EACH 0.005 *N*) SOLUTION

Clay mineral or soil clay	Origin	Size (μm)	Ca (meq/100)	K (kg)	kK/Ca (1/[mmol][a,b] × 10^{-2})
Clay Minerals					
Illite	Morris, IL	2—0.2	21.0	4.3	6.5
Muscovite	Ontario, Canada	<0.2	18.0	18.6	32.8
Montmorillonite	Otay, CA	<2	87.5	22.8	8.2
Montmorillonite	Wyoming	<2	71.5	7.8	3.4
Vermiculite	Libby, MT	<0.2	99.6	17.8	5.7
Vermiculite-treated	Libby, MT	<0.2	82.0	28.9	11.1
Vermiculite-Biotite	Libby, MT	20—5	57.0	3.6	14.7
Soil Clays					
Nason (A)	Virginia	2—0.2	22.0	16.8	24.3
Nason (C)	Virginia	2—0.2	20.3	20.9	32.6
Berks (Ap)	Virginia	2—0.2	8.6	7.5	27.6
Carrington (Ap)	Iowa	2—0.2	18.8	11.0	18.5
Putnam (B22)	Missouri	<2	42.0	21.5	16.2

[a] Based on calcium and potassium extracted from the wet samples by 1 *N* NH_4OAc, except for the untreated vermiculite, in which case the values for 1 *N* $Mg(OAc)_2$ were used (NH_4OAc removed only 15.8 meq).

[b] Al^{3+} saturated and boiled in H_2O for 2 h prior to K^+ selectivity determination.

From Rich, C. I. and Black, W. R., *Soil Sci.*, 97, 384, 1964. With permission.

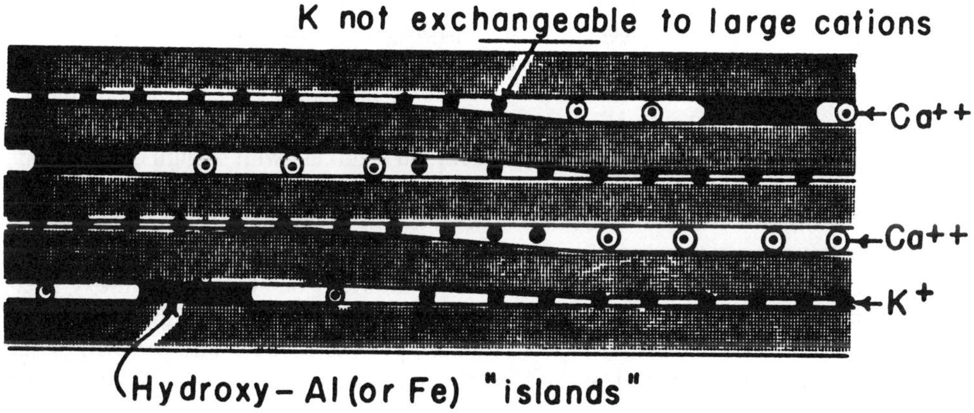

FIGURE 10. Proposed model of an expansible layer silicate with interlayers indicating an effect on potassium fixation. (From Rich, C. I., *The Role of Potassium in Agriculture,* Kilmer, V. J., Ed., American Society of Agronomy, Crop Science Society of America, and Soil Science Society of America, Madison, WI, 1968, 79. With permission.)

of this configuration in potassium selectivity.[171] Potassium is selected from among larger hydrated ions (e.g., Ca^{2+} or Mg^{2+}) because of the space limitations for diffusion of the latter into the wedge zone. If the wedge zone is near the edge of the particle, potassium-selectivity is minimal, but if it is deep within the particle, then the "catalytic" effect can result in considerably more potassium selectivity. As silicate layers close about the sorbed K^+ ions, a new wedge zone can form (see Figure 10). This progression depends on the concentration of K^+ ions, the structural alignment, the charge density of the mineral, the concentration of large hydrated ions, and the presence of hydroxy-aluminum or other "prop"-like interlayer materials. At very low concentrations of potassium, there is selection for potassium at the wedge site, but higher potassium concentrations promote collapse at the wedge site. With still further increases in potassium concentration, there is a tendency for the collapse of the silicate layers where the exchange is initiated (at the edge of the particles) with a consequent entrapment of cations deeper within the crystal. In the absence of wedge zones (e.g., true vermiculite), a large concentration of K^+ (or similar ions) is necessary to initiate the collapse of the vermiculite. Potassium selectivity would be expected to be low until the collapse was initiated.

Schachtschabel[172] studied the relative replacing power of cations in relation to mineral composition and concentration. He concluded that there was no constant order of the replacing power of cations, but their affinity depended upon both the concentration of the external solution and the mineral composition. Often the strong replacing position of H_3O^+ in the lyotropic series is a result of Al^{3+} confused for H_3O^+ by a reaction termed "interchange reaction" by Barshad.[70] Foscolos[173] determined the order of relative replacing power in the lyotropic series by cation-exchange equilibrium coefficients to be $H_3O^+ < Na^+ < Mg^{2+} < Ca^{2+} < K^+ < Al^{3+}$ in Colony, Wyoming montmorillonite (90 cmol/kg CEC); $H_3O^+ < Na^+ < Mg^{2+} < Ca^{2+} < Al^{3+} < K^+$ in Otay, California montmorillonite (125 cmol/kg CEC); and $Mg^{2+} < Ca^{2+} < H_3O^+ < Al^{3+} < Na^+$ in Jeffersite vermiculite (175 cmol/kg CEC). Bloom et al.[40] reported the preference of ions for a smectite surface (94 cmol/kg CEC) to follow the order $H_3O^+ < Ca^{2+} < Al^{3+} <$ hydroxy-aluminum.

Banin[174] measured a preference for calcium over sodium on montmorillonite with the selectivity coefficients varying with the relative fraction of the adsorbed ion. On completely separated montmorillonite platelets, these two ions are distributed randomly on the exchange sites. Calcium saturation in interlayer positions results in quasicrystal formation, thus increasing the preference of calcium over sodium on exchange sites.[174] The affinity of internal surfaces for calcium has been calculated to be 2 to 4 times greater than that for external surfaces.[175] Thus, quasicrystal formation produces ion demixing with sodium concentration on external surfaces and calcium on internal surfaces.[176] The presence of hydroxy-aluminum interlayers in montmorillonite further increases the selectivity of calcium over sodium, presumably because of increased quasicrystal formation.[147]

Specific adsorption of copper has been measured on synthetic aluminum-hydroxy interlayered montmorillonite[177] and hectorite,[178] which exceeded adsorption on microcrystalline gibbsite. Greater pH-dependency and more chemisorption to specific sites occurred on the aluminum-hydroxy interlayered clay than on the noninterlayered clays.[177,178] Much more Cu^{2+} also existed at discrete sites as monomeric Cu^{2+} with the aluminum-interlayered clays than with microcrystalline gibbsite.[178]

B. Physical

1. Swelling and Aggregate Stability

The adsorption of aluminum species on clays and soils greatly affects moisture-retention properties and aggregate stability of the former. Aluminum hydroxides, as well as iron hydroxides, are considered to improve the structure of lateritic and related soils through the interaction of their positive charge with the negative charge of other soil components.

Increasing evidence suggests that aluminum hydroxides are more important than iron hydroxides as bonding agents in these highly weathered, well-oxidized soils.[99,179-182] Most free iron oxides in these soils exist as small particles that have little effect on the physical properties of the soil. Tweneboah et al.[183] have shown that prior removal of aluminum oxides in such systems destabilizes soil pores and drastically reduces soil-water diffusivity. The effectiveness of aluminum-hydroxy species in promoting stable aggregates in heavy-textured paddy soils has also been demonstrated by Nakano and Kono[184] and Shiraishi.[185,186] Under paddy cultivation, macropores became clogged with clay particles, which reduced the permeability of the soil. When drainage becomes necessary for the cultivation of crops other than paddy rice, the stability of the soil structure must be enhanced in order for ditches and underdraining to work effectively.[187] Soil aggregation and permeability are commonly improved and maintained by applications of hydroxy-aluminum or $CaSO_4$. The former treatment has the advantage of partially irreversible binding to soil clays, with consequent long-lasting effects. However, hydroxy-aluminum may also increase the acidity of the soil dramatically.

Hydroxy-aluminum species are also more effective than iron hydroxides in reducing clay mineral swelling.[181,182,188] The greater control of expansion is likely the result of more uniform interlayering and stronger bonds between aluminum hydroxide and the clay surface.[88] Rich[15] and Carstea et al.[144] have shown that aluminum interlayers in montmorillonite and vermiculite are more stable than iron interlayers. The occurrence of swelling chlorite supports the mechanism of bonding instead of a blocking action.[189] The precipitation of aluminum hydroxide onto clay surfaces is planar in shape, whereas iron precipitates have spherical or acicular shapes.[99,182] Thus, aluminum hydroxides are more likely to be bonded directly over a larger area of contact with clay particles and inhibit expansion more effectively. Although clay swelling is greatly reduced with aluminum saturation, Karathanasis and Evangelou[190] suggested that such systems have a higher field-moisture capacity and higher moisture retention during drought.

2. *Flocculation and Colloidal Interactions*

Aluminum plays an important role in the control of soil flocculation and the colloidal stability of soils and suspensions. The presence of aluminum species on the surfaces of clays is thought to be responsible for some positive sites which prevent complete dispersion of clays as negatively charged colloids.[191,192] Nonhydrolyzable metal cations (e.g., K^+ or Ca^{2+}) coagulate negatively charged colloids by compression of the diffuse double layer. However, hydrolyzable metal cations such as aluminum reduce the double-layer charge to near zero and impart some positive charge to the clay, which enhances colloidal interactions and thus flocculation.[192-194] An excess of hydrolyzable metal ions on a clay surface may result in a reversal of the charge, leading to a stable suspension with a positive charge. Roberts et al.[193] demonstrated that the flocculation of kaolinite with polyacrylamide can be enhanced by the adsorption of hydrolyzed aluminum onto the clay. The positively charged aluminum species served as "anchor points" to attach negatively charged polyacrylamides to clay surfaces of negative charge. An excess of adsorbed hydroxy-aluminum on the clay, however, may flatten the polyacrylamide and hinder flocculation. Colloidal interactions of clay minerals with adsorbed aluminum and iron-hydroxides in the absence of drying was investigated by El-Swaify and Emerson.[99] Both aluminum and iron hydroxides provided equal enhancement of clay colloidal stability above their isoelectric point; however, the former hydroxide species was more effective in inducing clay flocculation and charge reversal at low pH values.

3. *Surface Area*

The addition of hydroxy-aluminum to the interlayers of expandable phyllosilicates will decrease internal and total surface area. Reductions in total surface area with the addition of hydroxy-aluminum to interlayer positions have been measured by ethylene glycol adsorption for montmorillonite[36,72] and vermiculite.[12,72] Generally, a linear decrease in the total surface area of montmorillonite occurs with increasing additions of hydroxy-aluminum because of linear decreases in internal surface area and slight increases in external surface area.[72] Although the total and external surface area of vermiculite are reduced drastically with small additions of hydroxy-aluminum material, systematic changes with further additions are complicated by structural considerations, providing further support for the "atoll" structure in vermiculite.[72]

VI. CONCLUSIONS

Aluminum occupies a somewhat anomalous position among the elements, in that it is a very common and important constituent of many inorganic materials of the biosphere, but is quite rare and usually an unimportant component of living matter. Although aluminum is found typically in low amounts in plants and animals, the aluminum content of most soils is quite high. Such aluminum is usually not readily available and is contained primarily in the structures of primary minerals and aluminosilicate clays. As soils become acidified through weathering and the leaching of nonhydrolyzable cations, these sources of aluminum solubilize, releasing aluminum in its readily available chemical forms. The newly released aluminum may be readsorbed onto the solid surfaces of soil colloids or remain in the soil solution.

The differentiation of the various forms of aluminum in a soil is especially difficult, because these forms have ranges of extractability in solution because of a wide range in particle size, crystallinity, and ionic composition. In addition to the exchangeable aluminum that is bound to the soil by electrostatic charge, each precipitate of aluminum has its own solubility product and equilibrium with soil solution. The aluminum in mineral soils can be categorized as either exchangeable or nonexchangeable, with the amount of exchangeable aluminum controlled by the solubility of the nonexchangeable aluminum and the adsorption characteristics of the soil.

On a theoretical, equilibrium basis, there is considerable support for a mononuclear $Al(OH)_3$ or mononuclear-dinuclear-$Al(OH)_3$-controlled hydrolysis scheme. Under equilibrium constraints, polynuclear hydroxy-aluminum ions would be considered as unstable, transient species on the way to becoming solid $Al(OH)_3$. In a real-world, steady-state system, there is general agreement that a monomer or monomer-dimer hydrolysis scheme is adequate to describe experimental results for aluminum solutions with $n < 0.5$ or concentrations $< 10^{-3}$ *M* (see Chapter 2). As solution concentrations and/or n values increase, it becomes necessary to assume the presence of polynuclear hydroxy-aluminum species (see Chapter 4). The degree of polymerization of these species increases as the n value increases, until the precipitation of solid $Al(OH)_3$ occurs. Based on structural considerations, kinetic studies of formation and dissolution, and light-scattering studies, it seems likely that polynuclear hydroxy-aluminum ions would have a structure composed of hexamer units or cyclic derivatives of a hexamer unit. Many species not conforming to these constraints are proposed and possible under specific conditions, but are probably based on nonsteady-state data or failure to account for precipitated solids. The stability and, thus, the "steady-state" hydrolysis quotients of the polynuclear hydroxy-aluminum species vary with the experimental conditions and methods (see Chapter 4 for a complete discussion).

Most studies prior to 1964 were conducted with exchangers that were assumed to be saturated with Al^{3+}. These studies failed to account for hydrolysis induced by the removal of excess salts. Nevertheless, important groundwork was established and mechanisms elucidated for the formation of aluminum hydroxy-interlayered vermiculites and montmorillonites. Studies of the forms of exchangeable aluminum indicated initially that only Al^{3+} was exchanged. More recently it has been recognized that hydrolyzed monomeric and polymeric aluminum can be exchanged from montmorillonite, although vermiculite and resins seem to exchange only Al^{3+}.

Nonexchangeable aluminum was assumed initially to be precipitated as $Al(OH)_3$ in the interlayer of clay minerals. This was disproved by workers attempting to synthesize aluminum-hydroxy interlayers in vermiculite and smectites when large decreases in the CEC of these minerals were observed. More detailed studies using sodium- or calcium-saturated exchangers indicated that polymer forms were being fixed. The OH/Al molar ratio of these polymers varies from 2.0 to as high as 2.7. The ability of different minerals to fix aluminum polymers with high n values is dependent on the charge and the expansibility of the interlayer. Thus, vermiculite, with high charge and low expansibility, fixes large amounts of Al^{3+} and polymer aluminum of $n \leq 2.0$. Montmorillonite preferentially fixes polymeric aluminum of higher n values of 2 to 2.5 because of its lower charge and highly expansible interlayer, while kaolinite seemed to prefer a species of ~ 2.0. Peat seems to display no distinct selectivity, but adsorbed aluminum with similar basicity to that initially added. These results indicate that the decrease in exchangeable aluminum with increasing organic matter content of soils results primarily from the greater exchange capacity of organic matter instead of the enhanced uptake attributable to specific adsorption or complex formation.

Montmorillonite exchanges more aluminum than other exchangers, whereas kaolinite fixes more aluminum more rapidly, and with greater reduction in CEC, than any other exchanger when compared on a unit-CEC basis. The reason for the unexpectedly strong reactions of kaolinite with aluminum must lie in the nature of the bonding sites. It is almost certain that most of the charge in kaolinite arises from edge sites, but the reason for such strong aluminum affinity is not apparent.

The addition of aluminum to exchangers at amounts less than or equivalent to the CEC has resulted most notably in (1) *increases* in the pH value of solutions relative to solutions with no exchanger (as compared to reported *decreases* when excess aluminum is added) and (2) the depolymerization of aluminum, added at initial OH/Al molar ratios of 2.0 and 2.4, by kaolinite when aluminum equivalent to 40 and 60% of the exchange capacity was added. When excessive aluminum was added to kaolinite samples, the pH value decreased compared to the pH value of the solutions with no kaolinite, the amount of exchangeable and nonexchangeable aluminum increased, and the basicity of nonexchangeable aluminum increased. Other effects of nonexcessive aluminum additions included decreased hydrolysis of aluminum adsorbed from solution of low initial n value, and decreased basicity of the exchangeable aluminum when compared with excessive additions of aluminum.

The use of dilute aluminum solutions resulted in decreased basicity of the initially adsorbed aluminum, but increases in the basicity of exchangeable aluminum and nonexchangeable aluminum when compared with a more concentrated aluminum solution. Increased percentages of exchangeable aluminum and less reduction in CEC are also noted in the more dilute solutions.

The adsorption of polynuclear and mononuclear aluminum onto soil constituents is reported to be affected by the surface properties of the constituents as well as the presence of associated systems that compete for aluminum. Some inorganic and organic anions present in the soil solution compete strongly with mineral and organic surfaces for the aluminum ion. The adsorbed aluminum has been reported to affect significantly both chemical and physical properties of clays and soils. Most noted chemical changes are CEC reduction, decreased cation fixation, and increased cation selectivity. Most noted physical changes are reduced swelling, increased aggregate stability, and enhanced colloidal interaction.

REFERENCES

1. **Frye, K.,** *Modern Mineralogy,* Prentice-Hall, Englewood Cliffs, NJ, 1974.
2. **Hem, J. D.,** Aluminum species in water, *Adv. Chem. Ser.,* 73, 98, 1968.
3. **Baes, C. F., Jr. and Mesmer, R. E.,** *The Hydrolysis of Cations,* John Wiley & Sons, New York, 1976.
4. **Hsu, P. H.,** Aluminum hydroxides and oxyhydroxides, in *Minerals in Soil Environments,* Dixon, J. B. and Weed, S. B., Eds., Soil Science Society of America, Madison, WI, 1977, 99.
5. **Smith, R. W.,** Relations among equilibrium and nonequilibrium aqueous species of aluminum hydroxy complexes, *Adv. Chem. Ser.,* 106, 250, 1971.
6. **Nair, V. D. and Prenzel, J.,** Calculations of equilibrium concentration of mono- and polynuclear hydroxyaluminum species at different pH and total aluminum concentrations, *Z. Pflanzenernaehr. Bodenkd.,* 141, 741, 1978.
7. **Smith, R. W. and Hem, J. D.,** Effect of aging on aluminum hydroxide complexes in dilute aqueous solutions, *U.S. Geol. Surv. Water-Supply Pap.,* 1827-D, 1972.
8. **Jardine, P. M. and Zelazny, L. W.,** Mononuclear and polynuclear aluminum speciation through differential kinetic reactions with ferron, *Soil Sci. Soc. Am. J.,* 50, 895, 1986.
9. **Turner, R. C.,** A second species of polynuclear hydroxyaluminum cation, its formation and some of its properties, *Can. J. Chem.,* 54, 1910, 1976.
10. **Turner, R. C.,** Effect of aging on properties of polynuclear hydroxyaluminum cations, *Can. J. Chem.,* 54, 1528, 1976.
11. **Tsai, P. P. and Hsu, P. H.,** Studies of aged OH-Al solutions using kinetics of Al-ferron reactions and sulfate precipitation, *Soil Sci. Soc. Am. J.,* 48, 59, 1984.
12. **Jenny, H.,** Reflections on the soil acidity merry-go-round, *Soil Sci. Soc. Am. Proc.,* 25, 428, 1961.
13. **Jackson, M. L.,** Aluminum bonding in soils: a unifying principle in soil science, *Soil. Sci. Soc. Am. Proc.,* 27, 1, 1963.
14. **Coleman, N. T. and Thomas, G. W.,** The basic chemistry of soil acidity, in *Soil Acidity and Liming,* Pearson, R. W. and Adams, F., Eds., American Society of Agronomy, Madison, WI, 1967, 1.
15. **Rich, C. I.,** Hydroxy interlayers in expansible layer silicates, *Clays Clay Miner.,* 16, 15, 1968.
16. **Coulter, B. S.,** The Chemistry of hydrogen and aluminum ions in soils, clay minerals and resins, *Soils Fert.,* 32, 215, 1969.
17. **Frink, C. R.,** Aluminum chemistry in acid sulfate soils, in *Acid Sulfate Soils,* Dost, H., Ed., Proc. Int. Symp. Acid Sulfate Soils, International Institute for Land Reclamation and Improvement, The Netherlands, 1972, 131.
18. **Barhisel, R. I.,** Chlorites and hydroxy interlayered vermiculite and smectite, in *Minerals in Soil Environment,* Dixon, J. B. and Weed, S. B., Eds., Soil Science Society of America, Madison, WI, 1977, 331.
19. **Thomas, G. W. and Hargrove, W. L.,** The chemistry of soil acidity, in *Soil Acidity and Liming,* 2nd ed., Adams, F., Ed., American Society of Agronomy, Crop Science Society of America, and Soil Science Society of America, Madison, WI, 1984, 3.
20. **Rich, C. I.,** Aluminum in interıayers of vermiculite, *Soil Sci. Soc. Am. Proc.,* 24, 26, 1960.
21. **Rich, C. I.,** Conductometric and potentiometric titration of exchangeable Al, *Soil Sci. Soc. Am. Proc.,* 34, 31, 1970.
22. **Kaddah, M. T. and Coleman, N. T.,** Salt displacement and titration of $AlCl_3$-treated trioctahedral vermiculites, *Soil Sci. Soc. Am. Proc.,* 31, 328, 1967.
23. **Kaddah, M. T. and Coleman, N. T.,** Salt displacement of acid-treated trioctahedral vermiculites, *Soil Sci. Soc. Am. Proc.,* 31, 333, 1967.
24. **Kissel, D. E., Gentzsch, E. P., and Thomas, G. W.,** Hydrolysis of nonexchangeable acidity in soils during salt extractions of exchangeable acidity, *Soil Sci.,* 111, 293, 1971.
25. **Veith, J. A. and Sposito, G.,** On the average equilibrium OH/Al molar ratio for aluminum adsorbed by a synthetic cation exchanger, *Soil Sci.,* 127, 161, 1979.
26. **Pratt, P. F. and Bair, F. L.,** A comparison of three reagents for the extraction of aluminum from soils, *Soil Sci.,* 91, 357, 1961.
27. **Coleman, N. T., Weed, S. B., and McCracken, R. J.,** Cation-exchange capacity and exchangeable cations in Piedmont soils of North Carolina, *Soil Sci. Soc. Am. Proc.,* 23, 146, 1959.
28. **Bache, B. W. and Sharp, G. S.,** Soluble polymeric hydroxy-aluminum ions in acid soils, *J. Soil Sci.,* 27, 167, 1976.
29. **McLean, E. O., Heddleson, M. R., Bartlett, R. J., and Holowaychuk, N.,** Aluminum in soils. I. Extraction methods and magnitudes in clays and Ohio soils, *Soil Sci. Soc. Am. Proc.,* 22, 382, 1958.
30. **Yuan, T. L. and Fiskell, J. G. A.,** Aluminum studies. II. The extraction of aluminum from some Florida soils, *Soil Sci. Soc. Am. Proc.,* 23, 202, 1959.
31. **Skeen, J. B. and Sumner M. E.,** Exchangeable aluminum. II. The effect of concentration and pH value of the extractant on the extraction of aluminum from acid soils, *S. Afr. J. Agric. Sci.,* 10, 303, 1967.

32. **Bloom, P. R., McBride, M. B., and Weaver, R. M.,** Aluminum organic matter in acid soils: salt-extractable aluminum, *Soil Sci. Soc. Am. J.*, 43, 813, 1979.
33. **McLean, E. O.,** Aluminum, in *Methods of Soil Analysis, Part II,* Black, C. A., Ed., American Society of Agronomy, Madison, WI, 978, 1965, 978.
34. **Lin, C. and Coleman, N. T.,** The measurement of exchangeable aluminum in soils and clays, *Soil Sci. Soc. Am. Proc.*, 24, 444, 1960.
35. **Skeen, J. B.,** Determination of Exchangeable Aluminum in Acid Soils, M.S. thesis, University of Natal, 1964, 100.
36. **Shen, M. J. and Rich, C. I.,** Aluminum fixation in montmorillonite, *Soil Sci. Soc. Am. Proc.*, 26, 33, 1962.
37. **Thomas, G. W.,** Forms of aluminum in cation exchangers, Trans., 7th Int. Congr. Soil Sci., Madison, WI, 2:364, 1960.
38. **Hsu, P. H.,** Heterogeneity of montmorillonite surface and its effect on the nature of hydroxy-aluminum interlayers, *Clays Clay Miner.*, 16, 303, 1968.
39. **Bache, B. W.,** Barium isotope method for measuring cation exchange capacity of soils and clays, *J. Sci. Food. Agric.*, 21, 169, 1970.
40. **Bloom, P. R., McBride, M. B., and Chadbourne, B.,** Adsorption of aluminum by a smectite. I. Surface hydrolysis during Ca^{2+}-Al^{3+} exchange, *Soil Sci. Soc. Am. J.*, 41, 1068, 1977.
41. **Veith, J. A.,** Selectivity and adsorption capacity of smectite and vermiculite for aluminum of varying basicity, *Clays Clay Miner.*, 26, 45, 1978.
42. **Brown, G. and Newman, A. C. D.,** The reactions of soluble aluminum with montmorillonite, *J. Soil Sci.*, 24, 339, 1973.
43. **Hodges, S. C. and Zelazny, L. W.,** Interactions of dilute, hydrolyzed aluminum solutions with clays, peat, and resin, *Soil Sci. Soc. Am. J.*, 47, 206, 1983.
44. **Hodges, S. C. and Zelazny, L. W.,** Influences of OH/Al ratios and loading rates on aluminum-kaolinite interactions, *Soil Sci. Soc. Am. J.*, 47, 221, 1983.
45. **Skeen, J. B. and Sumner, M. E.,** Measurement of exchangeable aluminum in acid soils, *Nature*, 207, 712, 1965.
46. **Skeen, J. B. and Sumner, M. E.,** Exchangeable aluminum. I. The efficiency of various electrolytes for extracting aluminum from acid soils, *S. Afr. J. Agric. Sci.*, 10, 3, 1967.
47. **Sivasubramaniam, S. and Talibudeen, O.,** Potassium-aluminum exchange in acid soils. I. Kinetics, *J. Soil Sci.*, 23, 163, 1972.
48. **Amedee, G. and Peech, M.,** The significance of KCl-extractable Al (III) as an index to lime requirement of soils of the humid tropics, *Soil Sci.*, 121, 227, 1976.
49. **Hsu, P. H. and Bates, T. F.,** Fixation of hydroxy-aluminum polymers by vermiculite, *Soil Sci. Soc. Am. Proc.*, 28, 763, 1964.
50. **Evans, C. E. and Kamprath, E. J.,** Lime response as related to percent aluminum saturation, solution aluminum, and organic matter content, *Soil Sci. Soc. Am. Proc.*, 34, 893, 1970.
51. **Thomas, G. W.,** The relationship between organic matter content and exchangeable aluminum in acid soil, *Soil Sci. Soc. Am. Proc.*, 39, 591, 1975.
52. **Hargrove, W. L. and Thomas, G. W.,** Effect of organic matter on exchangeable aluminum and plant growth in acid soils, in *Chemistry in the Soil Environment,* Stelly, M., Ed., American Society of Agronomy, Spec. Publ. No. 40, 1981, 151.
53. **Oates, K. M. and Kamprath, E. J.,** Soil acidity and liming. I. Effect of the extracting solution cation and pH on the removal of aluminum from acid soils, *Soil Sci. Soc. Am. J.*, 47, 686, 1983.
54. **Juo, A. S. R. and Kamprath, E. J.,** Copper chloride as an extractant for estimating the potentially reactive aluminum pool in acid soils, *Soil Sci. Soc. Am. J.*, 43, 35, 1979.
55. **Ragland, J. L. and Coleman, N. T.,** The hydrolysis of aluminum salts in clay and soil systems, *Soil Sci. Soc. Am. Proc.*, 24, 457, 1960.
56. **Frink, C. R. and Peech, M.,** Hydrolysis and exchange reactions of the aluminum ion in hectorite and montmorillonite suspensions, *Soil Sci. Soc. Am. Proc.*, 27, 527, 1963.
57. **Hsu, P. H. and Rich, C. I.,** Aluminum fixation in a synthetic cation exchanger, *Soil Sci. Soc. Am. Proc.*, 24, 21, 1960.
58. **Clark, J. S.,** Aluminum and iron fixation in relation to exchangeable hydrogen in soils, *Soil Sci.*, 94, 302, 1964.
59. **Turner, R. C. and Brydon, J. E.,** Factors affecting the solubility of $Al(OH)_3$ precipitated in the presence of montmorillonite, *Soil Sci.*, 100, 176, 1965.
60. **Turner, R. C. and Brydon, J. E.,** Effect of length of time of reaction on some properties of suspensions of Arizona bentonite, illite, and kaolinite in which aluminum hydroxide is precipitated, *Soil Sci.*, 103, 111, 1967
61. **Turner, R. C.,** Some properties of aluminum hydroxide precipitated in the presence of clays, *Can. J. Soil Sci.*, 45, 331, 1965.

62. **Turner, R. C.,** Aluminum removed from solution by montmorillonite, *Can. J. Soil Sci.*, 47, 217, 1967.
63. **Hsu, P. H. and Bates, T. F.,** Formation of X-ray amorphous and crystalline aluminum hydroxides, *Mineral Mag.*, 33, 749, 1964.
64. **Jackson, M. L.,** Structure role of hydronium in layer silicates during soil genesis, Trans. 7th Int. Congr. Soil Sci., Madison, WI, II:445, 1960.
65. **Stol, R. J., van Helden, A. K., and De Bruyn, P. L.,** Hydrolysis-precipitation studies of aluminum (III) solutions. II. A kinetic study model, *J. Colloid Interface Sci.*, 57, 115, 1976.
66. **Grim, R. E. and Johns, W. D.,** Clay mineral investigation of sediments in the northern Gulf of Mexico, *Clays Clay Miner.*, 2, 81, 1954.
67. **Chakravarti, S. N. and Talibudeen, O.,** Phosphate interaction with clay minerals, *Soil Sci.*, 92, 232, 1961.
68. **Dixon, J. B. and Jackson, M. L.,** Properties of intergradient chlorite-expansible layer silicates of soils, *Soil Sci. Soc. Am. Proc.*, 26, 358, 1962.
69. **Frink, C. R.,** Characterization of aluminum interlayers in soil clays, *Soil Sci. Soc. Am. Proc.*, 29, 379, 1965.
70. **Barshad, I.,** The effect of the total chemical composition and crystal structure of soil minerals on the nature of exchangeable cations in acidified clays and in naturally occurring acid soils, Trans., 7th Int. Congr. Soil Sci., Madison, WI, II:435, 1960.
71. **Harsh, J. B. and Doner, H. E.,** The nature and stability of aluminum hydroxide precipitated on Wyoming montmorillonite, *Geoderma*, 36, 45, 1985.
72. **Barnhisel, R. I.,** Changes in specific surface areas of clays treated with hydroxy-aluminum, *Soil Sci.*, 107, 126, 1969.
73. **Kozak, L. M. and Huang, P. M.,** Adsorption of hydroxy-Al by certain phyllosilicates and its relation to K/Ca cation exchange selectivity, *Clays Clay Miner.*, 19, 95, 1971.
74. **Sawhney, B. L.,** Aluminum interlayers in layer silicates: effect of OH/Al ratio of Al solution, time of reaction, and type of structure, *Clays Clay Miner.*, 16, 157, 1968.
75. **Novak, R. J., Motto, H. L., and Douglas, L. A.,** The effect of time and particle size on mineral alteration in several Quaternary soils in New Jersey and Pennsylvania, U.S.A., in *Paleopedology-Origin, Nature and Dating of Paleosols*, Yaahon, D. H., Ed., International Society of Soil Science and Israel University Press, Jerusalem, 1971, 211.
76. **Hsu, P. H.,** Formation of gibbsite from aging hydroxy-aluminum solutions, *Soil Sci. Soc. Am. Proc.*, 30, 173, 1966.
77. **Barnhisel, R. I. and Rich, C. I.,** Gibbsite, bayerite, and nordstrandite formation as affected by anions, pH and mineral surfaces, *Soil Sci. Soc. Am. Proc.*, 29, 531, 1965.
78. **Schoen, R. and Roberson, E. C.,** Structures of aluminum hydroxide and geochemical implication, *Am. Mineral.*, 55, 43, 1970.
79. **Lodding, W. W.,** Gibbsite vermiforms in the Pensauken formation of New Jersey, *Am. Mineral.*, 46, 394, 1961.
80. **Keller, W. D.,** The origin of high alumina clay minerals, a review, *Clays Clay Miner.*, 12, 129, 1964.
81. **Violante, A. and Jackson, M. L.,** Crystallization of nordstrandite in citrate systems in the presence of montmorillonite, Proc. 6th Int. Clay Conf. 1:517, 1979.
82. **Davis, C. E. and Hill, V. G.,** Occurrence of nordstrandite and its possible significance in Jamaica bauxites, *Travaux*, 11, 61, 1974.
83. **Coleman, N. T. and Harward, M. E.,** The heats of neutralization of acid clays and cation exchange resins, *J. Am. Chem. Soc.*, 75, 6045, 1953.
84. **Low, P. F.,** The role of aluminum in titration of bentonite, *Soil Sci. Soc. Am. Proc.*, 19, 135, 1955.
85. **Rich, C. I. and Obenshain, S. S.,** Chemical and clay mineral properties of a red-yellow podzolic soil derived from muscovite shist, *Soil Sci. Soc. Am. Proc.*, 19, 334, 1955.
86. **Chernov, V. A. and Maximova, W. S.,** Austauschreaktionen zwischen adsorbierten Wasserstoffionen und den-Al und Mg-ionen in tonen, Trans. Int. Soc. Soil Sci., Comm. II and IV. 2:179, 1958.
87. **Coleman, N. T. and Craig, D.,** The spontaneous alteration of hydrogen clay, *Soil Sci.*, 91, 14, 1961.
88. **Carstea, D.,** Formation of hydroxy-Al and -Fe interlayers in montmorillonite and vermiculite: influence of particle size and temperature, *Clays Clay Miner.*, 16, 231, 1968.
89. **Ross, G. J. and Turner, R. C.,** Effect of different anions on the crystallization of aluminum hydroxide in partially neutralized aqueous aluminum salt systems, *Soil Sci. Soc. Am. Proc.*, 35, 389, 1971.
90. **Serna, C. I., White, J. L., and Hem, S. L.,** Anion-aluminum hydroxide gel interactions, *Soil Sci. Soc. Am. J.*, 41, 1009, 1977.
91. **Akitt, J. W. and Farthing, A.,** Aluminum-27 nuclear magnetic resonance studies of the hydrolysis of aluminum (III). V. Slow hydrolysis using aluminum metal, *J. Chem. Soc. Dalton Trans.*, 1624, 1981.
92. **Brindley, G. W. and Sempels, R. E.,** Preparation and properties of some hydroxy-aluminum beidellites, *Clay Miner.*, 12, 129, 1977.

93. **Plee, D., Borg, F., Gatineau, L., and Fripiat, J. J.,** High-resolution solid-state ^{27}Al and ^{29}Si nuclear magnetic resonance study of pillared clays, *J. Am. Chem. Soc.,* 107, 2362, 1985.
94. **Veith, J. A.,** Basicity of exchangeable aluminum, formation of gibbsite, and composition of the exchange acidity in the presence of exchangers, *Proc. Soil Sci. Soc. Am. J.,* 41, 865, 1977.
95. **Lindsey, W. L.,** *Chemical Equilibria in Soils,* John Wiley & Sons, New York, 1979.
96. **Singh, S. S. and Brydon, J. E.,** Precipitation of aluminum by calcium hydroxide in the presence of Wyoming bentonite and sulfate ions, *Soil Sci.,* 103, 162, 1967.
97. **Jardine, P. M., Zelazny, L. W., and Parker, J. C.,** Mechanisms of aluminum adsorption on clay minerals and peat, *Soil Sci. Soc. Am. J.,* 49, 862, 1985.
98. **Brown, D. W. and Hem, J. D.,** Reactions of aqueous aluminum species at mineral surfaces, *U.S. Geol. Surv. Water-Supply Pap.,* 1827-F, 1975.
99. **El-Swaify, S. A. and Emerson, W. W.,** Changes in the physical properties of soil clays due to precipitated aluminum and iron hydroxides. I. Swelling and aggregate stability after drying, *Soil Sci. Am. Proc.,* 39, 1056, 1975.
100. **Jackson, M. L.,** Interlaying of expansible layer silicates in soils by chemical weathering, *Clays Clay Miner.,* 11, 29, 1963.
101. **Jardine, P. M., Parker, J. C., and Zelazny, L. W.,** Kinetics and mechanisms of aluminum adsorption on kaolinite using a two-site nonequilibrium transport model, *Soil Sci. Soc. Am. J.,* 49, 867, 1985.
102. **Huang, P. M. and Kozak, L. M.,** Adsorption of hydroxy-aluminum polymers by muscovite and biotite, *Nature,* 228, 1084, 1970.
103. **Columbera, P. M., Posner, A. M., and Quirk, J. P.,** The adsorption of aluminum from hydroxy-aluminum solutions onto Fithian illite, *J. Soil Sci.,* 22, 118, 1971.
104. **Edwards, D. G., Posner, A. M., and Quirk, J. P.,** Repulsion of chloride ions by negatively charged clay surfaces. III. Di- and tri-valent cation clays, *Trans. Faraday Soc.,* 61, 2920, 1965.
105. **Cotten, S. B.,** Hydrolysis of Aluminum in Synthetic Cation Exchange Resins and Dioctahedral Vermiculite, Ph.D. thesis, Virginia Polytechnic Institute and State University, Blacksburg, VA, 1965.
106. **Theng, B. K. G.,** *The Chemistry of Clay-Organic Reactions,* Adams Hilger Ltd., 1974.
107. **Schnitzer, M.,** Humic substances: chemistry and reactions, in *Soil Organic Matter,* Schnitzer, M. and Khan, S. U., Eds., Elsevier, New York, 1978, 1.
108. **Pionke, H. B. and Corey, R. B.,** Relations between acidic aluminum and soil pH, clay, and organic matter, *Soil Sci. Soc. Am. Proc.,* 31, 749, 1967.
109. **Spyridakis, D. E., Chesters, G., and Wilde, S. A.,** Kaolinization of biotite as a result of coniferous seedling growth, *Soil Sci. Soc. Am. Proc.,* 31, 203, 1967.
110. **Bloom, P. R., McBride, M. B., and Weaver, R. M.,** Aluminum organic matter in acid soils: buffering and solution aluminum activity, *Soil Sci. Soc. Am. J.,* 43, 488, 1979.
111. **White, R. E. and Thomas, G. W.,** Hydrolysis of aluminum on weakly acidic organic exchangers: implications for phosphate adsorption, *Fert. Res.,* 2, 159, 1981.
112. **Hargrove, W. L. and Thomas, G. W.,** Titration properties of Al-organic matter, *Soil Sci.,* 134, 216, 1982.
113. **Bloom, P. R. and McBride, M. B.,** Metal ion binding and exchange with hydrogen ions in acid-washed peat, *Soil Sci. Soc. Am. J.,* 43, 687, 1979.
114. **Bloom, P. R.,** Phosphorus adsorption by an aluminum-peat complex, *Soil Sci. Soc. Am. J.,* 45, 267, 1981.
115. **Arai, S. and Kumada, K.,** Factors controlling stability of Al-humate, *Geoderma,* 26, 1, 1981.
116. **Young, S. D. and Bache, B. W.,** Aluminum-organic complexation: formation constants and a speciation model for the soil solution, *J. Soil Sci.,* 36, 261, 1985.
117. **Jardine, P. M. and Zelazny, L. W.,** Influence of organic anions on the speciation of mononuclear and polynuclear aluminum by ferron, *Soil Sci. Soc. Am. J.,* 51, 885, 1987.
118. **Turner, R. C.,** Conditions in solution during the formation of gibbsite in dilute Al salt solutions. II. Effect of length of time of reaction on the formation of polynuclear hydroxy aluminum cations. The substitution of other anions for OH^- in amorphous $Al(OH)_3$ and the crystallization of gibbsite, *Soil Sci.,* 106, 338, 1968.
119. **Turner, R. C. and Ross, G. J.,** Conditions in solution during the formation of gibbsite in dilute Al salt solutions. IV. Effect of Cl concentration and temperature and a proposed mechanism for gibbsite formation, *Can. J. Chem.,* 48, 723, 1970.
120. **Violante, A. and Jackson, M. L.,** Clay influence on the crystallization of aluminum hydroxide polymorphs in the presence of citrate, sulfate, or chloride, *Geoderma,* 25, 199, 1981.
121. **Gastuche, M. C. and Herbillon, A.,** Etude des gels d'alumine: cristallisation en mi lieu disionise, *Bull. Soc. Chim. Fr.,* 1404, 1962.
122. **Huang, P. M. and Jackson, M. L.,** Fluoride interaction with clays in relation to third buffer range, *Nature (London),* 211, 779, 1966.
123. **Liu, J. C., Feldkamp, J. R., White, J. L., and Hem, S. L.,** Adsorption of phosphate by aluminum hydroxycarbonate, *J. Pharm. Sci.,* 73, 1355, 1984.

124. **White, R. E., Tiffin, L. O., and Taylor, A. W.,** The existence of polymeric complexes in dilute solutions of aluminum and orthophosphate, *Plant Soil,* 45, 521, 1976.
125. **Jardine, P. M. and Zelazny, L. W.,** Influence of inorganic anions on the speciation of mononuclear and polynuclear aluminum by ferron, *Soil Sci. Soc. Am. J.,* 51, 889, 1987.
126. **Kwong, K. F. and Huang, P. M.,** Influence of citric acid on the crystallization of aluminum hydroxides, *Clays Clay Miner.,* 23, 164, 1975.
127. **Kwong, K. F. and Huang, P. M.,** Influence of citric acid on the hydrolytic reactions of aluminum, *Soil Sci. Soc. Am. J.,* 41, 692, 1977.
128. **Kwong, K. F. and Huang, P. M.,** Nature of hydrolytic products of aluminum as influenced by low molecular weight complexing organic acids, in *Proc. 1978 Int. Clay Conf. (Oxford),* Mortland, M. M. and Farmer, V. C., Eds., Elsevier, Amsterdam, 1979, 527.
129. **Kwong, K. F. and Huang, P. M.,** The relative influence of low-molecular weight, complexing organic acids on the hydrolysis and precipitation of aluminum, *Soil Sci.,* 128, 337, 1979.
130. **Kwong, K. F. and Huang, P. M.,** Surface reactivity of aluminum hydroxides precipitated in the presence of low molecular weight organic acids, *Soil Sci. Soc. Am. J.,* 43, 1107, 1979.
131. **Nail, S. L., White, J. L., and Hem, S. L.,** Structure of aluminum hydroxide gel. I. Initial precipitate, *J. Pharm. Sci.,* 65, 1188, 1976.
132. **Goh, T. B. and Huang, P. M.,** Formation of hydroxy-Al-montmorillonite complexes as influenced by citric acid, *Can. J. Soil Sci.,* 64, 411, 1984.
133. **Vicente, M. A., Razzaghe, M., and Robert, M.,** Formation of aluminum hydroxy vermiculite (intergrade) and smectite from mica under acidic conditions, *Clay Miner.,* 12, 101, 1977.
134. **Clark, J. S.,** Some cation-exchange properties of soils containing free oxides, *Can. J. Soil Sci.,* 44, 203, 1964.
135. **Barnhisel, R. I. and Rich, C. I.,** Gibbsite formation from aluminum-interlayers in montmorillonite, *Soil Sci. Soc. Am. Proc.,* 27, 632, 1963.
136. **Klages, M. G. and White, J. L.,** A chlorite-like mineral in Indiana soils, *Soil Sci. Soc. Am. Proc.,* 21, 16, 1957.
137. **Sawhney, B. L.,** Aluminum interlayers in clay minerals, Trans., 7th Int. Congr. Soil Sci., Madison, WI, IV:476, 1960.
138. **Sawhney, B. L.,** Weathering and aluminum interlayers in a soil catena; Hollis-Charlton-Sutton-Leicester, *Soil Sci. Soc. Am. Proc.,* 24, 221, 1960.
139. **Rich, C. I. and Cook, M. G.,** Formation of dioctahedral vermiculite in Virginia soils, *Clays Clay Miner.,* 11, 96, 1963.
140. **Reneau, R. B. and Fiskell, J. G. A.,** Selective dissolution effects on cation-exchange capacity and specific surface of some tropical soil clays, *Soil Sci. Soc. Am. Proc.,* 34, 809, 1970.
141. **Lietzke, D. A. and Mortland, M. M.,** The dynamic character of a chloritized vermiculitic soil clay, *Soil Sci. Soc. Am. Proc.,* 37, 651, 1973.
142. **Brydon, J. E. and Kodama, H.,** The nature of aluminum hydroxide-montmorillonite complexes, *Am. Mineral.,* 51, 875, 1966.
143. **Huang, P. M. and Lee, S. Y.,** Effect of drainage on weathering transformations of mineral colloids of some Canadian prairie soils, *Proc. Int. Clay Conf. (Tokyo),* 1, 541, 1969.
144. **Carstea, D. D., Harward, M. E., and Knox, E. G.,** Comparison of iron and aluminum hydroxy interlayers in montmorillonite and vermiculite. I. Formation, *Soil Sci. Soc. Am. Proc.,* 34, 517, 1970.
145. **Carstea, D. D., Harward, M. E., and Knox, E. G.,** Comparison of iron and aluminum hydroxy interlayers in montmorillonite and vermiculite. II. Dissolution, *Soil Sci. Soc. Am. Proc.,* 34, 522, 1970.
146. **Tullock, R. J. and Roth, C. B.,** Stability of mixed iron and aluminum hydrous oxides on montmorillonite, *Clays Clay Miner.,* 23, 27, 1975.
147. **Keren, R.,** The effect of hydroxy-aluminum precipitation on the exchange properties of montmorillonite, *Clays Clay Miner.,* 27, 303, 1979.
148. **Keren, R.,** Effects of titration rate, pH, and drying process on cation exchange capacity reduction and aggregate size distribution of montmorillonite hydroxy-aluminum complexes, *Soil Sci. Soc. Am. J.,* 44, 1209, 1980.
149. **Keren, R.,** Reduction of the cation-exchange capacity of montmorillonite by take-up of hydroxy-Al polymers, *Clays Clay Miner.,* 34, 534, 1986.
150. **Keren, R., Gast, R. G., and Barnhisel, R. I.,** Ion exchange reactions in nondried Chambers montmorillonite hydroxy-aluminum complexes, *Soil Sci. Soc. Am. Proc.,* 41, 34, 1977.
151. **Kinniburgh, D. G., Syers, J. K., and Jackson, M. L.,** Specific adsorption of trace amounts of calcium and strontium by hydrous oxides of iron and aluminum, *Soil Sci. Soc. Am. Proc.,* 39, 464, 1975.
152. **Kwong, K. F. and Huang, P. M.,** Comparison of the influence of tannic acid and selected low-molecular-weight organic acids on precipitation products of aluminum, *Geoderma,* 26, 179, 1981.
153. **Rich, C. I.,** Ammonium fixation by two Red-yellow Podzolic soils as influenced by interlayer-Al, Trans., 7th Int. Congr. Soil Sci., Madison, WI, IV:468, 1960.

154. **Scott, A. D., Ahlrichs, J. L., and Stanford, G.,** Aluminum effect on potassium fixation by Wyoming bentonite, *Soil Sci.*, 84, 377, 1957.
155. **Cook, M. G. and Hutcheson, T. B., Jr.,** Soil potassium reactions as related to clay mineralogy of selected Kentucky soils, *Soil Sci. Soc. Am. Proc.*, 24, 252, 1960.
156. **Carter, D. L., Harward, M. E., and Young, J. L.,** Variation in exchangeable K and relation to intergrade layer silicate minerals, *Soil Sci. Soc. Am. Proc.*, 27, 283, 1963.
157. **Rich, C. I. and Lutz, J. A., Jr.,** Mineralogical changes associated with ammonium and potassium fixation in soil clays, *Soil Sci. Soc. Am. Proc.*, 29, 167, 1965.
158. **Somasiri, S. and Huang, P. M.,** Effect of hydrolysis of aluminum on competitive adsorption of potassium and aluminum by expansible phyllosilicates, *Soil Sci.*, 117, 110, 1974.
159. **Nagasawa, K., Brown, G., and Newman, A. C. D.,** Artificial alteration of biotite into a 14A layer silicate with hydroxy-aluminum interlayers, *Clays Clay Miner.*, 22, 241, 1974.
160. **Rich, C. I. and Black, W. R.,** Potassium exchange as affected by cation size, pH and mineral structure, *Soil Sci.*, 97, 384, 1964.
161. **Rich, C. I.,** Effect of cation size and pH on potassium exchange in Nason soil, *Soil Sci.*, 98, 100, 1964.
162. **Murdock, L. M. and Rich, C. I.,** Ion selectivity in three soil profiles as influenced by mineralogical characteristics, *Soil Sci. Soc. Am. Proc.*, 36, 167, 1972.
163. **Dolcater, D. L., Lotse, E. G., Syers, J. K., and Jackson, M. L.,** Cation exchange selectivity of some clay-sized minerals and soil materials, *Soil Sci. Soc. Am. Proc.*, 32, 795, 1968.
164. **Tamura, T. and Jacobs, D. G.,** Structural implications in cesium sorption, *Health Phys.*, 2, 391, 1960.
165. **Jacobs, D. G. and Tamura, T.,** The mechanisms of ion fixation using radioisotope techniques, Trans., 7th Int. Congr. Soil Sci., Madison, WI, II:206, 1960.
166. **Elprince, A. M.,** Effect of pH on the adsorption of trace radioactive cesium by sediments, *Water Resour. Res.*, 14, 696, 1978.
167. **LeRoux, J., Rich, C. I., and Ribbe, P. H.,** Ion selectivity by weathered micas as determined by electron probe analysis, *Clays Clay Miner.*, 18, 333, 1970.
168. **Rich, C. I.,** Potassium in minerals, *Proc. Colloq. Int. Potash Inst.*, 9, 15, 1972.
169. **Zelazny, L. W. and Rich, C. I.,** Temperature effects on potassium-calcium exchange selectivity in soils, clay minerals and resins, Agronomy Abstracts, American Society of Agronomy, Madison, WI, 1970, 91.
170. **Elprince, A. M., Rich, C. I., and Martens, D. C.,** Effect of temperature and hydroxy aluminum interlayer on the adsorption of trace radioactive cesium by sediments near water-cooled nuclear reactors, *Water Resour. Res.*, 13, 375, 1977.
171. **Rich, C. I.,** Mineralogy of soil potassium, in *The Role of Potassium in Agriculture,* Kilmer, V. J., Ed., American Society of Agronomy, Crop Science Society of America, and Soil Science Society of America, Madison, WI, 1968, 79.
172. **Schachtschabel, P.,** Untersuchungen iiber die sorption der Tonmineralien und organischen bodenkolloide und die bestimmung des anteils dieser kolloides an der sorption in boden, *Kolloid-Beih.*, 51, 199, 1940.
173. **Foscolos, A. E.,** Cation-exchange equilibrium constants of aluminum-saturated montmorillonite and vermiculite clays, *Soil Sci. Soc. Am. Proc.*, 32, 350, 1968.
174. **Banin, A.,** Ion exchange isotherms of montmorillonite and structural factors affecting them, *Isr. J. Chem.*, 6, 27, 1968.
175. **Shainberg, I. and Kemper, W. D.,** Electrostatic forces between clay and cations calculated and inferred from electrical conductivity, *Clays Clay Miner.*, 14, 117, 1966.
176. **Shainberg, I. and Otoh, H.,** Size and shape of montmorillonite particles saturated with Na/Ca ions (inferred from viscosity and optical measurements), *Isr. J. Chem.*, 6, 251, 1968.
177. **Harsh, J. B. and Doner, H. E.,** Specific adsorption of copper on an hydroxy-aluminum-montmorillonite complex, *Soil Sci. Soc. Am. J.*, 48, 1034, 1984.
178. **Harsh, J. B., Doner, H. E., and McBride, M. B.,** Chemisorption of copper on hydroxy-aluminum-rectorite: an electron spin resonance study, *Clays Clay Miner.*, 32, 407, 1984.
179. **Saini, G. R., MacLean, A. A., and Doyle, A. A.,** The influence of some physical and chemical properties on soil aggregation and response to VAMA, *Can. J. Soil Sci.*, 46, 155, 1966.
180. **Deshpande, T. L., Greenland, D. J., and Quirk, J. P.,** Changes in soil properties associated with the removal of iron and aluminum oxides, *J. Soil Sci.*, 19, 108, 1968.
181. **El-Rayah, H. M. E. and Rowell, D. L.,** The influence of iron and aluminum hydroxides on the swelling of Na-montmorillonite and the permeability of Na-soil, *J. Soil Sci.*, 24, 137, 1973.
182. **Oades, J. M.,** Interactions of polycations of aluminum and iron with clays, *Clays Clay Miner.*, 32, 49, 1984.
183. **Tweneboah, C. K., Kijne, J. W., and Greenland, D. J.,** Influence of active aluminum oxides on water movement in soils, *Aust. J. Soil Res.*, 20, 325, 1969.
184. **Nakano, K. and Kono, M.,** Influence of basic aluminum on physical soil properties (Abstr.), 1976 Meeting Soc. Sci. Soil Manure, Jpn., 22, 4, 1976.

185. **Shiraishi, K.,** Studies on the effect of hydroxyaluminum on soil physical properties. I. Effects of the formation of water stable aggregates and soil chemical properties, *Bull. Kyushu Natl. Agric. Exp. Stn.*, 20, 257, 1979.
186. **Shiraishi, K.,** Studies on the effect of hydroxyaluminum on soil physical properties. III. Effects of hydroxyaluminum applied as soil conditioner on the properties of clayey paddy field, *Bull. Kyushu Natl. Agric. Exp. Stn.*, 22, 203, 1982.
187. **Wada, K., Tsumori, Y., Nitawaki, Y., and Egashira, K.,** Effects of hydroxy-aluminum on flocculation and permeability of smectite clays, *Soil Sci. Plant Nutr. (Tokyo)*, 29, 305, 1983.
188. **Gerald, K. and Reed, L. W.,** Swelling characteristics of hydroxy-aluminum interlayered clays, *Clays Clay Miner.*, 20, 13, 1972.
189. **Stephen, I. and MacEwan, D. M. C.,** Some chloritic clay minerals of unusual type, *Clay Miner. Bull.*, 1, 157, 1951.
190. **Karathanasis, A. D. and Evangelou, V. P.,** Water sorption characteristics of aluminum- and calcium-saturated soil clays, *Soil Sci. Soc. Am. J.*, 50, 1063, 1986.
191. **Schofield, R. K. and Samson, H. R.,** The deflocculation of kaolinite suspensions and the accompanying change-over from positive to negative chloride adsorption, *Clay Miner. Bull.*, 2, 45, 1954.
192. **Lahav, N. and Shani, U.,** Cross-linked smectites. II. Flocculation and micro-fabric characteristics of hydroxy-aluminum-montmorillonite, *Clays Clay Miner.*, 26, 116, 1978.
193. **Roberts, K., Kowalewska, J., and Friberg, S.,** The influence of interactions between hydrolyzed aluminum ions and polyacrylamides on the sedimentation of kaolin suspensions, *J. Colloid Interface Sci.*, 48, 361, 1974.
194. **Rengasamy, P. and Oades, J. M.,** Interaction of monomeric and polymeric species of metal ions with clay surfaces. III. Aluminum (III) and Chromium (III), *Aust. J. Soil Res.*, 16, 53, 1977.

Chapter 7

THE SURFACE CHEMISTRY OF ALUMINUM OXIDES AND HYDROXIDES

James A. Davis and John D. Hem

TABLE OF CONTENTS

I. INTRODUCTION

The coordination behavior of Al^{3+} toward hydroxide ions explains the marked dependence of aluminum hydroxide solubility on pH. The ratio of the ionic radii favors octahedral coordination, with Al^{3+} at the center of an octahedron formed by six close-packed hydroxide ions (compare Figure 1 in Chapter 2). The water molecule, OH^-, and O^{2-} ions all have nearly the same physical dimensions and can replace each other in some crystal structures. A crystal structure is broken or cleaved along preferred crystal planes where the structure is weakest, creating the surface of the mineral. Thus, the initial surface structure can be related to the crystal structure and the way in which the surface was formed. Furthermore, the surface chemical behavior of aluminum oxides and hydroxides can be explained in part by surface structure expected for particular crystals. Complications may arise, however, if the surface is altered by exposure to water or changes in environmental conditions.

The general physical properties of various forms of aluminum oxides and hydroxides are reviewed in Chapter 3 and elsewhere.[1-3] In this chapter, we review the surface chemical behavior of these oxides and hydroxides, with an emphasis on adsorption studies and the electrochemical behavior of the alumina-water interface. When possible we have contrasted the observed behavior of different crystal forms in terms of what is known about their surface structure. The hydroxides and oxyhydroxides of aluminum are probably the more important phases in terrestrial systems.[3] However, most of the data on surface chemical behavior are available from studies of the crystalline hydroxide, gibbsite [α-$Al(OH)_3$], or the synthetic oxide, γ-Al_2O_3, although limited information is available for other phases. We also present a discussion of crystal formation for various phases and the ways in which surface chemical behavior may influence the final precipitate formed.

The phase discontinuity that occurs at the alumina-water interface may play an important role in the geochemical processes of soils. The composition of soil solutions and the flux of solutes through soils are often controlled by reactions at the mineral-water interface,[4] and the surfaces of aluminum oxides and hydroxides are particularly reactive in comparison to other minerals in the soil environment. Probably one of the most important topics in this context is the geochemical control of dissolved phosphate in soil solutions because of the commercial importance of phosphate as a nutrient. The rate of delivery of phosphate and other nutrients to plant roots may be controlled in some soils by processes which occur on the surfaces of aluminum oxides and hydroxides (see Chapter 8).

II. RELATIONSHIP OF CRYSTAL STRUCTURE TO SURFACE CHEMISTRY

A. Surface Functional Groups

All strongly dried aluminum oxides and hydroxides chemisorb at least a monolayer of water when exposed to moisture at room temperature. Figure 1 shows the idealized structure of a γ-Al_2O_3 surface when dried and when exposed to water.[5] When dry, the top layer contains only oxide ions, regularly arranged over aluminum ions in octahedral sites in the lower layer (Figure 1a). This upper layer contains only half of the oxide ions present in the next lower layer, which represents the *100* plane of a cubic, close-packed oxide lattice, where each oxide ion occupies an area of 0.08 nm^2. Aluminum ions are located in all interstices between oxide ions. The stoichiometry of the upper two layers combined corresponds to Al_2O_3. When the surface is hydrated, water is chemisorbed to convert the top layer of oxide ions to a filled, square lattice of hydroxyl ions (Figure 1b). Each hydroxyl ion is assumed to be directly over an aluminum ion in the next lower layer. In the hydrated form, the two upper layers correspond stoichiometrically to $Al_2O_3 \cdot H_2O$.

Hydroxyl ions coordinated in various ways with aluminum cations constitute the reactive functional groups of alumina surfaces. In general, more than one kind of surface OH group

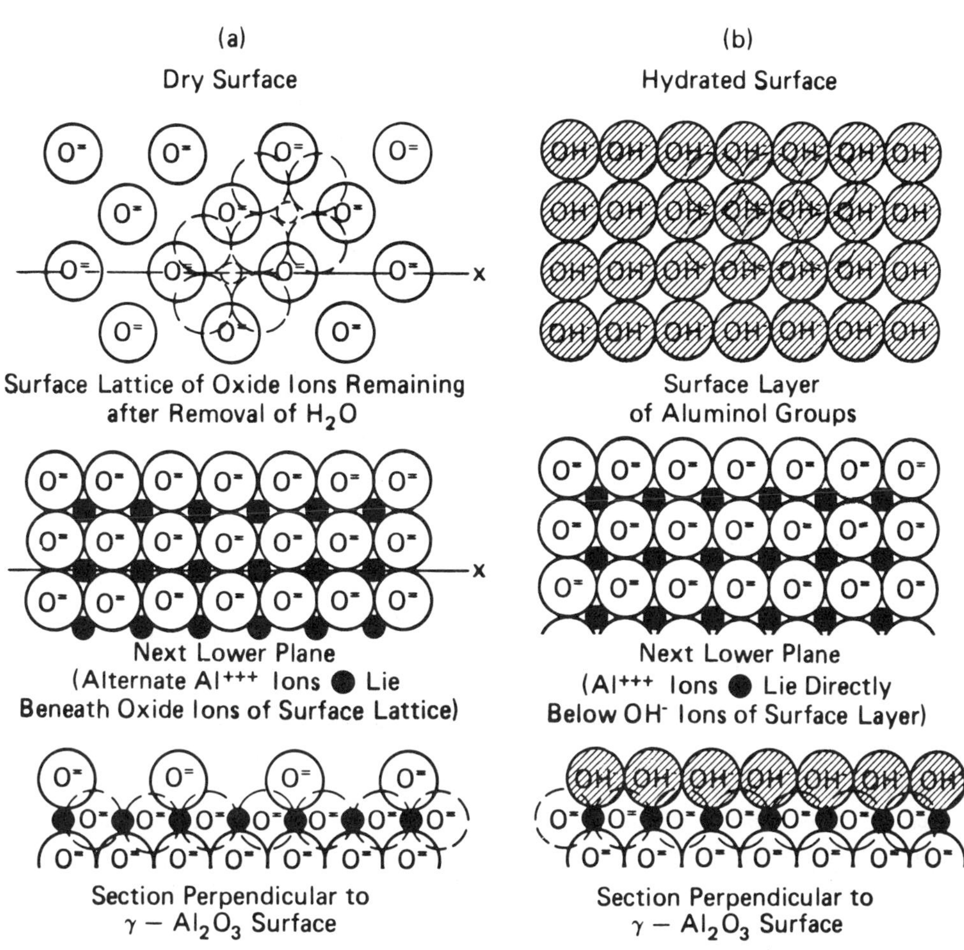

FIGURE 1. Idealized illustration of the dry and hydrated surface of γ-Al_2O_3. (Modified from Peri, J. B., *J. Phys. Chem.*, 69, 220, 1965.)

can be distinguished on the basis of stereochemical reasoning, and these different groups have properties (e.g., their IR absorption spectra) that set them apart from OH groups inside the bulk mineral structure.[6] The number and type of each surface OH group depend on which crystal planes are exposed preferentially and how aluminum ions are distributed at the surface. For example, cleavage of gibbsite crystals occurs preferentially along the *001* planes, since this involves only the breaking of relatively weak hydrogen bonds which hold the layers of the structure together. The surface hydroxyl ions in this plane would appear as a close-packed, pseudo-hexagonal array, with each OH coordinated to a pair of underlying aluminum ions. This basal plane makes up most of the surface of gibbsite. At the edge faces (*100* and *110* planes), aluminum ions with potentially unsatisfied positive charge may be exposed. At such sites, hydroxide ions can be present that are in effect, coordinated by only one aluminum ion and these groups are more reactive, as will be discussed below in more detail.

B. Gibbsite, α-$Al(OH)_3$

The crystal structure of gibbsite, the most common of the $Al(OH)_3$ minerals, has been modeled in a number of ways. One simple model uses spheres of two different sizes. The larger spheres, representing OH^- ions, are arranged in a close-packed layer structure with

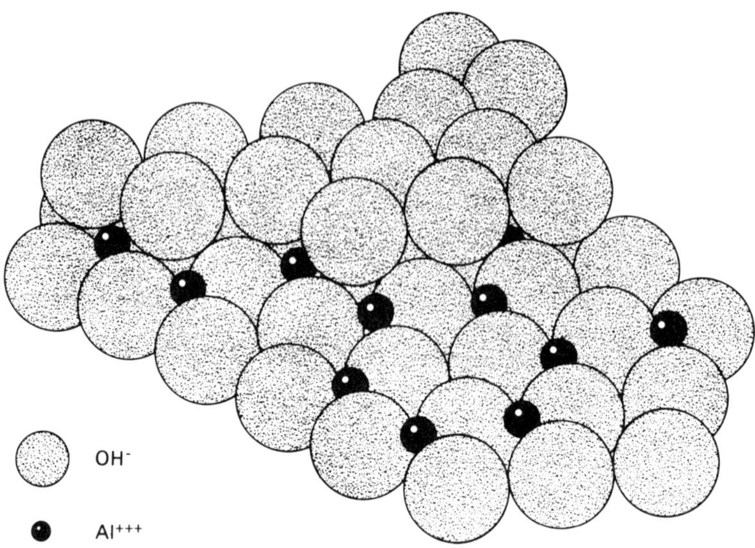

FIGURE 2. Sphere model of gibbsite structure. Large spheres represent OH^- ions, smaller spheres represent Al^{3+} ions in octahedral coordination sites. Part of upper OH^- layer removed to show arrangement of Al^{3+}

the smaller spheres, representing Al^{3+} ions, occupying interstices between two layers. Two types of interstices (i.e., cation sites) occur in this structure. The smaller, tetrahedral sites are surrounded by four of the large spheres, three in one layer and one in the other. The larger, octahedral sites are surrounded by six of the larger spheres, three in each layer. Figure 2 is a sphere model of a segment of the gibbsite structure with part of the upper hydroxide layer removed to show the arrangement of Al^{3+} ions. An electroneutral structure is achieved when Al^{3+} ions occupy only two thirds of the octahedral sites (i.e., dioctahedral arrangement). The Al^{3+} ions are coordinated to six OH^- ions with each OH^- ion shared with two Al^{3+} ions. This results in the orderly arrangement shown in Figure 2. Crystal growth occurs by extension in the plane of the layer structure (a and b directions) and by vertical stacking of these layers (c direction).

The surface hydroxyls of an ideal gibbsite basal plane may be coordinated to two Al^{3+} ions while edge plane hydroxyls are coordinated to one Al^{3+} ion. Parfitt et al.[7] estimated that the basal plane of gibbsite contains approximately 12 hydroxyl groups per square nanometer of surface while edge planes contains about 4 groups/nm^2.

The structure of gibbsite can also be viewed from the perspective of the Al^{3+} ions, which are arranged in a pattern of coalesced hexagonal rings. A single ring of six Al^{3+} ions, joined above and below by six pairs of bridging OH^- ions, is the smallest recognizable unit of the gibbsite structure. A sphere model of two six-membered rings is shown in Figure 3b. The shaded spheres in this model are OH^- ions and the clear ones are H_2O molecules. The excess positive charge on each Al^{3+} ion retains a pair of H_2O molecules, shown in the model to be inclined at an angle of 60° to the face of the ring structure. Two rings can be joined by replacing inclined pairs of H_2O molecules on adjacent rings with two bridging OH^- ions. Figure 3a represents the ring with the upper hydroxide layer removed to show the locations of Al^{3+} ions.

Alternatively, the $Al(OH)_6$ unit can be represented schematically as an octahedron, with each apex being at the center of an OH^- ion. Two octahedra are joined along one edge by sharing a pair of OH ions, i.e., $Al(OH)_2Al^{4+}$. Six octahedra, each sharing two edges, yield an $Al_6(OH)_{12}^{6+}$ ring. This model depicts the three-dimensional gibbsite crystal structure as an array of octahedra sharing three edges to form the pattern shown in Figure 4. The shared

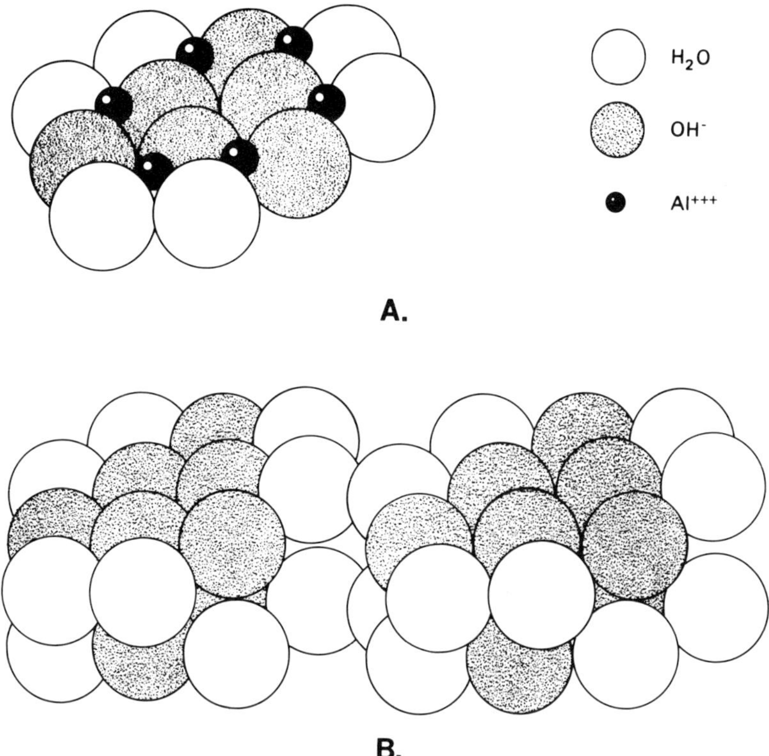

FIGURE 3. Sphere model of polymeric ion $Al_6(OH)_{12}(H_2O)_{12}^{6+}$. (A) Single ion with upper OH^- + H_2O layer removed to show arrangement of Al^{3+} ions (small dark spheres). (B) Two complete ions. White spheres represent H_2O molecules, shaded spheres represent OH^- ions.

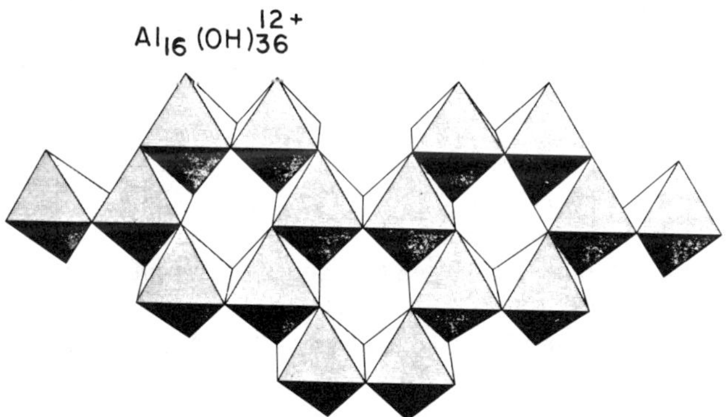

FIGURE 4. Octahedral model of gibbsite structure consisting of three coalesced $Al_6(OH)_{12}$ rings with two additional aluminum ions.

edges represent pairs of bridging OH^- ions. The apparent "holes" in the structure in Figure 4 stress the dioctahedral arrangement of Al^{3+} ions in the structure.

Another way of representing gibbsite structures in two dimensions is shown in Figure 5. The double OH bridges in Figure 5a are represented in the other structures by straight lines describing hexagons with Al^{3+} ions at each corner. The rings, in turn, are joined by formation of additional OH bridges, represented by dashed lines, joining the corner Al^{3+} ions and

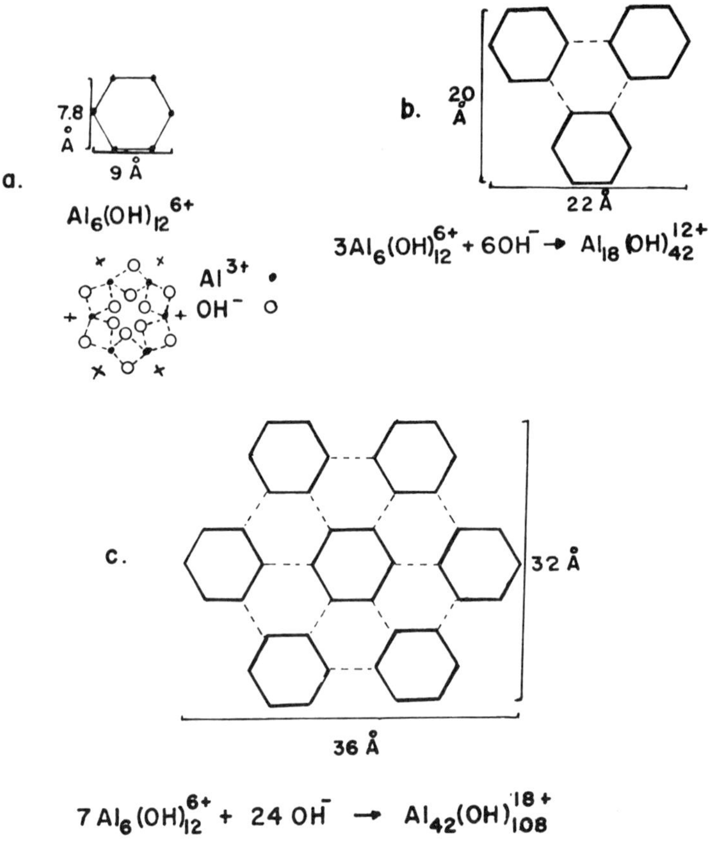

FIGURE 5. Two dimensional representations of aluminum hydroxide structures and approximate dimensions in Angstroms. (a) $Al_6(OH)_{12}$ ring shown in two different ways — with OH^- double bridges drawn in (not to scale) and with bridges represented by straight lines connecting Al^{3+} ions. (b) 4-ring polymer, formed from 3 independent $Al_6(OH)_{12}^{6+}$ units joined by 3 additional double OH^- bridges (dashed lines). (c) 13-ring polymer formed from 7 independent 6-membered rings joined by 12 double OH bridges (dashed lines).

giving polymeric ions with the formulas and physical dimensions noted in the illustration. If the polymer growth process is based on coalescence of $Al_6(OH)_{12}^{6+}$ rings (compare Chapter 4), the geometry of the units is constrained by the requirement that two complete ring units cannot combine with each other by sharing hexagonal edges, but must join at corners (i.e., octahedral edges) by establishing additional $(OH)_2$ bridges.

The structural model proposed for gibbsite (Figures 2 to 5) suggests that Al^{3+} ions located at the corners of $Al_6(OH)_{12}$ rings will have one unsatisfied positive charge. Two water molecules could be retained in coordination positions at such sites. Attachment of another ring at that site entails deprotonation of two of the four H_2O molecules involved and release of the other two into solution. At higher pH, the coordinated water molecules might be replaced by a single OH^-. This condition would tend to inhibit lateral growth of the gibbsite structure.

The corner charge sites described above constitute aluminol groups (AlOH) or surface hydroxyl sites that will be described below. Another possible form of edge Al^{3+} ions is represented diagrammatically in Figure 4. This structure comprises three $Al_6(OH)_{12}$ rings, with two additional aluminum octahedra that are attached to the structure through single $(OH)_2$ bridges. If the edges of gibbsite crystals actually include Al^{3+} ions that are not part of a ring structure, then these two Al^{3+} ions could represent potential sites with double positive charges.

Defects and elemental substitutions in the gibbsite crystal structure may lead to a net charge imbalance. At the surface, where the crystal structure is abruptly truncated, electroneutrality is achieved by adsorption of ions or molecules at empty or incompletely filled coordination sites. Both positive and negative charge sites may exist on the crystal faces. However, where defects or layer growth sites are absent, the charge density on crystal faces will be substantially less intense than those at crystal edges or corners. Electron micrographs of synthetic gibbsite crystals bearing adsorbed particles of a negatively charged gold sol illustrate this phenomenon rather strikingly.[8] The gold sol particles are generally attached to crystal edges, corners, and imperfections. Another example is shown in Figure 6a, an electron micrograph of gibbsite particles produced by slow addition of OH^- to a dilute $Al(ClO_4)_3$ solution maintained at pH 5.4 and 25°C. The particles display different amounts of layer stacking and various types of surface irregularities that could produce favorable sites for adsorption. McBride et al.[9] have also observed the occurrence of crystal steps at regular intervals on gibbsite plates.

C. γ-Al_2O_3

Although gibbsite is probably the most important aluminum-bearing oxide mineral in soils and sediments, more is known about the surface chemistry of γ-Al_2O_3 than gibbsite because of the significance of γ-Al_2O_3 as a catalyst in industry. Electron and X-ray diffraction studies indicate that γ-Al_2O_3 may preferentially expose the *100* plane of the defect spinel lattice. The surface of γ-Al_2O_3 is hydroxylated when water chemisorbs and combines with oxide ions which have broken bonds along the cleavage plane (Figure 1). The hydrated surface retains 12.5 molecules of water per square nanometer of surface after evacuation at 25°C for 100 h.[10] Although most of the water is chemisorbed and held in hydroxyl groups, IR studies suggest that some of the water strongly adsorbed at 25°C remains as molecular water. Kummert and Stumm[11] reported a value of 8.5 exchangeable protons per square nanometer on γ-Al_2O_3 using the isotopic exchange technique of Yates and Healy.[12] This value is about one third of that expected on the basis of results from Peri and Hannan.[10] Other authors have attempted hydroxyl group determinations from acid-base titrations in water. These methods yield significantly lower values (e.g., 1.3 protons per square nanometer by Kummert and Stumm[11]) because of incomplete ionization of hydroxyl groups.

In the idealized model of the hydrated alumina surface presented in Figure 1, two types of surface hydroxyl group can be distinguished: those coordinated with one aluminum ion and those coordinated by two aluminum ions. The singly coordinated groups are more basic and more likely to participate in ligand exchange reactions. Since the hydroxyl groups on the edge planes of gibbsite may also be singly coordinated, one may expect some correlations of the surface chemistry of γ-Al_2O_3 with that of the edge faces of gibbsite. It would be useful to know if this hypothesis is true, since a significant number of studies have been conducted with γ-Al_2O_3. These studies then would be useful in modeling the chemistry on the edge faces of gibbsite. Since the faces are more reactive than the basal planes of gibbsite, it is probably more important to understand the chemical behavior of the edge faces than that of the basal planes.

In comparing the behavior of one γ-Al_2O_3 sample with another, one should ostensibly normalize the reactivity in terms of surface area. For example, one may compare the maximum adsorption of a specific solute or its energy of adsorption per unit surface area. However, this approach is probably inappropriate when comparing the reactivity of one type of aluminum oxide with another. The unique nature of the crystal faces means that the reactivity of the surface cannot be normalized on the basis of the total surface area. The reactivity of each face should be considered separately.

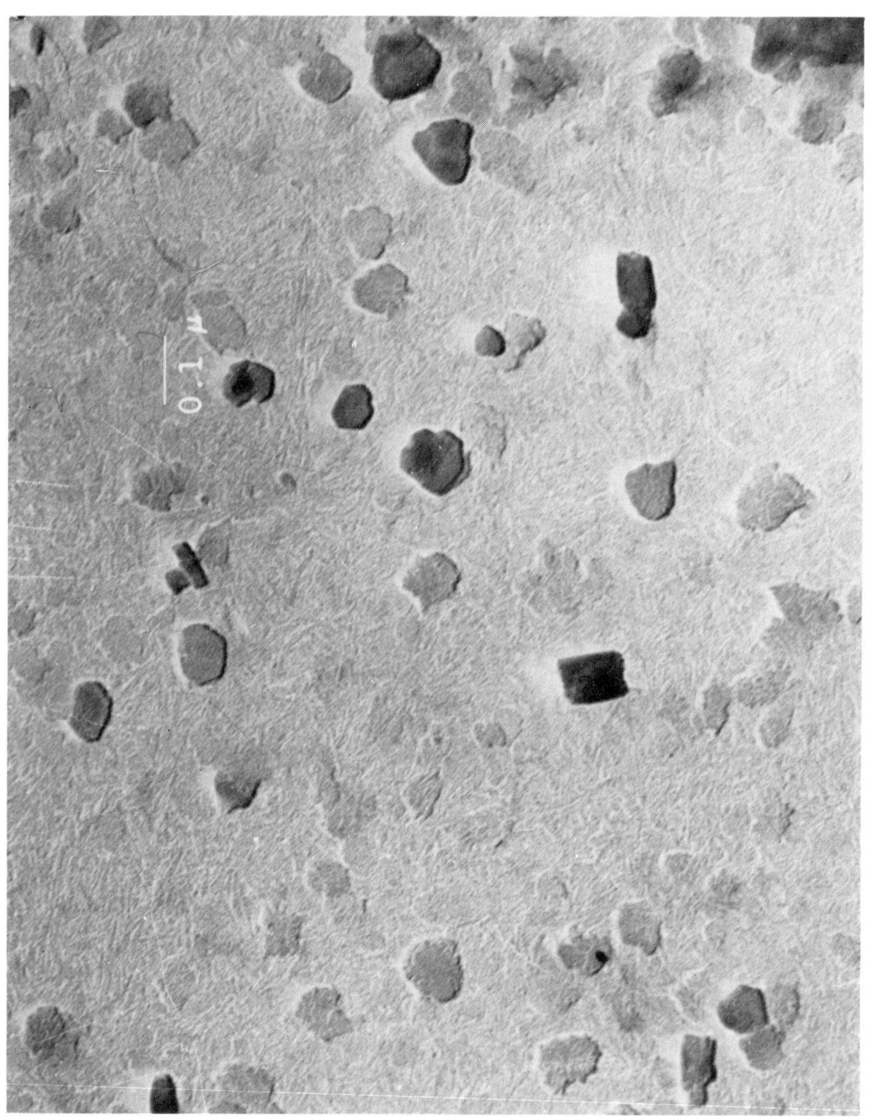

A

FIGURE 6. Transmission electron micrographs of aluminum hydroxide precipitates: (A) gibbsite, (B) bayerite, (C) bayerite and nordstrandite (rectangles).

FIGURE 6B

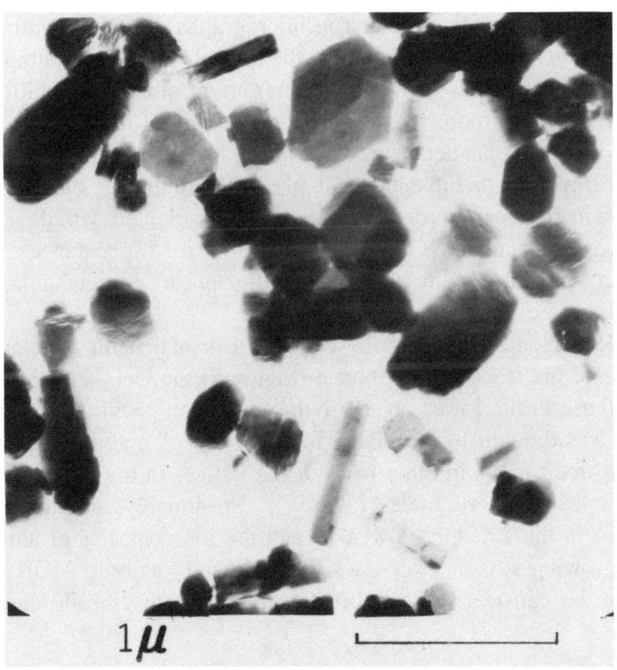

FIGURE 6C

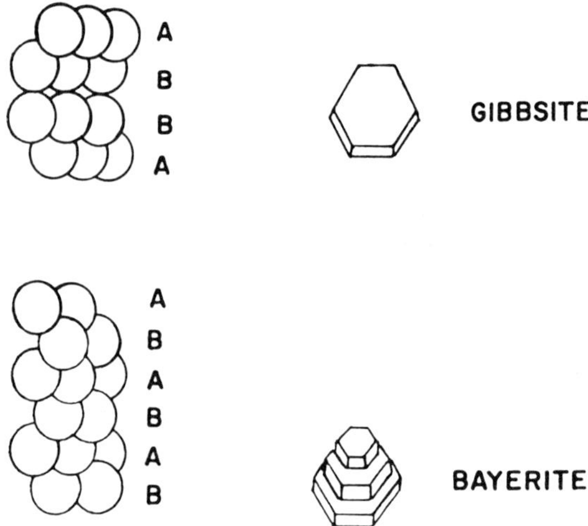

FIGURE 7. Gibbsite and bayerite layer stacking patterns.

D. Other $Al(OH)_3$ Polymorphs

Two other minerals, bayerite and nordstrandite, have the composition $Al(OH)_3$ and differ in structure from gibbsite only in the way the double layers of coordinated OH^- and Al^{3+} are stacked to form the three-dimensional crystal units. Bayerite and nordstrandite were first discerned in synthetic material, but were later found to occur naturally.[13] They are relatively rare, however, in natural material (see Chapter 3).

During the synthesis of gibbsite near pH 5.0, it appears that some critical size of individual sheet structures is attained before a significant amount of growth through stacking of layers takes place.[14] The crystals seen in electron micrographs are fairly uniform in size and the number of layers is generally not large; thus, the crystals appear predominantly as hexagonal plates. However, at pH values above that of minimum gibbsite solubility, growth in the c direction appears to occur more readily through stacking of the gibbsite sheets. As will be discussed later, this pH dependence may be caused by the development of electrical charge on the surface. Bayerite, precipitated near pH 9.0 (Figure 6b) displays a layered crystal habit. Units seen in electron micrographs are generally oriented with their c axis in the plane of the photograph and usually have a conical or barrel-shaped appearance. Another polymorph found at higher pH is nordstrandite, which generally appears as rectangular shapes in electron micrographs (Figure 6c).

Schoen and Roberson[13] explained the change in crystal habit as related to the way the layers of OH^- ions are stacked. In gibbsite (Figure 7) the OH^- ions of the upper layer are directly above those in the lower layer, giving an "open packing" arrangement. Bonding between layers was thought to be related to distortion of hydroxyl ions in the vicinity of Al^{3+} ions, compared to ions in other parts of the lattice. In the bayerite structure the OH^- ions of adjacent layers interdigitate (Figure 7). Presumably, at higher pH the effect of coordinated Al^{3+} in the structure is weaker and the ionic species of aluminum in solution that enters the growing structure becomes predominantly anionic $AlOH_4^-$. In nordstrandite the structure can be considered a combination of alternate gibbsite and bayerite layering patterns.

E. α-Al_2O_3 (Corundum)

Corundum occurs naturally as an igneous and metamorphic mineral and is synthesized artificially by various high temperature techniques. The crystal structure of corundum is described by Wefers and others[2,15] as a hexagonal close-packing of oxygen ions with Al^{3+}

ions occupying two thirds of the octahedral interstices. In effect this structure could be visualized as composed of alternating layers of O^{2-} ions and Al^{3+}. Thus, interlayer bonding between aluminum and oxygen is well established in the c direction as well as in the a and b directions. It is to be expected that the surface properties of different crystal faces of corundum will be similar. However, the mode of preparation of synthetic material may affect the distribution of crystal defects. The surface hydroxyl density of α-Al_2O_3 (corundum) has been estimated as 6 groups/nm^2 by Smit and Holten[16] from crystal lattice considerations.

F. γ-AlOOH (Boehmite)

Boehmite is a common mineral with the composition AlOOH. It is present in some bauxites, especially those of the Mediterranean region.[2] The processes by which boehmite is produced in bauxite deposits in preference to gibbsite are not well understood. Sphere models of the crystal structure of boehmite are given by Wefers and Bell,[15] who describe the structure as composed of double layers of oxygen and hydroxide in cubic packing. As in gibbsite, the double layers are joined by hydrogen bonds, and crystals exhibit preferential cleavage in the direction perpendicular to the general direction of the hydrogen bonding. Aluminum oxyhydroxide structures have some similarities with those of iron oxyhydroxides and, therefore, a significant amount of Fe^{3+} can replace Al^{3+} in the boehmite crystal lattice. DeBoer et al.[17] determined the hydroxyl group density on γ-AlOOH (boehmite) from infrared studies as 16.5 groups per square nanometer.

III. SURFACE ACIDITY AND THE ELECTRICAL DOUBLE LAYER

A. Acidity of the Dehydrated Alumina Surface

The surface acidity of aluminum oxides is of long-standing interest, for both practical and theoretical reasons. The acidic properties of γ-Al_2O_3 have been of particular interest, because it is widely believed that the catalytic properties of this material are related to the density and type of acidic sites.[18] A variety of methods has been used to investigate the relative acidity of hydroxyls on alumina surfaces, including direct titration with acid-base indicators,[18-22] IR spectroscopy,[21-26] and NMR spectroscopy.[27,28] However, these techniques generally have examined the acidity of functional groups on the alumina surface after dehydration at elevated temperatures in a vacuum. As shown by Peri,[5,29] most surface hydroxyls are removed by evacuation at temperatures greater than 400°C, but distinct types of hydroxyls are left on the surface instead of hydroxyls at random sites. The relationship between the acidity of groups remaining on the dehydrated surface and that of the hydrated surface remain obscure. ''Strain'' sites are created by the process of dehydration at high temperature and some of these sites exhibit considerable acidity. A portion of the sites are produced by the creation of strained Al–O–Al linkages, but other types of site must also develop during dehydration.[29]

B. Acidity of the Hydrated Alumina Surface

The hydroxylation of oxide surfaces is complete at 25°C for relative humidities well below 1%.[30] In the idealized model presented in Figure 1, two types of surface hydroxyl group were distinguished: those coordinated with one aluminum cation and those coordinated by two aluminum cations. The doubly coordinated hydroxyl groups are expected to be more acidic because of the strong polarization induced by the aluminum ions on the oxygen atom.[19] The singly coordinated hydroxyl groups should be less acidic in nature. The hydrated alumina surface exhibits amphoteric behavior in water, i.e., it is capable of adsorbing or releasing protons. This behavior has been attributed generally to amphoteric surface hydroxyl groups instead of independent basic and acidic surface groups.[31] For a hypothetical system consisting only of pure water and an immersed alumina surface, the ionization reactions of the surface hydroxyl (aluminol) groups can be described in the following way:

$$AlOH_2^+ = AlOH + H^+ \quad (1)$$

$$AlOH = AlO^- + H^+ \quad (2)$$

where $AlOH$ represents an aluminol group. These reactions lead to the development of electrical charge and potential on the surface. To preserve electrical neutrality, an equal amount of counterion charge must accumulate at the alumina-water interface. In the presence of electrolyte ions, the counterion charge may consist of a diffuse atmosphere of counterions or a compact layer of bound counterions or a combination of both. The complete interface, including the surface and the accumulated counterions, is often called the *electrical double layer*. Background theory on the electrical double layer can be found in standard reference books.[30,32-36]

The techniques used for characterizing the electrochemical properties of alumina surfaces include: (1) potentiometric acid-base titrations, (2) measurement of the dependence of electrokinetic potentials on pH and electrolyte concentration, (3) measurement of surface electrical potential with an ion-sensitive field effect transistor (ISFET), and (4) direct measurement of counterion adsorption.

1. Acid-Base Potentiometric Titrations

Proton or hydroxide ion consumption by a solid phase is determined by comparing the titration curve of a suspension of alumina in an electrolyte solution with that of the electrolyte alone. Details of the methodology are given by Huang.[37] The adsorption isotherm for the proton is obtained with the mass balance equation:

$$A\Gamma = (C_A - C_B) - (C_{H^+} - C_{OH^-}) \quad (3)$$

where $\Gamma = (\Gamma_{H+} - \Gamma_{OH-})$ is the net surface excess of hydrogen over hydroxide ions per square centimeter of surface, A is the total surface area of the alumina per unit volume (cm^2/dm^3), C_A and C_B are the concentrations of acid and base added to reach a given point in the titration curve, and C_{H+} and C_{OH-} are the concentrations of hydrogen and hydroxide ions, respectively. If the supporting electrolyte ions have no specific affinity for the surface, or if the positive and negative ions of the supporting electrolyte have equal specific affinity, then curves representing Γ as a function of pH for each of several electrolyte concentrations should intersect at a common pH value, the pH_{pzc},[38] where:

$$\Gamma_{H^+} - \Gamma_{OH^-} = 0 \quad (4)$$

Figure 8 shows plots of Γ as a function of pH and sodium chloride concentration for titrations of γ-Al_2O_3.[39] The intersection point indicates a pH_{pzc} of 8.5. Note that the protons released or consumed by the alumina increase with electrolyte concentration. The surface is positively charged at pH values below 8.5 and negatively charged when the pH is greater than 8.5. Other titration data for γ-Al_2O_3 can be found in Westall and Hohl,[40] Kummert and Stumm,[11] Healy and White,[31] and a comparison of these data at one electrolyte concentration can be found in Bousse and Meindl.[41] Westall and Hohl[40] and Kummert and Stumm[11] reported pH_{pzc} values of 8.3 and 8.7, respectively, for γ-Al_2O_3.

Kavanagh et al.[42] studied the proton balance in potentiometric titrations of gibbsite as a function of pH and $NaNO_3$ concentration (0.01 to 1 *M*). The data are somewhat unique in comparison to γ-Al_2O_3 and other oxides in that the quantity Γ is less sensitive to electrolyte concentration. The authors reported a pH_{pzc} value of 9.8 for gibbsite. The pH_{pzc} of gibbsite was quoted by Smith[43] as 11.2. Titration data for chromatographic Al_2O_3 (crystal form not reported) as a function of pH and electrolyte concentration were published by Tewari and

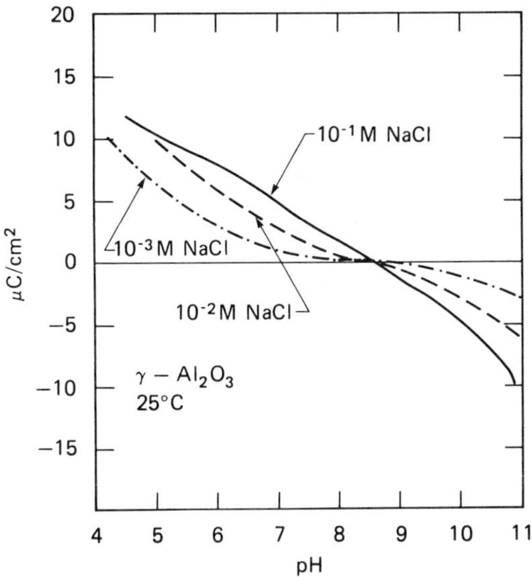

FIGURE 8. pH and ionic strength dependence of the net proton balance of γ-Al_2O_3. (Modified from James, R. O. and Parks, G. A., in *Surface and Colloid Science*, Vol. 12, Matijevic, E., Ed., Plenum Press, New York, 1982, 119; data from Huang, C.-P. and Stumm, W., *J. Colloid Interface Sci.*, 43, 409, 1973.)

McLean.[44] These authors found that pH_{pzc} decreased with increasing temperature from a value of 9.1 at 30°C to 8.4 at 90°C. The change in pH_{pzc} with temperature appeared to be explained solely by the change in the ionization constant of water as a function of temperature.

Smit and Holten[16] and Ahmed[45] have reported pH_{pzc} values for α-Al_2O_3 (corundum) of 4.5 and 4.8; the former investigators studied a large crystal while the latter author worked with a powder. The lower pH_{pzc} of α-Al_2O_3 can be expected on the basis of crystal structure. The dehydrated oxide has hydroxyl groups which are more polarized than those in the aluminum hydroxides, because of differences in the coordination number of the aluminum ion. This difference will result in a relatively acidic surface for the oxide and a more basic surface for $Al(OH)_3$.[46]

In contrast, Yopps and Fuerstenau[47] observed a pH_{pzc} of 9.1 in studies of a commercial alumina (Linde A) identified as α-Al_2O_3 by X-ray diffraction. The authors also cite several other publications which have found the value of pH_{pzc} to fall in this range. However, Smit and Holten[16] state that grinding α-Al_2O_3 results in a disturbed surface layer which, upon contact with water, hydrates to give a surface layer like that of gibbsite. This could result in artificially high values of pH_{pzc} for α-Al_2O_3 powders. A study by Smith[43] found that α-Al_2O_3 (Linde A) had a pH_{pzc} of 9.0 to 9.1 after 3 d of immersion in water, but after 3 months the pH_{pzc} had risen to pH 10.1 to 10.2. These results suggest that the α-Al_2O_3 surface may undergo alteration in water to give an external ''shell'' of gibbsite. Support for this hypothesis was given by electron micrographs of altered α-Al_2O_3 particles, which showed the presence of hexagonal microcrystals of gibbsite on the particle surfaces.[43] Slow alteration of the surfaces of other aluminum oxides and hydroxides may also occur, since gibbsite is probably the thermodynamically stable phase at 25°C.[48,49] The surfaces of gibbsite preparations may also vary in their degree of crystallinity, which may influence their solubility, surface area, concentration of structural defects, and rates of recrystallization.[50]

The increase in positive charge (as indicated by the increase in pH_{pzc}) can be attributed to localized strong positive charge on the edges of gibbsite crystals, as compared to the more diffuse distribution of surface charge expected for the dehydrated material. The distribution of charge sites per unit area of crystal faces of well-crystallized gibbsite, however, would be expected to be sparse. As noted previously, bayerite crystals grow preferentially in the c direction and therefore a substantial fraction of the surface of an individual bayerite crystal comprises the edges of individual layers, not the faces.

The relative acidity of oxide surfaces can be inferred from values of the pH_{pzc} and, therefore alumina is usually less acidic than silica or titania.[37] A quantitative evaluation of the acidity of aluminol groups on a surface, however, requires a model for the surface reactions and the introduction of several assumptions. Models which have been used to describe the surface reactions will be discussed below. The quantity, Γ, the net proton adsorption density defined in Equation 3, is usually called "surface charge" in the literature. It is important to remember that the experimental quantity determined is the net proton release or consumption of the surface, which is not an actual determination of surface charge. If not properly taken into account,[37,51] the dissolution of alumina during titrations may significantly affect the titration curve and hence the net surface proton balance, Γ.

It is only through the development of a model for the surface reactions that the net surface proton balance, Γ, can be assumed to be a determination of surface charge. This is particularly significant when titrations of alumina suspensions are conducted in the presence of specifically adsorbing ions. "Specific adsorption" is a general term which describes the chemical coordination process involving specific interactions between solutes and reactive surface groups. This may be contrasted with the nonspecific interaction between the charged surface and counterions located some distance away in the diffuse double layer. Breeuwsma and Lyklema[52] have shown that the pH_{pzc} is shifted to greater values in the presence of specifically adsorbing anions and to lower values in systems containing specifically adsorbing cations. An interpretation of the net proton balance, Γ, in systems containing strongly adsorbing ions, e.g., phosphate or lead, requires the division of proton consumption among several possible surface reactions. Thus, interpretation of the experimental quantity, Γ, and the definition of the term "surface charge", are *model-dependent* in these types of systems.

2. Electrokinetic Potential Measurements

Electrokinetic techniques for inferring the electrical potential difference across the alumina-water interface rely on the relative movement between surface and solution. Some water is strongly bound by the alumina surface and moves along with the particles when placed in an electric field. At some distance from the surface, a shear plane develops where the water movement is independent of the particles. The shear plane is assumed generally to occur within the electrical double layer. The relative motion of the alumina with respect to the bulk solution results in a separation of charge, and a potential difference is created across the shear plane; this potential difference is referred to as the "zeta potential".[30,53]

Wiese and Healy[54] have studied the effect of pH and electrolyte concentration on the zeta potential of γ-Al_2O_3 colloidal particles suspended in KNO_3 solutions (Figure 9). The isoelectric point, pH_{iep}, the pH value where the zeta potential is zero, was reported as 8.9, very close to the values of pH_{pzc} found in potentiometric titrations of γ-Al_2O_3 samples. Other values for the pH_{iep} of γ-Al_2O_3 have been reported in the range 8.75 to 9.2.[39,55] At a given pH value (except the pH_{iep}), the zeta potential increases in absolute value as the electrolyte concentration decreases (Figure 9). The absolute value of the zeta potential increases rapidly with pH near the pH_{iep} but the slope of the curves decreases considerably at pH values far from the pH_{iep}. Wiese and Healy[54] also studied the rate of coagulation of alumina suspensions as a function of pH and electrolyte concentration. The maximum rate of coagulation was observed in the pH region near the pH_{iep}. Rapid coagulation occurred when the absolute value of the zeta potential was less than 14 mV, regardless of the pH value or electrolyte concentration.

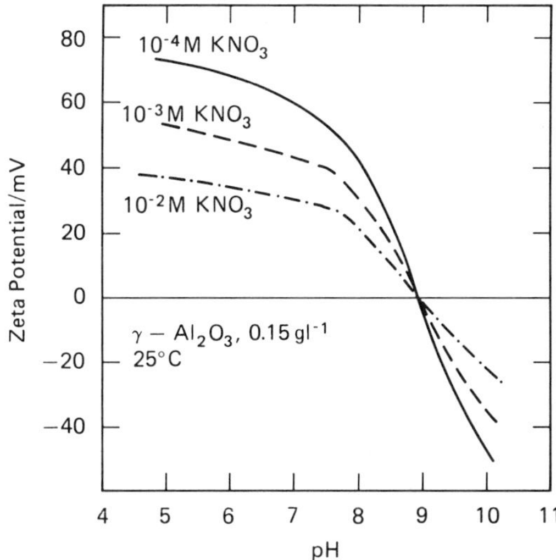

FIGURE 9. pH and ionic strength dependence of the zeta potential of γ-Al_2O_3. (Modified from Wiese, G. R. and Healy, T. W., *J. Colloid Interface Sci.*, 51, 427, 1975.)

The decrease in zeta potential with electrolyte concentration at a given pH is in contrast to the dependence of the surface proton balance on electrolyte concentration (Figure 8). If the surface proton balance in the absence of specifically adsorbed ions is in fact a measure of ''surface charge'', then potentiometric titration results suggest that the surface charge of γ-Al_2O_3 increases with electrolyte concentration at a given pH. Since the zeta potential decreases with increasing electrolyte concentration, one must conclude that either (1) a greater percentage of counterion charge is present inside the hydrodynamic shear plane at higher electrolyte concentrations or (2) the location of the hydrodynamic shear plane is farther away from the alumina surface at higher electrolyte concentrations. This point will be discussed below in the section on models for the alumina surface.

Anderson et al.[56] found a value of 8.5 for the pH_{iep} of freshly precipitated aluminum hydroxide (precipitated from a sulfate solution). Residual sulfate in the precipitate used by Anderson et al.[56] could have lowered the measured pH_{iep} in comparison to the pH_{iep} of a completely washed sample.[57] Yopps and Fuersteneau[47] studied the variation in zeta potential of α-Al_2O_3 as a function of pH and type of electrolyte medium. A pH_{iep} of 9.1 was reported, in agreement with their determination of pH_{pzc} for the same sample (Linde A). Like Wiese and Healy,[54] these authors found that the maximum rate of coagulation of the alumina suspensions occurred in the pH region of the pH_{iep}. Relatively large mobilities were observed in comparison to those measured under similar conditions for γ-Al_2O_3. Yopps and Fuersteneau[47] also compared the mobilities of α-Al_2O_3 particles suspended in 0.001 *M* solutions of $KClO_4$, KNO_3, and KCl. The mobilities of the particles were the same in all three electrolyte solutions when $pH > pH_{iep}$, i.e., when the particles were negatively charged. When $pH < pH_{iep}$ and the particles were positively charged, however, the mobility of the particles depended upon the anion present in the system.

Smit and Holten[16] found a pH_{iep} near 3.5 for single crystals of α-Al_2O_3 in NaBr solutions; a pH_{pzc} of 4.5 was reported for the same samples. The low value for pH_{iep} in comparison to published values for powdered samples was attributed to sample grinding. The fact that the pH_{iep} was lower than the pH_{pzc} was evidence for specific adsorption of bromide ions. The authors studied the binding of sodium and bromide ions on the crystal surface and found that both ions were indeed specifically adsorbed.

Bleam and McBride[58] reported a pH_{iep} of 10.4 for boehmite. The data reported by these authors were unique in that little dependence of particle mobility on electrolyte concentration was observed in very dilute solutions of $NaClO_4$.

Specific adsorption of ions changes the value of the isoelectric point; anions decrease pH_{iep} while cations cause an increase.[52] The shifts in pH_{pzc} and pH_{iep} are in opposite directions when specific adsorption occurs. Van Riemsdijk et al.[59] recently proposed that the pH_{pzc} at which no specific adsorption is indicated (and thus, $pH_{pzc} = pH_{iep}$) should be referred to as the *pristine* point-of-zero charge, or pH_{ppzc}, to distinguish this pH_{pzc} from other pH_{pzc} values (at which specific adsorption occurs). This appears to be an excellent nomenclature suggestion; the pH_{pzc} in the presence of specific adsorption represents the pH at which the surface proton balance equals zero, which may not be the pH at which there is zero surface charge. The condition of zero surface charge can only be indicated unambiguously by zero electrophoretic mobility.

3. Measurement of Surface Potential

The measurement of changes in electrical surface potential between an insulating material, e.g., alumina, and an aqueous solution can be made by incorporating a thin film of alumina in an electrolyte/alumina/silicon structure.[41] A field effect transistor (ISFET device) then can be fabricated to measure lateral current flow, and relative changes in the surface potential of alumina can be measured.[60,61]

Potential-determining ions (pdi), by definition, are ions which determine the interfacial electrical potential difference by virtue of their equilibrium distribution between solid and liquid phases.[62] The half-cell potentials of reversible electrodes, such as AgI, are established by transfer of charge in the form of pdi between the two bulk phases.[30] In the case of AgI, the pdi are Ag^+ and I^-, and the potential difference between the two phases is given by the Nernst equation,

$$E = E^o + RT/F[\ln\{Ag^+\} - \ln\{Ag^+\}^o] \quad (5)$$

where E^o is a reference potential and $\{Ag^+\}^o$ is the silver ion concentration at that reference potential. Many of the early research studies of oxide-water interfaces assumed that hydrogen and hydroxide ions were pdi. For example, for the solid Al_2O_3, it could be argued that Al^{3+} and O^{2-} are the pdi, e.g., where ψ_o is the surface potential. Since the activities of both Al^{3+} and O^{2-} may be small and highly pH-dependent in most solutions, the Nernst equation can be written more conveniently in terms of pH.

$$\psi_o = 2.3RT/3F[\log\{Al^{3+}\} - \log\{Al^{3+}\}_{pzc}] \quad (6)$$

$\{Al^{3+}\}$ is related directly to $\{H^+\}$ through the solubility product constant for alumina and thus, Equation 6 may be rewritten in terms of H^+:

$$\psi_o = 2.3RT/F\ (pH_{pzc} - pH) \quad (7)$$

In this way, H^+ and OH^- may be considered as pdi and, according to the Nernst equation, ψ_o will change 59.2 mV per pH unit at 25°C.

Bousse et al.[61] measured the pH dependence of the surface potential of γ-Al_2O_3 in 0.1*M* $NaNO_3$ with an ISFET device (Figure 10). The dependence of ψ_o on pH was not consistent with the Nernst equation, in that the slope was less than 59.2 mV per pH unit and the relationship was not quite linear.[41] The smallest slope (48 mV per pH) was measured near the pH_{pzc} (8.0). At pH values less than the pH_{pzc}, the slope increased to about 56 mV per pH unit; above the pH_{pzc}, the slope increased to values as high as 53 mV. The failure of

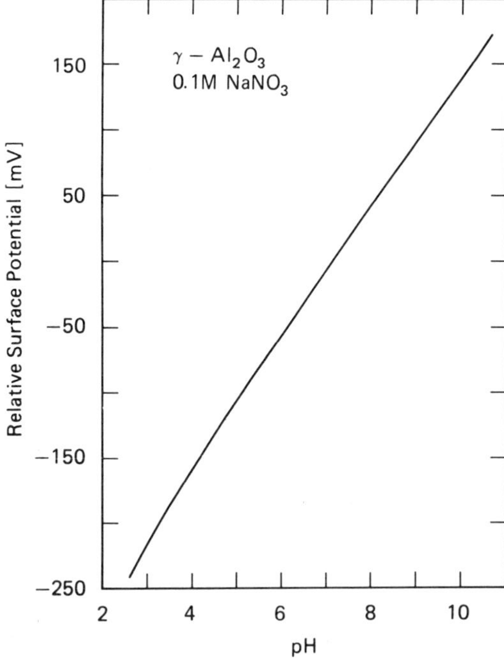

FIGURE 10. pH dependence of the surface potential, ψ_o, of γ-Al_2O_3 in 0.1 *M* $NaNO_3$ solution. (Modified from Bousse, L. and Meindl, J. S., in *Geochemical Processes at Mineral Surfaces,* Davis, J. A. and Hayes, K. F., Eds., ACS Symposium Series 323, American Chemical Society, Washington, D.C., 1986, chap. 5.)

the Nernst equation to describe the surface potential relationship is an indication that counterions form complexes (or ion pairs) with charged aluminol groups on the surface.[41]

4. Measurement of Counterion Adsorption

Breeuwsma and Lyklema[52] were among the first to study directly the ionic composition of counterion charge on a dispersion of a hydrous oxide (hematite). These authors found that the pH_{pzc} of hematite equalled the pH_{iep} in nitrate and chloride solutions of Na^+, K^+, and Cs^+. In potassium sulfate solution, the pH_{pzc} was shifted to a higher value by the specific adsorption of sulfate ions. In calcium and barium chloride solutions, the pH_{pzc} was shifted to lower values by the specific adsorption of calcium or barium ions. When specific adsorption occurs, the pH_{pzc} is no longer equal to the pH_{iep}, as noted above. Regardless of the value of pH_{pzc}, the adsorbed counterion charge was nearly equal to the surface proton balance as a function of pH in solutions containing the alkali and alkaline earth cations and chloride or nitrate anions. Near the pH_{pzc} the adsorption of electrolyte ions in these solutions was minimal. Similar measurements of the pH dependence of counterion adsorption on gibbsite have been made by Hingston and others.[63,64] These studies and others led to the general conclusion that many univalent ions (e.g., Na^+, Cl^-) were not specifically adsorbed but were held at the surface only by electrostatic forces just to balance the surface charge.

Recently, several experimental studies[16,65,66] and model calculations[38] have reopened the subject of counterion binding by oxide surfaces. Sprycha[65] found that the net charge from adsorption of Na^+ and Br^- ions on γ-Al_2O_3 was greater than the surface proton balance in the pH region near the pH_{pzc}. At the pH_{pzc}, adsorption of Na^+ was nearly equal to adsorption of Br^-, but both quantities were significantly greater than the surface proton balance.

Adsorption of Na^+ was significant at $pH < pH_{pzc}$ (when the surface was positively charged!) as was adsorption of Br^- at $pH > pH_{pzc}$. Supporting evidence for this behavior can be found in studies of γ-Al_2O_3 suspended in NaBr solutions,[67] of α-Al_2O_3 crystals suspended in NaBr solution,[16] and of colloidal samples of titanium dioxide suspended in uni-univalent electrolyte solutions.[68]

Yates et al.[69] suggested that counterions form weak ion pairs with charged surface hydroxyl groups, and this may explain the observations of specifically adsorbed counterions at the pH_{pzc}. Ion pair formation at alumina surfaces in NaBr solutions could be described by electrolyte binding reactions, i.e.,

$$AlO^- + Na^+ = AlO^-Na^+ \tag{8}$$

$$AlOH_2^+ + Br^- = AlOH_2^+Br^- \tag{9}$$

Equation 9 refers to ion pair formation with a Bronsted acid site, $AlOH_2^+$, but it is also possible that the reaction proceeds by complex formation at a Lewis acid site (by ligand exchange) as postulated by Smit and Holten,[16] i.e.,

$$AlOH_2^+ + Br^- = Al^+Br^- + H_2O \tag{10}$$

The formation of ion pairs with charged aluminol groups may explain other experimental observations: (1) the dependence of surface proton balance and electrokinetic potential on electrolyte concentration[38], and (2) the failure of the Nernst equation to describe the pH dependence of the surface potential.[41] More discussion of these observations will follow below in the section on modeling.

IV. ADSORPTION OF METAL IONS

Numerous studies of the adsorption of metal ions on aluminum oxides and hydroxides have been conducted. It has been demonstrated often that the mechanism of metal ion association with hydrous oxide surfaces involves an ion exchange process in which the adsorbed metal ions replace bound protons.[39,70-72] The adsorption of metal ions from dilute solution onto alumina surfaces is highly dependent on pH, as has been demonstrated by several investigators (Table 1). Figure 11 illustrates data of Benjamin and Leckie[73] for Pb^{2+} and Cd^{2+} adsorption on γ-Al_2O_3. These authors were among the first to demonstrate that the pH region in which adsorption increases rapidly (the ''adsorption edge'') depends on metal concentration. This is true even at very low concentrations of metals, far below those which would saturate the maximum adsorption capacity of γ-Al_2O_3. A consequence of this phenomenon is that adsorption isotherms for metal ions on aluminas usually are best described by the Freundlich equation:

$$\Gamma_M = aC_M^b \tag{11}$$

where Γ_M is the adsorption density of the metal ion, C_M is the equilibrium concentration of the metal ion, and a and b are empirical constants. The parameter b usually has a value in the range 0.5 to 0.7 for metal ion adsorption on alumina. A maximum in adsorption density for metal ions is often not evaluated because of experimental difficulties. Spectroscopic studies generally have led to the conclusion that there is a continuum of adsorption phenomena from monolayer adsorption to cluster formation or multilayer adsorption and, finally, surface precipitation.[9,58,74,75]

At very low concentrations, the adsorption of metal ions can be considered as a bimolecular

Table 1
ADSORPTION OF METAL IONS ON ALUMINUM OXIDES AND HYDROXIDES

Solid phase[a]	Cations	Type of investigation[b]	Ref.
γ-Al_2O_3	Pb, Cd, Zn, Cu	a, c, d, h, i	77
	Pb, Zn, Cu, Mn, Co	a, b, f, g, j	76
	Pb, Cd	a, d, h	73
	Mg, Ca, Sr, Ba	a, c, d, g	39
	Pb	a, c	80
	Cu	f	84
	Ni, Cd, Cu, Pb, Zn, Ca	a, i, l	108
	Cu	a, i	107
	Cu, Cd	a, i	109
	Cu, Cd	a, i	104
α-Al_2O_3	Cd	a, c, d	92
	Cd	a, c, d	81
	Zn	a, d	140
α-$Al(OH)_3$	Cu	a, f, j, k	9
α-$Al(OH)_3$, am-$Al(OH)_3$	Ca, Sr	a, d, e, j	141
α-$Al(OH)_3$, am-$Al(OH)_3$, γ-AlOOH	Cu	a, b, d, e, f, k	74
γ-AlOOH	Mg, Mn	a, f, g, k	58
am-$Al(OH)_3$	Cu, Pb, Zn, Ni, Co, Cd, Mg, Sr	a, e, h	142
	Ca, Mg, Mn, Cu,	d, e, f, h, i	143

[a] Solid phases: α-$Al(OH)_3$ = gibbsite, α-Al_2O_3 = corundum, γ-AlOOH = boehmite, and am-$Al(OH)_3$ = noncrystalline alumina.

[b] Type of investigation: (a) pH dependence, (b) kinetics of adsorption, (c) proton/hydroxyl stoichiometry, (d) dependence on cation concentration (isotherm), (e) co-precipitation or effects of aging precipitate, (f) spectroscopic, (g) electrokinetic potentials, (h) competitive adsorption, (i) effect of complexing ligand concentration, (j) desorption, (k) surface precipitation, and (l) temperature dependence.

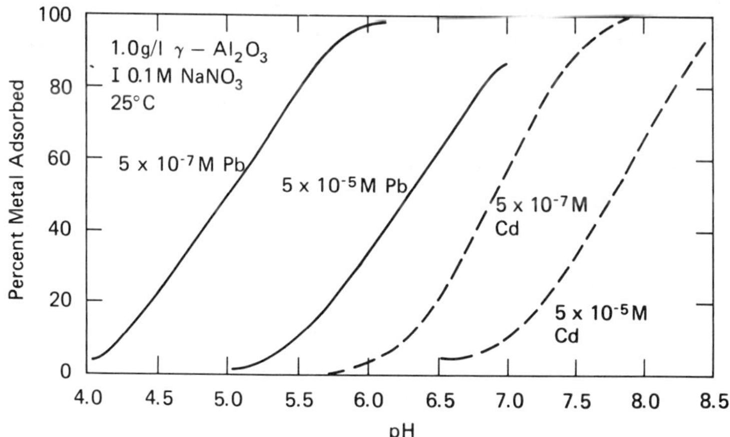

FIGURE 11. Adsorption of Pb^{2+} and Cd^{2+} on γ-Al_2O_3 from 0.1 *M* $NaNO_3$ solution as a function of pH and dilute metal ion concentration. (Data from Benjamin, M. M. and Leckie, J. O., in *Contaminants and Sediments,* Vol. 2, Baker, R. A., Ed., Ann Arbor Science, Ann Arbor, MI, 1980, 305.)

adsorption/desorption reaction in which a metal ion is bound by an aluminol group, releasing a proton to solution, i.e.,

$$AlOH + Pb^{2+} = AlOPb^{+} + H^{+} \tag{12}$$

Yasunaga and Ikeda[76] have shown by electric field pulse and pressure-jump methods that the adsorption and desorption reactions are very fast, with half-times considerably less than one second under typical experimental conditions. However, additional slow uptake of metal ions is often observed over hours or days, e.g., Cd^{2+} adsorption on γ-Al_2O_3.[77] The slow processes are very poorly understood. If the γ-Al_2O_3 surface has been altered to form a gibbsite-like layer, perhaps the slower process could be caused by diffusion through or around reorganized surface material to deeper layers.

As metal ions are adsorbed, protons are released (Equation 12). Most of these protons are readsorbed by the surface through ion-pair formation or the protonation of aluminol groups, but some remain in solution. The protons released during metal adsorption can be determined by back titration to constant pH[78,79] or by measuring the displacement of potentiometric titration curves.[39,80] The number of H^+ ions released per metal ion adsorbed is usually found to have a noninteger value between 1 and 2.[81] For example, Hohl and Stumm[80] found an average value of 1.5 protons released per Pb^{2+} ion adsorbed on γ-Al_2O_3. Because the overall proton release usually shows values much greater than one, many authors[72,78-80,82] have assumed that a second adsorption reaction must be considered in which two protons are released per metal ion adsorbed, e.g.,

$$AlOH + Pb^{2+} + H_2O = AlOPbOH + 2H^{+} \tag{13}$$

$$2AlOH + Pb^{2+} = \begin{matrix} AlO \diagdown \\ AlO \diagup \end{matrix} Pb + 2H^{+} \tag{14}$$

The fractional and nonequivalent values observed for the stoichiometry result from the fact that the equilibria of ion-pair and surface ionization reactions (Equations 1, 2, 8, and 9) are perturbed by the adsorption of metal ions. Few authors have taken these reactions into account when evaluating the stoichiometry of metal adsorption from dilute solution.[78,81] A recent experimental study and thermodynamic evaluation by Hayes[83] suggests that the microscopic adsorption reaction of Pb^{2+} on goethite results in the release of only one proton (as in Equation 12). Additional protons are released because of an increase in surface potential and the adjustment of the ion-pair and surface ionization reactions to a new equilibrium condition. The quantitative analysis by Hayes[83] was able to account for Pb^{2+} adsorption data and potentiometric titration of goethite in the presence and absence of Pb^{2+}, without invoking additional reactions analogous to those in Equations 13 and 14. Hohl and Stumm[80] also concluded that the bidentate complexation of Pb^{2+} (Equation 14) was insignificant in comparison to monodentate complexes formed on the γ-Al_2O_3 surface.

On the other hand, spectroscopic studies have typically found that the alumina surface acts as a bidentate ligand in binding Cu^{2+} (Figure 12).[9,74,84] Rudin and Motschi[84] found from a combination of EPR (electron paramagnetic resonance) and ENDOR (electron spin double resonance) spectra that Cu^{2+} formed inner-sphere surface complexes of predominantly pseudo-square planar geometry. The conclusion that a bidentate complex was formed was based on an interrelation between the surface stability constant and the values of EPR parameters[74] and a comparison of these parameters to those of aqueous Cu(II) complexes.[9] Ligands containing nitrogen groups formed ternary complexes at the surface by coordination of Cu^{2+} in the equatorial positions instead of the axial positions.[84]

Few studies have examined the effect of metal ion adsorption on the electrokinetic potential

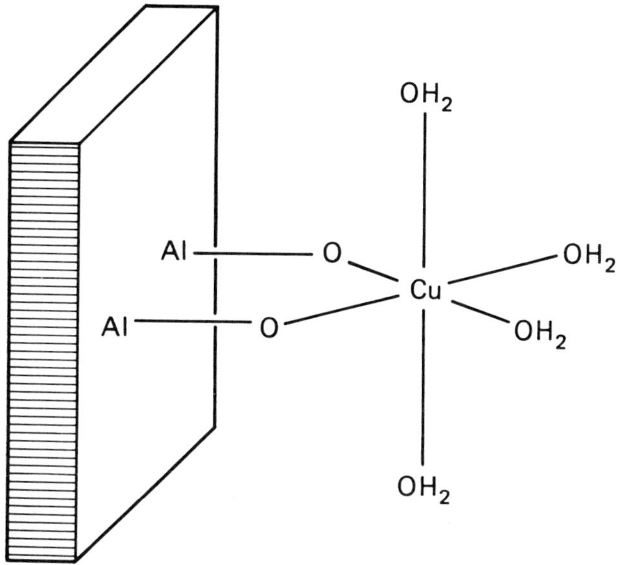

FIGURE 12. Suggested structure of the coordinative complex formed by Cu^{2+} with two aluminol groups of the γ-Al_2O_3 surface. (Modified from Rudin, M. and Motschi, H., *J. Colloid Interface Sci.*, 98, 285, 1984.)

of alumina particles. Huang and Stumm[39] have shown that the adsorption of Ca^{2+} increases the zeta potential of γ-Al_2O_3, increasing the pH_{iep} from 9.0 in NaCl solution to near 10.5 in 0.0001 *M* $CaCl_2$. At some pH values (e.g., pH 9.5), a reversal of charge was demonstrated upon adsorption of Ca^{2+}. The γ-Al_2O_3 particles were negatively charged at low concentrations of Ca^{2+}, but became positively charged as the concentration of Ca^{2+} was increased.

McBride[74] conducted one of the few studies that compared the adsorption behavior of different crystal forms of aluminum oxyhydroxides. He concluded that the strong adsorption sites for Cu^{2+} were surface hydroxyl groups coordinated to a single aluminum ion. The Cu^{2+} sorption capacities followed the order: noncrystalline alumina > boehmite > gibbsite, which is consistent with the relative number of singly coordinated hydroxyl groups on each surface. In the simplest interpretation of the gibbsite surface, the basal planes are unreactive and the adsorption sites of cations (and anions) occur only on the edges of crystals. However, in subsequent work, McBride et al.[9] showed by electron microscopy that many gibbsite plates have crystal steps at regular intervals. Strong adsorption sites are expected to be found at edge sites that occur at these step dislocations on the basal planes.

V. ADSORPTION OF ANIONS

Anions can adsorb on the surface of aluminum oxides by ion-pair formation with positively charged surface sites (Equation 9) or by ligand exchange with surface hydroxyls,[63,64,85-88] e.g.,

$$AlOH + H_2PO_4^- = AlH_2PO_4 + OH^- \quad (15)$$

$$2AlOH + H_2PO_4^- = \begin{matrix} Al \diagdown \\ Al \diagup \end{matrix} HPO_4 + H_2O + OH^- \quad (16)$$

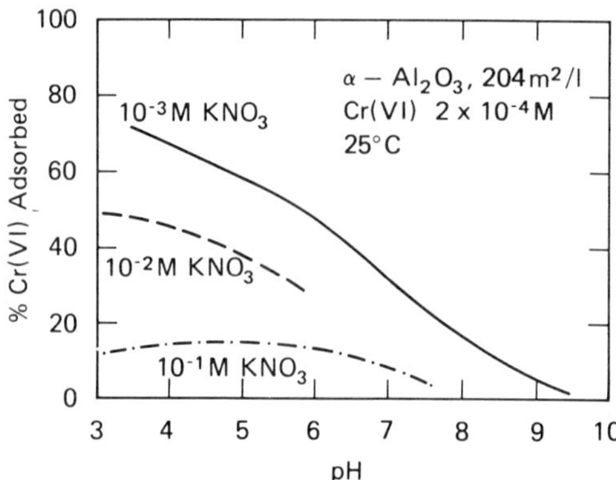

FIGURE 13. Adsorption of Cr(VI) on α-Al_2O_3 as a function of pH and KNO_3 concentration. (Modified from Davis, J. A. and Leckie, J. O., *J. Colloid Interface Sci.*, 74, 32, 1980; data from MacNaughton, M. G., in *Biological Implications of Metals in the Environment,* Ducker, H. and Wildung, R. E., Eds., CONF-75029, National Technical Information Service, Springfield, VA, 1977.)

Because these reactions are ligand exchange reactions which replace hydroxyls on the surface, the adsorption of anions usually increases with decreasing pH.[56,63,64,89-93] Figure 13 shows the adsorption of chromate on α-Al_2O_3 as a function of pH and background electrolyte concentration. Adsorption of Cr(VI) is greatest at low pH and decreases gradually over several pH units to zero adsorption. Adsorption was also significantly affected by the concentration of KNO_3. The adsorption of anions which are more strongly bound (e.g., $H_2PO_4^-$, SeO_3^{2-}) is less dependent on the ionic strength.[63,83] It is difficult to derive meaningful parameters for anion adsorption isotherms because of the slow approach to equilibrium and multiple processes occurring at the surface (see below). Nonetheless, many authors have attempted to analyze anion sorption data for a specified reaction time.[56,85,94-96] Typically a portion of the experimental data can be approximated by the Langmuir isotherm, and parameters such as the "energy of adsorption" and the "maximum adsorption capacity" are derived from these data. However, the maximum adsorption derived with one set of data frequently underestimates the actual adsorption densities measured at high adsorbate concentrations, or even may exceed an independent estimate of total adsorption sites.[11]

The studies of anion adsorption on aluminum oxides and hydroxides are numerous (Table 2). Many aspects of anion adsorption behavior have been examined, and it is difficult to make comparisons among these studies because of differences in adsorbent, kinetics of adsorption, and experimental procedures. Because of its importance as a nutrient, phosphate has been the subject of intense study. Yasunaga and Ikeda[76] showed that adsorption-desorption reactions of phosphate on γ-Al_2O_3 approach equilibrium rapidly. However, as was the case in cation adsorption, additional slow processes perturb the adsorption equilibria and reveal additional phosphate uptake in long-term kinetic studies.[87,94,97,98] In general, the slow processes are poorly understood. Chen et al.[97] attributed the slow uptake of phosphate on α-Al_2O_3 to the formation of a surface precipitate of aluminum phosphate. Phosphate desorption (and isotopic exchange) from the gibbsite surface is slow and incomplete;[64,87,94,99] this effect has been explained in terms of the high activation energy for dissociation of the binuclear-bidentate surface complex of phosphate with two aluminol groups (Equation 16) or in terms of surface precipitation of an aluminum phosphate phase at high phosphate concentrations.[99] However, Anderson et al.[100] suggested that the lack of reversibility in phosphate adsorption by a goethite suspension was related to the state of aggregation of the

Table 2
ADSORPTION OF ANIONS BY ALUMINUM OXIDES AND HYDROXIDES

Solid phases[a]	Anions	Type of investigation[b]	Ref.
γ-Al_2O_3	PO_4	a, c, d	85
	Organic acids	a, b, c, d	11
γ-Al_2O_3, am-$Al(OH)_3$	PO_4, AsO_4	a, d, h	93
γ-Al_2O_3	CrO_4	a, b, h	91
	PO_4	a, b, h	90
	NTA, other organic ligands	a, d	107
	EDTA, other polyacetic amino acids	a, c, d, h	105
	Natural organic acids	a, c, d, h, k, m	55
	Natural organic acids	a, c, d, h, l	104
α-Al_2O_3	CrO_4	a, c, d	92
	CrO_4	a, i	89
α-Al_2O_3, kaolinite	PO_4	a, b, d, e, j	97
α-$Al(OH)_3$	PO_4	b, d, g, j, m	94
	F, Cl, SO_4, SeO_3, MoO_4, H_4SiO_4	a, c	63
	PO_4	a, b, d, g	87
	PO_4	b, g, k	98, 144
	PO_4	b, d, g, k	99
α-$Al(OH)_3$, kaolinite	PO_4	a, b, d, m	95
α-$Al(OH)_3$	PO_4, F, SeO_3	a, b, d, g, i	64
	AsO_4, SeO_3, PO_4	a, d, m	86
α-$Al(OH)_3$, γ-AlOOH, am-$Al(OH)_3$	PO_4, organic acids	d, f, m	103
α-$Al(OH)_3$, imogolite	Natural organic acids	a, d, f	106
α-$Al(OH)_3$, bayerite	H_4SiO_4	a, b, c, d, e, i, j	145
α-$Al(OH)_3$, imogolite	PO_4, organic acids	a, c, d, f	7
γ-AlOOH, am-$Al(OH)_3$	Many	d, e, f	118
Bayerite	PO_4, SeO_3, SO_4	c, d, g	96
am-$Al(OH)_3$	AsO_4	a, b, d, h	56
	PO_4	b, d	102
	PO_4	a, b, c, d, e, g,	101

[a] Solid phases: α-$Al(OH)_3$ = gibbsite, α-Al_2O_3 = corundum, γ-AlOOH = boehmite, and am-$Al(OH)_3$ = noncrystalline alumina.

[b] Type of investigation: (a) pH dependence, (b) kinetics of adsorption, (c) proton/hydroxyl stoichiometry, (d) dependence on anion concentration (isotherm), (e) co-precipitation or effects of aging precipitate, (f) spectroscopic, (g) desorption, (h) electrokinetic potentials, (i) dependence on ionic strength, (j) dependence on temperature, (k) effects of metal ions, (l) dependence on molecular size, and (m) competitive adsorption.

dispersion. These authors hypothesized that phosphate ions formed bonds with more than one particle surface, increasing the aggregation of the dispersion and effectively burying phosphate within aggregates. This type of phenomenon could also explain the desorption behavior observed in alumina suspensions.

Nonequilibrium behavior is very evident in the studies of phosphate sorption by noncrystalline aluminum hydroxide.[101,102] Phosphate sorption and reversibility is influenced significantly by the age of the precipitate and the experimental procedures used. Phosphate uptake continues to increase even after one month of reaction. Arsenate adsorption also reaches equilibrium slowly.[56] Large differences in phosphate sorption were observed in comparing phosphate uptake on freshly precipitated aluminum hydroxide to the phosphate content of an aluminum precipitate formed in a phosphate solution.[101] The latter precipitate was reported to contain equimolar amounts of phosphorus and aluminum when formed at pH 7. If $AlPO_4$ is more stable than $Al(OH)_3$ under these conditions,[97] then the slow phosphate sorption observed in the former system could be caused by diffusion into the precipitate to convert $Al(OH)_3$ to $AlPO_4$. Similar processes may occur at low pH value on the surfaces of crystalline alumina.

Organic acids exhibit similar adsorption behavior on aluminum oxides and hydroxides.[7,11,55,103-106] Kummert and Stumm[11] reported on the adsorption of phthalate, salicylate, benzoate, and catechol on γ-Al_2O_3 as a function of pH. Adsorption of phthlate and salicylate was highly pH dependent; strong adsorption was observed at low pH. Benzoate was weakly adsorbed, indicating that the bidentate functionality of phthalate and salicylate molecules was an important factor in forming a strong bond at the surface. Catechol adsorption exhibited a different pattern, increasing with increasing pH. The authors attributed the adsorption behavior to ligand exchange reactions involving a single aluminol group and the deprotonated or singly protonated anion. In the case of catechol, it was presumed that the singly protonated anion formed the predominant surface complex. The general pH dependence for adsorption of these organic acids is in good agreement with observations of fulvic acid adsorption on aluminum oxides and hydroxides.[55,104,106]

McBride and Wesselink[103] recently compared adsorption of catechol on gibbsite, boehmite, and amorphous aluminum hydroxide and found that the dominant crystal faces of gibbsite and boehmite were unreactive. Only singly coordinated aluminol groups on the edge faces of gibbsite were reactive, similar to observations made on phosphate adsorption.[7] IR spectroscopy suggested that the surface complex formed was a 1:1 bidentate complex of deprotonated catecholate with aluminum ions at the surface. These spectroscopic results suggest that the model-dependent conclusions concerning surface structure made by Kummert and Stumm[11] may need to be reexamined. Parfitt et al.[7] concluded that oxalate also formed a bidentate complex with singly coordinated aluminol sites on the edge faces of gibbsite.

Huang and others[105,107,108] have conducted detailed studies of the adsorption of strongly complexing ligands (EDTA, NTA) and their metal complexes on γ-Al_2O_2. Both EDTA and NTA adsorption increase with decreasing pH, similar to the adsorption behavior of phthlate and salicylate.[11] In homogeneous solution, EDTA forms hexadentate bonds with metal ions, generally using all its functional groups and coordination positions of the metal ion. At the surface, this is not possible without dissolution and detachment of the metal ion from the solid lattice. Surface complex formation involving more than one carboxylic group seems possible; however, Huang and co-authors have explained their observations in terms of hydrogen bonding between aluminol groups and carboxylic groups of EDTA and NTA.[105,107] Metal adsorption in these systems is enhanced by adsorption of the metal-ligand complexes;[107,108] this effect has also been observed with fulvic acid.[104,109]

The electrical charge on particles is decreased by strong anion adsorption. Several studies have shown that the pH_{iep} of alumina suspensions is decreased by the adsorption of anions.[55,56,93,104,105,108] Anderson and Malotky[93] have derived interesting isotherms as a function of pH_{iep} for arsenate adsorption on $Al(OH)_3$ and for phosphate on γ-Al_2O_3. By separating out the influence of electrostatic forces, the authors attempted to derive the chemical energy of adsorption and maximum adsorption density from a linearized isotherm obtained at various pH_{iep} values. They also demonstrated that electrophoretic mobilities can be described by one curve when plotted as a function of the reduced pH ($pH-pH_{iep}$) in the presence of variable concentrations of arsenate and phosphate.

Few studies of competitive anion adsorption have been conducted. In one study by Hingston et al.,[86] the adsorption of arsenate and phosphate on gibbsite and goethite were compared and the effect of phosphate on arsenate adsorption was determined. Although the tendency of these two anions to adsorb on goethite was similar, phosphate adsorption was much stronger on gibbsite than arsenate. The effect of changes in surface charge and potential were not considered in the quantitative analysis of competitive adsorption.

VI. PRECIPITATION OF ALUMINUM OXIDES AND HYDROXIDES

The hydrolysis of aluminum in aqueous systems has received much attention in the past few decades as researchers have attempted to understand the complicated processes of polymerization and precipitation of aluminum oxides and hydroxides. These processes are discussed in detail in other chapters of this volume, especially in Chapter 4. A brief review also is presented here to emphasize the interrelated nature of surface chemistry and precipitation.

The hydrolysis and precipitation of aluminum from unseeded, acidic solutions has been shown to produce three general types of species: (1) monomeric aluminum hydroxide complexes, (2) polynuclear or polymeric aluminum hydroxide complex ions, and (3) crystalline or microcrystalline gibbsite.[3,8,110-113] The relative proportions of these species depends upon the conditions during hydrolysis (e.g., pH value, Al^{3+} concentration) and age of the hydrolysis products.[8,111-113] Under some conditions, much of the dissolved aluminum may be present as polymeric ions with a considerable range of composition and structure, and some of these species may persist for a long time.[8,112,114] A significant degree of supersaturation is required to produce microcrystalline gibbsite from unseeded solutions,[14] but the crystal growth reaction may proceed at lower reaction affinities in the presence of a surface which is favorable for nucleation. If the initial surface is gibbsite, a reversible solubility equilibrium controlled by a well-crystallized form of gibbsite may be attained at 25°C in 422 h or less.[49]

Van Straten and others[115,116] have studied the precipitation of aluminum hydroxide phases from basic aqueous solutions. A precipitation sequence was observed in which the thermodynamically least stable phase was usually the first to form, followed by transformation to other phases in stages, i.e., amorphous aluminum hydroxide, pseudoboehmite, bayerite, and gibbsite. The growth of bayerite was retarded by the presence of pseudoboehmite. The authors[115,116] concluded that pseudoboehmite formation was favored at relatively high (but still alkaline) pOH values, and that this phenomenon was related to the lower interfacial tension of pseudoboehmite in comparison to bayerite. A two-dimensional nucleation mechanism was proposed as the rate-determining step for bayerite formation. The growth rate of bayerite was proportional to the available surface area and the square of the degree of supersaturation. The rate of transformation of bayerite to gibbsite increased with increasing temperature or with decreasing pOH values. These studies clearly demonstrated thc role of surface chemical behavior in nucleation and crystal growth processes.

It should be apparent that the specific surface area of a precipitate will be influenced by the size of particles formed. Calculations of the surface energy of microcrystalline gibbsite suggest that the difference in solubility between this material and well-crystallized gibbsite can be accounted for by the difference in particle size and hence the difference in surface energy per unit mass of material.[117] Parthasarathy and Buffle[111] found that a substantial fraction of the microcrystalline gibbsite formed in their experiments was removed by an ultrafilter with pores 13 nm in diameter. The remaining material was removed by a filter having pores of 2 nm in diameter. These dimensions are within the general range of minimum particle size for microcrystalline gibbsite calculated by Smith and Hem.[8] Using the gibbsite structural model (Figure 5) and assuming all OH is present in bridging positions, a single sheet of $Al(OH)_6$ octahedra with this composition would be 50 to 100 Å units in diameter. Stacking of the layers might decrease the lateral dimension by a factor of two or more, but the particle size would still be within the general range indicated by the ultrafiltration experiments.

A. Influence of Adsorption on Precipitation

As discussed in detail in Chapter 4, a number of different anions may perturb the precipitation of aluminum and lead to the formation of noncrystalline products or poorly crystalline pseudoboehmite.[118] Hsu[119] found that several inorganic ligands inhibited the crystallization of aluminum hydroxide. Numerous investigations have also shown this effect when precipitation occurs in the presence of organic ligands.[120-123] These results demonstrate that interactions between adsorbing anions and the coalescing polymers may influence the type of precipitate formed.

The fluoride ion has nearly the same effective radius as the hydroxide ion, and aluminum fluoride complexes are sufficiently stable in solution to influence the solubility of aluminum substantially below pH 7.0 if the F^- activity is 10^{-4} or greater. (The hydroxyl complex $Al(OH)_4^-$ becomes completely dominant at pH values above 7.0.[124]) When fluoride is present during the polymerization of aluminum hydroxide, a substantial amount of substitution of F^- for OH^- may occur. Roberson and Hem[125] synthesized cryolite (Na_3AlF_6) in this way and prepared a form of ralstonite [$Al_2(F,OH)_6H_2O$]. More recently, Parthasarathy et al.[126] studied this synthesis and suggested treatment with aluminum hydroxide polymers as a means for decreasing the dissolved F^- concentration in industrial waste water. Although direct reaction of F^- at a preexisting $Al(OH)_3$ surface might be expected to be limited to substitution of F^- for exposed OH^- ions at crystal edges, the dissolution and reprecipitation that may occur at such surfaces could incorporate considerable quantities of F^- in the altered surficial layers of the solid.

Aqueous aluminum sulfate complexes are considerably less stable than the fluoride complexes. In systems with high sulfate concentrations at pH values less than about 4.0, however, aluminum hydroxysulfate species can occur and may control aluminum solubility by precipitation of sparingly soluble minerals such as alunite $\{KAl_3(OH)_6(SO_4)_2\}$.[125,127] DeHek et al.[128] found that the process of aluminum hydroxide polymerization at 25°C was catalyzed in the presence of sulfate. The final product at neutral or lower pH was microbayerite. At pH 9.7 a hydrated boehmite formed with an average particle size less than 20 nm. Little or no sulfate was present in the aluminum hydroxide precipitate. A strong association can also occur between aluminum polymers and phosphate ions. The interaction is of particular interest in connection with behavior of phosphate in soils. Studies of the interaction have been made by Hsu.[119,129]

If dissolved $H_4SiO_4(aq)$ is present under certain conditions during the polymerization of aluminum hydroxide and precipitation of gibbsite, a disordered 1:1 aluminosilicate with the approximate composition of halloysite, $Al_2Si_2O_5(OH)_4$, is formed.[130] Lind and Hem[131] found that in the presence of an organic ligand capable of forming complexes with Al^{3+}, small amounts of well-crystallized kaolinite were produced during aging for periods of a year or more at room temperature and near neutral pH. Several other investigators have reported low-temperature syntheses of kaolinite when organic complexing agents were present. Linares and Huertas[132] reported a synthesis of kaolinite using fulvic acid as a complexing agent. However, the synthesis of a well-crystallized clay mineral by interactions among $Al(OH)_3$ species and dissolved silica is not easily achieved in the absence of organic ligands. The fusing together of the tetrahedrally coordinated silicon - oxygen sheet and the octahedrally coordinated Al–OH sheet to attain the clay mineral structure entails establishment of Al–O–Si bonding along the c direction. This in turn, requires conversion of an OH^- ion to an O^{2-} ion in the octahedral sheet and deprotonation of one of the 4-coordinated OH^- ions associated with the aqueous silica. From a coordination chemistry point of view, it is obvious that the synthesis of kaolinite in an aqueous system should be very slow. At higher temperatures, however, the kinetics of the process can be much more favorable. The rate of synthesis may also be increased by seeding the solution with crystalline clay particles.[133] Difficulties in establishing interlayer bonding in this kind of system are an indirect indication of the rather inert surface chemical characteristics of the faces of the gibbsite sheet structure.

Zutic and Stumm[134] have conducted an interesting study of the influence of adsorption on the rates of aluminum oxide and hydroxide dissolution. These authors proposed that the rate of dissolution was controlled by a surface process in the presence of organic ligands and at low pH. The detachment of surface complexes was thought to be the rate-controlling step in the dissolution process. Fluoride ions increased the dissolution rate dramatically at concentrations as low as 10^{-6} *M*.

B. Coprecipitation and Adsorption in Natural Systems

In laboratory experiments it is possible to maintain conditions that will prevent extensive alterations and recrystallization of an $Al(OH)_3$ solid, but natural systems are not this simple. For that reason it may be necessary to allow for coprecipitation effects in evaluating the surface properties of naturally occurring $Al(OH)_3$ solids.

In all the specific interactions mentioned above it is necessary to consider relationships to the aluminum hydroxide polymerization process. If the polymerization has gone essentially to completion so that only a form of crystalline gibbsite and monomeric aluminum ions are present, the extent of incorporation of other ions into the solid phase will depend primarily on the extent to which the surface may be undergoing alteration. In natural systems there will generally be changes in pH, temperature, and concentrations of various solutes as water moves past the surface and dissolution and reprecipitation reactions can be expected.

VII. ADSORPTION MODELS FOR ALUMINUM OXIDES AND HYDROXIDES

Quantitative models for the surface chemistry of oxides have evolved substantially during the last 2 decades. The *surface complexation models*, which describe the formation of charge, potential, and the adsorption of ions at the oxide-water interface, are the most popular. The fundamental concept upon which all surface complexation models are based is that adsorption takes place at defined coordination sites (the surface hydroxyl groups, present in finite number) and that adsorption reactions can be described quantitatively via mass action equations (e.g., Equations 1, 2, 8 to 10, and 12 to 16).[135] Various models can be distinguished on the basis of different assumptions made for electrostatic interaction terms, or more correctly, surface activity coefficients.[136] However, all of these models reduce to a similar set of simultaneous equations that can be solved numerically, including: (a) mass action equations for all surface reactions, (b) a mass balance equation for surface hydroxyl groups, (c) equations for calculation of surface charge, and (d) a set of equations that describes the charge and potential relationships of the electrical double layer.[135]

One of the earliest models for adsorption on alumina surfaces was proposed by Huang and Stumm,[39] sometimes called the "constant capacitance model". In this model, all adsorbing species reside in a single surface plane (including H^+, OH^-, and specifically adsorbed cations and anions), and the resulting electrical charge is balanced by a diffuse layer of counterions in the adjacent water. The electrical double layer is considered to act as a parallel plate capacitor, with the surface charge linearly related to the surface potential. Huang and Stumm[39] illustrated how conditional acidity constants could be calculated for the γ-Al_2O_3 surface (for Equations 1 and 2) from acid-base potentiometric titration data. These acidity constants were determined for the condition of zero charge and potential at the surface, but the actual acidity constants used in the model varied as a function of pH due to the change in surface potential. The acidity constants found by Huang and Stumm[39] for the zero charge condition were pK (Equation 1) = 7.9 and pK (Equation 2) = 9.1. This simple model for the surface has also been used to describe cation[80] and anion[11,137,138] adsorption on alumina.

Some of the contradictions between the results of the constant capacitance model and actual experimental data have been reviewed by Bousse and Meindl.[41] One of the problems is that the constant capacitance model predicts the same degree of linearity in ψ_o/pH curves and σ_o/pH curves, where σ_o represents the surface proton balance, Γ_{H+}, in units of electrical charge per unit surface area. Although the ψ_o/pH relationship is reasonably linear, σ_o/pH relationships are usually very nonlinear, and this observation contradicts the model.[41]

Another problem of the constant capacitance model lies in its ability to determine charge and potential relationships as a function of ionic strength. Ideally one would hope that an adsorption model would quantitatively describe variations in adsorption as a function of ionic strength, such as that shown in Figure 13. Since the total number of sites is finite, the potential and charge cannot increase to infinity, and a certain degree of nonlinearity is introduced in the ψ_o/pH relationships in the constant capacitance model through the saturation of sites. Frequently the model is used with a numerical optimization technique to achieve the best fit of experimental data, including the total site density as an adjustable parameter.[40,135] The numerical optimization techniques typically yield values for the total site density which are considerably lower than found by isotopic exchange or expected from theoretical calculations of surface structure. This is because a lower site density is required by the constant capacitance model to achieve a good fit between the ψ_o/pH relationships at different ionic strengths. The model parameters are strongly interdependent, and a range of site concentrations may be used to model the data through the variation of other parameters, as has been shown by Westall and Hohl[40] and discussed recently by Bousse and Meindl.[41] This approach has been justified on the basis of simplicity of the model or uncertainty in the available techniques for measuring total site density.[135] However, the uncertainty in determining total site density should be weighed against the error created by numerical optimization of parameters of an inadequate model. There is sufficient experimental data, including adsorption isotherms for anions at high concentrations, to suggest that the total site density is greater than that determined by numerical optimization of σ_o/pH relationships. Moreover, the ψ_o/pH relationship measured by Bousse and Meindl[41] and their analysis suggest that the basic experimental data for the γ-Al_2O_3 surface are best described by the consideration of ion-pair formation with electrolyte ions.

Yates et al.[69] introduced the idea of ion-pair formation by electrolyte ions and hydroxyl groups as part of the surface complexation model. These authors showed that ion pair formation would explain the observed large discrepancy between "surface charge" determined by potentiometric titration and that determined indirectly from zeta potential measurements. In this model, the electrolyte ions were bound at a plane some distance from the surface; conceptually this plane is equivalent to the classical Stern plane.[135] This model is frequently referred to as the "triple layer model". Davis et al.[58] developed a procedure for estimating parameters for this model from potentiometric titrations at different ionic strengths. An important point, however, is that each model for the interface may produce different estimates for the acidity constants of the surface hydroxyl groups. These estimates are *model dependent*, which is why the reported values of these constants vary widely.[41]

Bousse and Meindl[41] have shown that the consideration of ion-pair formation with surface hydroxyls provides a superior description of the charge and potential relationships at the alumina-water interface as a function of ionic strength. These authors have shown that ψ_o/pH and σ_o/pH relationships are sensitive to different aspects of the surface chemistry. The proton balance data only allow the determination of parameters which describe ion pair formation by electrolyte ions at the surface. The acidity constants are best determined from the measurements of ψ_o as a function of pH. As has been noted by many authors, the σ_o/pH relationships are sometimes not symmetrical about the pH_{pzc}, and σ_o may depend on the anion or cation used in the electrolyte solution. Figure 14 shows a schematic illustration of the model for the surface proposed by Bousse and Meindl.[41] These authors have modified the triple layer model to consider different values of capacitance on each side of the pH_{pzc}. The authors illustrate that different values of the capacitance are required for different electrolyte ions because some ions are able to approach the surface more closely than others. In general, cations may approach the surface more closely than anions, as shown in Figure 14, resulting in a higher capacitance for the surface when $pH > pH_{pzc}$.

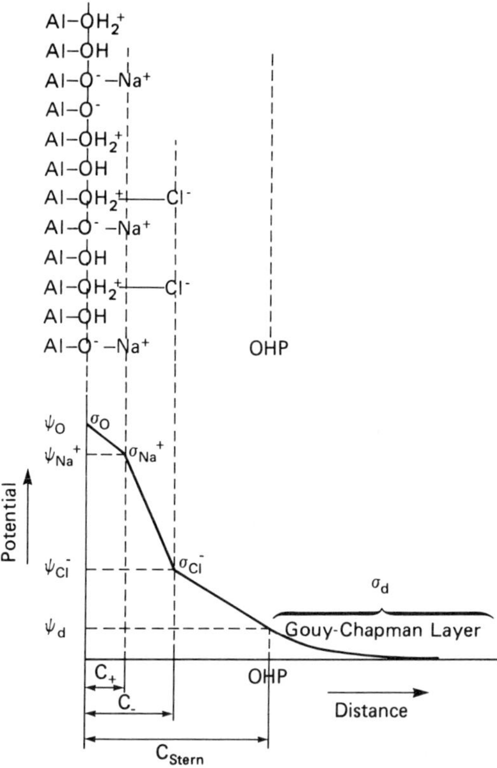

FIGURE 14. Illustration of the species of the generalized site-binding model for the alumina-water interface in NaCl solution, the possible location of ions, and the distribution of electrical charge and potential. (Modified from Bousse, L. and Meindl, J. S., in *Geochemical Processes at Mineral Surfaces,* Davis, J. A. and Hayes, K. F., Ed., ACS Symposium Series 323, American Chemical Society, Washington, D.C., 1986, chap 5.)

Numerous investigators have employed the triple layer model to describe the adsorption of cations and anions from dilute solutions. An advantage of this model is its ability to describe adsorption of weakly bound ions at the surface as a function of ionic strength.[88] In most cases, this model has been employed by formulating mass action equations with electrostatic correction factors consistent with the location of specifically adsorbed ions in a single Stern plane, as advocated by Davis and Leckie.[82,88] This approach predicts a significant variation in the adsorption of strongly adsorbed ions as a function of ionic strength as has been observed by Vuceta[139] for adsorption of Pb^{2+} and Cu^{2+} on quartz. However, the recent results of Hayes[83] show a lack of ionic strength dependence for the adsorption of Pb^{2+} and SeO_3^{2-} on goethite. The results were best modeled by formulating the electrostatic correction factors as if the ions were located in the surface plane rather than in the Stern plane. Moreover, Hayes[83] found that only one proton was released per Pb^{2+} adsorbed, which is more consistent with the location of Pb^{2+} in the surface plane of the triple layer model. Adsorption of weakly bound ions, e.g., Ba^{2+} and SeO_4^{2-}, was dependent on the ionic strength and best modeled by the location of these ions in the Stern plane. Very little work has been done on the adsorption of anions or cations on aluminum oxides and hydroxides from dilute solutions as a function of ionic strength, so it is difficult to speculate on the best modeling approach for these solid phases. Future investigations on the adsorption behavior of these solids should focus on the ionic strength dependence to increase our basic understanding of the surface complexation process and the applicability of surface complexation models to aluminum oxides and hydroxides.

VIII. CONCLUDING REMARKS

Knowledge of the surface chemical behavior of aluminum oxides and hydroxides is still advancing. More is known in terms of systematic laboratory studies about the synthetic material, γ-Al_2O_3, than any other phase. The results of experimental studies of the gibbsite surface are consistent with what is expected in terms of the exposure of different crystal faces as predicted from the crystal structure. Little is known regarding the effects of surface alteration on the observed surface chemical behavior. More systematic and comprehensive studies of gibbsite are needed.

Recent advances, such as the measurement of surface potential as a function of pH and studies of the specific adsorption of counterions such as Na^+ and Br^-, are affecting dramatically the evolution of models for the alumina-water interface. However, models for the adsorption of solutes from dilute solutions have not yet been combined in a consistent way with spectroscopic studies of the surface. Much more research also is needed on the kinetics of adsorption reactions, the determination of maximum adsorption densities, the stoichiometry of surface reactions, and the ionic strength dependence of adsorption reactions. Although model development and the experimental studies conducted to date represent a considerable increase in knowledge, quantitative application of this knowledge to the natural environment still appears problematic. Furthermore, a greater understanding of the role of adsorption in the precipitation process and the rates at which the surfaces of metastable minerals are altered by reorganization is required to improve geochemical models.

REFERENCES

1. **Lippens, B. C. and Steggerda, J. J.,** Active alumina, in *Physical and Chemical Aspects of Adsorbents and Catalysts,* Linsen, B. G., Ed., Academic Press, New York, 1970, chap. 4.
2. **Wefers, K. and Misra, C.,** Oxides and hydroxides of aluminum, Tech. Paper No. 19, Rev., Alcoa Research Laboratories, Alcoa Center, PA, 1987.
3. **Hsu, P. H.,** Aluminum hydroxides and oxyhydroxides, in *Minerals in Soil Environments,* Dixon, J. B., Weed S. B., et al., Eds., Soil Science Society of America, Madison, WI, 1977, 99.
4. **Davis, J. A. and Hayes, K. F.,** Geochemical processes at mineral surfaces: an overview, in *Geochemical Processes at Mineral Surfaces,* ACS Symposium Series 323, Davis, J. A. and Hayes, K. F., Eds., American Chemical Society, Washington, D.C., 1986, chap. 1.
5. **Peri, J. B.,** A model for the surface of γ-alumina, *J. Phys. Chem.,* 69, 220, 1965.
6. **Sposito, G.,** *The Surface Chemistry of Soils,* Oxford University Press, New York, 1984, 16.
7. **Parfitt, R. L., Fraser, J. D., and Farmer, V. C.,** Adsorption on hydrous oxides. II. Oxalate, benzoate and phosphate on gibbsite, *J. Soil Sci.,* 28, 40, 1977.
8. **Smith, R. W. and Hem, J. D.,** Effect of aging on aluminum hydroxide complexes in dilute aqueous solutions, *U.S. Geol. Surv. Water-Supply Pap.,* 1827D, D1, 1972.
9. **McBride, M. B., Fraser, A. R., and McHardy, W. J.,** Cu^{2+} interaction with microcrystalline gibbsite. Evidence for oriented chemisorbed copper ions, *Clays Clay Miner.,* 32, 12, 1984.
10. **Peri, J. B. and Hannan, R. B.,** Surface hydroxyl groups on γ-alumina, *J. Phys. Chem.,* 64, 1526, 1960.
11. **Kummert, R. and Stumm, W.,** The surface complexation of organic acids on hydrous γ-Al_2O_3, *J. Colloid Interface Sci.,* 75, 373, 1980.
12. **Yates, D. E. and Healy, T. W.,** The structure of the silica/electrolyte interface, *J. Colloid Interface Sci.,* 44, 9, 1976.
13. **Schoen, R. and Roberson, C. E.,** Structures of aluminum hydroxide and geochemical implications, *Am. Mineral.,* 55, 43, 1970.

14. **Hem, J. D.,** Aluminum hydroxide polymerization and synthesis of microcrystalline gibbsite in solutions near pH 5.0 (Abstr.), *EOS - Transactions,* American Geophysical Union, 67, 931, 1986.
15. **Wefers, K. and Bell, G. M.,** Oxides and hydroxides of aluminum, Tech. Paper No. 19, Alcoa Research Laboratories, Alcoa Center, PA, 1972, 51.
16. **Smit, W. and Holten, C. L. M.,** Zeta-potential and radiotracer adsorption measurements on EFG α-Al_2O_3 single crystals in NaBr solutions, *J. Colloid Interface Sci.,* 78, 1, 1980.
17. **DeBoer, J. H., Fortuin, J. M. H., Lippens, B. C., and Meijs, W. H.,** Study of the nature of surfaces with polar molecules. II. The adsorption of water on aluminas, *J. Catal.,* 2, 1, 1963.
18. **Tanabe, K.,** *Solid Acids and Bases,* Academic Press, New York, 1970.
19. **Boehm, P.,** Acidic and basic properties of hydroxylated metal oxide surfaces, *Discuss. Faraday Soc.,* 52, 264, 1971.
20. **Medema, J., Van Bokhoven, J. J. G. M., and Kuiper, A. E. T.,** Adsorption of bases on γ-Al_2O_3, *J. Catal.,* 25, 238, 1972.
21. **Hughes, T. R., White, H. M., and White, R. J.,** Bronsted and Lewis acid site concentrations in fluorided alumina from the infrared spectra of adsorbed pyridine species, *J. Catal.,* 13, 58, 1969.
22. **Scokart, P. O. and Rouxhet, P. G.,** Characterization of the basicity of oxides through the infrared study of pyrrole adsorption, *J. Chem. Soc. Farady Trans. 1,* 76, 1476, 1980.
23. **Basila, M. R. and Kantner, T. R.,** The nature of the acidic sites on silica-alumina. A reevaluation of the relative adsorption coefficients of chemisorbed pyridine, *J. Phys. Chem.,* 70, 1681, 1981.
24. **Scokart, P. O. and Rouxhet, P. G.,** Comparison of the acid-base properties of various oxides and chemically treated oxides, *J. Colloid Interface Sci.,* 86, 96, 1982.
25. **Peri, J. B.,** Infrared study of adsorption of ammonia on dry γ-alumina, *J. Phys. Chem.,* 69, 231, 1965.
26. **Parfitt, G. D. and Rochester,** Surface characterization: chemical, *Characterization of Powder Surfaces,* Academic Press, San Francisco, 1976, chap. 2.
27. **Kiviat, F. E. and Petrakis, L.,** Surface acidity of transition metal modified aluminas. Infrared and nuclear magnetic resonance investigation of adsorbed pyridine, *J. Phys. Chem.,* 77, 1232, 1973.
28. **Dawson, W. H., Kaiser, S. W., Ellis, P. D., and Inners, R. R.,** Carbon-13 cross-polarization magic-angle-spinning NMR study of *n*-butylamine adsorbed on γ-alumina: characterization of surface acid sites, *J. Am. Chem. Soc.,* 103, 6780, 1981.
29. **Peri, J. B.,** Infrared and gravimetric study of the surface hydration of γ-alumina, *J. Phys. Chem.,* 69, 211, 1965.
30. **James, R. O. and Parks, G. A.,** Characterization of aqueous colloids by their electrical double-layer and intrinsic surface chemical properties, in *Surface and Colloid Science,* Vol. 12, Matijevic, E., Ed., Plenum Press, New York, 1982, 119.
31. **Healy, T. W. and White, L. R.,** Ionizable surface group models of aqueous interfaces, *Adv. Colloid Interface Sci.,* 9, 303, 1978.
32. **Overbeek, J. Th. G.,** Electrochemistry of the double layer, in *Colloid Science,* Vol. 1, Kruyt, H. R., Ed., Elsevier, Amsterdam, 1952, chap. 4.
33. **Bockris, J. O'M. and Reddy, A. K. N.,** *Modern Electrochemistry,* Vols. 1 and 2, Plenum Press, New York, 1970.
34. **Sparnaay, M. J.,** *The Electrical Double Layer,* Pergamon Press, Elmsford, NY, 1972.
35. **Hiemenz, P. C.** *Principles of Colloid and Surface Chemistry,* Marcel Dekker, New York, 1977.
36. **van Olphen, H.,** *An Introduction to Clay Colloid Chemistry,* 2nd ed., Wiley-Interscience, New York, 1977.
37. **Huang, C.-P.,** The surface acidity of hydrous solids, in *Adsorption of Inorganics at Solid-Liquid Interfaces,* Anderson, M. A. and Rubin, A. J., Eds., Ann Arbor Science, Ann Arbor, MI, 1981, chap. 5.
38. **Davis, J. A., James, R. O., and Leckie, J. O.,** Surface ionization and complexation at the oxide/water interface. I. Computation of electrical double layer properties in simple electrolytes, *J. Colloid Interface Sci.,* 63, 480, 1978.
39. **Huang, C.-P. and Stumm, W.,** Specific adsorption of cations on hydrous γ-Al_2O_3, *J. Colloid Interface Sci.,* 43, 409, 1973.
40. **Westall, J. and Hohl, H.,** A comparison of electrostatic models for the oxide/solution interface, *Adv. Colloid Interface Sci.,* 12, 265, 1980.
41. **Bousse, L. and Meindl, J. S.,** Surface potential-pH characteristics in the theory of the oxide-electrolyte interface, in *Geochemical Processes at Mineral Surfaces,* Davis, J. A. and Hayes, K. F., Eds., ACS Symposium Series 323, American Chemical Society, Washington, D.C., 1986, chap. 5.
42. **Kavanagh, B. V., Posner, A. M., and Quirk, J. P.,** Effect of polymer adsorption on the properties of the electrical double layer, *Faraday Discuss. Chem. Soc.,* 59, 242, 1975.
43. **Smith, R. W.,** The State of Al(III) in Aqueous Solution and Adsorption of Hydrolysis Products on α-Al_2O_3, Ph.D. thesis, Stanford University, Palo Alto, CA, 1969.
44. **Tewari, P. H. and McLean, A. W.,** Temperature dependence of point of zero charge of alumina and magnetite, *J. Colloid Interface Sci.,* 40, 267, 1972.

45. **Ahmed, S. M.,** Studies of the double layer at oxide-solution interface, *J. Phys. Chem.,* 73, 3546, 1969.
46. **Wnek, W. and Davies, R.,** An analysis of the dependence of the zeta potential and surface charge on surfactant concentration, ionic strength, and pH, *J. Colloid Interface Sci.,* 60, 361, 1977.
47. **Yopps, J. A. and Fuersteneau, D. W.,** The zero point of charge of alpha-alumina, *J. Colloid Sci.,* 19, 61, 1964.
48. **Robie, R. A., Hemingway, B. S., and Fisher, J. R.,** Thermodynamic properties of minerals and related substances at 298.15 K and 1 bar (10^5 Pascals) pressure and at higher temperatures, *U.S. Geol. Surv. Bull.,* 1452, 1977.
49. **May, H. M., Helmke, P. A., and Jackson, M. L.,** Gibbsite solubility and thermodynamic properties of hydroxy — aluminum ions in aqueous solution at 25°C, *Geochim. Cosmochim. Acta,* 43, 861, 1979.
50. **Bloom, P. R. and Weaver, R. M.,** Effect of removal of reactive surface material on the solubility of synthetic gibbsites, *Clays Clay Miner.,* 30, 281, 1982.
51. **Schulthess, C. P. and Sparks, D. L.,** Backtitration technique for proton isotherm modeling of oxide surfaces, *Soil Sci. Soc. Am. J.,* 50, 1046, 1986.
52. **Breeuwsma, A. and Lyklema, J.,** Physical and chemical adsorption of ions in the electrical double layer on hematite (α-Fe_2O_3), *J. Colloid Interface Sci.,* 43, 437, 1973.
53. **Napper, D. H. and Hunter, R. J.,** Hydrosols, in *Surface Chemistry and Colloids,* Kerker, M., Ed., (Vol. 7 of MPT International Review of Science, Series One, Physical Chemistry, Buckingham, A. D., Ed.), University Park Press, Baltimore, 1972, 241.
54. **Wiese, G. R. and Healy, T. W.,** Coagulation and electrokinetic behavior of TiO_2 and Al_2O_3 colloidal dispersions, *J. Colloid Interface Sci.,* 51, 427, 1975.
55. **Davis, J. A.,** Adsorption of natural dissolved organic matter at the oxide/water interface, *Geochim. Cosmochim. Acta,* 46, 2381, 1982.
56. **Anderson, M. A., Ferguson, J. F., and Gavis, J.,** Arsenate adsorption on amorphous aluminum hydroxide, *J. Colloid Interface Sci.,* 54, 391, 1976.
57. **Stol, R. J., Van Helden, A. K., and DeBruyn, P. L.,** Hydrolysis-precipitation studies of aluminum (III) solutions. II. A kinetic study and model, *J. Colloid Interface Sci.,* 57, 115, 1976.
58. **Bleam, W. F. and McBride, M. B.,** Cluster formation versus isolated-site adsorption. A study of Mn(II) and Mg(II) adsorption on boehmite and goethite, *J. Colloid Interface Sci.,* 103, 124, 1985.
59. **Van Riemsdijk, W. H., Bolt, G. H., Koopal, L. K., and Blaakmeer, J.,** Electrolyte adsorption on heterogeneous surfaces: adsorption models, *J. Colloid Interface Sci.,* 109, 219, 1986.
60. **Bousse, L., De Rooij, N. F., and Bergveld, P.,** The influence of counter-ion adsorption on the ψ_o/pH characteristics of insulator surfaces, *Surf. Sci.,* 135, 479, 1983.
61. **Bousse, L., De Rooij, N. F., and Bergveld, P.,** Operation of chemically sensitive field-effect sensors as a function of the insulator-electrolyte interface, *IEEE Trans. Electron Devices,* 30, 1263, 1983.
62. **Stumm, W. and Morgan, J. J.,** *Aquatic Chemistry,* 2nd ed., John Wiley & Sons, New York, 1981.
63. **Hingston, F. J., Posner, A. M., and Quirk, J. P.,** Anion adsorption by goethite and gibbsite. I. The role of the proton in determining adsorption envelopes, *J. Soil Sci.,* 23, 177, 1972.
64. **Hingston, F. J., Posner, A. M., and Quirk, J. P.,** Anion adsorption by goethite and gibbsite. II. Desorption of anions from hydrous oxide surfaces, *J. Soil Sci.,* 25, 16, 1974.
65. **Sprycha, R.,** Attempt to estimate σ_β charge components on oxides from anion and cation adsorption measurements, *J. Colloid Interface Sci.,* 96, 550, 1983.
66. **Foissy, A., M'Pandou, A., LaMarche, J. M., and Jaffrezic-Renault, N.,** Surface and diffuse-layer charge at the TiO_2-electrolyte interface, *Colloids Surf.,* 5, 363, 1982.
67. **Shiao, S.-Y. and Meyer, R. E.,** Adsorption of inorganic ions on alumina from salt solutions: correlations of distribution coefficients with uptake of salt, *J. Inorg. Nucl. Chem.,* 43, 3301, 1981.
68. **Sprycha, R.,** Surface charge and adsorption of background electrolyte ions at anatase/electrolyte interface, *J. Colloid Interface Sci.,* 102, 173, 1984.
69. **Yates, D. E., Levine, S., and Healy, T. W.,** Site-binding model of the electrical double layer at the oxide/water interface, *J. Chem. Soc., Faraday Trans.* 1, 70, 1807, 1974.
70. **Kurbatov, M. H., Wood, G. B., and Kurbatov, J. D.,** Isothermal adsorption of cobalt from dilute solutions, *J. Phys. Chem.,* 55, 1170, 1951.
71. **Dugger, D. L., Stanton, J. H., Irby, B. N., McConnel, B. L., Cummings, W. W., and Maatman, R. W.,** The exchange of twenty metal ions with the weakly acidic silanol group of silica gel, *J. Phys. Chem.,* 68, 757, 1964.
72. **Schindler, P. W.,** Surface complexes at oxide-water interfaces, *Adsorption of Inorganics at Solid-Liquid Interfaces,* Anderson, M. A. and Rubin, A. J., Eds., Ann Arbor Science, Ann Arbor, MI, 1981, chap. 1.
73. **Benjamin M. M. and Leckie, J. O.,** Adsorption of metals at oxide interfaces: effects of the concentrations of adsorbate and competing metals, in *Contaminants and Sediments,* Vol. 2, Baker, R. A., Ed., Ann Arbor Science, Ann Arbor, MI, 1980, 305.
74. **McBride, M. B.,** Cu^{2+}-adsorption characteristics of aluminum hydroxide and oxyhydroxides, *Clays Clay Miner.,* 30, 21, 1982.

75. **Tewari, P. H. and Lee, W.,** Adsorption of Co(II) at the oxide/water interface, *J. Colloid Interface Sci.*, 52, 77, 1975.
76. **Yasunaga, T. and Ikeda, T.,** Adsorption-desorption kinetics at the metal-oxide-solution interface studied by relaxation methods, *Geochemical Processes at Mineral Surfaces,* Davis, J. A. and Hayes, K. F., Eds., ACS Symposium Series 323, American Chemical Society, Washington, D.C., 1986, chap. 12.
77. **Benjamin, M. M.,** Effects of Competing Metals and Complexing Ligands on Trace Metals Adsorption at the Oxide/Solution Interface, Ph.D. thesis, Stanford University, Palo Alto, CA, 1978.
78. **Kinniburgh, D. G.,** The H^+/M^{2+} exchange stoichiometry of calcium and zinc adsorption on ferrihydrite, *Soil Sci.*, 34, 759, 1983.
79. **Forbes, E. A., Posner, A. M., and Quirk, J. P.,** The specific adsorption of divalent Cd, Co, Cu, Pb, and Zn on goethite, *J. Soil Sci.*, 154, 1976.
80. **Hohl, J. and Stumm, W.,** Interaction of Pb^{2+} with hydrous γ-Al_2O_3, *J. Colloid Interface Sci.*, 55, 281, 1976.
81. **Honeyman, B. D. and Leckie, J. O.,** Macroscopic partitioning coefficients for metal ion adsorption, in *Geochemical Processes at Mineral Surfaces,* Davis, J. A. and Hayes, K. F., Eds., ACS Symposium Series 323, American Chemical Society, Washington, D.C., 1986, chap. 9.
82. **Davis, J. A. and Leckie, J. O.,** Surface ionization and complexation at the oxide/water interface. II. Surface properties of amorphous iron oxhydroxide and adsorption of metal ions, *J. Colloid Interface Sci.*, 67, 90, 1978.
83. **Hayes, K. F.,** Equilibrium, Spectroscopic, and Kinetic Studies of Ion Adsorption at the Oxide/Aqueous Interface, Ph.D. thesis, Stanford University, Palo Alto, CA, 1987.
84. **Rudin, M. and Motschi, H.,** A molecular model for the structure of copper complexes on hydrous oxide surfaces: an ENDOR study of ternary Cu(II) complexes on δ-alumina, *J. Colloid Interface Sci.*, 98, 285, 1984.
85. **Huang, C. P.,** Adsorption of phosphate at the hydrous γ-Al_2O_3 electrolyte interface, *J. Colloid Interface Sci.*, 53, 178, 1975.
86. **Hingston, F. J., Posner, A. M., and Quirk, J. P.,** Competitive adsorption of negatively charged ligands on oxide surfaces, *Discuss. Faraday Soc.*, 52, 334, 1971.
87. **Kyle, J. H., Posner, A. M., and Quirk, J. P.,** Kinetics of isotopic exchange of phosphate adsorbed on gibbsite, *J. Soil Sci.*, 26, 32, 1975.
88. **Davis, J. A. and Leckie, J. O.,** Surface ionization and complexation at the oxide/water interface. III. Adsorption of anions, *J. Colloid Interface Sci.*, 74, 32, 1980.
89. **MacNaughton, M. G.,** Adsorption of chromium (VI) at the oxide/water interface, in *Biological Implications of Metals in the Environment,* Ducker, H. and Wildung, R. E., Eds., CONF-75029, National Technical Information Service, Springfield, VA, 1977.
90. **Mikami, N., Sasaki, M., Hachiya, K., Astumian, R. D., Ikeda, T., and Yasunaga, T.,** Kinetics of the adsorption-desorption of phosphate on the γ-Al_2O_3 surface using the pressure-jump technique, *J. Phys. Chem.*, 87, 1454, 1983.
91. **Mikami, N., Sasaki, M., Kikuchi, T., and Yasunaga, T.,** Kinetics of adsorption — desorption of chromate on γ-Al_2O_3 surfaces using the pressure-jump technique, *J. Phys. Chem.*, 87, 5245, 1983.
92. **Honeyman, B. D.,** Cation and Anion Adsorption at the Oxide/Solution Interface in Systems Containing Binary Mixtures of Adsorbents: an Investigation of the Concept of Adsorptive Additivity, Ph.D. thesis, Stanford University, Palo Alto, CA, 1984.
93. **Anderson, M. A. and Malotky, D. T.,** The adsorption of protolyzable anions on hydrous oxides at the isoelectric pH, *J. Colloid Interface Sci.*, 72, 413, 1979.
94. **Kuo, S. and Lotse, E. G.,** Kinetics of phosphate adsorption and desorption by hematite and gibbsite, *Soil Sci.*, 116, 400, 1974.
95. **Chen, Y.-S. R., Butler, J. N., and Stumm, W.,** Adsorption of phosphate on alumina and kaolinite from dilute aqueous solutions, *J. Colloid Interface Sci.*, 43, 421, 1973.
96. **Rajan, S. S. S.,** Adsorption of selenite, phosphate and sulphate on hydrous alumina, *J. Soil Sci.*, 30, 709, 1979.
97. **Chen, Y.-S. R., Butler, J. N., and Stumm, W.,** Kinetic study of phosphate reaction with aluminum oxide and kaolinite, *Environ. Sci. Technol.*, 7, 327, 1973.
98. **Helyar, K. R., Munns, D. N., and Burau, R. G.,** Adsorption of phosphate by gibbsite. I. Effects of neutral chloride salts of calcium, magnesium, sodium, and potassium, *J. Soil Sci.*, 27, 307, 1976.
99. **Van Riemsdijk, W. H. and Lyklema, J.,** The reaction of phosphate with aluminum hydroxide in relation with phosphate bonding in soils, *Colloids Surf.*, 1, 33, 1980.
100. **Anderson, M. A., Tejedor-Tejedor, M. I., and Stanforth, R. R.,** Influence of aggregation on the uptake kinetics of phosphate by goethite, *Environ. Sci. Technol.*, 19, 632, 1985.
101. **Lijklema, L.,** Interaction of orthophosphate with iron(III) and aluminum hydroxides, *Environ. Sci. Technol.*, 14, 537, 1980.

102. **Bolan, N. S., Barrow, N. J., and Posner, A. M.,** Describing the effect of time on sorption of phosphate by iron and aluminium hydroxides, *J. Soil Sci.*, 36, 187, 1985.
103. **McBride, M. B. and Wesselink, B.,** Chemisorption of catechol on gibbsite, boehmite, and amorphous alumina surfaces, *Environ. Sci. Technol.*, in press.
104. **Davis, J. A. and Gloor, R.,** Adsorption of dissolved organics in lake water by aluminum oxide. Effect of molecular weight, *Environ. Sci. Technol.*, 15, 1223, 1981.
105. **Bowers, A. R. and Huang, C. P.,** Adsorption characteristics of polyacetic amino acids onto hydrous γ-Al_2O_3, *J. Colloid Interface Sci.*, 105, 197, 1985.
106. **Parfitt, R. L., Fraser, A. R., and Farmer, V. C.,** Adsorption of hydrous oxides. III. Fulvic acid and humic acid on goethite, gibbsite and imogolite, *J. Soil Sci.*, 28, 289, 1977.
107. **Elliott, H. A. and Huang, C. P.,** The adsorption characteristics of Cu(II) in the presence of chelating agents, *J. Colloid Interface Sci.*, 70, 29, 1979.
108. **Bowers, A. R. and Huang, C. P.,** Adsorption characteristics of metal-EDTA complexes onto hydrous oxides, *J. Colloid Interface Sci.*, 110, 575, 1986.
109. **Davis, J. A.,** Complexation of trace metals by adsorbed natural organic matter, *Geochim. Cosmochim. Acta*, 48, 679, 1984.
110. **Hsu, P. H.,** Formation of gibbsite from aging hydroxy-aluminum solutions, *Proc. Soil Sci. Soc. Am.*, 30, 173, 1966.
111. **Parthasarathy, N. and Buffle, J.,** Study of polymeric aluminum (III) hydroxide solutions for application of waste water treatment: properties of the polymer and optimal conditions of preparation, *Water Res.*, 19, 25, 1985.
112. **Vermeulen, A. C., Geus, J. W., Stol, R. J., and De Bruyn, P. L.,** Hydrolysis-precipitation studies of aluminum (III) solutions. I. Titration of acidified aluminum nitrate solutions, *J. Colloid Interface Sci.*, 51, 449, 1975.
113. **Stol, R. J., Van Helden, A. K., and De Bruyn, P. L.,** Hydrolysis-precipitation studies of aluminum (III) solutions. II. A kinetic study and model, *J. Colloid Interface Sci.*, 57, 115, 1976.
114. **Hem, J. D.,** Kinetics of dissolution and structure of aluminum hydroxide polymers, in *Leaching and Diffusion in Rocks and Their Weathering Products,* Augustithis, S. S., Ed., Theophrastus Publications S. A., Athens, 1983, 51.
115. **van Straten, H. A., Holtkamp, B. T. W., and de Bruyn, P. L.,** Precipitation from supersaturated aluminate solutions. I. Nucleation and growth of solid phases at room temperature, *J. Colloid Interface Sci.*, 98, 342, 1984.
116. **van Straten, H. A. and De Bruyn, P. L.,** Precipitation from supersaturated aluminate solutions. II. Role of temperature, *J. Colloid Interface Sci.*, 102, 260, 1984.
117. **Smith, R. W.,** Relations among equilibrium and nonequilibrium aqueous species of aluminum hydroxide complexes, in *Nonequilibrium Systems in Natural Water Chemistry,* Advances in Chemistry Series 106, American Chemical Society, Washington, D.C., 1971, 250.
118. **Violante, A. and Huang, P. M.,** Influence of inorganic and organic ligands on the formation of aluminum hydroxides and oxyhydroxides, *Clays Clay Miner.*, 33, 181, 1985.
119. **Hsu, P. H.,** Effect of phosphate and silicate on the crystallization of gibbsite from OH-Al solutions, *Soil Sci.*, 127, 219, 1979.
120. **Kwong, N.-K. K. F. and Huang, P. M.,** The relative influence of low-molecular-weight, complexing organic acids on the hydrolysis and precipitation of aluminum, *Soil Sci.*, 128, 337, 1979.
121. **Violante, A. and Violante, P.,** Influence of pH, concentration, and chelating power of organic anions on the synthesis of aluminum hydroxides and oxyhydroxides, *Clays Clay Miner.*, 28, 425, 1980.
122. **Kodama, H. and Schnitzer, M.,** Effect of fulvic acid on the crystallization of aluminum hydroxides, *Geoderma,* 24, 195, 1980.
123. **van Straten, H. A., Schoonen, A. A., Verheul, R. C. S., and De Bruyn, P. L.,** Precipitation from supersaturated aluminate solutions. IV. Influence of citrate ions, *J. Colloid Interface Sci.*, 106, 175, 1985.
124. **Hem, J. D.,** Geochemistry and aqueous chemistry of aluminum, *Kidney Int.*, 29(Suppl 18), S3, 1986.
125. **Roberson, C. E. and Hem, J. D.,** Solubility of aluminum in the presence of hydroxide, fluoride, and sulfate, *U.S. Geol. Surv. Water-Supply Pap.*, 1827C, C1, 1969.
126. **Parthasarathy, N., Buffle, J., and Haerdi, W.,** Study of interaction of polymeric aluminum hydroxide with fluoride, *Can. J. Chem.*, 64, 24, 1986.
127. **Nordstrom, D. K.,** The effect of sulfate on aluminum concentrations in natural waters: some stability relations in the system Al_2O_3-SO_3-H_2O at 298°K, *Geochim. Cosmochim. Acta,* 46, 681, 1982.
128. **DeHek, H., Stol, R. J., and DeBruyn, P. L.,** Hydrolysis — precipitation studies of aluminum (III) solutions. III. The role of the sulfate ion, *J. Colloid Interface Sci.*, 64, 72, 1978.
129. **Hsu, P. H.,** Precipitation of phosphate from solution using aluminum salts, *Water Res.*, 9, 1155, 1975.
130. **Hem, J. D., Roberson, C. E., Lind, C. J., and Polzer, W. L.,** Chemical interactions of aluminum with aqueous silica at 25°C, *U.S. Geol. Surv. Water-Supply Pap.*, 1827, E1, 1973.

131. **Lind, C. J. and Hem, J. D.,** Effects of organic solutes on chemical reactions of aluminum, *U.S. Geol. Surv. Water-Supply Pap.*, 1827G, G1, 1975.
132. **Linares, J. and Huertas, F.,** Kaolinite — synthesis at room temperature, *Science,* 171, 896, 1971.
133. **Kittrick, J. A.,** Precipitation of kaolinite at 25°C and 1 atm, *Clays Clay Miner.*, 18, 261, 1970.
134. **Zutic, V. and Stumm, W.,** Effect of organic acids and fluoride on the dissolution kinetics of hydrous alumina. A model study using the rotating disc electrode, *Geochim. Cosmochim. Acta,* 48, 1493, 1984.
135. **Dzombak, D. A. and Morel, F. M. M.,** Adsorption of inorganic pollutants in aquatic systems, *J. Hydraul. Eng.*, 113, 430, 1987.
136. **Sposito, G.,** On the surface complexation model of the oxide-aqueous solution interface, *J. Colloid Interface Sci.*, 91, 329, 1983.
137. **Goldberg, S. and Sposito, G.,** A chemical model of phosphate adsorption by soils. I. Reference oxide minerals, *Soil Sci. Soc. Am. J.*, 48, 772, 1984.
138. **Goldberg, S.,** Chemical modeling of arsenate adsorption on aluminum and iron oxide minerals, *Soil Sci. Soc. Am. J.*, 50, 1154, 1986.
139. **Vuceta, J.,** Adsorption of Pb(II) and Cu(II) on α-Quartz from Aqueous Solutions: Influence of pH, Ionic Strength, and Complexing Ligands, Ph.D. thesis, California Institute of Technology, Pasadena, 1976.
140. **James, R. O. and MacNaughton, M. G.,** The adsorption of aqueous heavy metals on inorganic materials, *Geochim. Cosmochim. Acta,* 41, 1549, 1977.
141. **Kinniburgh, D. G., Syers, J. K., and Jackson, M. L.,** Specific adsorption of trace amounts of calcium and strontium by hydrous oxides of iron and aluminum, *Soil Sci. Soc. Am. J.*, 39, 464, 1975.
142. **Kinniburgh, M. L., Jackson, M. L., and Syers, J. K.,** Adsorption of alkaline earth, transition, and heavy metal cations by hydrous oxide gels of iron and aluminum, *Soil Sci. Soc. Am. J.*, 40, 796, 1976.
143. **McBride, M. B.,** Retention of Cu^{2+}, Ca^{2+}, Mg^{2+}, and Mn^{2+} by amorphous alumina, *Soil Sci. Soc. Am. J.*, 42, 27, 1978.
144. **Helyar, K. R., Munns, D. N., and Burau, R. G.,** Adsorption of phosphate by gibbsite. II. Formation of a surface complex involving divalent cations, *J. Soil Sci.*, 27, 315, 1976.
145. **Hingston, F. J. and Raupach, M.,** The reaction between monosilicic acid and aluminum hydroxide. I. Kinetics of adsorption of silicic acid by aluminum hydroxide, *Aust. J. Soil Res.*, 5, 295, 1967.

Chapter 8

THE SOLUBILITY OF ALUMINUM IN SOILS

W. L. Lindsay and P. M. Walthall

TABLE OF CONTENTS

I. INTRODUCTION

Aluminum is one of the more abundant elements in soils, making up approximately 7.1% of the solid matter in an average soil.[1] For this reason, the chemical reactions and solubility relationships of soil aluminum need to be understood in order to predict and overcome environmental problems caused by excess concentrations. A major environmental problem of aluminum is its toxicity to plants growing in highly weathered acid soils. The agricultural practice of liming can eliminate this problem by raising the pH value above 6.0. However, pH is not the only factor that affects the solubility of aluminum. Since aluminum combines with silica to form aluminosilicates, the solubility of silica directly affects that of aluminum. During weathering, primary silicates and aluminosilicates dissolve as secondary aluminosilicates and aluminum hydroxides precipitate. Thus the solubility of aluminum is largely the result of numerous weathering transformations.[1]

During the past decade considerable attention has been focused on acid rain and its impact on soils, plants, and aquatic life. Aluminum is central to this concern for two major reasons. First, added acidity from rainfall hastens the weathering of soil minerals and the release of bases contained therein. Second, the increased acidity that impacts on highly weathered soils solubilizes and mobilizes aluminum, increasing its toxicity to plants and aquatic organisms.[2] This chapter will describe how solid-phase controls and the solution speciation of aluminum govern its solubility and, in turn, its toxicity.

II. SELECTION OF ALUMINUM EQUILIBRIUM CONSTANTS

One way of representing aluminum solubility relationships in soils is through the use of equilibrium relationships. Consider the reaction

$$aA + bB = cC + dD \tag{1}$$

where A and B are reactants, C and D are products, and a, b, c, and d are the number of moles of each involved in the reaction. The thermodynamic equilibrium constants (K°) at standard temperature and pressure for the above reaction is defined as:

$$K^\circ = \frac{[C]^c\,[D]^d}{[A]^a\,[B]^b} \tag{2}$$

where products and reactants are expressed in terms of activities, denoted by [], based on molar concentrations and raised to a power equal to the number of moles of each reactant or product in the equilibrium reaction. Complex equilibria can be developed by combining individual reactions.

The chemical reactions and equilibrium constants used to illustrate the solubility relationships in this chapter are summarized in Table 1. The equilibrium constants were carefully screened from the literature and represent our best selections at this time. All figures and equations included in this chapter were derived from these equilibrium data. Either original sources or previously documented literature sources are given in the "reference" column. Where thermodynamic data were used to calculate the equilibrium constants, care was taken to combine only internally consistent data sources. The equilibrium constant values appearing in Table 1 are in almost every case in agreement with the definitive compilations given in Chapters 2 and 3. Differences that may occur reflect normal variability among the choices of experts in a rapidly developing field of environmental chemistry.

Table 1
EQUILIBRIUM CONSTANTS FOR DISSOCIATION REACTIONS OF ALUMINUM MINERALS AND COMPLEXES AND FOR ASSOCIATED REACTIONS

Reaction no.	Reaction	log K°	Ref.
	Oxides and Hydroxides		
1	$Al(OH)_3$(am) + $3H^+$ = Al^{3+} + $3H_2O$	9.66	1, 3
2	$Al(OH)_3$(bayerite) + $3H^+$ = Al^{3+} + $3H_2O$	8.51	1, 3
3	$AlOOH$(boehmite) + $3H^+$ = Al^{3+} + $3H_2O$	8.13	1, 3
4	$Al(OH)_3$(norstrandite) + $3H^+$ = Al^{3+} + $3H_2O$	8.13	1, 3
5	$Al(OH)_3$(gibbsite) + $3H^+$ = Al^{3+} + $3H_2O$	8.04	1, 3
6	$AlOOH$(diaspore) + $3H^+$ = Al^{3+} + $2H_2O$	7.92	1, 3
	Sulfates		
7	$KAl_3(SO_4)_2(OH)_6$(alunite) + $6H^+$ = K^+ + $3Al^{3+}$ + $2SO_4^{2-}$ + $6H_2O$	−1.04	4
8	$Al_2(SO_4)_3 \cdot 17H_2O$(alunogen) = $2Al^{3+}$ + $3SO_4^{2-}$ + $17H_2O$	−7.00	5
9	$Al_4(OH)_{10}SO_4 \cdot 5H_2O$(basaluminite) + $10H^+$ = $4Al^{3+}$ + SO_4^{2-} + $15H_2O$	22.30	4
10	$Al_4(OH)_{10}SO_4 \cdot 5H_2O$(am. basaluminite) + $10H^+$ = $4Al^{3+}$ + SO_4^{2-} + $15H_2O$	24.00	4
11	$AlSO_4(OH) \cdot 5H_2O$(jurbanite) + H^+ = Al^{3+} + SO_4^{2-} + $6H_2O$	−3.80	5
	Aluminosilicates		
12	$Al_2O_3SiOH(OH)_3$(imogolite) + $6H^+$ = $2Al^{3+}$ + $H_4SiO_4^{\circ}$ + $3H_2O$	11.57	6
13	$Al_2Si_2O_5(OH)_4$(halloysite) + $6H^+$ = $2Al^{3+}$ + $2H_4SiO_4^{\circ}$ + H_2O	8.72	1, 3
14	$Al_2Si_2O_5(OH)_4$(kaolinite) + $6H^+$ = $2Al^{3+}$ + $2H_4SiO_4^{\circ}$ + H_2O	5.45	1, 3
	K-Al-Si		
15	$KAl_2(AlSi_3O_{10})(OH)_2$(muscovite) + $10H^+$ = K^+ + $3Al^{3+}$ + $3H_4SiO_4^{\circ}$	13.44	1, 3
	Mg-Al-Si		
16	$(Si_{2.97}Al_{1.03})(Al_{1.44}Fe^{3+}{}_{0.07}Fe^{2+}{}_{0.99}Mg_{3.24})O_{10}(OH)_8$(Vermont chlorite) + $16.08H^+$ = $2.97H_4SiO_4^{\circ}$ + $2.47Al^{3+}$ + $3.24\ Mg^{2+}$ + $0.99Fe^{2+}$ + $0.07Fe^{3+}$ + $6H_2O$	48.2	7
17	$K_{0.6}Mg_{0.25}Al_{2.3}Si_{3.5}O_{10}(OH)_2$(illite) + $8H^+$ + $2H_2O$ = $0.6K^+$ + $0.25Mg^{2+}$ + $2.3Al^{3+}$ + $3.5H_4SiO_4^{\circ}$	10.35	1, 3
18	$0.4^+(Si_{3.81}Al_{1.71}Fe(III)_{0.22}Mg_{0.29})O_{10}(OH)_2$(montmorillonite) + $6.76H^+$ + $3.24H_2O$ = $0.29Mg^{2+}$ + $1.71Al^{3+}$ + $0.22Fe^{3+}$ + $3.81H_4SiO_4^{\circ}$	2.68	1, 3
19	$1.02^+(Si_{7.40}Al_{0.60})(Al_{2.69}Mg_{0.54}Fe^{3+}{}_{0.81})O_{20}(OH)_4$(Houston Black smectite) + $5.60H_2O$ + $14.40H^+$ = $7.40H_4SiO_4^{\circ}$ + $3.29Al^{3+}$ + $0.54Mg^{2+}$ + $0.81Fe^{3+}$	3.21	8
	Hydrolysis Reactions		
20	Al^{3+} + H_2O = $AlOH^{2+}$ + H^+	−4.99	9
21	Al^{3+} + $2H_2O$ = $Al(OH)_2^+$ + $2H^+$	−10.13	9
22	Al^{3+} + $4H_2O$ = $Al(OH)_4^-$ + $4H^+$	−22.16	9
23	$2Al^{3+}$ + $2H_2O$ = $Al_2(OH)_2^{4+}$ + $2H^+$	−7.69	1, 3

Table 1 (continued)
EQUILIBRIUM CONSTANTS FOR DISSOCIATION REACTIONS OF ALUMINUM MINERALS AND COMPLEXES AND FOR ASSOCIATED REACTIONS

Reaction no.	Reaction	log K°	Ref.
	Fluoride Complexes		
24	$Al^{3+} + F^- = AlF^{2+}$	6.98	1, 3
25	$Al^{3+} + 2F^- = AlF_2^+$	12.60	1, 3
26	$Al^{3+} + 3F^- = AlF_3^{\circ}$	16.65	1, 3
27	$Al^{3+} + 4F^- = AlF_4^-$	19.03	1, 3
	Sulfate Complexes		
28	$Al^{3+} + SO_4^{2-} = AlSO_4^+$	3.20	1, 3
29	$Al^{3+} + 2SO_4^{2-} = Al(SO_4)_2^-$	1.90	1, 3
30	$2Al^{3+} + 3SO_4^{2-} = Al_2(SO_4)_3^{\circ}$	−1.88	1, 3
	Nitrate Complexes		
31	$Al^{3+} + 3NO_3^- = Al(NO_3)_3^{\circ}$	0.12	1, 3
	Iron Reactions		
32	$Fe(OH)_3(soil) + 3H^+ = Fe^{3+} + 3H_2O$	2.70	1, 3
33	$1/2\ Fe_2O_3(hematite) + 3H^+ = Fe^{3+} + 3/2\ H_2O$	0.09	1, 3
34	$Fe_3(OH)_8(ferrosic\ hydroxide) + 8H^+ + 2e^- = 3Fe^{2+} + 8H_2O$	43.75	1, 3
35	$Fe^{3+} + e^- = Fe^{2+}$	13.04	1, 3
	Silica Reactions		
36	$SiO_2(am) + 2H_2O = H_4SiO_4^{\circ}$	−2.74	1, 3
37	$SiO_2(soil) + 2H_2O = H_4SiO_4^{\circ}$	−3.10	1, 3
38	$SiO_2(quartz) + 2H_2O = H_4SiO_4^{\circ}$	−4.00	1, 3
	Phosphate Reactions		
39	$H_6K_3Al_5(PO_4)_8 \cdot 18H_2O(K\text{-}taranakite) + 10H^+ = 3K^+ + 5Al^{3+} + 8H_2PO_4^- + 18H_2O$	−22.30	1, 3
40	$H_6(NH_4)_3Al_5(PO_4)_8 \cdot 18H_2O(NH_4\text{-}taranakite) + 10H^+ = 3NH_4^+ + 5Al^{5+} + 8H_2PO_4^- + 18H_2O$	−19.10	1, 3
41	$AlPO_4 \cdot 2H_2O(variscite) + 2H^+ = Al^{3+} + H_2PO_4^- + 2H_2O$	−2.50	1, 3

III. SOLUBILITY OF ALUMINUM MINERALS

A. Oxides and Hydroxides

The solubility of aluminum in soils is controlled initially by minerals present in significant amounts that have the highest solubility. Pedogenic processes slowly remove the thermodynamically unstable minerals, so that the more stable minerals supporting the lowest activity of Al^{3+} ultimately control the solubility of aluminum. At any given time, the solubility of aluminum may be controlled by either the dissolving or the precipitating minerals, depending on the rates of dissolution and precipitation of each. By examining the solution ion activity product (IAP) for various minerals, one can determine the minerals most likely involved in solubility control.

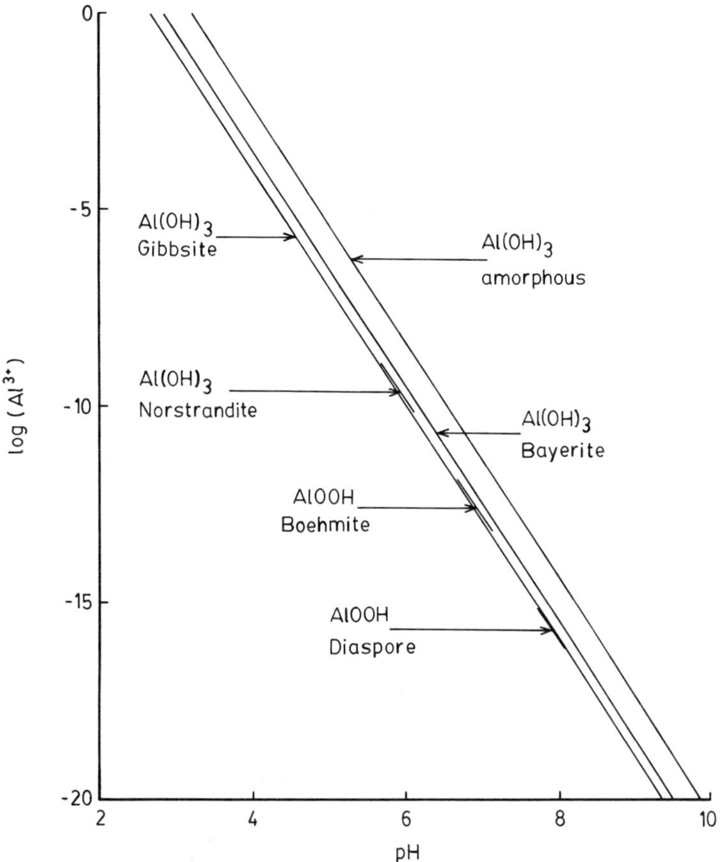

FIGURE 1. Solubility control by varius aluminum oxides and hydroxides.

Figure 1 shows the Al^{3+} activities that can be maintained by various oxides and hydroxides of aluminum. The most soluble product is $Al(OH)_3$(amorphous), followed by $Al(OH)_3$(bayerite), $Al(OH)_3$(norstrandite), AlOOH(boehmite), $Al(OH)_3$(gibbsite), and AlOOH(diaspore).[1,3] The last four minerals have nearly the same solubility. Gibbsite is the most commonly found hydrous aluminum oxide in soils. It occurs mainly in highly weathered soils where more soluble forms of silica have been removed by weathering. The presence of gibbsite imparts a strong pH dependence of the Al^{3+} activity, which decreases 1000-fold for each unit increase in pH value.

B. Sulfates and Oxysulfates

Aluminum forms several oxysulfate minerals in acid soils and aqueous environments.[2,4,5,10,11] In such environments, the aluminum oxysulfate minerals are more stable than gibbsite and kaolinite. The solubilities of several such minerals are described by Reactions 7 to 11 of Table 1, and the corresponding Al^{3+} activities are plotted in Figure 2.

The mineral $Al_2(SO_4)_3 \cdot 17H_2O$(alunogen) is too soluble to form in natural environments,[5] so it is not included in Figure 2. Gibbsite is more stable than any of the oxysulfate minerals at pH values above 5.5 to 6.0 depending on the SO_4^{2-} activity. As the pH is lowered and the SO_4^{2-} activity is increased, $KAl_3(SO_4)_2(OH)_6$(alunite) becomes more stable than gibbsite. At still lower pH values and higher SO_4^{2-} activities, $AlOHSO_4 \cdot 5H_2O$(jurbanite) becomes more stable than alunite.

Both crystalline and amorphous $Al_4(OH)_{10}SO_4 \cdot 5H_2O$(basaluminite) depicted in Figure 2

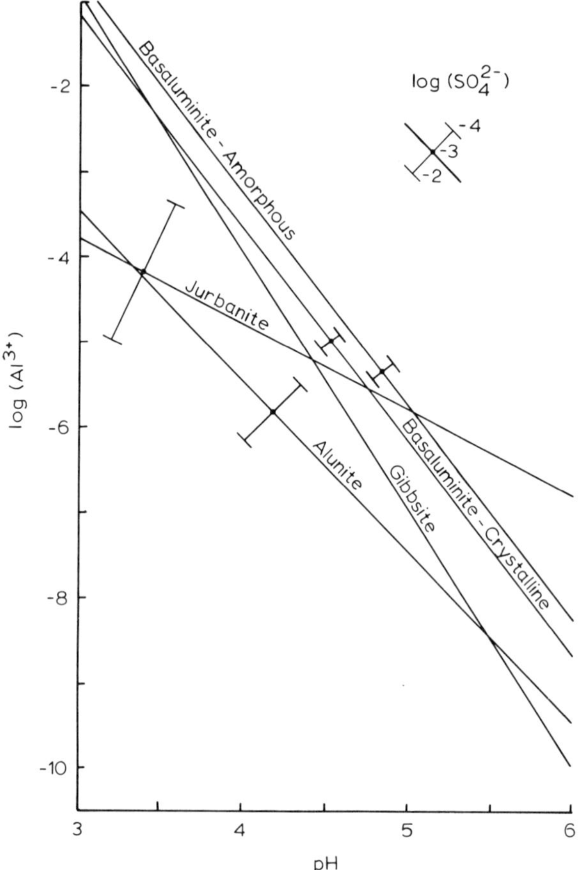

FIGURE 2. Solubility control by various aluminum oxysulfate minerals when the K^+ and SO_4^{2-} activities are 10^{-3}. Shifts are shown for activities of 10^{-2} and 10^{-4}.

are never stable minerals. However, they have been known to form and act as solubility controls under certain conditions.[4,5] Many of the solubility relationships involving the aluminum oxysulfates shown here have been considered previously[2,4,5,10,11] in describing acidic environments. Other oxysulfate minerals whose solubilities have not been sufficiently well-characterized cannot be included at this time.[5]

C. Phosphates

Aluminum forms insoluble phosphates at low pH values[1,12] Figure 3 shows the $H_2PO_4^-$ activities that can be maintained by NH_4-taranakite ($H_6(NH_4)_3Al_5(PO_4)_8 \cdot 18H_2O$), K-taranakite ($H_6K_3Al_5(PO_4)_8 \cdot 18H_2O$), and variscite ($AlPO_4 \cdot 2H_20$). These comparisons are made when quartz (SiO_2) controls $H_4SiO_4^\circ$, kaolinite ($Al_2Si_2O_5(OH)_4$) controls $[Al^{3+}]$, and the activities of K^+ and NH_4^+ are taken equal to 10^{-3}. The taranakites are more soluble than variscite. As the activity of Al^{3+} increases from that supported by kaolinite-quartz equilibrium to that supported by gibbsite, the solubility of variscite decreases, as shown by the arrow in Figure 3.

Although $AlPO_4 \cdot 2H_2O$ (variscite) is a stable aluminum mineral in acid soils, its presence is not expected to affect aluminum solubility. The reason is that soils contain much more aluminum than phosphate, so even if all the phosphate is bound with aluminum, other aluminum minerals such as hydroxides, aluminosilicates, or oxysulfates would still form and control the Al^{3+} activity.

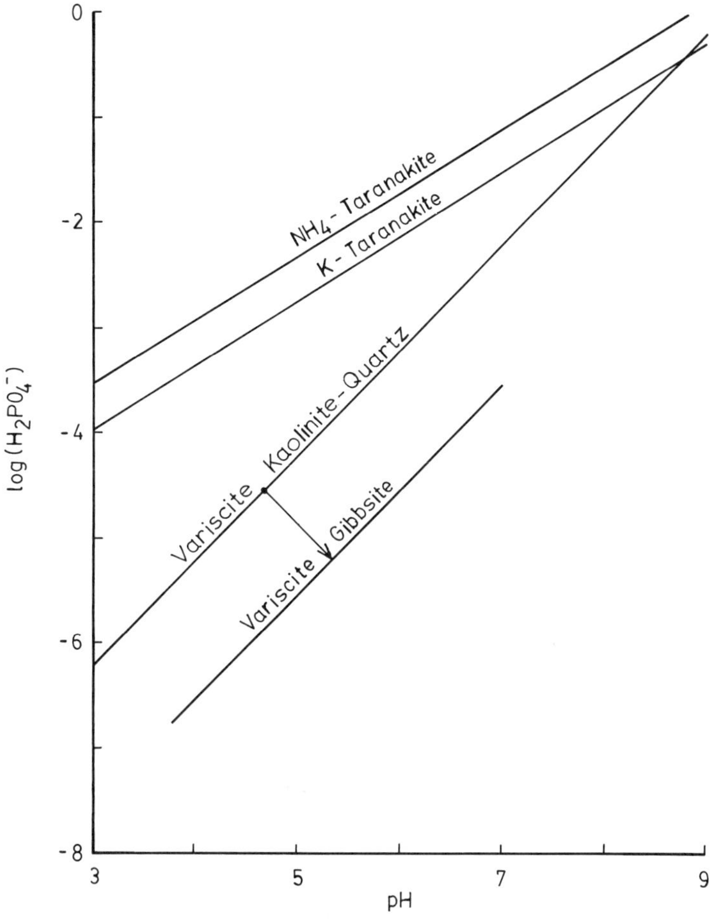

FIGURE 3. Solubility control by various aluminum phosphates in equilibrium with kaolinite and quartz. Note how the variscite line changes when aluminum solubility is controlled by equilibrium with gibbsite.

D. Aluminosilicates

The widespread presence of aluminosilicates in soils is accompanied by a reciprocal solubility relationship between aluminum and silica. When the $H_4SiO_4^\circ$ activity is high, the activity of Al^{3+} is depressed because of the increased stability of the aluminosilicates. On the other hand, if the H_4SiO_4 activity is lowered, that of Al^{3+} is increased. For illustrative purposes (see Figure 4) we combine the activity of Al^{3+} with pH, yielding the parameter of log $[Al^{3+}]$ + 3pH, often referred to as the *aluminum potential.*[13] This allows us to view pH and $[Al^{3+}]$ as functions of $[H_4SiO_4^\circ]$. It should be noted that a number of the aluminosilicates have a pH dependence which is not entirely accounted for by the aluminum potential parameter. This additional pH dependence must be accounted for by indicating the residual shifts in the pH value for each specific mineral where necessary. As shown in Figure 4, the value of log$[Al^{3+}]$ + 3pH supported by $Al(OH)_3$(am) is 9.66 and that of $Al(OH)_3$(gibbsite) is 8.04. Under the acid conditions (pH 4.0) shown, illite is the most soluble aluminosilicate mineral. Only at high $H_4SiO_4^\circ$ activity does illite depress $[Al^{3+}]$ below that supported by amorphous $Al(OH)_3$. As the pH value increases, illite becomes more stable.

Imogolite represented by the formula $Al_2O_3Si(OH)(OH)_3$,[6] supports an Al^{3+} activity very near that of gibbsite, being slightly more stable at high $H_4SiO_4^\circ$ activities and less stable at low activities. Imogolite constitutes an initial weathering product of volcanic ash, with

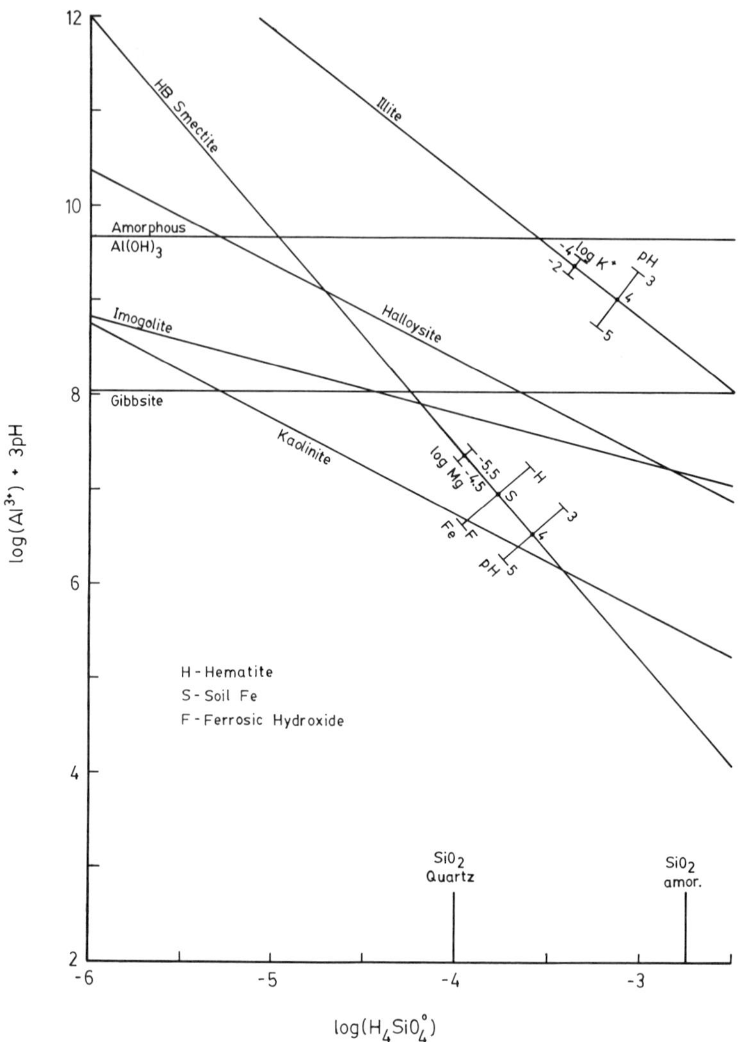

FIGURE 4. Solubility relationships of various aluminum hydroxides and aluminosilicates at pH 4, assuming equilibrium with soil-iron and $[Mg^{2+}]$ and $[K^+] = 10^{-4.5}$. Shifts for changes in K^+ and Mg^{2+} activities are shown.

subsequent transformations to halloysite, kaolinite, and gibbsite.[14] At extremely high $H_4SiO_4^\circ$ activities, in the vicinity of that supported by amorphous silica (see Figure 4), halloysite is more stable than imogolite, thus confirming the weathering sequence described above, in that the more ordered crystallinity of halloysite is more stable than the paracrystalline imogolite. As silica levels decrease and halloysite shifts to a more unstable position with respect to imogolite, the transformation to the more stable kaolinite results because of dehydration of the halloysite to produce kaolinite. Under acidic soil conditions, kaolinite is fairly stable. However, the 2:1 clay mineral, Houston Black smectite,[8] is more stable under high silica, lowering the activity of Al^{3+} below that of kaolinite. This smectite mineral, occurring extensively throughout the coastal plains near the Gulf of Mexico, also contains an appreciable amount of iron, so the activity of aluminum which it supports is somewhat sensitive to the solid phase controlling the activity of Fe^{3+}. The smectite line in Figure 4 was drawn in equilibrium with $Fe(OH)_3$(soil).[15] Shifts with respect to Fe_2O_3(hematite) and $Fe_3(OH)_8$(ferrosic hydroxide) are also shown. Solubility measurements for iron in soils suggest that soil-Fe and in some cases ferrosic hydroxide[16] are more likely to control iron solubility than hematite. If this is the case, Houston Black smectite is least stable when hematite[7] controls the solubility of iron. It can be seen from Figure 4 that, if ferrosic hydroxide controls iron solubillity, the Houston Black smectite, kaolinite, and quartz all can coexist in acidic soils, even at pH 4.0. Higher silica and pH value shift aluminum solubility control to smectite, whereas lower pH value, silica, and iron shift the stability control to kaolinite. These solubility relationships help to demonstrate the complexity of factors that affect the solubility of aluminum in soils.

As weathering continues and the activity of $H_4SiO_4^\circ$ drops to $10^{-5.3}$, kaolinite and gibbsite can coexist. If the activity of $H_4SiO_4^\circ$ drops even lower, gibbsite becomes more stable than kaolinite. Under advanced weathering conditions, the presence of gibbsite provides an upper limit on the solubility of aluminum in soils.

The stability relationships of several aluminosilicates at pH 7.0 are shown in Figure 5. In general these minerals decrease in solubility as pH value rises. In the pH value range near 7.0, illite is comparable in solubility to gibbsite, imogolite, and halloysite when quartz controls silica. Vermont chlorite is also included, and its strong dependence on pH value $[Mg^{2+}]$, and iron are evident. Again, these minerals are of intermediate solubility and are expected to weather to kaolinite and smectite with time. At pH 7.0, the montmorillonite in equilibrium with soil-iron can also exist in equilibrium with kaolinite and quartz. The high stability of the Houston Black smectite is shown by the fact that, at high pH values and high iron solubility, it can coexist not only with kaolinite but also with gibbsite. Natural conditions favoring the occurrence of low $H_4SiO_4^\circ$ activity under the high pH value conditions imposed in Figure 5 are very unlikely. Weathering conditions which lower $H_4SiO_4^\circ$ activity also lower pH value, making Figure 4 more realistic for depicting smectite stability under conditions of low $H_4SiO_4^\circ$ activity.

IV. ALUMINUM SPECIATION IN THE SOIL SOLUTION

A. Hydrolysis

The Al^{3+} aquo-ion hydrolyzes readily in aqueous media giving the hydrolysis species defined by Reactions 20 to 23 in Table 1. (See Chapters 2 and 4 for a definitive discusson.) The activities of these species in equilibrium with gibbsite are shown in Figure 6. Only below pH 5 is Al^{3+} the dominant aluminum species. The hydrolysis species dominate soluble aluminum at higher pH values.

If the Al^{3+} activity is controlled by aluminosilicates instead of gibbsite, then the activities of Al^{3+} and all the monomeric hydrolysis species shown in Figure 6 drop by one log unit for each log unit drop in $[Al^{3+}]$. The $[Al_2(OH)_2{}^{4+}]$ will drop by two log units because it contains two aluminums. Similarly, if the activity of Al^{3+} were controlled by $Al(OH)_3$(am) instead of gibbsite, the activities of the monomeric hydrolysis species in Figure 6 would rise by $9.66 - 8.04 = 1.64$ log units and that of the dimeric species by 3.28 log units. Thus the hydrolysis species of Al^{3+} can be related readily to the solid phase controlling Al^{3+} and pH value. The slopes of the lines representing the hydrolysis species are related directly to the charge of the ionic species. Total soluble inorganic aluminum at any given pH value can be obtained by summing the concentrations of each species after conversion of the activities to concentrations, as shown in Section IV.E below.

B. Fluoride Complexes

Fluoride forms strong complexes with Al^{3+} (see Chapter 2) and these complexes contribute to total soluble aluminum in soils, particularly in acid soils, where $[Al^{3+}]$ is highest. Figure 7 was developed to show the importance of aluminum fluoride complexes as a function of the activity of F^-. This diagram was developed on the basis that $[Al^{3+}] = 10^{-10}$, the value

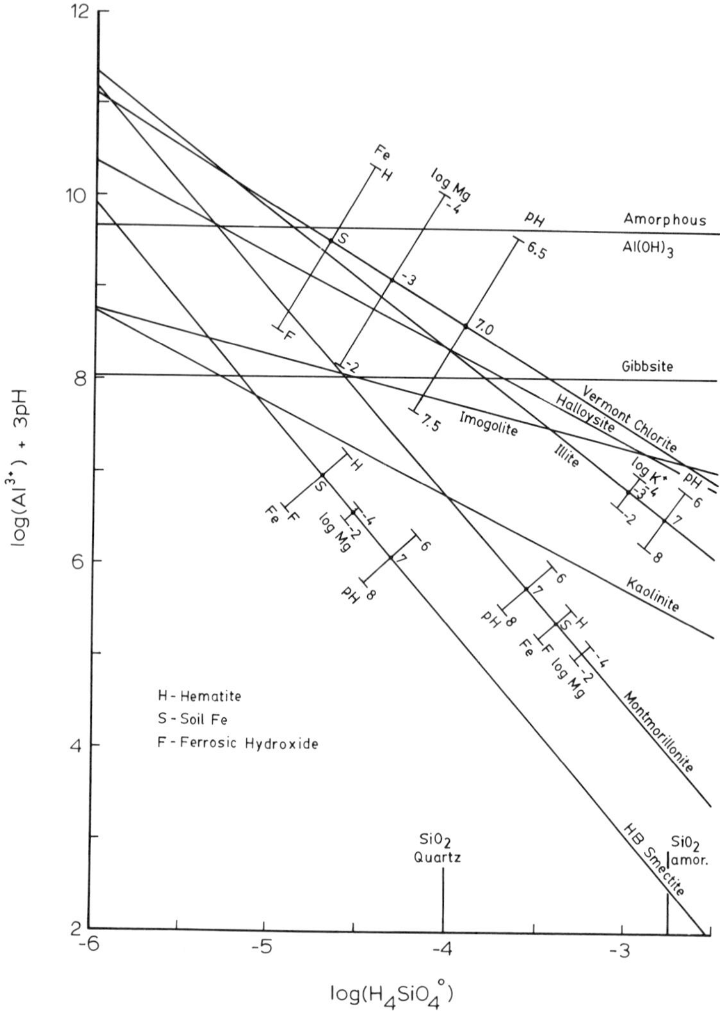

FIGURE 5. Solubility relationships of various aluminum hydroxides and aluminosilicates at pH 7, assuming equilibrium with soil-Fe, and $[Mg^{2+}]$ and $[K^+]$ = 10^{-3}. Shifts are shown for changes in the K^+ and Mg^{2+} activities.

supported by gibbsite at pH 6.0 (see Reaction 5, Table 1). This diagram can be easily modified to show the importance of aluminum fluoride complexes at any Al^{3+} activity. For example, if we were concerned about a gibbsitic soil of pH 5 instead of 6, $[Al^{3+}]$ would be 10^{-7} rather than 10^{-10}, so all the lines in Figure 7 would be raised three log units while their relative positions would remain unchanged.

It is apparent from Figure 7 that the importance of fluoride complexes of aluminum increases as the F^- activity increases. Below $[F^-] = 10^{-8}$, the fluoride complexes can be largely ignored, whereas at $[F^-] = 10^{-4}$, the activity of aluminum fluoride complexes would be 10^6 times greater than that of Al^{3+}.

C. Other Complexes

The Al^{3+} aquo-ion also forms soluble complexes with anions such as sulfate and nitrate. The stability constants for these complexes accompany Reactions 28 to 31 in Table 1. Their importance relative to Al^{3+} is shown in Figure 8. Only $AlSO_4^+$ contributes significantly. For example, when the SO_4^{2-} activity is equal to $10^{-3.2}$ (see Reaction 28, Table 1), the

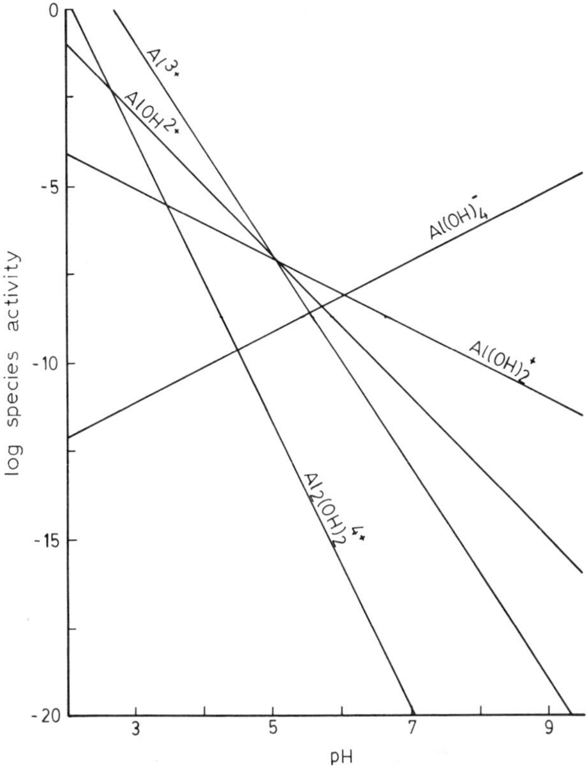

FIGURE 6. The activity of Al^{3+} and its hydrolysis species in equilibrium with gibbsite.

activities of $AlSO_4^+$ and Al^{3+} are equal. Tenfold changes in SO_4^{2-} activity result in corresponding changes in the ratio of complexed to free Al^{3+}. Other ions whose formation constants are known can be similarly included in Figure 8 to ascertain their significance in soils.

D. Organic Complexes

Aluminum forms fairly strong complexes with different functional groups in organic ligands (see Chapters 2 and 5). The extent of complexation varies from soil to soil. The synthesis and breakdown of the complexes is expected to change with time, temperature, biological activity, and other factors.[17] Soluble alumino-organic complexes increase total soluble aluminum in soils and contribute to its mobility. There are problems in describing naturally occurring alumino-organic complexes. First, one needs to know which chemical species are present, their formation constants, and the formation constants of organic complexes with other competing metal ions. This maze of uncertain parameters is handled in detail in Chapter 5.[17]

We offer two approaches around this difficult problem in quantifying aluminum solubility relationships. Separation of solid phases from the soil solution by filtration permits an estimate of total soluble aluminum. This fraction of aluminum provides the parameter for estimating the total mobility of aluminum for both diffusive and convective movement. For many environmental problems, this is the parameter of interest. On the other hand, it is not this parameter that is needed to examine toxicity relationships and ion activity products (see Chapters 1 and 2). The activity of Al^{3+} is needed for this purpose. Several attempts have been made to resolve this dilemma by adding complexing agents such as 8-hydroxyquinoline[18,19]

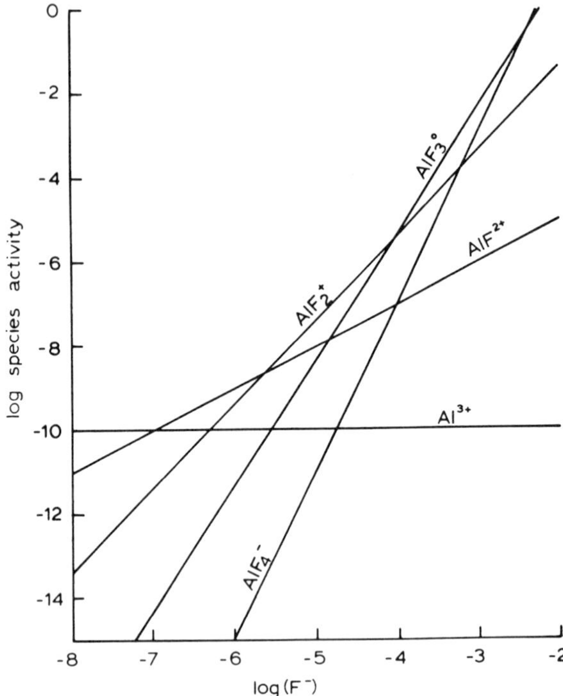

FIGURE 7. Fluoride complexes[1] of aluminum in equilibrium with $[Al^{3+}] = 10^{-10}$. (From Lindsay, W. L., *Chemical Equilibria in Soils*, Wiley-Interscience, New York, 1979. With pemission.)

or crown ether[20] to a soil extract for short periods, hoping that only the inorganic aluminum will be complexed. Another approach makes use of exchange resins.[21] An indirect determination involving the measurement of fluoride has also been reported.[22] There is still some skepticism about achieving a clean separation of the inorganic aluminum. An added chelating agent or resin that combines with Al^{3+} lowers its activity and causes all aluminum complexes, including organic complexes, to dissociate to some degree. The result is an overestimation of inorganic aluminum. See Chapter 1 for a complete discussion.

Another suggestion for separating the organic-metal complexes of aluminum from inorganic aluminum would be to add well-characterized, activated carbon to soil suspensions before equilibration. Both inorganic and organic aluminum would then equilibrate to attain equilibrium relationships in the suspension. During filtration, soluble organic matter, including the alumino-organic complexes, is adsorbed by the activated carbon and is retained on the soil solids. Thus organic aluminum can be separated from the inorganic aluminum which passes through the filter. Speciation of inorganic aluminum in the filtrate would permit a more reliable estimation of the Al^{3+} activity, which then can be used to examine ion activity products and relationships involving primary aluminum minerals and their weathering products, both minerals and soluble complexes.

A second approach for estimating the Al^{3+} activity would be to make use of a chelation method similar to that used to measure Fe^{3+} activity.[15] In this method, a chelating agent containing different mole fractions of bound Al^{3+} and a second ion are equilibrated with soil for a few days. By measuring the molar ratio of the two chelated ions along with the activity of the second ion, the activity of Al^{3+} can be estimated. This estimate is independent of organic complexes or polymeric aluminum species that may be present.

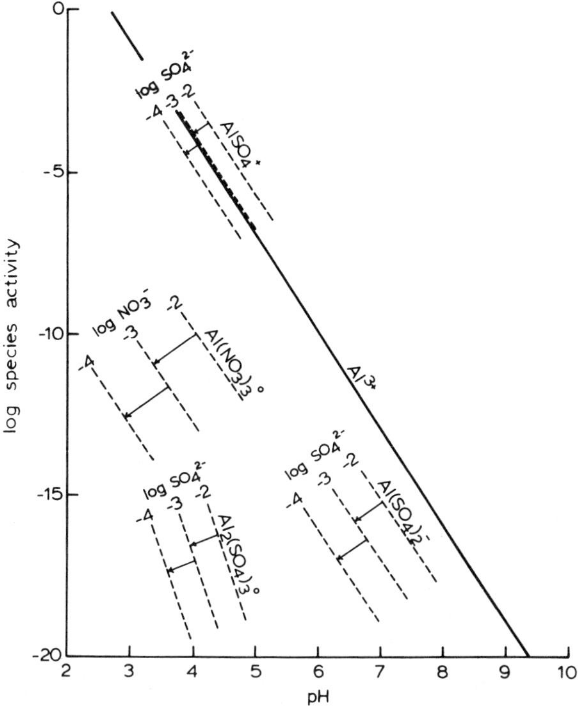

FIGURE 8. Nitrate and sulfate complexes of aluminum in equilibrium with gibbsite.[1] (From Lindsay, W. L., *Chemical Equilibria in Soils*, Wiley-Interscience, New York, 1979. With pemission.)

E. Speciation Models

If soluble organic aluminum can be separated from inorganic aluminum by the procedure suggested above or by another reliable procedure, then the inorganic fraction can be speciated as follows:

$$\text{Total inorganic aluminum} = \{Al^{3+}\} + \{AlOH^{2+}\} + \{Al(OH)_2^+\} + \{Al(OH)_4^-\} + \{AlF^{2+}\} + \{AlF_2^+\} + \{AlF_3^\circ\} + \{AlSO_4^+\} \quad (3)$$

In terms of activities, Equation 3 becomes

$$\text{Total inorganic aluminum} = \frac{[Al^{3+}]}{\gamma Al^{3+}} + \frac{[AlOH^{2+}]}{\gamma AlOH^{2+}} + \frac{[Al(OH)_2^+]}{\gamma Al(OH)_2^+} + \frac{[Al(OH)_4^-]}{\gamma Al(OH)_4^-} + \frac{[AlF_2^+]}{\gamma AlF^{2+}} + \frac{[AlF_2^+]}{\gamma AlF_2^+} + \frac{[AlF_3^\circ]}{\gamma AlF_3^\circ} + \frac{[AlSO_4^+]}{\gamma AlSO_4^+} \quad (4)$$

The reactions in Table 1 can be used to express the activity of each of the soluble species in terms of H^+, Al^{3+}, F^-, and SO_4^{2-} activities, giving

$$\text{Total inorganic aluminum} = \frac{[Al^{3+}]}{\gamma Al^{3+}} + \frac{10^{-4.99}[Al^{3+}]}{\gamma AlOH^{2+}(H^+)}$$

$$+ \frac{10^{-10.13}[Al^{3+}]}{\gamma Al(OH)_2^+(H^+)^2} + \frac{10^{-22.16}[Al^{3+}]}{\gamma Al(OH)_4^-[H^+]^4} + \frac{10^{6.98}[Al^{3+}][F^-]}{\gamma AlF^{2+}}$$

$$+ \frac{10^{12.60}[Al^{3+}][F^-]^2}{\gamma AlF_2^+} + \frac{10^{16.65}[Al^{3+}][F^-]^3}{\gamma AlF_3^\circ} + \frac{10^{3.20}[Al^{3+}][SO_4^{2-}]}{\gamma AlSO_4^+} \tag{5}$$

By rearranging Equation 5

$$[Al^{3+}] = \frac{\text{Total inorganic aluminum}}{\text{REM}} \tag{6}$$

where

$$\text{REM} = \frac{1}{\gamma Al^{3+}} + \frac{10^{-4.99}}{\gamma AlOH^{2+}[H^+]} + \frac{10^{-10.13}}{\gamma Al(OH)_2^+[H^+]^2}$$

$$+ \frac{10^{-22.16}}{\gamma Al(OH)_4^-[H^+]^4} + \frac{10^{6.98}[F^-]}{\gamma AlF^{2+}}$$

$$+ \frac{10^{12.60}[F^-]^2}{\gamma AlF_2^+} + \frac{10^{16.65}[F^-]^3}{\gamma AlF_3^\circ} + \frac{10^{3.20}[SO_4^{2-}]}{\gamma AlSO_4^+} \tag{7}$$

Thus, Al^{3+} activities can be calculated from Equation 6 using REM defined by Equation 7. For this calculation, the following parameters are necessary: pH value, electrolytic conductivity for estimating activity coefficients (γ)[1], F^- activity, and SO_4^{2-} activity. The total inorganic aluminum concentration used in the calculation must not include organic complexes. These calculations are limited to aqueous systems in which polymeric aluminum complexes do not contribute significantly to inorganic aluminum in solution[23] (see Chapter 5).

V. ALUMINUM SOLUBILITY MEASUREMENTS IN SOILS

A. Gibbsite and Variscite

An early attempts was made to measure Al^{3+} activities in soils and relate them to solid phases.[12] In this study, monocalcium phosphate monohydrate (MCP) was used to fertilize six New York acid glacial-till soils. The soils were allowed to wet and dry in a greenhouse for periods up to 18 months. No plants were grown. At various times, 0.1-kg samples of the soil were removed and shaken with 200 ml of 10 mol m^{-3} $CaCl_2$ for 24 h. The filtrates were analyzed for pH value, total phosphorus, and total aluminum. The activities of Al^{3+} and $H_2PO_4^-$ were calculated using appropriate speciation equations. The results are summarized in Figure 9.[1,12]

The pH values of these six soils in 10 mol m^{-3} $CaCl_2$ extract ranged from 3.7 to 5.0. The unfertilized soils fell very near or slightly below the $AlPO_4{\cdot}2H_2O$(variscite) solubility line, indicating possible equilibrium with this phosphate mineral. Adding 220 mg/kg P as MCP raised the $H_2PO_4^-$ activity in these soils by approximately 1.0 to 1.5 log units. During the ensuing 18 months, the solubility points moved down toward the variscite line. These results suggest that the initial solid reaction products of phosphate fertilization were 10 to 30 times more soluble than variscite and may have been an amorphous aluminum phosphate. During the ensuing 18 months, the solubility of phosphate decreased and approached that

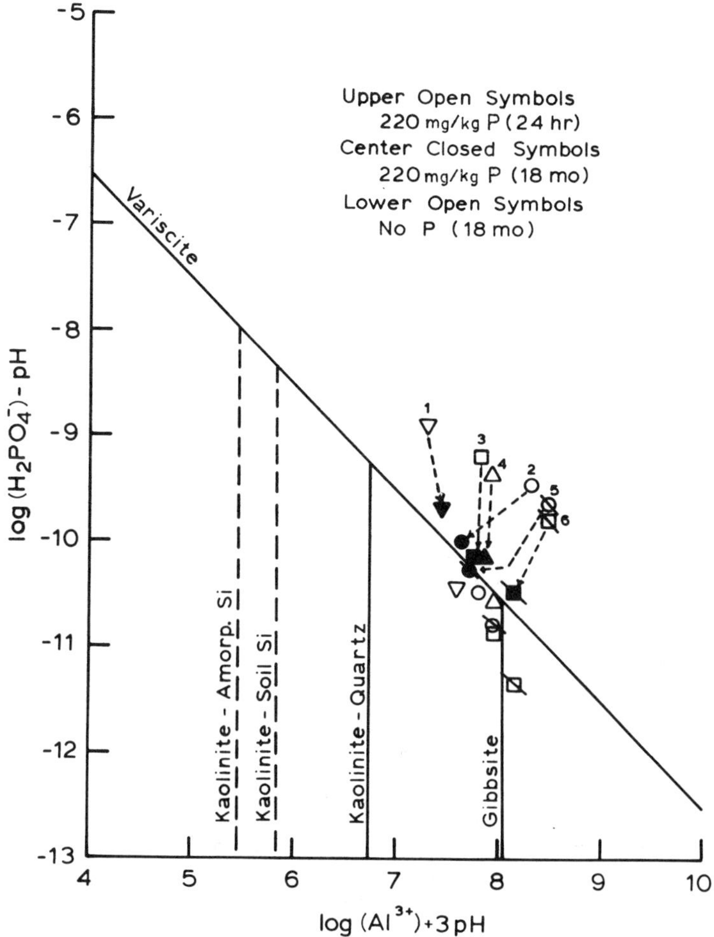

FIGURE 9. Changes in the solubilities of aluminum and phosphate in six acidic New York soils during an 18-month period following the addition of 220 mg/kg phosphorus as MCP.[1,12] (From Lindsay, W. L., *Chemical Equilibria in Soils*, Wiley-Interscience, New York, 1979. With pemission.)

of variscite. The values of $\log[Al^{3+}] + 3pH$ for these soils changed slightly during the 18-month period, but approached a value of approximately 7.8, which is only slightly below that supported by gibbsite (8.04). Thus it would appear that the aluminum solubility was controlled by gibbsite. Certainly kaolinite-quartz, kaolinite-soil Si[1], or kaolinite-amorphous Si mineral combinations were not controlling Al^{3+} activity, for otherwise $\log [Al^{3+}] + 3pH$ would have been much lower.

B. Aluminosilicate Equilibria

Walthall[24] examined soils from the Loch Vale Watershed (LVWS) in Rocky Mountain National Park, Colorado, and plotted $\log [Al^{3+}] + 3pH$ vs. $\log [H_4SiO_4^\circ]$ for these soils. The results are shown in Figure 10. Because the clay mineralogy of these soils was found to be dominated by smectite, the Houston Black smectite was chosen as a reference mineral because of its high stability under acid conditions. Solution conditions for this mineral were imposed to cover the pH range found in LVWS (pH 3.3 to 5.0), and the mean measured Mg^{2+} activity of $10^{-4.6}$. The smectite solubility lines were constructed under the condition that soil-Fe[15] (S) was used to establish the activity of Fe^{3+} with accompanying shifts if the more soluble ferrosic hydroxide[16] (F) should control iron solubility.

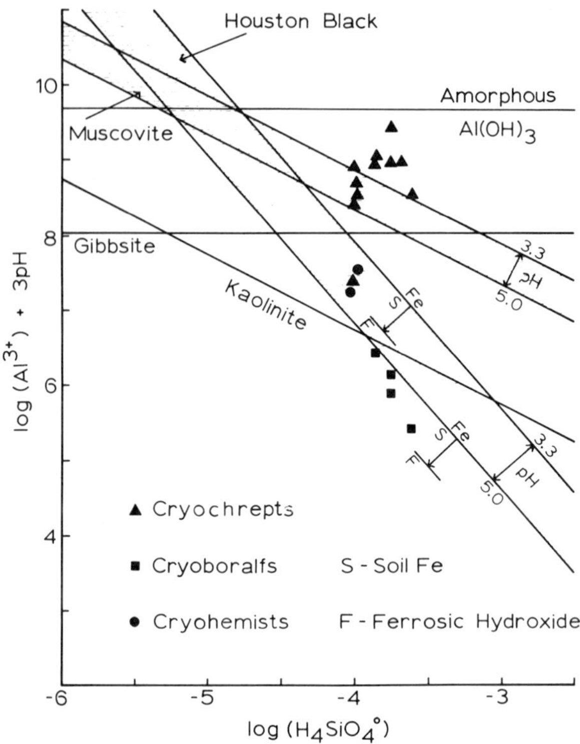

FIGURE 10. Aluminum and silica solubility relationships measured in soils compared to the solubilities of known minerals. Equilibrium was assumed with soil-iron (S), ferrosic hydroxide (F), pe + pH 13.0, $[Mg^{2+}] = 10^{-4.6}$, and the pH range of 3.3 to 5.0, which brackets the pH values measured in the soils.[23]

Two general clusters of points can be seen in Figure 10: the Inceptisols (with one exception) fall on or above the muscovite solubility line, while the Cryoboralfs and Cryohemists fall on or slightly below the smectite stability region. Soil solutions from the Inceptisols suggest that aluminum solubility may be controlled by rapid weathering of primary minerals, such as muscovite. This is possible considering the abundance of chlorite and vermiculite in the clay fractions of these soils. Both of these minerals are highly unstable under acid conditions. The presence of hydroxy-interlayered vermiculite in a number of these soils suggests that an amorphous aluminum hydroxide phase forming in the interlayer could also be an active component in these systems, since some of the data fall near the corresponding higher solubility line. Soil solutions from the Cryoboralfs show good agreement with the smectite solubility line. When ferrosic hydroxide was assumed to control the solubility of Fe, the solubility points fell within the solubility region for smectite. The Cryohemists — although the data are limited in number — also fall in the smectite solubility region.

These data suggest that two distinct environments exist among the soils of LVWS which reflect the degree of pedogenic weathering that they have undergone. The Inceptisols represent the initial conditions of weathering in which a range of unstable aluminosilicates are dissolving and producing a high and somewhat variable Al^{3+} activity. The Alfisols represent a steady-state weathering environment in which smectite is a stable secondary clay mineral controlling Al^{3+} activity. This extremely acidic environment favors the rapid transition of the unstable clay minerals initially present in the Inceptisols to the more stable secondary smectite phase in the more highly weathered Alfisols. In the Cryohemists, on the other hand, either aluminum chemistry is dominated by drainage waters from the Cryoboralfs situated upslope or inorganic constituents dominated by smectite control the solubility of aluminum.

C. Exchangeable Aluminum

Exchangeable aluminum constitutes an important buffered reserve of labile aluminum that can be solubilized readily by other exchangeable cations.[25] An example of this exchange was demonstrated by the senior author in an experiment in which the concentration of $CaCl_2$ was increased from 1 to 100 mol m^{-3} with a Mardin silt loam.[26] The results showed that both H^+ and Al^{3+} were displaced by exchange with Ca^{2+}, yet the parameter $\log[Al^{3+}] + 3pH$ remained constant at 7.55. This experiment reflects the tendency of soils to maintain a constant $\log[Al^{3+}] + 3pH$, as predicted by the Boltzmann equation,[1] independently of solid phases that may be present. Equilibrium with gibbsite gives $\log[Al^{3+}] + 3pH = 8.04$, whereas equilibrium with $Al(OH)_3$(am) gives a value of 9.66 (see Reactions 1 and 5, Table 1). Thus the Mardin silt loam was undersaturated with respect to all of the minerals shown in Figure 1, yet $\log[Al^{3+}] + 3pH$ for this soil tended to remain constant over a wide range of displaced Al^{3+}. The fact that $\log[Al^{3+}] + 3pH$ in Mardin silt loam was lower than the value supported by gibbsite suggests that more insoluble aluminosilicates are likely controlling the Al^{3+} activity.

Although exchangeable cations tend to maintain fixed activity ratios, as demonstrated here by the constancy of $\log[Al^{3+}] + 3pH$ as $[Ca^{2+}]$ was increased, exchangeable ions do not have ultimate control on the activity of these ions in soil solutions. The solubility of Al^{3+} ultimately is controlled by the dissolution and precipitation of solid phases containing aluminum. Thus exchangeable Al^{3+} is only a buffer and not a final determinant of aluminum solubility; the solubility controls of aluminum ultimately rest with solid phases.

VI. SUMMARY AND FUTURE CHALLENGES

The complex pattern of aluminum solubility relationships in soils is gradually beginning to unfold. Many details remain to be worked out before observed aluminum solubility relationships can be fully reconciled with expected mineral equilibria. Until this is accomplished, many concerns about aluminum solubility relationships in soil environments will remain unresolved.

Soils consist largely of an assemblage of aluminosilicate and oxide minerals that undergo weathering. In the early stages of weathering, the dissolution of primary aluminosilicates, the precipitation of amorphous $Al(OH)_3$ or intermediate reaction products dominates the solubility control of aluminum. At later stages of weathering, the solubility control of aluminum shifts to the more slowly formed secondary clay minerals. The particular minerals involved depend largely on the solubility of silica in the weathering environment. If large quantities of silica are released during weathering, and the silica is not removed by leaching, a high activity of silica persists. Under these conditions, the precipitation of 2:1 type clay minerals occurs and the solubility of aluminum is depressed far below that present during the latter stages of weathering. Examples of such minerals include montmorillonite, Houston Black smectite, and other smectites with different isomorphous substitutions in their structures.

As weathering continues and soluble silica is removed by leaching, the 2:1 clay minerals become unstable relative to the 1:1 minerals. This results in an increase in the solubility of aluminum. As the $H_4SiO_4^\circ$ activity drops below $10^{-5.3}$, kaolinite becomes unstable relative to gibbsite. The formation of gibbsite then establishes a solubility limit on aluminum, which occurs in highly weathered soils largely depleted of silica.

Special soil conditions exist when other minerals control aluminum solubility. For example, in acid sulfate soils, the release of sulfuric acid favors the formation of alunite or other hydroxy aluminum sulfates. The solubility of aluminum in soils, however, is controlled ultimately by the dissolution and precipitation of solid phases, not by exchangeable aluminum.

Problems which need to be addressed before aluminum solubility in soil can be understood better include the following.

1. Better identification of minerals that are involved in aluminum and better quantification of their solubility. These minerals include many aluminosilicates, along with their isomorphically substituted counterparts, i.e., the smectites, montmorillonite, nontronite, and beidellite.
2. More accurate methods of obtaining Al^{3+} activity values in soils, uncomplicated by the presence of organic matter complexes. These data are needed so that meaningful thermodynamic equilibrium relationships can be examined.
3. More attention needs to be given to intricate mineral transformations and solubility relationships that occur in the weathering reactions of microenvironments. These findings then can be extended to more general soil conditions.

With this information, environmental chemists will be equipped better to monitor dynamic aqueous environments involving aluminum and to identify the solubility controls that govern aluminum mobility and toxicity.

REFERENCES

1. **Lindsay, W. L.,** *Chemical Equilibria in Soils,* Wiley-Interscience, New York, 1979.
2. **Reuss, J. O. and Johnson, D. W.,** Acid deposition and the acidification of soils and waters, *Ecological Studies 59,* Springer-Verlag, New York, 1986.
3. **Sadiq, M. and Lindsay, W. L.,** Selection of standard free energies of formation for use in soil chemistry, *Colo. State Univ. Exp. Stn. Tech. Bull.,* 134, 1979.
4. **Adams, F. and Rawajfih, Z.,** Basaluminite and alunite: a possible cause of sulfate retention by acid soils, *Soil Sci. Soc. Am. J.,* 41, 1977.
5. **Nordstrom, D. K.,** The effect of sulfate on aluminum concentrations in natural waters: some stability relations in the system Al_2O_3-SO_3-H_2O at 298 K, *Geochim. Cosmochim. Acta,* 46, 681, 1982.
6. **Farmer, V. C., Smith, B. F. L., and Tait, J. M.,** The stability, free energy and heat of formation of imogolite, *Clay Miner.,* 14, 103, 1979.
7. **Kittrick, J. A.,** Solubility of two high-Mg and two high–Fe chlorites using multiple equilibria, *Clays Clay Miner.,* 30(3), 167, 1982.
8. **Carson, C. D., Kittrick, J. A., Dixon, J. B., and McKee, T. R.,** Stability of soil smectite from a Houston Black clay, *Clays Clay Miner.,* 24, 151, 1976.
9. **May, H. M., Helmke, P. A., and Jackson, M. L.,** Gibbsite solubility and thermodynamic properties of hydroxy-aluminum ions in aqueous solution at 25°C, *Geochim. Cosmochim. Acta,* 43, 861, 1979.
10. **van Breemen, N.,** Dissolved aluminum in acid sulfate soils and in acid mine waters, *Soil Sci. Soc. Am. J.,* 37, 694, 1973.
11. **Adams, F. and Hajek, B. F.,** Effects of solution sulfate, hydroxide, and potassium concentrations on the crystallization of alunite, basaluminite, and gibbsite from dilute aluminum solutions, *Soil Sci.,* 126, 169, 1978.
12. **Lindsay, W. L., Peech, M., and Clark, J. S.,** Solubility criteria for the existence of variscite in soils, *Soil Sci. Soc. Am. Proc.,* 23, 357, 1959.
13. **Kittrick, J. A.,** Soil minerals in the Al_2O_3-SiO_2-H_2O system and a theory of their formation, *Clays Clay Miner.,* 17, 157, 1969.
14. **Wada, K.,** Allophane and imogolite, in *Minerals in Soil Environments,* Dixon, J. B. and Weed, S. B., Eds., Soil Science Society of America, Madison, WI, 1977.
15. **Norvell, W. A. and Lindsay, W. L.,** Estimation of the concentration of Fe^{3+} and the $(Fe^{3+})(OH)^3$ ion product from equilibria of EDTA in soil, *Soil Sci. Soc. Am. J.,* 46, 710, 1982.

16. **Schwab, A. P. and Lindsay, W. L.,** Effect of redox on the solubility and availability of iron, *Soil Sci. Soc. Am. J.,* 47, 201, 1983.
17. **Stevenson, F. J. and Vance, G. F.,** Naturally occurring aluminum-organic complexes, in *The Environmental Chemistry of Aluminum,* Sposito, G., Eds., CRC Press, Boca Raton, FL, 1989, chap. 5.
18. **Bloom, P R. and Erich, M. S.,** The quantitation of aqueous aluminum, in *The Environmental Chemistry of Aluminum,* Sposito, G., Ed., CRC Press, Boca Raton, FL, 1989, chap. 1.
19. **Bloom, P. R., Weaver, R. M., and McBride, M. B.,** The spectrophotometric and fluorometric determination of aluminum with 8-hydroxyquinoline and butyl acetate extraction, *Soil Sci. Soc. Am. J.,* 42, 713, 1978.
20. **Evans, A. and Zelazny, L. W.,** Determination of inorganic monomeric aluminum by selected chelation using crown ethers, *Soil Sci. Soc. Am. J.,* 50, 910, 1986.
21. **Driscoll, C. T.,** A procedure for the fractionation of aqueous aluminum in dilute acidic waters, *Int. J. Environ. Anal. Chem.,* 16, 267, 1984.
22. **Hodges, S. C.,** Aluminum speciation: a comparison of five methods, *Soil Sci. Soc. Am. J.,* 51, 57, 1987.
23. **Bertsch, P. M.,** Aqueous polymeric aluminum species, in *The Environmental Chemistry of Aluminum,* Sposito, G., Ed., CRC Press, Boca Raton, FL, 1989, chap. 4.
24. **Walthall, P. M.,** Acidic Deposition and the Soil Environment of Loch Vale Watershed in Rocky Mountain National Park, Ph.D. dissertation, Colorado State University, Fort Collins, 1985.
25. **Zelazny, L. W. and Jardine, P. M.,** Surface reactions of aqueous aluminum species, in *The Environmental Chemistry of Aluminum,* Sposito, G., Ed., CRC Press, Boca Raton, FL, 1989, chap. 6.
26. **Lindsay, W. L., Peech, M., and Clark, J. S.,** Determination of aluminum ion activity in soil extracts, *Soil Sci. Soc. Am. Proc.,* 23, 266, 1959.

Chapter 9

THE CHEMISTRY OF ALUMINUM IN SURFACE WATERS

Charles T. Driscoll

TABLE OF CONTENTS

I. INTRODUCTION

A. The Aluminum Cycle

Aluminum is the most abundant metallic element within the lithosphere.[1,2] Although quantitatively important, the aluminum cycle is complicated and poorly understood. Aqueous, particulate (soil/sediment), and biological transformations of aluminum may be conceptualized through a schematic representation of the aluminum cycle (see Figure 1). This cycle is a series of pools (e.g., aqueous, particulate, and living biomass) with many fluxes that regulate pool sizes. Some information is available on aluminum pools within the natural environment, although it is largely based on operationally defined methods. Unfortunately, there is virtually no information available on the relative importance of the pathways linking the pools or rates of these transformations.

Soil represents the largest pool of aluminum at the earth's surface.[2] Within the lithosphere, aluminum is associated largely with aluminosilicate minerals, most commonly as feldspars in metamorphic and igneous rocks, and as clay minerals in well-weathered soils (see Chapter 3). In this form, aluminum is generally unavailable (unreactive) for chemical and biological reactions. Through soil development, highly crystalline aluminosilicate minerals are decomposed and a small fraction of their aluminum becomes available to participate in biogeochemical processes. Soil (sediment) available pools of aluminum are characterized generally by operationally defined extraction techniques and are frequently termed "free" (nonsilicate bound) aluminum. Many processes contribute to free soil (sediment) aluminum pools; e.g., the weathering of aluminosilicate minerals by carbonic acid dissolution, strong acid dissolution, or neutral hydrolysis,[3] followed by the secondary precipitation of aluminum, produces amorphous mineral forms. Soil aluminum is also retained on charged surfaces[4] associated with soil organic matter and/or clay minerals (see Chapter 6). Finally, free soil (sediment) aluminum may be bound with organic matter through microbial/plant transformations[5,6] or formed by precipitation with organic solutes[7,8] within soil solutions (see Chapter 5).

All forms of soil/sediment aluminum potentially may control its solution concentration (see Chapter 8). However, the release of aluminum from highly crystalline minerals is very slow.[9] Although the mechanisms regulating its concentration are not established clearly, it is likely that aqueous aluminum is derived largely from free-soil/sediment pools (see Figure 1), including exchange from cation-exchange sites, dissolution from amorphous mineral phases, or decomposition/mineralization of organic forms.

Solution aluminum is the most chemically and biologically available form, although this pool represents an extremely small fraction of total aluminum in the environment. Aluminum is a strongly hydrolyzing metal and is relatively insoluble in the neutral pH range (6.0 to 8.0).[10] Under acidic ($pH < 6.0$) or alkaline ($pH > 8.0$) conditions, and/or in the presence of complexing ligands, the solubility of aluminum is enhanced, making it more available for biogeochemical transformations (see Chapter 8). Within the aqueous phase, aluminum may be associated with a variety of inorganic and organic complexes (see Section III.B and Chapters 2 and 5). The extent of complexation depends on the availability of soil aluminum, solution pH value, concentrations of complexing ligands, ionic strength, and temperature. Aqueous aluminum may be redeposited to free-soil (sediment) pools, assimilated by living biomass, or transported from the system.

The biological cycling of aluminum is assumed generally to be unimportant quantitatively. Aluminum is not a plant or animal nutrient[4] and, therefore, it generally does not accumulate in living tissue. Indeed, elevated concentrations of aluminum are often toxic to a wide variety of organisms.[11-14] Wood,[15] among others, has hypothesized that the selection and use of elements in the biochemical evolution of organisms was dictated by their abundance in the earth's crust, as well as their solubility under the anaerobic conditions that occurred during that period, about 4×10^9 years ago. Although abundant, aluminum was very insoluble

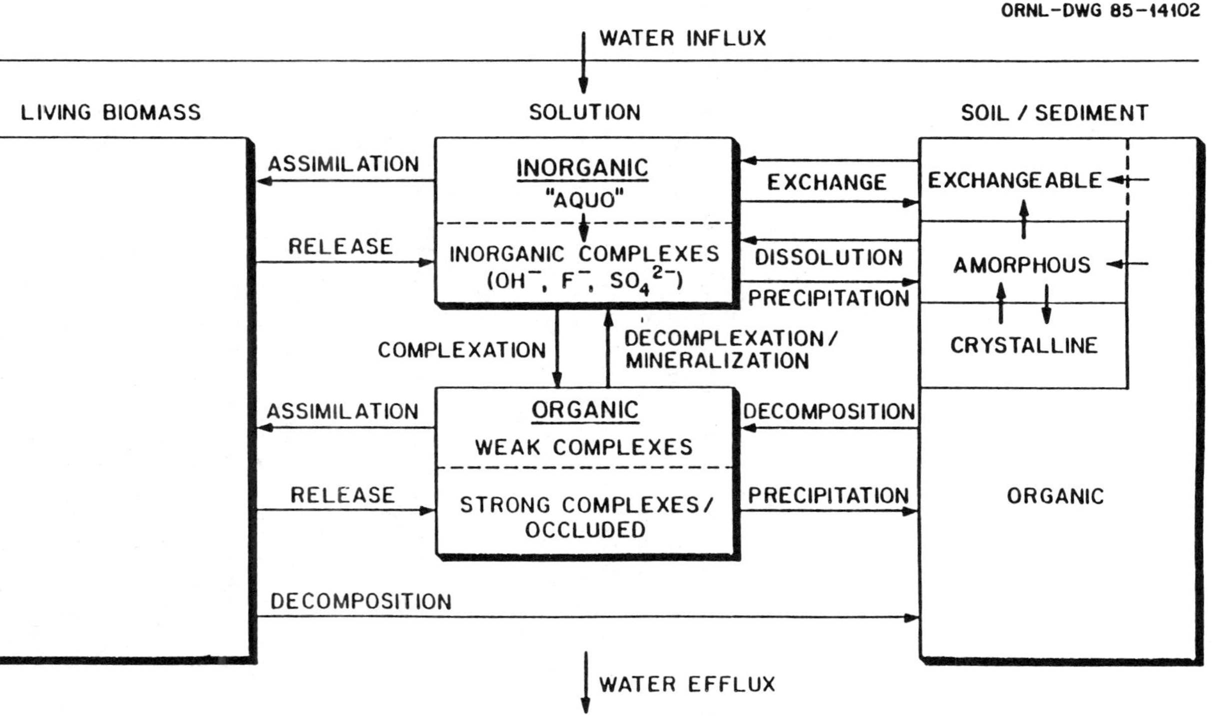

FIGURE 1. Schematic representation of the aluminum cycle. Within the environment aluminum can exist in pools of living biomass, solution, or nonliving particulate matter. A number of transformations link the individual pools and regulate aluminum concentrations.

under the conditions in which life evolved and, therefore, probably unavailable for transport into primitive anaerobic bacteria. As a result, most organisms have not adapted to elevated concentrations of bioavailable aluminum and cannot survive large intracellular accumulation. Nevertheless, organisms do assimilate limited quantities of aluminum, with some plants capable of accumulating high concentrations.[16] Biological uptake and release, therefore, represents a potential pathway of aluminum transfer to soil and aqueous environments (see Figure 1).

B. Scope of Chapter

The objective of this chapter is to provide a summary of the chemistry, transformations, and effects of aluminum in surface waters. This summary includes (1) a description of the mechanisms regulating aluminum inputs to surface waters, (2) a summary of concentrations of aluminum and speciation in surface waters, (3) a description of patterns of surface-water aluminum concentrations and mechanisms regulating temporal and spatial variations in the concentrations, and (4) an evaluation of the environmental consequences of elevated concentrations of aqueous aluminum.

Many of the concepts discussed in this chapter are illustrated by thermodynamic calculations. The thermochemical data used in these calculations are summarized in Appendix 1. They are in accord with the definitive compilations in Chapter 2 for inorganic species. Thermodynamic calculations were made with the chemical equilibrium model, ALCHEMI.[17] ALCHEMI facilitates calculations of the distribution of aluminum, F^-, SO_4^{2-}, dissolved inorganic carbon (DIC), and dissolved organic carbon (DOC) in natural solutions. Solutions can be equilibrated with hydroxide, silicate, and hydroxysulfate mineral phases, as well as without solid-phase constraints. Calculations can also be made in equilibrium with a given partial pressure of CO_2 or in a closed atmosphere system. A series of thermodynamic equations are solved with ALCHEMI using a Newton-Raphson method. Thermochemical calculations are corrected for the effects of temperature using enthalpy values (see Appendix 1) and for ionic strength using the Davies equation.[10] The symbols used in this chapter to summarize the distribution of aluminum are presented in Appendix 2.

II. PROCESSES REGULATING INPUTS OF ALUMINUM TO SURFACE WATERS

Inputs of aluminum to surface waters may occur through a variety of processes and exhibit many forms. On the global scale, aluminum inputs to the aqueous environment appear to be associated largely with particulate matter. For example, Garrels et al.[1] estimate the global stream flux of aluminum as 760 mol/ha-year. Virtually all this aluminum is associated with sediment transport, since aluminum is abundant in particulate minerals and relatively insoluble in most fresh waters. There have been very few studies comparing particulate and dissolved transport of aluminum in drainage areas unimpacted by anthropogenic activity. However, useful information is available from a small headwater stream at the Hubbard Brook Experimental Forest (HBEF) in New Hampshire.[17] This site exhibits very low erosion and transport of particulate matter (33 ± 13 kg/ha), even though the catchment has a steep slope (12 to 13°) as well as elevated precipitation inputs (132 ± 7 cm/year) and runoff (83 ± 7 cm/year). Moreover, stream water from the HBEF characteristically has elevated concentrations of dissolved aluminum because of acidic conditions which facilitate leaching losses.[18,19] From an 8-year study, Likens et al.[18] concluded that the total aluminum exported from the HBEF was 125 mol/ha-year, with 41% associated with particulate matter and 59% in a dissolved form. Therefore, particulate transport would appear to be quantitatively important as a component of total aluminum input to surface waters, even for those areas with minimal sediment loading and elevated concentrations of dissolved aluminum.

As mentioned previously, however, aluminum associated with mineral matter is relatively unavailable to participate in chemical and biological transformations. Therefore, this review will focus on processes regulating labile forms of surface-water aluminum.

A. Regulation of Aluminum Transport in the Absence of Strong Acid Inputs

The mobilization and subsequent deposition of aluminum is an integral component in the development of well-drained soils. A variety of interrelated factors are critical to soil development and associated transport of aluminum, including climate, watershed elevation and slope, temperature, vegetation, and the nature of soil minerals, as well as the movement of drainage water. In northern temperate regions, the process of podzolization strongly influences the transport of aluminum within a soil profile. Podzolization is thought traditionally to involve the mobilization of aluminum (and iron) from upper to lower mineral-soil horizons by organic acids leached from foliage as well as from decomposition in the forest floor.[21-26] As alumino-organic complexes are transported through the mineral soil, processes of complexation and/or microbial oxidation decrease the organic carbon-to-aluminum ratio of these solutes, thereby reducing their solubility.[7,8] In pristine environments, aluminum appears to be largely retained within the lower (B horizon) mineral soil. For example, Ugolini et al.[24] reported that there was little transport of aluminum from soil to surface water associated with soil development at Findley Lake in Washington. See Chapter 5 for additional discussion.

Recently, several researchers have hypothesized that CO_2 may have an important role in the solubilization and transport of aluminum.[27-29] Under elevated partial pressures of CO_2 that accumulate within soil (up to 10^{-1} atm),[4] the dissociation of H_2CO_3 acidifies the soil solution and may facilitate leaching losses of aluminum.

Within soil, the dissolution of aluminum occurs generally under acidic conditions and/or in the presence of complexing ligands. The dissociation of organic acids and/or H_2CO_3 results in the formation of temporarily mobile (transient) anions, which may allow for the transport of aluminum by serving as complexing ligands and/or counterions. Concentrations of these transient anions are dependent upon a variety of solution conditions, which may vary over time and between different soil horizons. When a transient anion is removed from solution, its associated aluminum (and iron) is deposited within the soil profile. The formation of transient anions in the upper soil, their transport and removal with associated aluminum (and iron) in lower soil horizons, is central to soil development.

The concept of transient transport of aluminum is probably most easily illustrated with reactions involving the hydration and dissociation of CO_2 (see Figure 2). In the soil environment, elevated partial pressures of CO_2 cause the formation and dissociation of H_2CO_3. This acidity may solubilize soil aluminum, whereas the production of HCO_3^- serves as a counterion for the transport of cationic aluminum through the soil. When the solution emerges from the soil to enter surface water, it equilibrates with atmospheric conditions (CO_2 partial pressure of $10^{-3.5}$ atm), CO_2 degasses, resulting in the removal of HCO_3^- and the hydrolysis and precipitation of aluminum (see Figure 2). To illustrate the range of conditions under which the mobilization of aluminum by CO_2 can occur, some hypothetical calculations were made using ALCHEMI (see Figure 3). Over a range of acid-neutralizing capacity (ANC) and partial pressure of CO_2, the amount of aluminum solubilized by CO_2 was determined. This "CO_2-mobilized aluminum" is calculated with ALCHEMI by allowing the solution to come into equilibrium with $Al(OH)_3$ for a given set of conditions (ANC and partial pressure of CO_2), then reequilibrating the system with atmospheric CO_2 ($10^{-3.5}$ atm). The difference in aqueous aluminum concentrations caused by precipitation represents the aluminum solubilized by elevated soil CO_2.

Mobilization of aluminum by CO_2 is most significant at values of ANC near zero (pH near 5.0) and increases with increasing partial pressure of CO_2 (see Figure 3). The extent

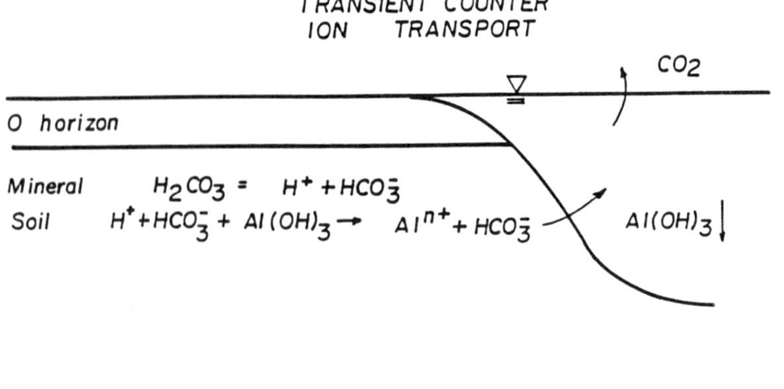

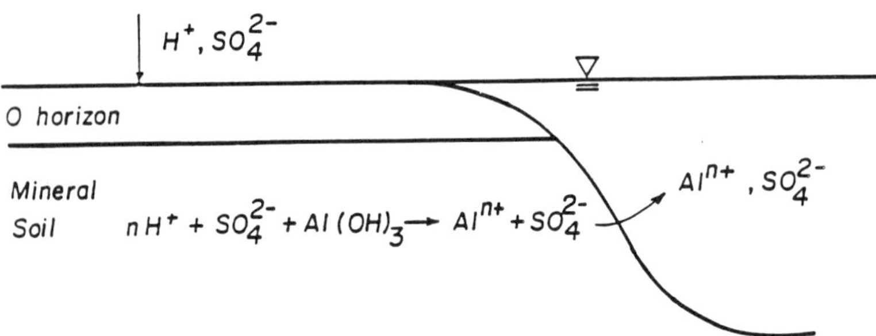

FIGURE 2. Conceptual model depicting the transport of aluminum by a transient anion, a (HCO_3^-) and a mobile anion, b (SO_4^{2-}). Dissociation of H_2CO_3 in soil under elevated partial pressures of CO_2 produces a H^+, solubilizing aluminum, and HCO_3^-, which can serve as a counterion in aluminum transport. When the solution equilibrates with atmospheric conditions, for example, upon discharge to a surface water, CO_2 degasses, resulting in aluminum hydrolysis and precipitation. Acid inputs associated with a mobile anion (e.g., SO_4^{2-}) can solubilize and transport aluminum to surface water.

of this phenomenon is closely coupled to the acid-base characteristics of dissolved inorganic carbon and aluminum. At low ANC and pH values, significant concentrations of aluminum are available because of the pH-dependent solubility of $Al(OH)_3$. However, acidic conditions restrict the hydration and dissociation of CO_2 to H^+ and HCO_3^- thereby limiting the availability of H^+ to solubilize aluminum as well as decreasing the concentration of HCO_3^-, which serves as a counterion in aluminum mobilization. Conversely, at high ANC and pH values, elevated concentrations of HCO_3^- occur, but the solubility of aluminum is low, so dissolution is limited. Because CO_2 is a weak Brønsted acid, elevated concentrations do not depress pH values substantially. Therefore, the concentrations of aluminum mobilized under elevated partial pressures of CO_2 are relatively low.

Concentrations of dissolved aluminum are generally low in most natural waters because of the relatively low solubility of natural aluminum minerals under circumneutral pH values and low concentrations of complexing ligands. Stumm and Morgan[10] report a median value of 0.4 μmol/l for terrestrial waters, whereas Bowen[30] gives an average concentration of 9 μmol/l for fresh waters, including bogs.

B. Role of Strong Acid Inputs in the Transport of Aluminum to Surface Waters

Superimposed on the natural process of soil development is the introduction of strong acids (e.g., H_2SO_4 and HNO_3) from atmospheric deposition in parts of eastern North America[31]

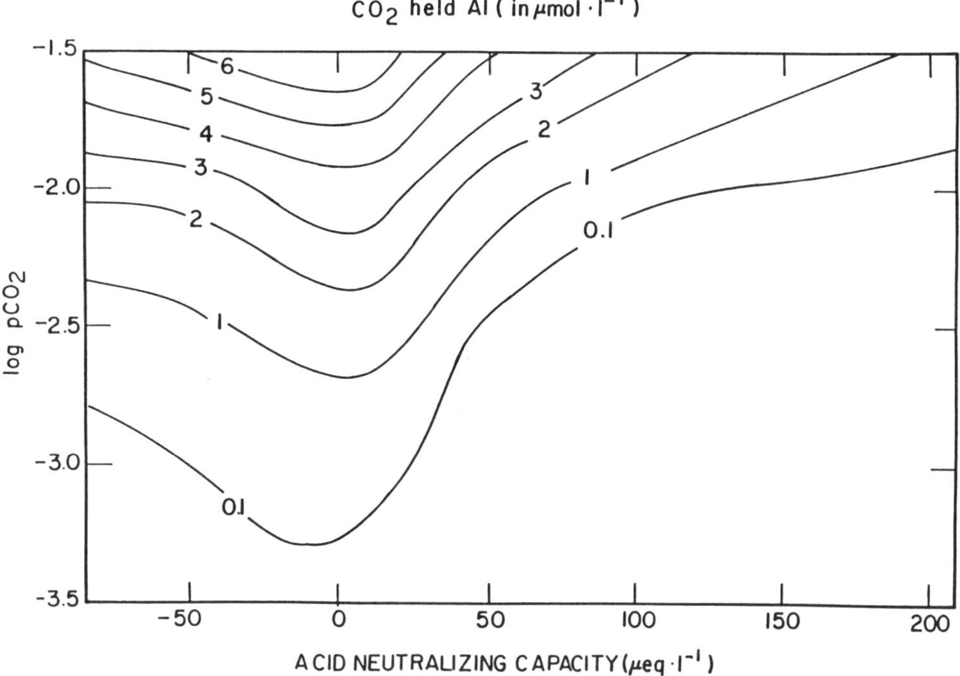

FIGURE 3. Diagram indicating aluminum mobilization by CO_2 (in μmol/l) under different values of acid neutralizing capacity (ANC) and partial pressure of CO_2 (pCO_2) for a system in equilibrium with natural $Al(OH)_3$.

and Europe.[32] There are many regions that are "sensitive" to inputs of strong acids. Acid-sensitive watersheds are generally underlain by granitic bedrock and have shallow acidic soil with small pools of "readily available" (i.e., exchangeable, easily weatherable) basic cations (C_B: Ca^{2+}, Mg^{2+}, Na^+, and K^+). These systems may also be characterized by an inability to retain inputs of strong acid anions (SO_4^{2-}, NO_3^-, Cl^-) through soil adsorption or biological retention. Therefore, high loadings of strong acids are not attenuated within the terrestrial environment, but instead are exported from the watershed with drainage water. Complete neutralization of strong acids is accomplished by dissolution or exchange of basic cations, and/or retention of strong acid anions within lake/watershed systems. However, when these processes are limited, neutralization is incomplete and acidic cations (H^+, Al^{n+}) are transported from soil to surface waters (see Figure 2).

Cronan and Schofield[31] have hypothesized that strong acids from atmospheric deposition have altered the natural process of podzolization in northern temperate soils by facilitating the transport of aluminum from soil to surface waters. Evidence to support this contention is available in water-chemistry data collected from acid-sensitive regions that are receiving a high atmospheric loading of strong acids.[33] When the charge concentration of strong acid anions (e.g., SO_4^{2-} or NO_3^-) approaches or exceeds the charge concentration of basic cations, high concentrations of acidic cations (H^+ or Al^{n+}) are observed in surface waters (see Figure 4).[34]

In order to understand better the aluminum cycle and to develop predictive models to assess the effects of acidic deposition, it is critical to identify the source of mobile aluminum (see Figure 1). Researchers have suggested several hypotheses for a solid phase regulating aluminum concentrations in dilute natural water systems, including poorly crystallized 1:1 layer-type clays,[35] kaolinite ($Al_2Si_2O_5[OH]_4$),[36] $Al(OH)_3$,[19,37] hydroxy sulfate minerals (e.g., basaluminite, $Al_4OH_{10}SO_4{\cdot}5H_2O$; alunite, $KAl_3OH[SO_4]_2$; or jurbanite, $Al[OH]SO_4{\cdot}5H_2O$),[38,39] and dissociation from soil organic matter.[40,41]

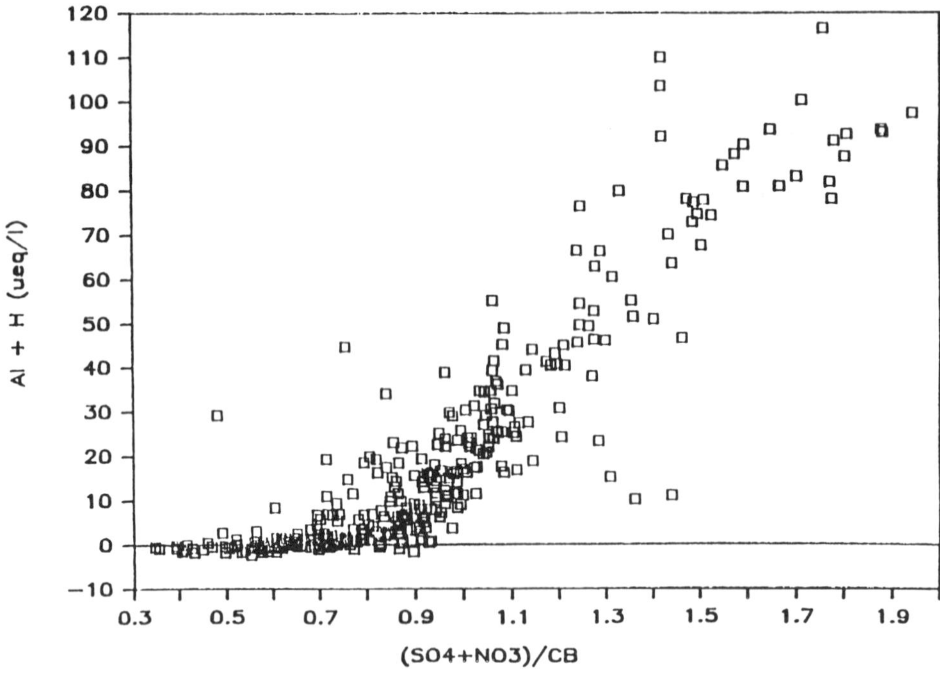

FIGURE 4. Concentration of acidic cations (H^+ and Al^{n+}) as a function of the ratio of SO_4^{2-} and NO_3^- concentration to basic cation concentration (C_B, Ca^{2+}, Mg^{2+}, Na^+, and K^+) on an equivalence basis for Adirondack, New York surface waters.[34] As this ratio approaches one, increases in the concentrations of acidic cations (H^+ and Al^{n+}) are observed. The equivalence of aluminum was computed with ALCHEMI.

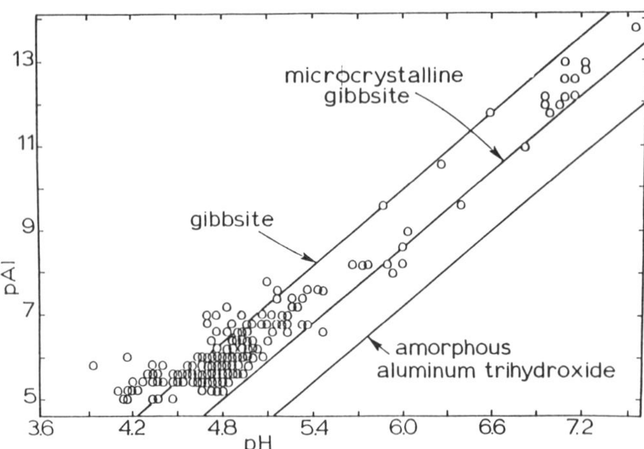

FIGURE 5. The logarithm of activity of aquo-aluminum as a function of pH in the Adirondack surface waters.[37] The activity of Al^{3+} was calculated from measurements of inorganic monomeric aluminum, pH, F^-, and SO_4^{2-} using a chemical-equilibrium model.

Johnson et al.[19] and Driscoll et al.[37] evaluated the chemistry of surface waters in New Hampshire and in the Adirondack region of New York, respectively, and reported that total solution aluminum concentrations were generally close to the theoretical solubility of $Al(OH)_3$ (see Figure 5). As a result, aluminum release is often depicted on the basis of $Al(OH)_3$ solubility in computer models developed to simulate aluminum geochemistry.[42,43] Deviations from this pattern, however, are evident in the literature. For example, solutions draining organic horizons in forest soils[44] and in peaty soils[34] are often highly undersaturated with respect to $Al(OH)_3$. Gherini et al.[43] have suggested that, under these conditions, the release of aluminum is no longer in equilibrium with $Al(OH)_3$, but rather is kinetically controlled. Other researchers,[40,41] however, contend that in the presence of soil organic matter, solution aluminum is regulated by cation exchange, which is a function of the concentration of aluminum on exchange surfaces. Finally, in regions of Europe experiencing very high SO_4^{2-} loading, a number of investigators have reported that the activity of aquo-aluminum in ground waters[45] and soil solutions[39] appears to be regulated by the solubility of jurbanite ($Al[OH]SO_4 \cdot 5H_2O$).

Thermodynamic calculations are used widely to interpret solution aluminum data, but the determination of a mineral phase regulating aqueous concentrations by this approach is tenuous. Calculations of aquo-aluminum activity are made using measured pH values. As a result of autocorrelation, strong trends are observed generally which do not necessarily reflect true solid-phase equilibrium. Moreover, results vary considerably depending on the source of thermochemical data, fractionation procedures used to determine the speciation of aluminum, and whether ionic strength, temperature corrections, or complexing ligands (e.g., F^- or DOC) are considered in the calculations (see Chapters 1, 2, and 4).

Additional information on processes regulating the aqeuous concentration of aluminum is available through soil solution and soil chemistry data (see Chapter 8). Studies of soil solution and soil aluminum from the Adirondack region of New York[44] and from New Hampshire[46] have yielded comparable results. Solutions draining organic soil horizons contained remarkably high concentrations of aluminum, which appeared to be associated largely with organic solutes. These solutions were highly undersaturated with respect to the solubility of readily forming mineral phases. In the lower mineral soil (B horizon), inorganic aluminum concentrations were elevated, and solutions approached the solubility of natural gibbsite. David and Driscoll[44] also evaluated differences in aluminum chemistry between soils supporting coniferous and deciduous vegetation. Solutions draining soils supporting coniferous vegetation contained higher aluminum concentrations in all soil horizons than those draining soils supporting deciduous stands. This enrichment of aluminum was attributed to complexation by organic solutes and coincided with elevated DOC concentrations (see also Chapter 5).

Results of field studies in the northeastern U.S.[44,46] have led investigators to conclude that the organic(o) horizon is an important souce of mobile aluminum. Aluminum transported from organic horizons is largely in an organic form and probably the by-product of biological decomposition processes. These observations are to some extent inconsistent with the traditional views of soil development (podzolization theory) and suggest that biocycling (vegetation assimilation and microbial decomposition) may be an important component of the aluminum cycle in forested ecosystems. Elevated concentrations of inorganic aluminum were generally observed only in solutions draining from the mineral soil. Therefore, it is likely that organic and inorganic fractions of aluminum observed in surface waters originate from different processes (either organic decomposition or dissolution by strong acids, respectively) and at different locations within the soil profile (either organic or mineral horizons, respectively).

It is apparent that there are inconsistencies in the literature on the processes regulating aluminum concentrations in drainage waters. These inconsistencies are, to some extent, the result of researchers relying on total aluminum concentration measurements to interpret processes. Fractionation procedures to evaluate the speciation of aluminum in solution have only recently been developed and applied to natural waters.[47] Unfortunately, there is no accepted aluminum fractionation procedure (see Chapter 1). These techniques, for the most part, are poorly defined; results obtained from a study using a given procedure may reflect the bias of the procedure rather than actual trends in solution characteristics.[47,48]

III. CONCENTRATION AND SPECIATION OF ALUMINUM IN SURFACE WATERS

Researchers often rely on operationally defined procedures to evaluate the concentration and forms of aluminum in natural waters. Unfortunately, the accuracy of many of these methods is difficult to evaluate, and there is considerable uncertainty as to what forms of aluminum are actually represented. The inability to differentiate precisely and accurately between aqueous and particulate aluminum, as well as between inorganic and organic forms of aqueous aluminum, have hampered studies of the environmental chemistry of this element.

Determinations of total (unfiltered) aluminum in natural waters are made using a wide variety of digestion procedures as well as extraction periods, resulting in a range of operational methods and considerable uncertainty in measured values (see Chapter 1). A number of studies employed filtration in an effort to distinguish between dissolved and particular aluminum forms. However, because particulate materials exhibit a continuous size distribution, no absolute distinction between dissolved and particulate forms can be made, and results show a strong dependence on filter pore size. For instance, in examining the dissolved and particulate fractions in a California stream, Kennedy et al.[49] found that the concentration of aluminum in the filtrate decreased as filter-pore size decreased from 0.45 to 0.1 μm.

In an attempt to eliminate problems and uncertainty associated with filtration, many researchers have relied on the addition of a complexing agent (e.g., 8-hydroxyquinoline or pyrocatechol violet) followed by either rapid detection or extraction in an organic solvent (e.g., methyl isobutyl ketone [MIBK] or butyl acetate), to partition between aqueous and particulate aluminum.[50-55] Rapid extraction techniques, however, are not without problems. For example, results vary with reaction time;[55] the greater the contact time the larger the measured concentrations of aluminum. Also, it is unclear to what extent the complexing agent desorbs aluminum from particulate matter or competes with naturally occurring ligands for aluminum. In view of the wide range and operational nature of collection and procedures, comparisons of studies of aluminum in natural waters should be made with caution. For a comprehensive discussion of these issues, see Chapter 1.

A. Concentrations of Aluminum in Surface Waters

A wide range of aluminum concentrations have been reported for surface waters. Elevated concentrations of aluminum generally have been found in acidic waters in regions receiving high inputs of acidic substances, such as Sweden,[56] Norway,[57] Belgium,[58] the Netherlands,[32] West Germany,[59] Canada,[60-63] and the U.S.[37,64-66] (see Table 1). Many investigators have observed an exponential increase in aluminum concentrations with decreasing solution pH values.[37,57-59,66] This phenomenon is characteristic of the theoretical solubility of a number of aluminum minerals. Unfortunately, synoptic surveys have used a variety of methods to determine solution aluminum, so only broad patterns should be interpreted.

In a synoptic survey of 1612 lakes in acid-sensitive regions of the eastern U.S., Linthurst et al.[66] measured two forms of aluminum. Total aluminum was determined on an unfiltered sample which was acid digested (pH < 2) and measured by atomic absorption spectrophotometry (AAS). To provide an estimate of concentrations of aqueous aluminum, extractable aluminum was determined by filtration (0.4 μm polycarbonate) followed by 8-hydroxyquinoline complexation, extraction using MIBK and detection by AAS. Lakes sampled as part of the Eastern Lake Survey were divided into regions and subregions within the eastern U.S. to facilitate geographical assessments. Moreover, the survey was designed statistically to allow for population estimates of lake characteristics.

Results of the Eastern Lake Survey indicated that lakes in acid-sensitive regions were generally circumneutral (see Table 2). Linthurst et al.[66] defined acidic lakes as those systems with pH values below 5.0. A relatively small number of lakes had pH values below 5.0.

Table 1
ALUMINUM CONCENTRATIONS IN SURFACE WATERS

Location	Description	pH range	Total aluminum range (μmol/l)	Ref.
	Lakes			
Europe				
Belgium	Moorland pools, 1975—1979	3.5—8.5	11—300	58
Norway	Regional survey, 1974—1977	4.2—7.8	0—27	57
Sweden	West coast, 1976	4.0—7.4	0.4—27	56
Scotland	Southwest Scotland	4.4—6.4	1—11	57
North America				
Canada	Ontario, 1980	4.1—6.5	0.2—32	60
	Sudbury, Ontario	4.3—7.0	5—42	61
	Quebec	4.7—7.0	0.8—13	63
U.S.	Adirondacks, 1977—1978	3.9—7.0	0.1—31	37
	New England, 1978—1981	4.0—8.2	0—21	64
	New England	4.2—7.0	0—16	65
	Streams			
Europe				
Sweden	Central Sweden, three brooks	4.6—5.2	5.9—15	67
West Germany	Snowmelt	3.7—7.5	0—740	60
North America				
U.S.	Adirondacks, 1977—1978	4.0—7.6	3—43	37
	New England, 1978—1981	4.1—7.7	0.5—14	64

For example, in the northeast region only 3.4% of the lakes were acidic. Within the Adirondack subregion, 10% of the lakes were acidic; in southern New England, 5%; all other northeast subregions had less than 2% acidic lakes. In the upper midwest, only 1.5% of the lakes surveyed had pH values below 5.0, with the Upper Peninsula of Michigan having 9% and all other upper midwest subregions having 2% or less. In the southeast region, the southern Blue Ridge contained no acidic lakes, while in the Florida subregion 12% of the lakes had pH values below 5.0.

As a result of the relatively low number of acidic lakes, very few of the waters sampled contained elevated aluminum concentrations. Lakes generally exhibited much higher concentrations of total aluminum than extractable aluminum (see Table 2), suggesting that water-column aluminum was largely in a particulate form. Because extractable aluminum is thought to be more bioavailable, Linthurst et al.[66] focused their analysis of aluminum on this measurement. Moreover, they evaluated only clearwater lakes (those with true color less than 30 platinum cobalt units) because of their concern that aluminum leaching by organic solutes would complicate the analysis.

Population estimates of clearwater lakes with elevated concentrations of extractable aluminum were consistent with pH trends (see Table 3). Generally, concentrations of extractable aluminum were below 1 μmol/l for waters with pH values above 6.0 and increased exponentially with decreases in pH below that value. Elevated concentrations of extractable aluminum were evident only in the subregions showing a significant number of acidic lakes, including the Adirondacks, the Poconos and Catskills, southern New England, the Upper Peninsula of Michigan, and Florida (see Table 3). The Adirondack region of New York clearly contained the largest number of clearwater lakes with elevated concentrations of extractable aluminum.

Table 2
SUMMARY OF ALUMINUM CONCENTRATIONS IN LAKES FROM ACID-SENSITIVE REGIONS OF THE EASTERN U.S.[66]

Location	n	pH	Total aluminum	Extractable aluminum
Northeast				
Adirondack	203	6.3 (4.2—9.4)	5.1 (0.3—28)	1.6 (0—12)
Poconos/Catskills	153	6.7 (4.9—9.5)	2.4 (0—24)	0.5 (0—9.6)
Central New England	212	6.7 (4.4—9.5)	3.3 (0.2—19)	0.5 (0—5.5)
Southern New England	124	6.5 (4.5—7.8)	2.2 (0.2—11)	0.4 (0—6.9)
Maine	183	6.9 (4.3—8.4)	3 (0.3—30)	0.5 (0—10)
Upper Midwest				
Northeast Minnesota	168	6.8 (5.6—8.3)	2.8 (0.1—19)	0.6 (0—4.7)
Upper Peninsula	165	6.5 (4.4—8.6	2.2 (0—16)	0.7 (0—7.9)
North Central Wisconsin	186	6.5 (4.7—8.0)	1.4 (0—30)	0.4 (0—0.6)
Upper Great Lakes	141	6.9 (5.4—8.7)	1.4 (0—15)	0.3 (0—6.7)
Southeast				
Southern Blue Ridge	111	6.9 (5.9—8.3)	4.3 (0.3—40)	0.1 (0—1.5)
Florida	168	6.3 (3.8—9.0)	3.3 (0.3—50)	0.8 (0—16)

Note: Mean value and range are given in μmol/l.

Table 3
POPULATION ESTIMATES OF PERCENT OF CLEARWATER LAKES WITH ELEVATED CONCENTRATIONS OF EXTRACTABLE ALUMINUM

Location	Above 1.8 (μmol/l)	Above 3.7 (μmol/l)	Above 5.5 (μmol/l)
Northeast	5.2	3.0	2.0
Adirondacks	14.4	12.1	9.9
Poconos/Catskills	6.0	2.9	0.3
Central New England	1.8	0.0	0.0
Southern New England	4.0	0.8	0.8
Maine	0.0	0.0	0.0
Upper Midwest	0.5	0.0	0.0
Northeast Minnesota	0.0	0.0	0.0
Upper Peninsula, Michigan	2.1	0.4	0.4
North Central Wisconsin	0.9	0.0	0.0
Upper Great Lakes	0.0	0.0	0.0
Southeast			
Southern Blue Ridge	0.0	0.0	0.0
Florida	7.4	4.4	1.5

B. Speciation of Aluminum in Surface Waters

In assessments of the environmental chemistry of aluminum, it is particularly useful to distinguish among species of aqueous aluminum. It is well established that aluminum forms strong complexes with OH^-, F^-, and SO_4^{2-},[68] as well as strong organic complexes.[69] Fortunately, thermodynamic data for OH^-, F^-, and SO_4^{2-} complexation are well established[17] and allow for precise estimates of the speciation of inorganic aluminum in solution (see Chapters 2 and 8).

The speciation of aluminum is pH-dependent. At low pH values, dissolved aluminum is almost entirely present in an aquo form. As pH values increase, aluminum undergoes hydrolysis, resulting in a series of OH^- complexes ($Al[OH]^{2+}$ and $Al[OH]_2^+$) and decreases in solubility. The solubility of aluminum is at a minimum near pH 6.5 and increases at higher pH values because of the formation of $Al(OH)_4^-$ (see Chapter 8).

Fluoride is similar in size and therefore readily substitutes for OH^- in metallic complexes. Under acidic conditions, F^- has a strong affinity for aluminum. The stoichiometry of Al-F complexes varies with the total F^- concentration. As F^- concentrations increase, H_2O or OH^- ligands are displaced in a step-wise manner, resulting in higher ligand-number complexes until all octahedral positions of aluminum are occupied by F^- (e.g., AlF_6^{3-}).[68] Concentrations of F^- in natural waters are typically low (median concentration of fresh waters is 10 μmol/l).[10] At pH values below 5.5, molar concentrations of dissolved aluminum generally exceed concentrations of F^-, and low ligand-number complexes predominate (e.g., AlF^{2+}, AlF_2^+). The concentration of Al-F complexes in natural acidic solutions is therefore generally limited by the total concentration of F^- available for complexation.[37] At high pH values (>7), OH^- concentrations are elevated and it becomes increasingly difficult for F^- to compete with OH^- for aqueous aluminum. As a result, OH^- complexes predominate over F^- complexes under alkaline conditions.

Like F^-, SO_4^{2-} forms complexes with aluminum under acidic conditions and is displaced by OH^- complexes under alkaline conditions. The predominant Al-SO_4 complex is $AlSO_4^+$ at low SO_4^{2-} concentrations and shifts to $Al(SO_4)_2^-$ at higher SO_4^{2-} concentrations. Sulfate complexes of aluminum are not as strong as F^- complexes. Although SO_4^{2-} concentrations are typically higher than F^- in fresh waters (median concentration 400 μmol/l),[10] Al-SO_4 complexes are significant only at high SO_4^{2-} concentrations and low pH values.

There is some qualitative evidence in the literature to suggest the importance of Al-Si species.[70-72] Unfortunately, there are no reliable thermodynamic data or aqueous Al-Si reactions. Given the concentration range of dissolved silicon in fresh waters (median concentration 300 μmol/l),[10] it might be reasonable to hypothesize the existence of Al-Si complexes, if only as metastable templates for clay formation. Because of the absence of thermochemical data, current studies on the distribution of inorganic aluminum in natural waters have generally ignored Al-Si complexes. However, if Al-Si complexation proves to be significant, it could dramatically alter current views of the speciation of aluminum in natural solutions. Other inorganic anions, found in natural waters (e.g., HCO_3^-, Cl^-, or NO_3^-) do not form significant complexes with aluminum (see Chapter 2).

Naturally occurring organic solutes may have a marked effect on the solubility and speciation of aluminum. Unfortunately, the chemistry of naturally occurring organic solutes is poorly understood (see Chapter 5). The complicated and diverse nature of dissolved organic matter leads to special problems in understanding and modeling aqueous aluminum chemistry. Not only is it an inherently complex ligand mixture that exhibits a distribution of binding sites with varying affinity for aluminum, but electrostatic interactions between ligand sites on the same molecule, configurational changes, and aggregation reactions arise,[73-76] making rigorous thermochemical descriptions difficult.

Because of uncertainty in thermodynamic data, particularly involving naturally occurring

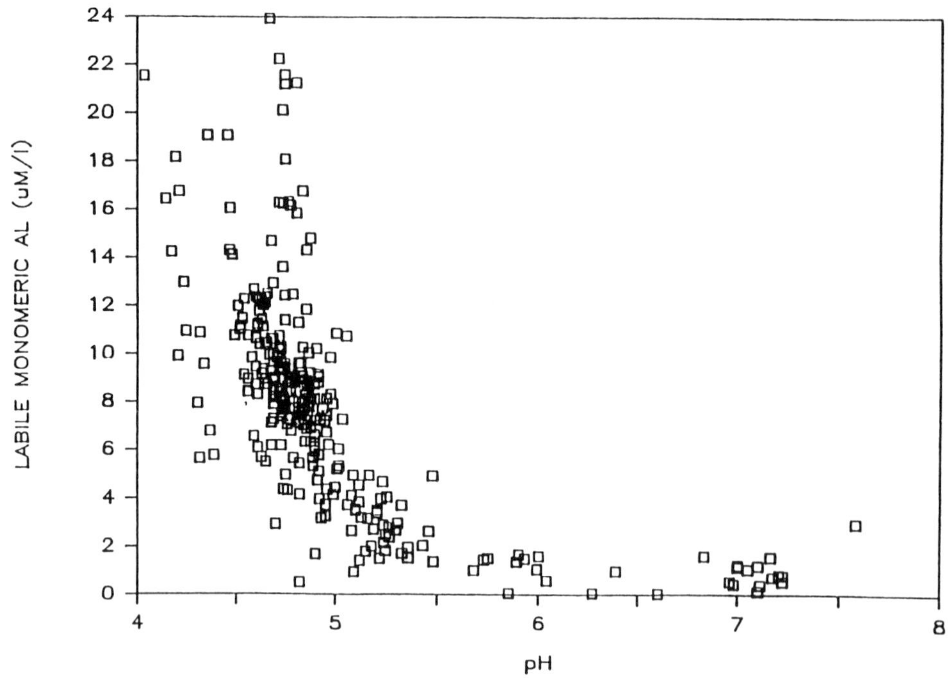

FIGURE 6. Concentrations of inorganic monomeric aluminum as a function of pH in Adirondack surface waters.[37]

organic solutes, it is useful to distinguish analytically among forms of aqueous aluminum. This information is critical to interpret process-level research on mechanisms controlling the concentration, transport, and bioavailability of aluminum. A number of fractionation procedures have been developed in recent years to attempt to distinguish among aqueous forms of aluminum. These techniques include rapid reaction with complexing agents,[52,54] dialysis,[62] batch chelating resin,[63] column cation exchange,[77] fluorine ion selective electrode,[77] and morin addition followed by fluorimetric detection.[78,79] All these procedures are operationally defined and undoubtedly detect somewhat different forms of aluminum.[47,48] See Chapter 1 for a complete discussion.

Until recently, there have been few studies of the speciation of aluminum in natural waters. Most of the available data are obtained from regions with waters that have been acidified by atmospheric deposition and contain elevated concentrations of aluminum. Studies of aluminum chemistry in surface waters in the Adirondack region of New York[37] and in New Hampshire,[19] using the cation-exchange column fractionation procedure, have yielded similar results. Concentrations of inorganic monomeric aluminum increased exponentially with decreases in solution pH below 5.5 (see Figure 6). Concentrations of organic monomeric aluminum were strongly correlated with organic carbon concentrations in both Adirondack ($Al_o = 3.36 \times 10^{-6} + 0.0204$ TOC; where Al_o is organic, monomeric aluminum in μmol/l and TOC is total organic carbon concentration in μmol C/l, $n = 322$, $r^2 = 0.76$, and $p < 0.0001$; see Figure 7), and New Hampshire ($Al_o = -7.3 \times 10^{-7} + 0.0155$ DOC; where DOC is the dissolved organic carbon concentration in μmol C/l, $n = 69$, $r^2 = 0.85$, and $P < 0.0001$) surface waters.[77]

Using the column ion-exchange fraction procedure, Driscoll et al.[37] observed that inorganic monomeric aluminum was the predominant form of aluminum in acidic Adirondack surface waters, although concentrations of organic monomeric aluminum were also significant (see Table 4). These authors applied values of inorganic monomeric aluminum, pH, F^-, and

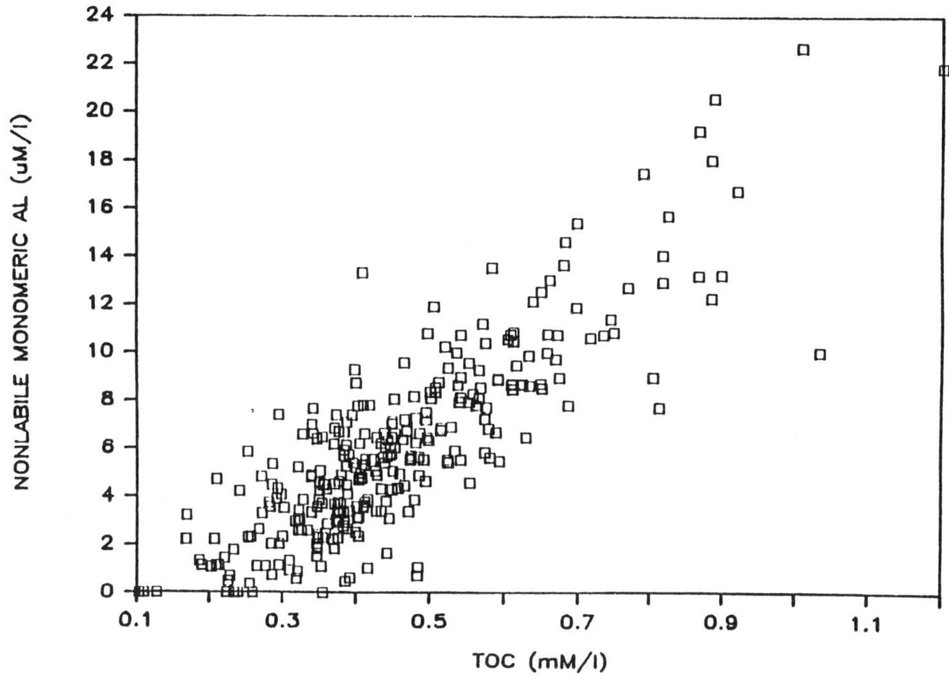

FIGURE 7. Concentrations of organic monomeric aluminum as a function of total organic carbon concentration (μmol C/l) in Adirondack surface waters.[37]

Table 4
THE DISTRIBUTION OF AQUEOUS ALUMINUM IN ADIRONDACK SURFACE WATERS (N = 321) USING THE CATION-EXCHANGE COLUMN FRACTIONATION PROCEDURE[37]

	Aluminum conc.		Aluminum distribution	
Aluminum form	Value (μmol/l)	SD	As total	As monomeric
Total	16.1	6.9	—	—
Inorganic monomeric	7.7	4.7	0.48	0.56
Organic monomeric	6.2	4.1	0.37	0.44
Acid soluble	2.1	1.8	0.14	—
Al^{3+}	1.6	1.3	0.10	0.11
Al-OH	1.9	1.9	0.25	0.14
Al-F	4.0	1.9	0.25	0.29
$Al\text{-}SO_4$	0.2	0.2	0.01	0.01

SO_4^{2-} to a chemical-equilibrium model to calculate the distribution of inorganic aluminum. Fluoride-complexed aluminum (aluminum-fluorine) was the predominant form of inorganic monomeric aluminum in acidic surface waters, and at pH values below 5.5 essentially all aqueous fluorine was complexed with alumunim. Concentrations of aquo-aluminum and Al-OH were less significant than Al-F. Concentration of the sulfate complexes of aluminum ($Al\text{-}SO_4$) were low in magnitude.

LaZerte[62] and Campbell et al.[63] have studied the speciation of aluminum in Canadian surface waters, using the dialysis and batch-chelating resin techniques, respectively. In these waters, pH values were generally higher than highly acidic waters in the northeastern U.S., so aluminum concentrations were generally lower and largely in an organic form. However, as surface-water pH values decreased, the distribution of aluminum shifted with increasing predominance of inorganic forms. Moreover, LaZerte[62] reported that in Ontario waters inorganic aluminum was predominantly complexed with fluorine.

Lee[80] evaluated the speciation of aluminum in the Lake Gardsjon catchment of Sweden, using the cation-exchange column procedure. At this study site, surface waters were highly acidic (pH < 4.5) and aluminum concentrations ranged from 11 to 33 μmol/l. Monomeric aluminum was predominantly in an inorganic form and largely associated with aquo (30 to 55%) or fluorine complexes (10 to 30%). Organic forms ranged from 0 to 43% of monomeric aluminum.

C. Models of Aluminum Speciation in Natural Waters

The speciation of aqueous aluminum is not determined routinely in synoptic surveys or monitoring programs of water chemistry. However, this information is critical for water quality assessments because the effects of aqueous aluminum are not only due to elevated concentrations, but also a function of species distribution[11] (see Section V). Researchers often rely on models to calculate the speciation of aluminum when direct measurements are not available. Calculations of the speciation of inorganic aluminum are relatively straightforward because of the availability of good thermodynamic data.[17] However, natural waters contain organic solutes that bind with aluminum, and difficulty arises in estimating alumino-organic concentrations (see Section III.B). A few models are available in the literature which allow for the calculation of aluminum species given measurements of pH, total or monomeric aluminum, and ligand concentrations.

Bakes and Tipping[81] developed two models using laboratory dialysis and acid/base-titration data to describe the partitioning of aluminum in solution between inorganic and organic forms. The first model is an empirical relationship:

$$\nu = \alpha[Al^{3+}]^{\beta}\,[H^{+}]^{\gamma}$$

where ν is the number of moles of aluminum bound per gram of organic material and α, β, and γ are empirical constants with values of 1.32×10^{-4} mol/g organic matter, 0.718, and -1.054, respectively.

The second model is thermodynamic in nature and considers aluminum complexation by two types of sites on an organic polyelectrolyte. Model parameters include a proton-dissociation constant, intrinsic-equilibrium constants for each binding site, an electrostatic interaction factor, as well as the number of sites per mole of organic carbon. These parameters were determined using laboratory data.

Bakes and Tipping[81] indicate that the fit of both models to laboratory observations were excellent ($r^2 = 0.86$). The empirical model is simple to apply directly to surface waters. The site-binding model has the advantage of allowing calculation of proton release associated with aluminum complexation, which may be important in studies of the pH buffering of aluminum (see Section V.A). However, it would be useful if this model were coupled with an inorganic-speciation model. These researchers suggest caution when applying the models to natural waters without further evaluation of model parameters.

LaZerte[62] developed a simple empirical model to predict the concentration of inorganic aluminum based on field studies of aluminum speciation using the dialysis method. This model relies on measurements of TOC and monomeric aluminum (inorganic monomeric aluminum $= -10 + 0.69$·[monomeric aluminum] -0.03·[TOC]·[monomeric aluminum]; $r^2 = 0.90$, $n = 537$; where monomeric aluminum and TOC are expressed in mg/l. LaZerte[62] indicates that the model was developed from data obtained from streams and lakes in Ontario, Canada during snowmelt and cautions that the model may not be applicable to surface waters that are highly acidic with low DOC concentrations.

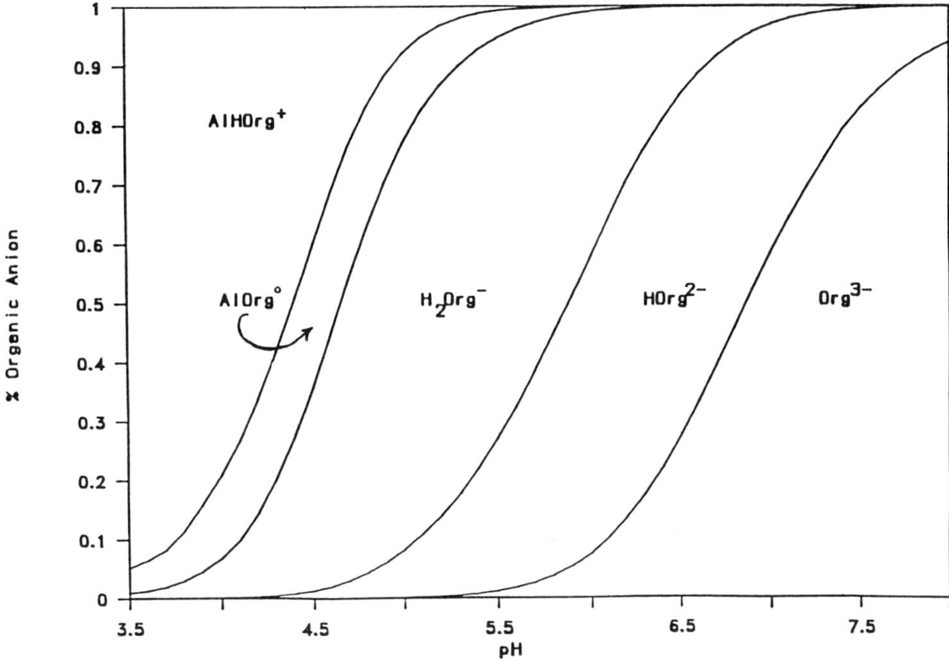

FIGURE 8. Distribution of model organic solute as a function of pH for a system in equilibrium with $Al(OH)_3$. Organic acid is represented as a triprotic solute which can complex aluminum.

Schecher[83] has coupled a chemical-equilibrium model for the speciation of inorganic aluminum (ALCHEMI), with a series of proton- and aluminum-binding reactions thought to be applicable to acid-sensitive waters. This model was developed using field data collected from the Adirondack region of New York, in which aluminum was fractionated using the cation-exchange column method. (Thermodynamic data used in this model are summarized in Appendix 1.) In ALCHEMI, organic solutes are represented as a triprotic acid, while two species of alumino-organic solute are depicted. Thermodynamic constants were calibrated using field observations, and a good fit of the data was obtained ($r^2 = 0.86$ and $n = 31$).

Using ALCHEMI, hypothetical calculations were made to illustrate the speciation of the model organic acid and aluminum for conditions that are typical of acid-sensitive waters in the Adirondack region of New York. Model calculations were made using F^- concentrations of 5 μmol/l and SO_4^{2-} concentrations of 100 μmol/l, and in equilibrium with the solubility of natural $Al(OH)_3$ (see Appendix 1). Two concentrations of DOC (100 and 1000 μmol C/l) indicative of clear and colored waters, respectively, were used to demonstrate the role of organic solutes in the speciation of aluminum.

For both concentrations of DOC, calculations show that a shift in the speciation of the model organic acid occurs with changes in pH value (see Figure 8). In the neutral pH range, the triprotic organic solute is essentially fully deprotonated. As pH values decrease below 7.0, the organic anion protonates, producing a series acidic species. At low pH values complexation of the organic anion by aluminum is evident, and below pH 4.5 these species represent the dominant organic form. At very low pH values a cationic alumino-organic complex is evident, suggesting that surface-water acidification may enhance the mobility of organic, as well as inorganic, forms of aluminum.

Hypothetical calculations on acid-sensitive waters demonstrate that the speciation of aluminum is highly dependent on pH value, as well as on the total ligand concentration (see Figure 9). In acidic waters, aquo-aluminum is generally the dominant form because the pH-

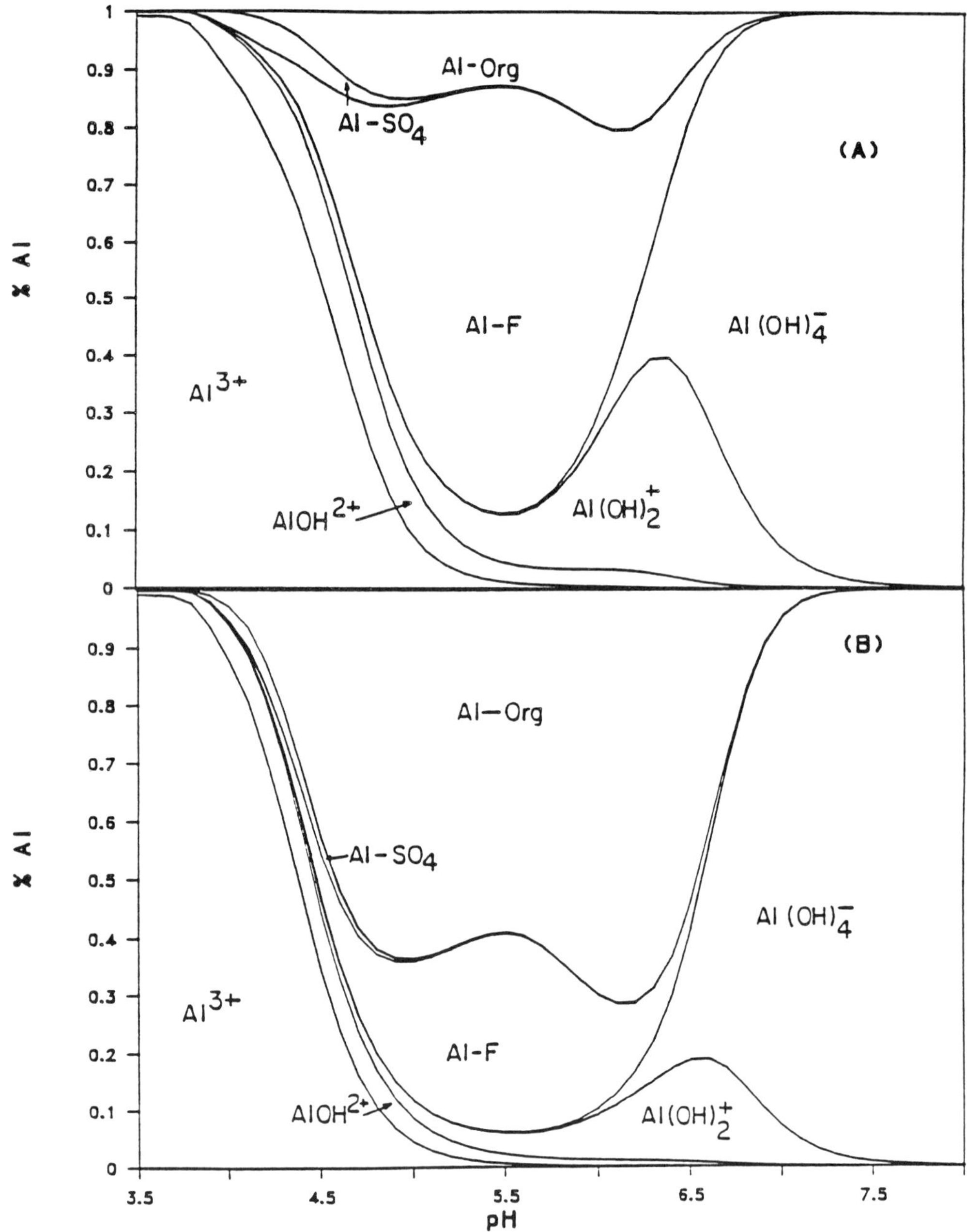

FIGURE 9. Distribution of aqueous aluminum as a function of pH at two concentrations of DOC. The low concentration, 100 μmol C/l (a), is indicative of clear waters, while the higher concentration, 1000 μmol C/l (b), is used to represent colored waters. Calculations were made in equilibrium with $Al(OH)_3$ and with 5 μmol/l fluorine and 100 μmol/l SO_4^{2-}, which are typical of acid-sensitive waters in the Adirondack region of New York.

dependent solubility of $Al(OH)_3$ results in aluminum concentrations that greatly exceed the concentrations of available ligands. At higher pH values, total aluminum concentrations decrease and the relative importance of complexing ligands increases. In low-DOC waters (100 μmol C/l; see Figure 9a) Al-F represents the dominant fraction of aqueous aluminum between pH 5.0 and 6.0. Increased concentrations of DOC (1000 μmol C/l; see Figure 9b) allow additional complexation of aluminum, greatly increasing the relative significance of these species. At neutral pH values, hydrolysis is evident and $Al(OH)_4^-$ is the predominant form of aqueous aluminum.

Models describing the aqueous chemistry of aluminum have been incorporated as subcomponents of detailed models of solute transport.[28,29,42,43] These solute-transport models have been developed to simulate soil and drainage water acidification in forested watersheds. In order to make assessments of surface water quality, an important output of acidification models is the concentration and speciation of aluminum.[83] Acidification models are highly variable in their depiction of aluminum chemistry; ranging from simple $Al(OH)_3$ dissolution and hydrolysis reactions[28,29] to detailed representations of soil and drainage water, including kinetically controlled release of solid-phase aluminum and comprehensive aqueous speciation.[43] Although the representation of aluminum chemistry is currently quite sophisticated in some acidification models, it is important that these models be evaluated continually and modified to reflect ongoing advances in the biogeochemistry of aluminum.

IV. PATTERNS AND PROCESSES OF ALUMINUM IN SURFACE WATERS

A. Chemistry and Transport of Aluminum in Streams

Johnson et al.[19] observed marked longitudinal variations in the chemistry of aluminum in a low-order woodland stream at the HBEF in New Hampshire. They reported that pH values increased and aluminum concentrations decreased, coinciding with increased concentrations of basic cations as stream order (and drainage area) increased. They suggested a two-step process for the neutralization of acidic deposition. Strong acids entering the watershed from atmospheric deposition were converted to a mixture of H^+ and aluminum acidity (H^+-Al-acidity) in the shallow acidic soils draining headwater areas. This acidity was subsequently neutralized by the release of basic cations as waters migrated through a greater drainage area with thicker mineral soil and glacial till. Johnson et al.[19] also observed a shift in aluminum speciation with increasing stream order. Aquo-aluminum and Al-OH concentrations decreased substantially with increasing drainage area, while Al-F remained constant throughout the experimental reach.

Lawrence et al.[20] also studied longitudinal variations in the aluminum chemistry of two upland streams at the HBEF, using the cation-exchange column fractionation procedure. Unlike Johnson et al.,[19] their study sites drained coniferous (spruce and fir) stands at high elevations while a deciduous (beech, birch, and maple) vegetation was evident at lower elevations. Solutions in high-elevation reaches contained a mixture of strong inorganic acids (SO_4^{2-} and NO_3^-) as well as organic acids derived from coniferous vegetation. Stream water within this elevational zone was characterized by low pH values and high concentrations of organic aluminum. With decreasing elevation, vegetation patterns changed from coniferous to deciduous, resulting in a marked shift in stream chemistry. Stream water acidity decreased within high-elevation deciduous reaches coinciding with a decrease in concentrations of organic acids (DOC). Nevertheless, solutions remained acidic because of high concentrations of strong inorganic acids (SO_4^{2-} and NO_3^-) relative to concentrations of basic cations. Within this elevational zone, aluminum concentrations were elevated, but a shift in speciation was evident with inorganic forms predominating. With subsequent increases in drainage area through the deciduous zone of the watershed, patterns of aluminum chemistry were very similar to results reported by Johnson et al.[19] Concentrations of both inorganic and organic forms of aluminum decreased with increasing drainage area as soil depth and stream concentrations of basic cations increased (see also Chapter 10).

Driscoll et al.[84] evaluated longitudinal variations in the chemistry of aluminum in an Adirondack stream, Pancake-Hall Creek. The drainage patterns of this catchment were considerably different from those of the upland streams of the HBEF. Within Pancake-Hall watershed, upland streams drain into a wetland area (formed by beaver activity) at lower elevations. In the upper reaches of Pancake-Hall Creek, stream water was highly acidic because of elevated concentrations of strong acid anions (SO_4^{2-} and NO_3^-) relative to concentrations of basic cations. Concentrations of aluminum were elevated and predominantly in an inorganic form (as determined using the column cation-exchange fractionation procedure). With increasing drainage area, stream acidity was neutralized and aluminum concentrations decreased. Neutralization of stream acidity was accomplished in part by increased release of basic cations from thicker deposits of glacial till at lower elevations in the watershed. Increases in ANC and associated aluminum hydrolysis/retentions were also facilitated by biological reduction processes occurring within the lower-elevation wetlands. In this drainage zone, marked retention of SO_4^{2-} was evident, particularly during low-flow summer conditions. Sulfate retention served to neutralize acidic solutions, restricting leaching losses of aluminum from the watershed. Within the wetland zone, only retention of inorganic aluminum was evident, and no change in concentrations of organic aluminum was observed.

In addition to spatial variations, temporal patterns are also evident in stream-water aluminum chemistry (see also Chapter 10). In Scandinavia[85-87] and eastern North America,[11,62,84,88] episodic acidification and pulsed increases in aluminum concentrations generally coincide with autumn rainfall events and spring snowmelt. Episodic acidification of surface water has been attributed to a variety of processes. For example, Johannessen[85] observed that snowmelt acidification of a stream in Norway coincided with increased SO_4^{2-} concentrations that were derived from snowpack release. Sullivan et al.[86] recently reported episodic aluminum release during both autumn and spring hydrologic events in Norway. These acidification events resulted from a dilution of major inorganic solutes, coupled with increases in TOC concentrations. Moreover, increased concentrations of aluminum were attributed to organic forms (by the cation-exchange column method). It would appear that, at this site, episodic acidification and elevated aluminum inputs to stream water were linked closely to increased release of organic solutes from soil.

In the Adirondack region of New York,[84] attributed snowmelt acidification of stream water to a combination of processes, dilution of basic cation concentrations, and increased nitric acid inputs. In acidic Adirondack waters, concentrations of basic cations and SO_4^{2-} are relatively invariant, so marked changes in NO_3^- concentrations are often accompanied by stoichiometric changes in concentrations of acidic cations (H^+, Al^{3+})[84] (see Figure 10). Although short-term changes in NO_3^- concentrations are often important in the regulation of aluminum leaching, the source of this input is uncertain. Possible mechanisms to explain seasonal variations in NO_3^- concentrations include fluctuations in biological assimilation of nitrogen over the annual cycle,[18,89] snowpack release of NO_3^-[90], or microbial mineralization of soil organic nitrogen during the winter period.[91]

Short-term changes in water chemistry associated with hydrologic events may be due to temporal variations in the flowpaths of water[92] as well as biological processes. During the low-flow summer period, transpiration reduces water movement through the soil and biological uptake decreases solution NO_3^- concentrations.[18] The resulting drainage water contains higher concentrations of basic cations, because of increased contact with the mineral soil, as well as low NO_3^- and DOC concentrations (see Figure 10). These transformations decrease acidity, lowering concentrations of aluminum. During winter, precipitation in northern temperate regions generally accumulates as snowpack. This water is discharged over a short period in the spring. For example, Likens et al.[18] reported that 54% of the annual runoff at the HBEF occurred during the months of March, April, and May, with more than 30% occurring in April alone. Elevated inputs of water to soil cause a shift in hydrologic flowpaths.[92] During hydrologic events, inputs from rain or snowmelt may exceed the ability of the lower mineral soil to transmit water, resulting in a short-circuiting of water through the upper soil horizons prior to entering the stream channel. The chemical composition of this interflow water is similar to that observed in organic horizon and upper mineral-soil

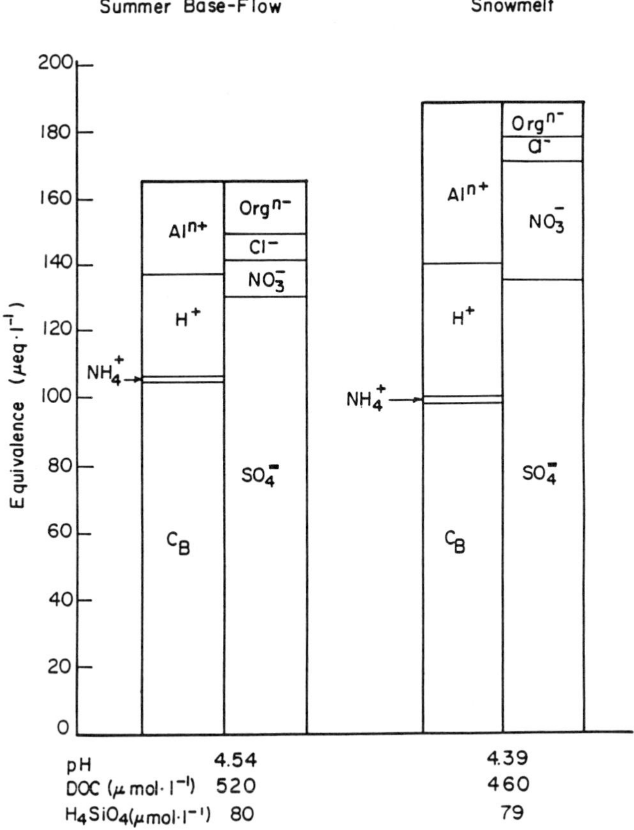

FIGURE 10. The charge distribution of solutes in an Adirondack lake outlet (Merriam Pond) under spring high flow (February, March, and April; eight observations) and during summer baseflow (June, July, August, and September; ten observations) conditions. C_B is the sum of the basic cation equivalence (Ca^{2+}, Mg^{2+}, Na^+, K^+). The equivalence of inorganic aluminum (Al^{n+}) is calculated with ALCHEMI by considering the various inorganic complexes. Free organic anions (Org^-) are calculated as the difference between measured inorganic cations and inorganic anions.

solutions, containing low pH values and basic cation concentrations, as well as elevated NO_3^-, DOC, and aluminum concentrations. During hydrologic events, high discharge of acidic interflow mixes with the ground water, draining from the lower mineral soil in the stream channel. This process greatly dilutes ground-water chemistry, producing stream water with low pH values, ANC, and basic cation concentrations, as well as elevated NO_3^-, DOC, and aluminum concentrations (see Figure 10).

During high runoff periods, stream waters characteristically have high concentrations of both inorganic and organic forms of aluminum.[62,86,93] Moreover, the solutions tend to depart from saturation with respect to mineral-phase solubility, becoming highly undersaturated.[41,84] The source of aluminum inputs to stream water is not clear. Organic horizon leachates typically contain low concentrations of inorganic aluminum.[44,46] Therefore, the elevated inorganic aluminum concentrations may be released from the mineral soil or possibly the stream bed. Norton and Henriksen[27] indicated that degassing of CO_2 from soil solutions emerging to surface waters results in aluminum hydrolysis and deposition of particulate aluminum on stream sediments. Freshly formed aluminum precipitates on stream substrate may be readily remobilized during episodic events.[27,94,95] Hall et al.[95] reported that aluminum

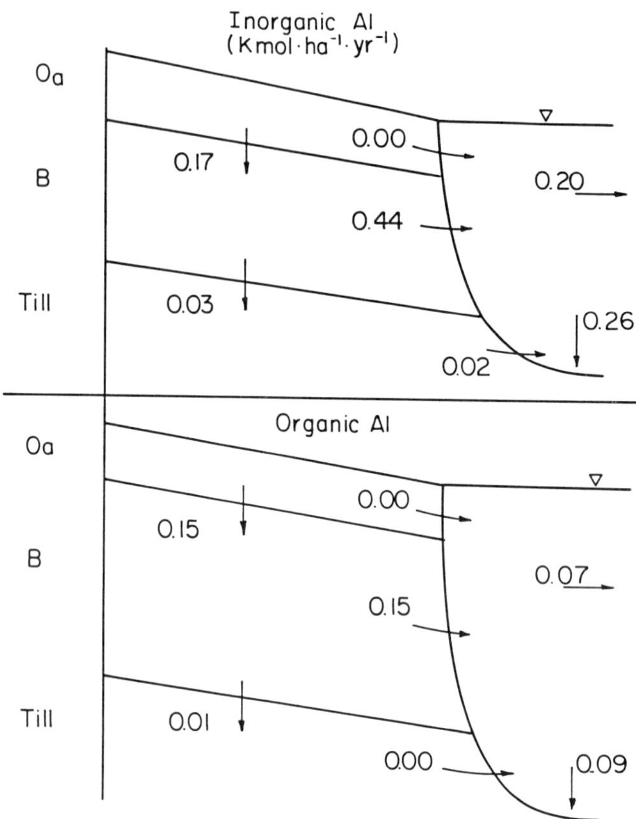

FIGURE 11. Fluxes of inorganic and organic monomeric aluminum within compartments of Pancake-Hall watershed (in kmol/ha-year).[96]

was released directly from stream sediments during experimental acidification of a HBEF stream by H_2SO_4 addition.

In an attempt to evaluate transformations of aluminum along the soil/stream interface, Driscoll et al.[96] calibrated a hydrologic model to the Pancake-Hall watershed in the Adirondack region of New York. Calculations of water flux through soil horizons were coupled with measurements of soil solution, ground water, and stream chemistry to assess the transport of solutes through the catchment. From this analysis inorganic aluminum was mobilized largely from the mineral soil, while organic aluminum was derived predominantly from the organic horizon (see Figure 11). In addition, immobilization of both inorganic and organic forms of aluminum appeared to occur in the near-stream/in-stream zone of the watershed. Retention of organic aluminum coincided with the loss of DOC while removal of inorganic aluminum was coincident with immobilization of NO_3^- along the soil/stream interface. Evidently, losses of NO_3^- by microbial/plant activity resulted in near-stoichiometric hydrolysis/retention of aluminum. The exact location of the zone of aluminum retention could not be determined from the study, although it appeared to occur somewhere along the near-stream/in-stream region. The immobilization of aluminum within this zone has potentially important implications for stream ecosystems (see Section V.B).

B. Chemistry and Transport of Aluminum in Lakes

There have been relatively few studies detailing the transport and cycling of aluminum in lake systems. Driscoll and Schafran[97] evaluated the temporal and spatial patterns of aluminum in Dart's Lake, an acidic Adirondack lake. They observed that aluminum chemistry and acidity associated with H^+ and aluminum (H^+-Al-acidity) were coupled closely with short-term variations in NO_3^- (H^+-Al-acidity $= 2.4 + 0.94\ NO_3^-$; μeq/l, $n = 172$, $r^2 = 0.54$, $p < 0.001$). This empirical correlation has a slope near one, suggesting a stoichiometric response. No significant relationship was evident with SO_4^{2-}, Cl^-, or organic anion concentrations. Short-term fluctuations in basic cations, SO_4^{2-} or Cl^-, are minor, so watershed/lake processes which result in temporal and spatial variations in NO_3^- strongly influence H^+ and aluminum concentrations in acidic lakes.[97-99]

In Dart's Lake, Schafran and Driscoll[98] observed distinct patterns in NO_3^-, pH values, and inorganic aluminum (see Figure 12). During autumn these parameters exhibited an orthograde vertical distribution following turnover. After ice development, NO_3^- and inorganic monomeric aluminum concentrations increased and pH values decreased in the upper waters in response to acidic inputs from the inlet stream. Concentrations of NO_3^- and inorganic monomeric aluminum decreased, while pH values increased with increasing water-column depth. Although the lake was well oxygenated, neutralization of acidity was attributed to sediment reduction of NO_3^- and transport across the sediment/water interface.

During the snowmelt period, the inlet water was acidic and elevated in NO_3^- concentrations. The large influx of water during snowmelt introduced low pH-value water enriched in NO_3^- and aluminum throughout the water column. After ice break-up, the lake quickly stratified. Summer stratification coincided with NO_3^- depletion in both the upper mixed waters, because of algal assimilation and denitrification in epilimnetic sediments, and lower hypolimnion because of sediment reduction. Both pH values and inorganic monomeric aluminum concentrations followed trends consistent with these patterns in NO_3^- (see Figure 12).

In Dart's Lake, a substantial fraction (30%) of monomeric aluminum was associated with organic solutes.[98] Within the water column there were significant spatial and temporal trends in DOC and organic monomeric aluminum (see Figure 13). An orthograde distribution in DOC was observed after autumn turnover. During the ice-cover season, DOC concentrations increased below the ice and decreased with increasing depth. In the hypolimnion, DOC increased slightly through the winter season. During spring snowmelt and prior to the break-up of ice cover, elevated concentrations of DOC were introduced throughout most of the water column. With summer stratification, DOC concentrations were depleted in the epilimnetic waters. This low-DOC water migrated with the thermocline to increasing lake depths during the summer season. In the hypolimnion DOC concentrations were enriched during summer stratification.

Patterns of the temporal and spatial variations in the concentrations of organic monomeric aluminum were generally similar to DOC (see Figure 13b). After autumn turnover and into the ice-cover season, organic monomeric aluminum concentrations decreased in the inlet, outlet, and water column. During much of the ice-cover period, organic monomeric aluminum displayed a slight negative heterograde distribution. Water-column concentrations increased markedly during the early summer. However, in late summer organic monomeric aluminum greatly increased in the hypolimnion. This accumulation peaked in September and then decreased until turnover. No significant increase was observed in the upper waters as the thermocline eroded, and, at turnover, concentrations throughout the water column were not substantially higher than inlet concentrations.

Lakes generally appear to be sinks for inputs of aluminum.[100-102] This trend may be attributed largely to in-lake generation of ANC through NO_3^- and SO_4^{2-} retention,[103] as well as the release of basic cations[104] in sediments. Production of ANC facilitates hydrolysis and deposition of aluminum for those solutions that are near saturation with respect to mineral-phase solubility. From mass-balance calculations, Driscoll[102] concluded that over the annual cycle, stream inputs of aluminum to Dart's Lake exceeded outputs, resulting in a net deposition of aluminum to the sediments (see Table 5). Moreover, distinct seasonal patterns

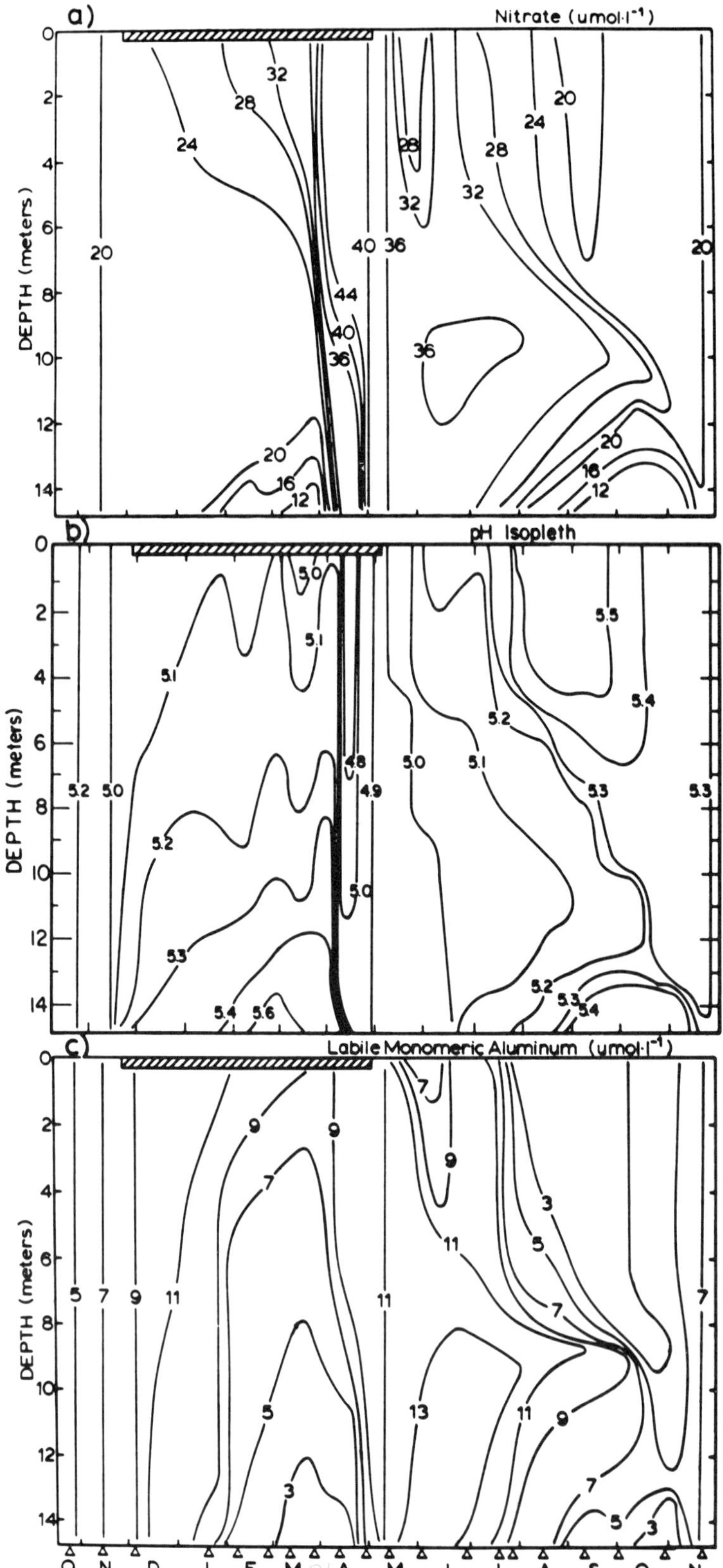

FIGURE 12. Temporal and spatial variations in (a) NO_3^-, (b) pH, and (c) inorganic monomeric aluminum within the water column of Dart's Lake.[98]

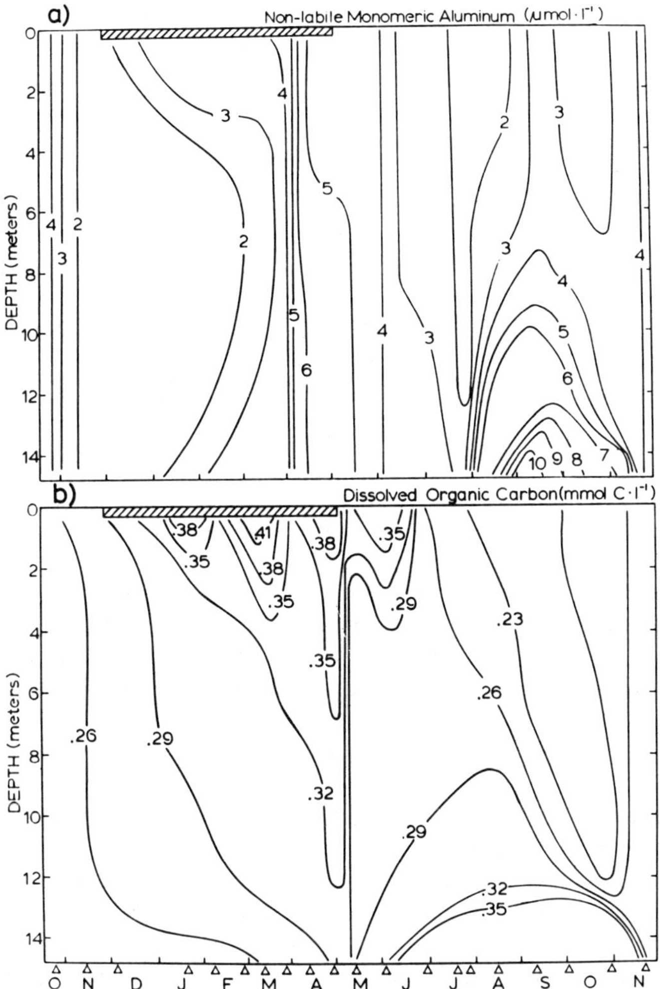

FIGURE 13. Temporal and spatial variations in (a) organic monomeric aluminum and (b) DOC in the water column of Dart's Lake.[98]

Table 5
MASS FLUX OF TOTAL ALUMINUM IN, OUT, AND RETAINED IN THE SEDIMENTS OF DART'S LAKE (IN mmol/m²-d)[99]

Period	Influx	Efflux	Retention
Annual average (October 1981—October 1982)	2.5	2.4	0.3
Spring high flow (March 28—June 28, 1982)	5.2	5.2	−0.1
Summer base flow (July 28—October 29, 1982)	1.2	0.9	0.4

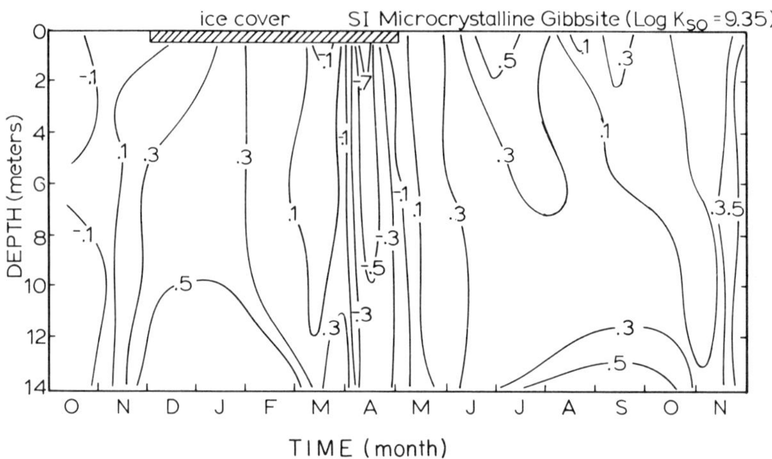

FIGURE 14. Temporal and spatial variations in the saturation index (SI) with respect to the solubility of microcrystalline gibbsite in the water column of Dart's Lake.[98]

were evident in lake-aluminum retention. During the high-flow snowmelt period, essentially all the aluminum entering Dart's Lake at the inlet stream was transported from the outlet. The conservative nature of aluminum within the lake during the snowmelt period was illustrated further by mineral-saturation index calculations (SI; see Figure 14):

$$SI = \log Q_P/K_p$$

where Q_p is the ion-activity product of the solution and K_p is the thermodynamic solubility-product constant of the mineral phase of interest. A SI value greater than zero suggests that the solution is oversaturated with respect to the solubility of the mineral phase of interest. A SI value near zero suggests equilibrium, and a negative value indicates undersaturation.

During the high-flow conditions associated with snowmelt, Dart's Lake solutions were highly undersaturated with respect to the solubility of readily forming mineral phases (e.g., microcrystalline gibbsite; see Figure 14). These calculations suggest that in-lake formation of particulate aluminum was not favored thermodynamically, and probably did not occur to any extent during snowmelt. Conditions of undersaturation may be attributed to relatively slow dissolution kinetics of aluminum minerals and/or minimal contact of solutions with the mineral soil because of the short-circuiting of water from the forest floor directly to the stream.

During the summer stratification period, NO_3^- concentrations in the water column were reduced through a combination of lower inlet concentrations, algal assimilation of NO_3^-, and denitrification (see Figure 12). Retention of NO_3^- by these latter two processes resulted in the production of ANC within the lake and a water column that was oversaturated with respect to the solubility of microcrystalline gibbsite (see Figure 14). Positive SI values suggest in-lake formation of particulate aluminum, and mass-flux calculations confirm that a considerable amount of aluminum was retained in the lake during the summer stratification period (see Table 5).

In addition to reduction processes, transformations of inorganic carbon can affect in-lake cycling of aluminum. In Dart's Lake, elevated concentrations of DIC from respiration reactions, accumulated in the lower waters during both winter and summer stratification.[105] These solutions were highly oversaturated with respect to the solubility of atmospheric CO_2. As discussed previously, the dissociation of carbon acid produces acidity capable of solu-

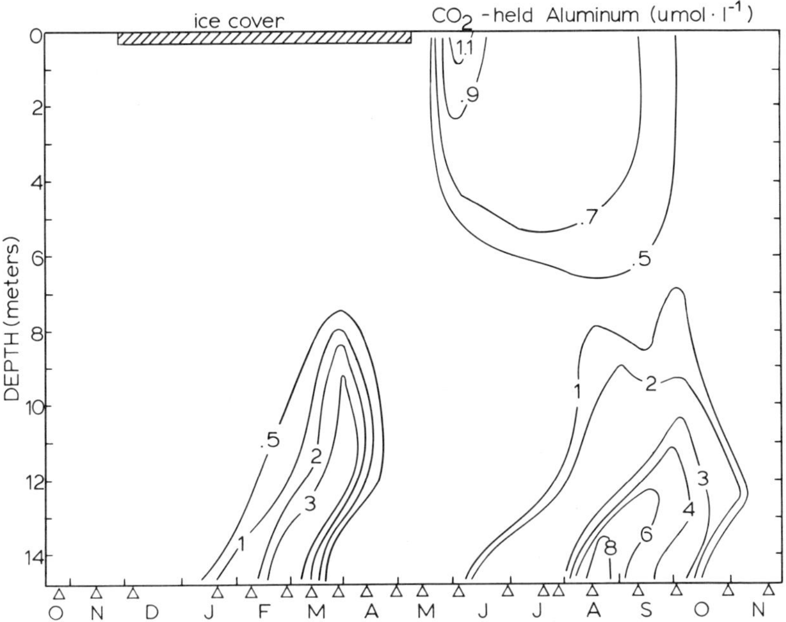

FIGURE 15. Temporal and spatial variations in the concentration of CO_2-mobilized aluminum (in μmol/l) for the water column of Dart's Lake.[105]

bilizing aluminum, as well as HCO_3^-, which can serve as a counter ion for aluminum transport (see Section II.A). As in soil solutions, this solubilization of aluminum is transient.

To quantify the extent to which accumulation of CO_2 facilitated the solubilization of aluminum, "CO_2-mobilized aluminum" was calculated for Dart's Lake (see Figure 15). CO_2-mobilized aluminum represents the concentration of aluminum that would be lost from solution if the system were allowed to equilibrate with the solubility of atmospheric CO_2 and microcrystalline gibbsite (Al[OH]$_3$). Significant concentrations of CO_2-mobilized aluminum were observed in the hypolimnitic waters of Dart's Lake during stratification (see Figure 15). Apparently the seasonal accumulation and loss of CO_2 can be an important mechanism of aluminum solubilization and deposition, respectively, in acidic lakes.

Trends in CO_2-mobilized aluminum, to some extent, run counter to changes in aluminum concentration induced by variations in NO_3^-. Nitrate inputs to the upper waters of Dart's Lake during autumn, winter, and spring coincided with in-lake accumulation of aluminum, while CO_2-mobilized aluminum was significant only in the lower waters during stratification. In the hypolimnitic waters during stratification, NO_3^- reduction facilitated in-lake deposition of aluminum.[97,98] However, production of CO_2 in the lower waters apparently attenuated the hydrolysis and deposition of aluminum to some extent.

Patterns in the depositon of aluminum are not always attributed to variations in inorganic solutes. Nilsson[101] made mass balances of aluminum for Lake Gardsjon in Sweden. He observed that the lake retained inputs of aluminum (11 to 31% retention). However, using the cation-exchange column fractionation procedure, he found little retention of inorganic monomeric aluminum. Rather, in-lake immobilization aluminum was attributed largely to a marked retention of alumino-organic solutes (63 to 81% retention). From this study it is not clear why organic forms of aluminum were retained in Lake Gardsjon, although microbial oxidation or photooxidation of organic matter may have contributed. However, the conservative transport of inorganic monomeric aluminum through the lake system was qualitatively consistent with solution conditions that were highly undersaturated with respect to readily forming mineral-phase solubility.

There have been very few studies of patterns in aluminum chemistry over large drainage basins. Driscoll et al.[93] studied the migration of aluminum along a chain-lake system draining 14,000 ha in the west-central Adirondack region of New York. Headwater lakes draining acidic subcatchments contained elevated concentrations of both inorganic monomeric (6 to 14 μmol/l) and organic monomeric aluminum (5 to 7 μmol/l), as measured using the cation-exchange column fractionation procedure. Inputs of this acidic water to a large lake (Big Moose Lake) resulted in a marked loss of both inorganic and organic forms of aluminum. Immobilization of aluminum coincided with a loss of water-column DOC and associated organic acidity.

With increasing drainage area, the neutralization of this acidic water was accomplished by successive inputs of water which were enriched in ANC and basic cations from adjacent sub-basins. These trends coincided with reduced concentrations of inorganic aluminum in water chemistry through the chain-lake basin, either from hydrolysis and deposition and/or from dilution. At the lowest reaches of the drainage basin, the export of aluminum was low and restricted largely to the high-flow snowmelt period.

V. ECOLOGICAL EFFECTS OF ELEVATED CONCENTRATIONS OF ALUMINUM IN SURFACE WATERS

A. Aluminum as a pH Buffer

Acidic surface waters are characteristically low in DIC because of the limited dissolution of soil minerals and low solubility of CO_2. Because dilute waters are inherently low in DIC, they are limited with respect to their inorganic carbon-buffering capacity. Consequently, noninorganic carbon acid/base reactions, such as the hydrolysis of aluminum and the protonation/deprotonation of natural organic carbon, may be important in the pH buffering of acidic waters.

Several researchers have investigated noninorganic carbon and weak-acid/base buffering of dilute waters. Dickson[56] observed that elevated aluminum concentrations increased the acidity of Swedish lakes. Waters were strongly buffered by aluminum in the pH range 4.5 to 5.5. The acidity of aluminum was particularly evident when acidic lakes were treated with base (i.e., limed). The aluminum component of total acidity was comparable in magnitude to the acidity attributed to H^+ and inorganic carbon; therefore, the presence of aluminum substantially increased base dose requirements and the cost associated with the treatment of acidic lakes.

Johannessen[106] investigated noninorganic carbon buffering in Norwegian waters. While reiterating the importance of aluminum as a buffer in dilute acidic waters, she also evaluated the role of natural organic acids. Natural organic solutes reduced the degree of aluminum hydrolysis in the pH range 5.0 to 5.5, presumably from complexation reactions, attenuating the pH buffering of aluminum. Natural organic acids also participated in proton-donor/acceptor reactions. Johannessen[106] concluded that organic carbon was the most important weak-acid/base system in acidic Norwegian waters because of the high organic carbon concentrations relative to aluminum.

Glover and Webb[107] evaluated the acid/base chemistry of surface waters in the Tovdal region of southern Norway. The acidity attributed to strong Brønsted acids was small compared to the acidity of weak-acid systems. These investigators suggested that of the total weak-acid acidity, 40 to 60 μeq/l could be attributed to dissolved aluminum and silicon, while 20 to 50 μeq/l could be attributed to natural organic acids.

In a comparable study, Henriksen and Seip[108] evaluated the strong- and weak-acid concentrations of surface waters in southern Norway and southwestern Scotland. In addition to a titrametric analysis, the aluminum, silicon, and TOC concentrations of water samples were determined. Weak-acid concentrations, determined by Gran plot analysis, were evaluated

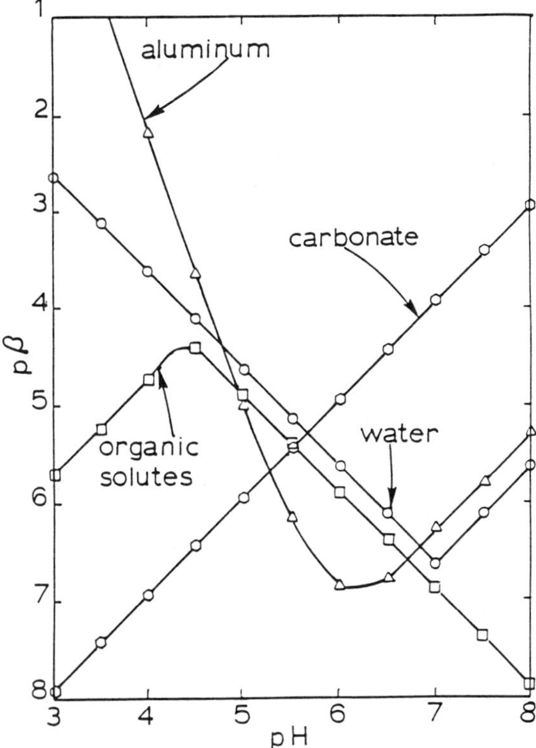

FIGURE 16. Buffer-intensity diagram for Adirondack surface-water systems. Equilibrium with $Al(OH)_3$ (p*Kso = 8.49), organic solutes (C_T 2 × 10^{-5} mols/l; pK_a = 4.4), and atmospheric CO_2 (PCO_2 = $10^{-3.5}$ atm) was assumed.[109]

by multiple-regression analysis. Most of the variance in the weak-acid concentration could be explained by the aluminum and TOC concentrations of the waters. Thus, it was concluded that the weak-acid content of acidic lakes in southern Norway and Scotland was largely a mixture of aluminum and natural organic acids.

Driscoll and Bisogni[109] quantitatively evaluated the weak-acid/base systems buffering dilute acidic waters in the Adirondack region of New York. Natural organic acids were fit to a monoprotic proton-dissociation constant model (pKa = 4.4), and proton-dissociation sites were determined from titration data. Aquo-aluminum activity, calculated from field observations, appeared to fit an $Al(OH)_3$ solubility model (see Figure 5). The calculated buffering capacity (β) is plotted as a function of pH in Figure 16 for a hypothetical system with properties characteristic of Adirondack waters.[109] *Buffering capacity* is defined as the quantity of strong acid or base (in mol/l) which would be required to change the pH value of a liter of solution by one log unit. Aluminum species may dominate the buffer system at low pH values if these conditions are fulfilled, suggesting that the lower limit of pH values observed in acidic waters with elevated aluminum concentrations may be controlled by the dissolution and hydrolysis of aluminum. Buffering capacity is minimal in the pH range 5 to 6, whereas at circumneutral pH values, buffering is regulated by the dissociation of inorganic carbon.

B. Interaction of Aluminum with Element Cycles

At certain times of the year and in certain regions within a stream or lake, acidic solutions may become oversaturated with respect to the solubility of readily forming aluminum minerals. As discussed previously, the processes generally responsible for this condition are retention of NO_3^- or organic anions, and degassing of CO_2. During conditions of oversaturation, aluminum will hydrolyze, forming particulate aluminum oxyhydroxide. These materials would include the microcrystalline hydroxide particles and their polymeric hydroxycation precursors. Smith and Hem[55] observed that during the polymerization process aluminum hydroxide units displayed metastable ionic solute behavior until they contained from 100 to 400 aluminum atoms. When particles developed to that size their behavior was characteristic of a suspended colloid. Microcrystalline precipitates may readily sorb both organic and inorganic solutes. The cycling of orthophosphate,[56,110] DOC,[56,111,112] and trace metals[113,114] within acidic surface waters may be altered by adsorption on aluminum oxyhydroxides. However, few studies have addressed this specific hypothesis.

Huang[110] studied the adsorption of orthophosphate on particulate aluminum oxide. He observed an adsorption maximum at pH 4.5. While Huang evaluated high concentrations of orthophosphate (100 to 1000 μmol/l), his observations of orthophosphate-aluminum interactions may be generally applicable to the lower phosphorus concentrations observed in acidic natural waters. Dickson[56] observed that when acidic lake water, elevated in aluminum, was amended with orthophosphate (1.6 and 3.2 μmol/l), dissolved phosphorus was removed from solution. This removal of phosphorus was most pronounced at pH 5.5. Dickson suggested that aqueous aluminum may substantially alter phosphorus cycling within acidic surface waters through adsorption or precipitation reactions. This hypothesis is noteworthy because phosphorus is often the nutrient that limits plant growth in fresh waters.[115] A decrease in aqueous phosphorus, induced by adsorption on aluminum oxyhydroxides, may decrease the net production of aquatic plants and diminish biologically generated ANC. Any reduction in ANC inputs would result in an aquatic ecosystem more susceptible to further acidification.

Aluminum forms strong complexes with natural organic matter,[69] and complexation alters substantially the character of natural organic acids (see Chapter 5). Driscoll et al.[116] observed that DOC was removed from the water column of an acidic lake following $CaCO_3$ addition. They hypothesized that DOC sorbed to the hydrolyzed aluminum that had formed within the water column shortly after base addition. Schafran and Driscoll[98] and Effler et al.[117] observed decreases in water-column DOC (see Figure 13) during conditions of oversaturation with respect to the solubility of $Al(OH)_3$ in an acidic lake (see Figure 14). They hypothesized that natural organic carbon was coagulated by particulate aluminum formed in the water column. Davis[111] has studied the adsorption of natural dissolved organic matter at the aluminum oxide/water interface. He observed that natural organic matter adsorbs by complex formation between the surface hydroxyl groups of aluminum oxide and acidic functional groups of organic matter. Davis[111] indicated that DOC adsorption was maximum near pH 5. Davis and Gloor[112] reported that DOC associated with relative molecular mass fractions greater than 1000 formed strong complexes with the aluminum oxide surface, but low relative molecular mass fractions were weakly adsorbed. Davis[111] suggested that under conditions typical for natural waters, almost complete surface coverage by adsorbed organic matter can be anticipated for aluminum oxide particles. Organic coatings may affect the extent of trace metal and anion adsorption on metal oxyhydroxide surfaces (see Chapter 7).

Hall et al.[12] observed a decrease in the DOC concentrations of a third-order stream in New Hampshire after manipulation by $AlCl_3$ addition. Moreover, a reduction in surface tension occurred at the air-water interface and was attributed to a decrease in the solubility of DOC due to interactions with aluminum.

The loss of DOC in acidic waters is significant in several respects. DOC represents a weak-acid/base system that serves to buffer solutions against changes in pH value[81,106,108] (see Figure 16). Moreover, in oligotrophic lakes, DOC may be primarily responsible for the attenuation of light and patterns of water-column heating. Effler et al.[117] reported that temporal variations in the DOC of an acidic Adirondack lake accounted for much of the change in attenuation of light ($K_d = -1.2 + 0.00063$ DOC; where K_d is the light-attenuation coefficient in m^{-1} and DOC is in μmol C/l, $r^2 = 0.86$, and $p < 0.01$). These authors also observed that variations in DOC within the photic zone were correlated with aluminum concentrations (aluminum $= -15.8 + 0.094$ DOC; where aluminum and DOC are in μmol/l, $n = 29$, $r^2 = 0.45$, and $p < 0.0001$). They suggested that in-lake hydrolysis and precipitation of aluminum facilitated the removal of DOC from the photic zone (by sorption, co-precipitation, or coagulation processes) and increased the clarity of the lake through the summer season. Decreases in the attenuation of light may allow heating of the lower waters, thereby decreasing the thermal stability of lakes. Changes in the heating and the thermal stratification of lakes may have profound effects on these ecosystems by altering vertical transport of solutes and limiting cold-water fisheries.

White and Driscoll[114] reported that the vertical deposition of lead was strongly correlated with aluminum deposition in an acidic Adirondack lake. These researchers hypothesized that in-lake hydrolysis and precipitation of aluminum enhanced the vertical deposition of lead through sorption reactions followed by sedimentation of particulate matter. This process would appear to have implications for trace-metal cycling as well as the interpretation of patterns of lead deposition to the sediments of acidic lakes.

VI. CONCLUSIONS

Aluminum may be mobilized in acidic soils by weak- or strong-acid leaching. Organic forms of aluminum appear to originate in the organic horizon of soil, while inorganic aluminum is derived largely from the mineral soil. Only leaching by strong acids results in the transport of ecologically significant concentrations of inorganic aluminum to surface waters. Within the aqueous environment, aluminum exhibits a variety of forms that vary markedly in their chemical and biological activity. Solution aluminum may be associated with particulate matter, or complexed with inorganic (H_2O, OH^-, F^-, or SO_4^{2-}) or organic ligands. The speciation of aluminum in natural water solutions is a function of many factors, including solution contact with mineral phases, pH value, concentrations of complexing ligands, temperature, and ionic strength.

Aluminum is mobilized largely to acidic surface water from shallow, acidic soil in upland regions. As waters migrate over larger drainage areas, stream acidity is neutralized by the release of basic cations from thicker deposits of soil and glacial till and/or retention of strong-acid anions (e.g., SO_4^{2-} or NO_3^-) from biological reduction processes. Both mechanisms result in the immobilization of aluminum. Under high-flow conditions, surface waters are acidic and aluminum concentrations are elevated. Conversely, under low-flow conditions, pH values are higher and aluminum concentrations are lower. Stream sediments may release aluminum directly to the water column under high-flow conditions in waters that are undersaturated with respect to mineral-phase solubility, while serving as zones of aluminum deposition under low-flow periods. Lakes are generally net sinks of aluminum, because of in-lake processes that generate ANC and facilitate the hydrolysis of aluminum.

Elevated concentrations of aluminum in surface waters may influence water quality through pH buffering, interactions with critical element cycles (e.g., phosphorus, organic carbon, and trace metals), and toxicity to aquatic organisms. An understanding of the pathways linking the aluminum cycle, as well as the speciation of aluminum, is critical to assess the effects of elevated concentrations in surface waters.

ACKNOWLEDGMENTS

We would like to thank the following journals and publishers for the use of figures for this chapter: Marcel Dekker, Butterworths, Martinus Nijhoff/Dr. W. Junk, *Environmental Science and Technology,* and *Environmental Health Perspectives.*

Appendix 1
AQUEOUS THERMOCHEMICAL DATA

Reaction	log K[a]	Ref.	ΔH_r[b]	Ref.
$Al^{3+} + H_2O = Al(OH)^{2+} + H^+$	−4.99	118	49.790	118
$Al^{3+} + 2H_2O = Al(OH)_2^+ + 2H^+$	−10.10	118		
$Al^{3+} + 4H_2O = Al(OH)_4^- + 4H^+$	−23.0	118	184.35	118
$Al^{3+} + F^- = AlF^{2+}$	7.02	119	4.602	120
$Al^{3+} + 2F^- = AlF_2^+$	12.76	119	8.368	120
$Al^{3+} + 3F^- + AlF_3$	17.03	119	10.46	120
$Al^{3+} + 4F^- = ALF_4^-$	19.73	119	9.20	120
$Al^{3+} + 5F^- = AlF_5^{2-}$	20.92	119	7.53	120
$Al^{3+} + SO_4^{2-} = AlSO_4^{2-}$	3.020	118	9.00	118
$Al^{3+} + 2SO_4^{2-} = Al(SO_4)_2^-$	4.92	118	11.88	118
$H_2CO_3^* = H^+ + HCO_3^-$	−6.35	118	9.401	118
$HCO_3^- = H^+ + CO_3^{2-}$	−10.33	118	−15.133	118
$CO_2(g) = CO_2(aq)$	−1.452	118	−20.92	118
$H^+ + F^- = HF$	3.169	118	14.48	118
$H_3Org = H^+ + H_2Org^-$	−1.76	82[c]		
$H_2Org^- = H^+ + H\ Org^{2-}$	−5.90	82		
$H\ Org^{2-} = H^+ + Org^{3-}$	−6.83	82		
$Al^{3+} + Org^{3-} = Al\ Org$	8.39	82		
$Al^{3+} + H^+ + Org^{3-} = AlH\ Org^+$	13.09	82		
Solubility:				
$Al(OH)_3 + 3H^+ = Al^{3+} + 3H_2O$				
Natural gibbsite	8.77	121		
Microcrystalline gibbsite	9.35	119	−95.40	119

[a] At 25°C.
[b] In kJ/mol.
[c] Org is a naturally occurring organic anion; ALCHEMI uses 0.43 mol Org/mol DOC.

Appendix 2
SYMBOLS USED IN CHAPTER

Parameter	Symbol
Aluminum hydroxide complexes	$Al\text{-}OH = \{Al(OH)^{2+}\} + \{Al(OH)^+\} + \{Al(OH)_4^-\}$
Aluminum fluoride complexes	$Al\text{-}F = \{AlF^{2+}\} + \{AlF_2^+\} + \{AlF_3\} + \{AlF_4^-\} + \{AlF_5^{2-}\} + AlF_6^{3-}\}$
Aluminum sulfate complexes	$Al\text{-}SO_4^{2-} = \{AlSO_4^+\} + \{Al(SO_4)_2^-\}$
Aluminum equivalence	$nAl^{n+} = 3\{Al^{3+}\} + 2\{(Al\text{-}OH)^{2+}\} + \{Al(OH)^+\} + 2\{AlF^{2+}\} + \{AlF_2^+\} + \{AlSO_4^+\} - \{Al(OH)_4^-\} - \{AlF_4^-\} - 2\{AlF_5^{2-}\} - 3\{AlF_6^{3-}\} - \{Al(SO_4)_2^-\}$
Aluminum acidity	$Al\text{-acidity} = 3\{Al^{3+}\} + 3\{Al\text{-}F\} + 3\{Al\text{-}SO_4\} + 2\{Al(OH)^{2+}\} + \{Al(OH)_2^+\}$
Hydrogen ion-aluminum acidity	$H^+\text{-}Al\text{-acidity} = \{H^+\} + Al\text{-acidity}$
Aluminum organic complexes	$Al\text{-}Org = \{Al\ Org\} + \{Al\ H\ Org^+\}$
Organic anions	$nOrg^{n-} = \{H_2Org^-\} + 2\{HOrg^{2-}\} + 3\{Org^{3-}\}$
Sum of basic cations	$C_B = 2\{Ca^{2+}\} + 2\{Mg^{2+}\} + \{Na^+\} + \{K^+\}$

REFERENCES

1. **Garrels, R. M., MacKenzie, F. T., and Hunt, C.,** *Chemical Cycles and the Global Environment,* William Kaufmann, Inc., Los Altos, CA, 1975, 59.
2. **Hem, J. D.,** Geochemistry and aqueous chemistry of aluminum, in *Kidney International,* Coburn, J. W. and Alfrey, A. C., Eds., Springer-Verlag, New York, 1986, 29, S3.
3. **Johnson, N. M.,** Acid rain neutralization by geologic materials, in *Geological Aspects of Acid Deposition,* Bricker, O. P., Ed., Butterworths, Boston, 1984, 37.
4. **Bohn, H. L., McNeal, B. L., and O'Connor, G. A.,** *Soil Chemistry,* 2nd ed., Wiley-Interscience, New York, 1985.
5. **Messenger, A. S.,** Climate, time and organisms in relation to podzol development in Michigan sands. II. Relationships between chemical element concentrations in mature tree foliage and upper humic horizons, *Soil Sci. Soc. Am. J.,* 39, 698, 1975.
6. **Messenger, A. S., Kline, J. R., and Wildrotter, D.,** Aluminum biocycling as a factor in soil change, *Plant Soil,* 49, 703, 1978.
7. **Schnitzer, M. and Skinner, S. I. M.,** Organic-metallic interactions in soils. I. Reactions between a number of metal ions and the organic matter of a podzol Bh horizon, *Soil Sci.,* 96, 86, 1963.
8. **Schnitzer, M. and Skinner, S. I. M.,** Organic-metallic interactions in soils. II. Reactions between forms of iron and aluminum and organic matter of a podzol Bh horizon, *Soil Sci.,* 96, 181, 1963.
9. **Bloom, P. R.,** The kinetics of gibbsite dissolution in nitric acid, *Soil Sci. Soc. Am. J.,* 47, 164, 1983.
10. **Stumm, W. and Morgan, J. J.,** *Aquatic Chemistry,* Wiley-Interscience, New York, 1970.
11. **Driscoll, C. T., Baker, J. P., Bisogni, J. J., and Schofield, C. L.,** Effect of aluminum speciation on fish in dilute acidified waters, *Nature,* 284, 161, 1980.
12. **Hall, R. J., Driscoll, C. T., Likens, G. E., and Pratt, J. M.,** Physical, chemical and biological consequences of episodic aluminum additions to a stream, *Limnol. Oceanogr.,* 30, 212, 1985.
13. **Helliwell, S., Batley, G. E., Florence, T. M., and Lumsden, B. G.,** Speciation and toxicity of aluminum in a model fresh water, *Environ. Technol. Lett.,* 4, 141, 1983.
14. **Pavan, M. A., Bingham, F. T., and Pratt, P. F.,** Toxicity of aluminum to coffee in Ultisols and Oxisols amended with $CaCO_3$, $MgCO_3$ and $CaSO_4{\cdot}2H_2O$, *Soil Sci. Soc. Am. J.,* 46, 1201, 1982.
15. **Wood, J. M.,** Effects of acidification on the mobility of metals and metalloids: an overview, *Environ. Health Perspect.,* 63, 115, 1985.
16. **Foy, C. D.,** Effects of aluminum on plant growth, in *The Plant Root and its Environment,* Carson, E. W., Ed., University Press of Virginia, Charlottesville, 1974, 601.
17. **Schecher, W. D. and Driscoll, C. T.,** An evaluation of uncertainty associated with Al equilibrium calculations, *Water Resour. Res.,* 23, 525, 1987.
18. **Likens, G. E., Bormann, F. H., Pierce, R. S., Eaton, J. S., and Johnson, N. M.,** *Biogeochemistry of a Forested Ecosystem,* Springer-Verlag, New York, 1977, 60.
19. **Johnson, N. M. Driscoll, C. T., Eaton, J. S., Likens, G. E., and McDowell, W. H.,** Acid rain, dissolved aluminum and chemical weathering at the Hubbard Brook Experimental Forest, New Hampshire, *Geochim. Cosmochim. Acta,* 45, 1421, 1981.
20. **Lawrence, G. B., Fuller, R. D., and Driscoll, C. T.,** Spatial relationships of aluminum chemistry in streams of the Hubbard Brook Experimental Forest, New Hampshire, *Biogeochemistry,* 2, 115, 1986.
21. **Bloomfield, C.,** The possible significance of polyphenols in soil formation, *J. Sci. Food Agric.,* 8, 389, 1957.
22. **Coulson, C. B., Davies, R. I., and Lewis, D. A.,** Polyphenols in leaves, litter and superficial humus from mull sites, *J. Soil Sci.,* 11, 20, 1960.
23. **Coulson, C. B., Davies, R. I., and Lewis, D. A.,** Polyphenols in plant, humus and soil. II. Reduction and transport by polyphenols of iron in model soil, *J. Soil Sci.,* 11, 30, 1960.
24. **Ugolini, F. C., Minden, R., Dawson, H., and Zachara, J.,** An example of soil processes in the Abies Amabilis zone of Central Cascades, Washington, *Soil Sci.,* 124, 291, 1977.
25. **Johnson, A. H. and Siccama, T. G.,** Effect of vegetation on the morphology of Windsor soils, Litchfield, Connecticut, *Soil Sci. Soc. Am. J.,* 43, 1199, 1979.
26. **DeConnick, F.,** Major mechanisms in formation of spodic horizons, *Geoderma,* 24, 101, 1980.
27. **Norton, S. A. and Henriksen, A.,** The importance of CO_2 in evaluation of effects of acidic deposition, *Vatten,* 4, 346, 1983.
28. **Reuss, J. O. and Johnson, D. W.,** Effect of soil processes on the acidification of water by acid deposition, *J. Environ. Qual.,* 14, 26, 1985.
29. **Reuss, J. O. and Johnson, D. W.,** *Acid Deposition and the Acidification of Soils and Waters,* Springer-Verlag, New York, 1986, 61.
30. **Bowen, H. J. M.,** *Trace Elements in Biochemistry,* Academic Press, New York, 1966.
31. **Cronan, C. S. and Schofield, C. L.,** Aluminum leaching response to acid precipitation: effects on high elevation watersheds in the northeast, *Science,* 204, 304, 1979.

32. **van Breemen, N., Burrough, P. A., Velthorst, E. J., van Dobben, H. F., de Wit, T., Ridder, T. B., and Reijnders, H. F. R.,** Soil acidification from atmospheric ammonium sulfate in forest canopy throughfall, *Nature,* 299, 548, 1982.
33. **Baker, J. P.,** Effects on fish of metals associated with acidification, in *Acid Rain/Fisheries,* Johnson, R. E., Ed., American Fisheries Society, Bethesda, MD, 1982, 165.
34. **Driscoll, C. T. and Newton, R. M.,** Chemical characteristics of acid-sensitive lakes in the Adirondack region of New York, *Environ. Sci. Technol.,* 19, 1018, 1985.
35. **Hem, J. D., Roberson, C. E., Lind, C. J., and Polzer, W. L.,** Chemical interactions of aluminum with aqueous silica at 25°C, *U.S. Geol. Surv. Water-Supply Pap.,* 1827-E, 1973.
36. **Norton, S. A.,** Changes in chemical processes in soils caused by acid precipitation, in Proc. 1st Int. Symp. on Acid Precipitation, Dochinger, L. S. and Seliga, T. A., Eds., U.S. Department of Agriculture, Forest Service General Tech. Rep. NE-2, 1976, 711.
37. **Driscoll, C. T. and Bisgoni, J. J.,** Weak acid/base systems in dilute acidified lakes and streams in the Adirondack region of New York State, in *Modeling of Total Acid Precipitation Impacts,* Schnoor, J. L., Ed., Butterworths, Boston, 1984, 53.
38. **Meiwes, K. J., Khanna, P. K., and Ulrich, B.,** Retention of sulfate by an acid brown earth and its relationship with atmospheric input of sulphur to forest vegetation, *Z. Pflanzenernaehr. Bodenkd.,* 143, 402, 1984.
39. **Nilsson, S. I. and Bergkvist, B.,** Aluminum chemistry and acidification processes in a shallow podzol on the Swedish west coast, *Water Air Soil Pollut.,* 20, 311, 1983.
40. **Bloom, P. R., McBride, M. C., and Weaver, R. M.,** Aluminum organic matter in acid soils: buffering and solution aluminum activity, *Soil Sci. Soc. Am. J.,* 43, 488, 1979.
41. **Cronan, C. S., Walker, W. J., and Bloom, P. R.,** Predicting aqueous aluminum concentrations in natural waters, *Nature,* 324, 140, 1986.
42. **Cosby, B. J., Hornberger, G. M., and Galloway, J. N.,** Modeling the effects of acid deposition: assessment of a lumped parameter model of soil water and streamwater chemistry, *Water Resour. Res.,* 21, 51, 1985.
43. **Gherini, S. A., Mok, L., Hudson, R. J. M., Davis, G. F., Chen, C. W., and Goldstein, R. A.,** The ILWAS model: formulation and application, *Water Air Soil Pollut.,* 26, 425, 1985.
44. **David, M. B. and Driscoll, C. T.,** Aluminum speciation and equilibria in soil solutions of a haplorthod in the Adirondack Mountains, New York, *Geoderma,* 33, 297, 1984.
45. **Erickson, E.,** Aluminum in groundwater possible solution equilibria, *Nord. Hydrol.,* 12, 43, 1981.
46. **Driscoll, C. T., van Breemen, N., and Mulder, J.,** Aluminum chemistry in a forested Spodosol, *Soil Sci. Soc. Am. J.,* 49, 437, 1985.
47. **Bloom, P. R. and Erich, M. S.,** The quantitation of aqueous aluminum, in *The Environmental Chemistry of Aluminum,* Sposito, G., Ed., CRC Press, Boca Raton, FL, 1989, chap. 1.
48. **Hodges, S. C.,** Aluminum speciation: a comparison of five methods, *Soil Sci. Soc. Am. J.,* 51, 57, 1987.
49. **Kennedy, V. C., Zellweger, G. W., and Jones, B. F.,** Filter pore-size effects on the analysis of Al, Fe, Mn and Ti in water, *Water Resour. Res.,* 10, 785, 1974.
50. **Turner, R. C.,** Three forms of aluminum in aqueous systems determined by 8-hydroxyquinoline extraction methods, *Can. J. Chem.,* 47, 2521, 1969.
51. **Turner, R. C.,** Kinetics of reaction of 8-quinolinol and acetate with hydroxyaluminum species in aqueous solutions. II. Initial solid phases, *Can. J. Chem.,* 49, 1688, 1971.
52. **Barnes, R. B.,** The determination of specific forms of aluminum in natural water, *Chem. Geol.,* 15, 177, 1975.
53. **Dougan, W. K. and Wilson, A. L.,** The absorptiometric determination of aluminum in water: a comparison of some chromogenic reagents and the development of an improved method, *Analyst,* 99, 413, 1974.
54. **James, R. B., Clark, C. J., and Riha, S. J.,** An 8-hydroxyquinoline method for labile and total aluminum in soil extracts, *Soil Sci. Soc. Am. J.,* 47, 893, 1983.
55. **Smith, R. W. and Hem, J. D.,** Effect of aging on aluminum hydroxide complexes in dilute aqueous solutions, *U.S. Geol. Surv. Water-Supply Pap.,* 1827-D, 1972.
56. **Dickson, W.,** Some effects of the acidification of Swedish lakes, *Verh. Int. Ver. Limnol.,* 20, 851, 1978.
57. **Wright, R. F., Dale,T., Henriksen, A., Hendrey, G. R., Gjessing, E. T., Johannessen, M., Lysholm, C., and Storen, E.,** Regional Surveys of Small Norwegian Lakes, SNSF Project IR33/77, Oslo, 1977.
58. **Vangenechten, J. H. D. and Vanderborght, O. L. J.,** Acidification of Belgian moorland pools by acid sulphur-rich rainwater, in *Ecological Impact of Acid Precipitation,* Drablos, D. and Tollan, A., Eds., SNSF Project, Oslo, 1980, 246.
59. **Schoen, R.,** Water acidification in the Federal Republic of Germany proved by simple chemical models, *Water Air Soil Pollut.,* 31, 187, 1986.
60. **Kramer, J. R.,** Aluminum: Chemistry, Analysis and Biology, Environmental Geochemistry Report 1981/82, Department of Geology, McMaster University, Hamilton, Ontario, 1981.

61. **Scheider, W. A., Adamski, J., and Paylor, M.,** Reclamation of Acidified Lakes Near Sudbury Ontario by Neutralization and Fertilization, Ontario Ministry of The Environment, Rexdale, Ontario.
62. **LaZerte, B. D.,** Forms of aqueous aluminum in acidified catchments of Central Ontario: a methodological analysis, *Can. J. Fish. Aquat. Sci.,* 41, 766, 1964.
63. **Campbell, P. G. L., Bisson, M., Brougie, R., Tessier A., and Villenuve, J. P.,** Speciation of aluminum in acidic freshwaters, *Anal. Chem.,* 55, 2246, 1983.
64. **Haines, T. A. and Akielaszek, J. J.,** A regional study of the chemistry of headwater lakes and streams in New England: vulnerability to acidification, in Air Pollution and Acid Rain, Rep. No. 15, Fish and Wildlife Service, U.S. Department of the Interior, FWS/OBS-80/40/15, 1983.
65. **Norton, S. A., Davis, R. B., and Brakke, D. F.,** Responses of Northern New England Lakes to Atmospheric Inputs of Acid and Heavy Metals, Completion Report Project A-048-ME, Land and Water Resources Center, University of Maine, 1981.
66. **Linthurst, R. A., Landers, D. H., Eilers, J. M., Brakke, D. F., Overton, W. S., Meier, E.P., and Crowe, R. E.,** Characteristics of Lakes in the Eastern United States, Vol. 1, Population Descriptions and Physico-Chemical Relationships, EPA/600/4-86/007a, U.S. Environmental Protection Agency, Washington, D.C., 1986.
67. **Rosen, K.,** Supply, loss and distribution of nutrients in three coniferous forest watersheds in central Sweden, Reports in Forest Ecology and Forest Soils 41, Department of Forest Soils, Swedish University of Agricultural Sciences, Uppsala, 1982.
68. **Roberson, C. E. and Hem, J. D.,** Solubility of aluminum in the presence of hydroxide, fluoride, and sulphate, *U.S. Geol. Surv. Water-Supply Pap.,* 1827-C, 1967.
69. **Lind, C. J. and Hem, J. D.,** Effects of organic solutes on chemical reactions of aluminum, *U.S. Geol. Surv. Water-Supply Pap.,* 1827-G, 1975.
70. **Luciak, G. M. and Huang, P. M.,** Effect of monosilicic acid on hydrolytic reactions of aluminum, *Soil Sci. Soc. Am. J.,* 38, 235, 1974.
71. **Farmer, V. C., Russell, J. D., and Berrow, M. L.,** Imogolite and proto-imogolite allophane in spodic horizons: evidence for a mobile aluminum silicate complex in podzol formation, *J. Soil Sci.,* 31, 673, 1980.
72. **Farmer, V. C. and Fraser, A. R.,** Chemical and colloidal stability of soils in the Al_2O_3-Fe_2O_3-SiO_2-H_2O system: their role in podzolization, *J. Soil Sci.,* 33, 737, 1982.
73. **Lakatos, B. J., Meisel, J., and Mady, J.,** Biopolymer-metal complex systems. I. Experiments for the preparation of high purity peat humic substances and their metal complexes, *Acta Agron. Acad. Sci. Hung.,* 26, 259, 1977.
74. **Sipos, S., Sipos, E., Dekany, I., Deer, A., Meisel, J., and Lakatos, B.,** Biopolymer-metal complex systems. II. Physical properties of humic substances and their metal complexes, *Acta Agron. Acad. Sci. Hung.,* 27, 31, 1978.
75. **Gosh, K. and Schnitzer, M.,** Macromolecular structure of humic substances, *Soil Sci.,* 129, 126, 1981.
76. **Ritchie, G. S. P. and Posner, A. M.,** The effect of pH and metal binding on the transport properties of humic acids, *J. Soil Sci.,* 33, 233, 1982.
77. **Driscoll, C. T.,** A procedure for the fractionation of aqueous Al in dilute acidic waters, *Int. J. Environ. Anal. Chem.,* 16, 267, 1984.
78. **Browne, B. A., McColl, J. C., and Driscoll, C. T.,** Speciation of aqueous aluminum in dilute acidic waters using morin. I. The chemistry of morin and its complexes with aluminum, *J. Environ. Qual.,* in press.
79. **Browne, B. A., Driscoll, C. T., and McColl, J. C.,** Speciation of aqueous aluminum in dilute acidic waters using morin. II. Principles and procedures, *J. Environ. Qual.,* in press.
80. **Lee, Y. H.,** Aluminum speciation in different water types, in *Lake Gardsjon: An Acid Forest Lake and its Catchment,* Andersson, F. and Olsson, B., Eds., Ecological Bulletin No. 37, Stockholm, 1985, 109.
81. **Bakes, C. A. and Tipping, E.,** Aluminum complexation by an aquatic humic fraction under acidic conditions, *Water Res.,* 21, 211, 1987.
82. **Schecher, W. D.,** Chemical Equilibrium Calculations and Uncertainty Within Drainage Water Acidification Models, Ph.D. dissertation, Syracuse University, Syracuse, New York, 1988.
83. **Schecher, W. D. and Driscoll, C. T.,** An evaluation of equilibrium calculations within acidification models: the effect of uncertainty in measured chemical components, *Water Resour. Res.,* 24, 533, 1988.
84. **Driscoll, C. T., Wyskowski, B. J., Cosentini, C. C., and Smith, M. E.,** Processes regulating temporal and longitudinal variations in the chemistry of a low-order woodland stream in the Adirondack region of New York, *Biogeochemistry,* 3, 255, 1987.
85. **Johannessen, M., Skartveit, A., and Wright, R. F.,** Stream water chemistry before, during and after snowmelt, in *Ecological Impact of Acid Precipitation,* Drablos, D. and Tollan, A., Eds., SNSF Project, Oslo, 1980, 224.
86. **Sullivan, T. J., Christophersen, N., Muniz, I. P., Seip, H. M., and Sullivan, P. D.,** Aqueous aluminum chemistry response to episodic increases in discharge, *Nature,* 323, 324, 1986.

87. **Henriksen, A., Skogheim, O. K., and Rosseland, B. O.,** Episodic changes in pH and aluminum-speciation kill fish in a Norwegian salmon river, *Vatten,* 40, 255, 1984.
88. **Schofield, C. L. and Trojnar, J. R.,** Aluminum toxicity to fish in acidified waters, in *Polluted Rain,* Toribara, T. Y., Miller, M. W., and Morrow, P. E., Eds., Plenum Press, New York, 1980, 347.
89. **Virtousek, P. M.,** The regulation of element concentrations in mountain streams in the northeastern United States, *Ecol. Monogr.,* 47, 65, 1977.
90. **Galloway, J. N., Schofield, C. L., Hendrey, G. R., Peters, N. E., and Johannes, A. H.,** Sources of acidity in three lakes acidified during snowmelt, in *Ecological Impact of Acid Precipitation,* Drablos, D. and Tollan, A., Eds., SNSF Project, Oslo, 1980, 264.
91. **Rascher, C. M., Driscoll, C. T., and Peters, N. E.,** Concentration and flux of solutes from snow and forest floor during snowmelt in the west-central Adirondack region of New York, *Biogeochemistry,* 3, 209, 1987.
92. **Chen, C. W., Dean, J. D., Gherini, S. A., and Goldstein, R. A.,** Acid rain model: hydrological module, *ASCE J. Environ. Eng. Div.,* 108, 455, 1982.
93. **Driscoll, C. T., Yatsko, C. P., and Unangst, F. J.,** Longitudinal and temporal trends in the water chemistry of the North Branch of the Moose River, *Biogeochemistry,* 3, 37, 1987.
94. **Hooper, R. P. and Shoemaker, C. A.,** Aluminum mobilization in an acidic headwater stream: temporal variation and mineral dissolution disequilibria, *Science,* 229, 463, 1985.
95. **Hall, R. J., Likens, G. E., Fiance, S. B., and Hendrey, G. R.,** Experimental acidification of a stream in the Hubbard Brook Experimental Forest, New Hampshire, *Ecology,* 61, 976, 1980.
96. **Driscoll, C. T., Wyskowski, B. J., DeStaffan P., and Newton, R. M.,** Processes regulating the chemistry and transfer of aluminum in a forested watershed, in *The Environmental Chemistry of Aluminum,* Lewis, T., Ed., Lewis Publishing, in press.
97. **Driscoll, C. T. and Schafran, G. C.,** Characterization of short term changes in the base neutralizing capacity of an acidic Adirondack, New York, lake, *Nature,* 310, 308, 1984.
98. **Schafran, G. C. and Driscoll, C. T.,** Spatial and temporal variations in aluminum chemistry of a dilute acidic lake, *Biogeochemistry,* 3, 105, 1987.
99. **Schofield, C. L., Galloway, J. N., and Hendrey, G. R.,** Surface water chemistry in the ILWAS basins, *Water Air Soil Pollut.,* 26, 403, 1985.
100. **LaZerte, B.,** Metals and acidification: an overview, *Water Air Soil Pollut.,* 31, 569, 1986.
101. **Nilsson, S. I.,** Budgets of aluminum species, iron and managanese in the Lake Gardsjon catchment in SW Sweden, in *Lake Gardsjon an Acid Forest Lake and its Catchment,* Andersson, F. and Olsson, B., Eds., Ecological Bulletin No. 37, Stockholm, 1985, 120.
102. **Driscoll, C. T.,** Aluminum in acidic surface waters chemistry, transport and effects, *Environ. Health Perspect.,* 63, 93, 1985.
103. **Kelly, C. A., Rudd, J. W. M., Hesslein, R.H., Schindler, D. W., Dillon, P. J., Driscoll, C. T., Gherini, S. A., and Hecky, R. E.,** Prediction of biological acid neutralization in acid-sensitive lakes, *Biogeochemistry,* 3, 129, 1987.
104. **Schiff, S. L. and Anderson, R. F.,** Alkalinity production in epilimnetic sediments: acidic and non-acidic lakes, *Water Air Soil Pollut.,* 31, 941, 1986.
105. **Driscoll, C. T. and Schafran, G. C.,** An Evaluation of Aluminum in Acidic Lake Ecosystems: Sources, Fate and Role in Nutrient Cycling, Final Report for USEPA/NCSU Acid Deposition Program Project: APP-0310-1983, Syracuse University Syracuse, New York, 1985.
106. **Johannessen, M.,** Aluminum, a buffer in acidic waters, in *Ecological Impact of Acid Precipitation,* Drablos, D. and Tollan, A., Eds., SNSF Project, Oslo, 1980, 222.
107. **Glover, G. M. and Webb, A. H.,** Weak and strong acids in the surface waters of the Tovdal region in southern Norway, *Water Res.,* 13, 781, 1979.
108. **Henriksen, A. and Seip, H. M.,** Strong and weak acids in surface waters of southern Scotland, *Water Res.,* 14, 809, 1980.
109. **Driscoll, C. T. and Bisogni, J. J.,** Weak acid/base systems in dilute acidified lakes and streams in the Adirondack region of New York State, in *Modeling of Total Acid Precipitation Impacts,* Schnoor, J. L., Ed., Butterworths, Boston, 1984, 53.
110. **Huang, C. P.,** Adsorption of phosphate on the hydrous γ-Al_2O_3 electrolyte interface, *J. Colloid Interface Sci.,* 53, 178, 1975.
111. **Davis, J. A.,** Adsorption of natural dissolved organic matter at the oxide/water interface, *Geochim. Cosmochim. Acta,* 46, 681, 1982.
112. **Davis, J. A. and Gloor, R.,** Adsorption of dissolved organic in lake water by aluminum oxide, *Environ. Sci. Technol.,* 15, 1223, 1981.
113. **Hohl, H. and Stumm, W.,** Interactions of Pb with hydrous γ-Al_2O_3, *J. Colloid Interface Sci.,* 53, 178, 1976.
114. **White, J. R. and Driscoll, C. T.,** Lead cycling in an acidic Adirondack lake, *Environ. Sci. Technol.,* 19, 1182, 1985.

115. **Schindler, D. W.,** Evolution of phosphorus limitations in lakes, *Science,* 195, 260, 1977.
116. **Driscoll, C. T., White, J. R., Schafran, G. C., and Rendall, J. D.,** Calcium carbonate neutralization of acidified surface waters, *ASCE Environ. Eng. Div.,* 108, 1128, 1982.
117. **Effler, S. W., Schafran, G. C., and Driscoll, C. T.,** Partitioning light attenuation in an acidic lake, *Can. J. Fish. Aquat. Sci.,* 42, 1707, 1985.
118. **Ball, J. W., Nordstrom, D. K., and Jenne, E. A.,** Additional and Revised Thermochemical Data for WATEQ-2 Computerized Model for Trace and Major Element Speciation in Mineral Equilibra of Natural Waters, U.S. Geological Survey Water Resource Investigations, Menlo Park, CA, 1980.
119. **Hem, J. D.,** Graphical methods for studies of aqueous aluminum hydroxide, fluoride and sulfate complexes, *U.S. Geol. Surv. Water-Supply Pap.,* 1827-B, 1968.
120. **Martell, A.E. and Smith, R. M.,** *Critical Stability Constants,* Plenum Press, New York, 1974.
121. **May, H. M., Helmke, P. A., and Jackson, M. L.,** Gibbsite solubility and thermodynamic properties of hydroxyaluminum ions in aqueous solutions at 25 C, *Geochim. Cosmochim. Acta,* 43, 861, 1979.

Chapter 10

ALUMINUM GEOCHEMISTRY AT THE CATCHMENT SCALE IN WATERSHEDS INFLUENCED BY ACIDIC PRECIPITATION

Dean S. Jeffries and William H. Hendershot

TABLE OF CONTENTS

I. INTRODUCTION

Aluminum is the most abundant metal in the earth's crust, occurring in a host of primary aluminosilicate minerals. ''Normal'' chemical weathering of these minerals by carbonic acid or various organic acids (i.e., abundantly occurring weak acids) generally causes only a short-lived mobilization of aluminum. The high surface affinity of dissolved aluminum species (see Chapter 6) and the relative insolubility of both primary aluminosilicate minerals and secondary aluminum-oxide minerals (see Chapter 3) ensure that the quantity of aluminum released to natural waters is always very small; most of it is retained within the B soil horizon. However, since the pool of aluminum within the soil environment is so large, even a small perturbation of normal weathering (i.e., soil acidification) may cause a large increase in aluminum mobility and profoundly increase the concentration of this element observed in some natural waters, particularly those close to the weathering site. Acidic precipitation is one such perturbation.[1,2] Numerous publications have detailed the occurrence of elevated aluminum concentrations in natural waters as the effect of acidic deposition (for example, see Dillon et al.,[3] Tri-academy Committee,[4] and Schindler[5]). Environmental concern for this increased aluminum mobility rests with its ultimate toxic effect on both terrestrial[6] and aquatic biota.[7-11]

The purpose of this chapter is to present and discuss the occurrence and variability of aluminum concentrations and species in surface and subsurface waters at the catchment scale, placing particular emphasis on discerning the biogeochemical factors that control the variability. For illustrative purposes, we shall use results from studies on watersheds located in central Ontario and southern Québec. Consideration of aluminum behavior on a watershed scale requires mainly a definition of how aluminum moves from a soil-weathering site into surface waters. Thus the hydrology of the watershed becomes very important in determining both the timing and chemistry of the water reaching streams. Given that the water encounters different materials on its path through soils to ground water, streams, or lakes, soil chemistry is important also. Therefore, we feel that it is useful to present brief summaries of the hydrological characteristics and soil properties that bear on the development of water quality within an ''acid-susceptible'' catchment. These summaries will also serve to establish the terminology that is used throughout this chapter.

A. The Relationship of Hydrology to Water Quality

The flowpath by which water reaches a stream channel is important. Most of the water reaching a stream during low-flow periods (i.e., baseflow) is moving through deeper soil horizons or porous bedrock. Under baseflow conditions, the water will normally reach the stream from the saturated zone because of the gradient in the water table. Flow from the vadose (unsaturated) zone may be important also, since water held in soil pores in excess of field capacity will drain vertically through this zone to the water table. Baseflow water has a lengthy contact time with the matrix through which it flows and, therefore, its composition reflects closely the chemical reactivity of the matrix material. It almost invariably has a higher pH value than the input precipitation and a low aluminum concentration. tration.

''Quickflow'' describes the situation in which runoff water from a precipitation or snowmelt event, or water already in the soil is displaced by incoming water and transmitted rapidly through the soil matrix, yielding a sharp rise in the storm hydrograph. The generation of quickflow requires two conditions: (1) a matrix that permits a high lateral hydraulic conductivity, and (2) a hydraulic gradient large enough to cause high flow rates through (or across) soil. In order to produce the high flow rates, the soil matrix must be saturated or close to it; hydraulic conductivities for unsaturated soils are too low to generate quickflow.

In humid regions, three major sources of precipitation-induced quickflow have been identified: (1) "direct precipitation runoff" from saturated soils, (2) "return flow", and (3) "subsurface stormflow".[12] In most stream basins, all three sources will contribute to the storm hydrograph, but the relative importance of each depends on the characteristics of the drainage basin, the antecedent soil moisture content, the position of the water table, and, finally, the nature of the input precipitation or snowmelt event.

Near the stream channel, where the water table approaches the soil surface, direct precipitation runoff from this saturated area will begin almost simultaneously with the start of the precipitation (or snowmelt) event. If there is a shallow, unsaturated zone, it will become saturated before incoming precipitation can flow overland to a stream, thereby creating a short lag in the flow response. When soils are dry, the area contributing to this type of flow is quite small; however, the area will increase during the precipitation event as the surface occupied by saturated soils increases.

Return flow is caused also by a rise in the water table. It occurs in response to the increased hydrostatic pressure exerted by water penetrating an unsaturated zone that is adjacent to areas which are saturated to the land surface. The water flows out of the saturated area and joins water flowing over saturated soil. Since return flow is moving toward the soil surface in the direction of least hydrostatic pressure (either horizontally or vertically), it effectively moves upward through the soil profile from the C horizon through the B and E or A horizons, and finally through the O horizon. The runoff-water composition will be affected accordingly by these different contacts with the soil matrix.

Subsurface stormflow is less well defined than the two preceding runoff mechanisms, but it is probably the most important for generating quickflow. Subsurface stormflow refers to water moving through the soil at any depth,which means that it can be moving just under the organic surface soil horizons or through permeable C horizons. The classical concept of subsurface stormflow[13] considers that water flows downward from the surface through the unsaturated zone, and then rapidly through the saturated zone directly into a stream channel. Increased hydrostatic pressure causes displacement of "old" subsurface water (i.e., water that may have spent a long time in the soil), an action sometimes called piston flow. In order for this mechanism to contribute significantly to quickflow, certain limiting conditions must exist. First, the water retention capacity of the unsaturated zone at the time of infiltration must be low relative to the amount of input, or else the incoming water will remain in the vadose zone. Second, the hydraulic conductivity of the saturated zone must be relatively high. Therefore, this mechanism will be most effective when the saturated zone is close to the surface and has a very high saturated hydraulic conductivity, as would be found in glacio-fluvial gravel, for example.

There is another less "classical" occurrence of subsurface stormflow that should be considered. When shallow soils overlay impermeable bedrock or other unconsolidated layers with restricted permeability, part of the soil profile will become saturated, but not up to the surface. Rapid subsurface flow then can be produced in two ways: (1) if part of the profile contains interconnecting macropores, such as dead root channels, very rapid "pipe" flow can be expected through them; and (2) more permeable near-surface horizons, such as O or eluviated E horizons, begin to transmit water to streams as they become saturated.[14-16] The input water to such a system will mix initially with the capillary water trapped previously in the unsaturated zone, so that the output stream water will appear older than the input water; as subsurface stormflow continues, the relative ages of the input and output waters will merge.

Horton overland flow[17] occurs when the water input rate exceeds the infiltration capacity. It is generally considered to be an insignificant factor in undisturbed, humid, forested watersheds. However, when snowmelt events are taken into consideration, it is possible that water input rates may exceed the infiltration capacity of the soils. Price and Hendrie[18] found that the formation of a compact ice layer in the snowpack or at the soil surface resulted in overland flow during one spring snowmelt period. During the intensive watershed studies presented below, no occurrence of Horton overland flow was observed during spring melt.

The flowpath of runoff water generated by each of these mechanisms can profoundly affect the water quality. In the case of direct precipitation on saturated soils and Horton overland flow, the water flows over the surface and will contact mainly the surface organic soil horizons. Return flow will move from lower to upper soil horizons and then along the surface. Subsurface stormflow can move through any soil horizon, depending on the relative permeability of the soil layers involved. However, since soil permeability is usually greatest nearer the surface than at depth (see below), the water quality of this type of subsurface runoff generally reflects the soil chemistry of near-surface horizons. The most striking occurrence of aluminum movement in a watershed is related to periods of peak stream flow (snowmelt and/or heavy rainfall) when the quickflow component is maximum. Runoff waters at this time have reduced pH values and elevated aluminum concentrations.

B. The Relationship of Soils to Water Quality

Areas where the effect of acidic deposition has been manifest as either long-term or episodic acidification of surface waters (usually with elevated aluminum concentrations, as well) are most commonly situated on acidic soils with natural forest vegetation. The role of these soils in the acidification process is important.

Four soil orders are usually associated with watersheds influenced by acidic deposition in eastern North America. They are the Inceptisols, Spodosols, Alfisols, and Ultisols, all of which have certain characteristics in common. First, their hydraulic conductivity tends to be greater near the surface and decreases with depth. In the case of the Inceptisols and Spodosols, the high hydraulic conductivity near the surface is related to the presence of amorphous metal-oxide material and organic matter that coat the soil particles and promote low bulk density and abundant pore space. Deeper in these same soils there are, occasionally compact layers of low permeability that may be related to compact glacial till, ortsteins, duripans, or fragipans. In Alfisols and Ultisols, the near-surface horizons have lost clay by eluviation, which creates a zone of relatively high hydraulic conductivity. Lower in the profile, in the illuvial zone or argillic horizon, clay has accumulated and the hydraulic conductivity is much lower. The underlying horizons may also have low permeability because of cementation or consolidation.

The second characteristic common to all four soil orders is a general increase in pH value with depth. The surface horizons tend to be more acidic because of the intense leaching that takes place at the surface and to the decomposition of organic matter. Thus the surface horizons are not only more acidic, they also contain more organic matter that can decompose to produce organic ligands capable of complexing aluminum.

In summary, the impact of acidic deposition on a surface water system will be influenced greatly by the flowpath by which the water reaches a stream channel. Given that ion exchange and adsorption/desorption reactions are extremely rapid, it is clear that water leaving the terrestrial portion of a catchment will have pH values and aluminum concentrations that generally reflect the soil horizon through which it last flowed. Water emerging from C horizons will be low in organic acids and have both a high pH value and a low aluminum content. In contrast, water flowing from the surface soil horizons will have high concentrations of organic acids, a low pH value, and high dissolved aluminum content.

II. STUDY SITES

A. The Turkey Lakes Watershed

The Turkey Lakes Watershed (TLW) is an undisturbed, remote catchment located approximately 50 km north of Sault Ste. Marie, Ontario (see Figure 1). Its total drainage area

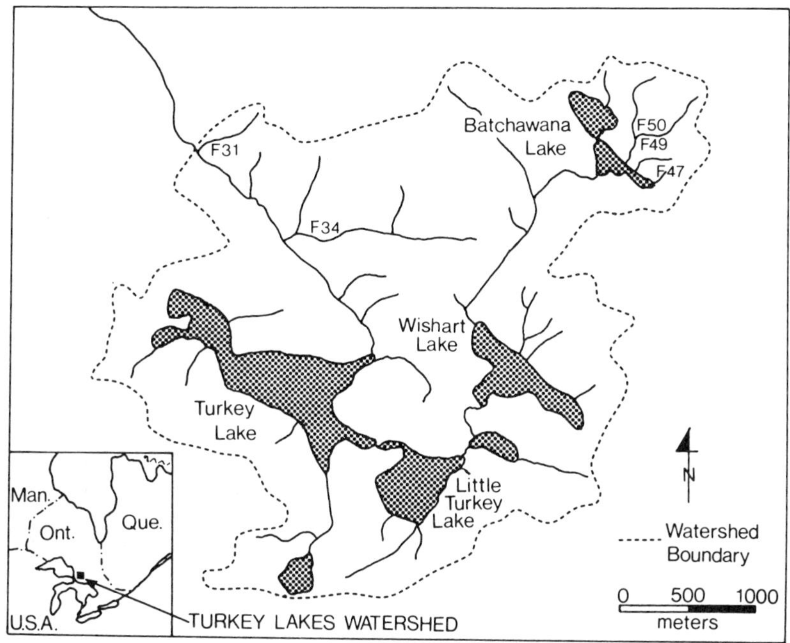

FIGURE 1. Location map of the Turkey Lakes watershed, central Ontario. The streams included in Table 1 are specifically noted.

is 10.5 km^2. The basin contains a chain of four lakes (Batchawana, Wishart, Little Turkey, and Turkey Lakes) that exhibit a gradient in their chemical composition and response to acidic deposition. The terrestrial portion of the TLW is completely forested by an "old-growth" mixed hardwood stand dominated by sugar maple (*Acer saccharum*) and yellow birch (*Betula alleghaniensis*). The closest point-source emitter of air pollutants is the steel industry located at Sault Ste. Marie. Annual atmospheric deposition of SO_4 and NO_3 is approximately 70 and 37 meq/m^2, respectively.

The elevation at the lowermost stream gauging station is 340 m AMSL, while the highest point is Batchawana Mountain (630 m) on the northern edge of the basin, giving an overall relief of 290 m. The large variations in elevation and the leeward position relative to Lake Superior influence the quantity of precipitation received. Semkin and Jeffries[19] recorded an average annual precipitation of 1212 mm from 1981 to 1984 at low elevations and up to 15% more at higher altitudes. In comparison, Sault Ste. Marie, which lies south of the highlands containing the TLW, receives only 935 mm/year (long-term average).

The watershed is almost entirely underlain by Precambrian silicate greenstone (i.e., a metamorphosed basalt), with only small outcrops of more felsic igneous rock occurring north of Batchawana Lake and near the main inflow to Little Turkey Lake.[20] Regional fault systems tending in approximately NW—SE and SW—NE directions have exerted some control over water drainage, giving the rather angular drainage pattern observed in Figure 1. A two-component glacial till overlies the bedrock. It is composed of an upper, relatively permeable ablation till and a lower, less permeable basal till. Till thickness varies from < 1 m at high-elevation locations (with frequent surface exposure of bedrock) to 1 to 2 m at lower elevations, with the occasional occurrence of extremely deep till sequences (up to 70 m) where valleys in the bedrock have been entirely filled. The mineralogy of the till is more felsic than the underlying bedrock, showing that the material was likely derived primarily from the large granitic intrusions that occur just north of the basin.[21] Finally, the tills contain a small but measurable amount of $CaCO_3$ (0 to 2%) that increases with depth,

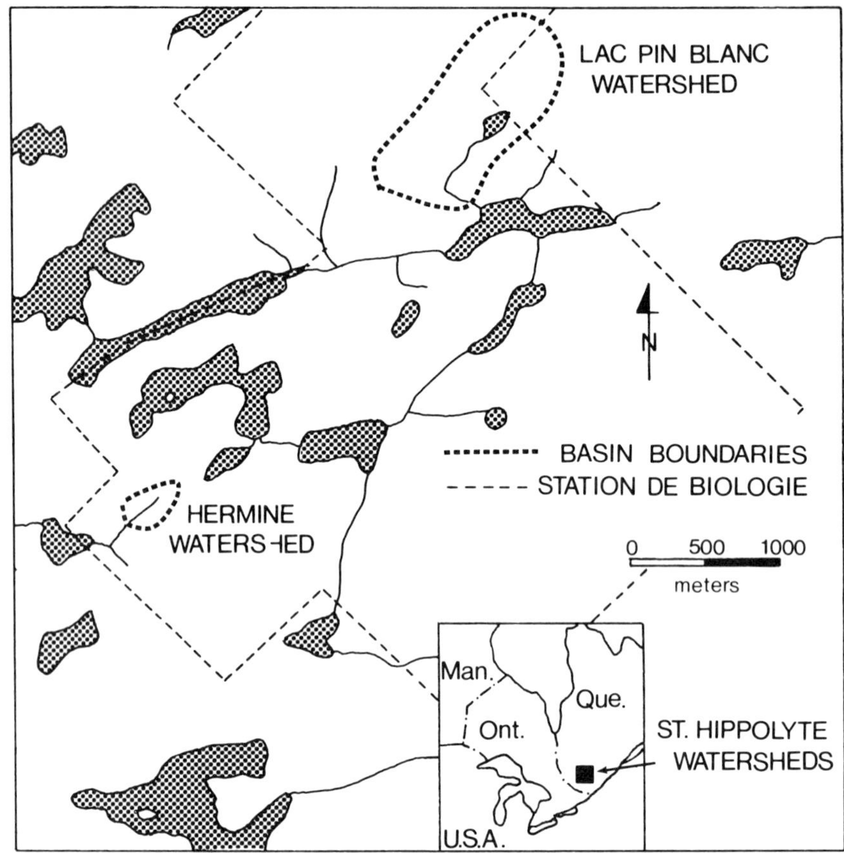

FIGURE 2. Location map of the Station de Biologie, Université de Montréal, Québec. The Lac Pin Blanc and Hermine watersheds are specifically noted.

and is higher (on average) at lower-elevation locations.[22] Soils developing within the till matrix are classified as Typic Haplorthods. More information on basin characteristics is presented by Jeffries et al.[23]

B. The St. Hippolyte Watersheds

1. Lac Pin Blanc Watershed

The Lac Pin Blanc watershed is located in the Station de Biologie de l'Université de Montréal near St. Hippolyte, Québec, about 80 km north of Montréal (see Figure 2). The catchment has a surface area of 0.56 km^2. The drainage system consists of a first-order stream. A depression in the bedrock formed by glacial action is now occupied by a small pond (Lac Pin Blanc) about 0.01 km^2. The area receives total precipitation of approximately 1100 mm/year with 30% as snow.[24] The annual atmospheric deposition of SO_4 and NO_3 is approximately 66 and 34 meq/m^2, respectively.

The vegetation on well-drained sites is a dominantly hardwood forest of sugar maple (*Acer saccharum*), white birch (*Betula papyrifera*), balsam fir (*Abies balsamea*), and trembling aspen (*Populus tremuloides*). The poorly drained sites are on organic soils, which have a vegetation cover of white cedar (*Thuja occidentalis*) and speckled alder (*Alnus rugosa*). The pond is edged with a mat of sphagnum moss.

The bedrock of this area is Precambrian anorthosite. Except for outcrops along many of the topographic divides between watersheds, the bedrock is covered by till with a mineralogical composition similar to that of the anorthosite. In most parts of the watershed the till is thin (<2 m), and the soils are Typic Haplorthods. In the depression near the pond, organic soils have become established — either Saprists or Hemists, depending on the degree of decomposition of the material. A more detailed description of the soils can be found in Hendershot et al.[25] and Dufresne.[26] This basin was monitored during 1984.

2. Hermine Watershed

During 1985 and 1986, the smaller Hermine watershed was studied (see Figure 2). This basin has a surface area of 0.06 km^2 and vegetation dominated by sugar maple (*Acer saccharum*, 78%), with lesser amounts of American beech (*Fagus grandifolia*, 9%) and yellow birch (*Betula alleghaniensis*, 6%). It is located about 3 km from the Lac Pin Blanc watershed and exhibits a similar geology. The soils are Typic Haplorthods, although differences in morphology and chemical properties have developed because of slope position and drainage.[27]

III. MATERIALS AND METHODS

A. Turkey Lakes Watershed

Atmospheric deposition was estimated from weekly bulk deposition samples while precipitation quantity was measured daily using appropriate gauges.[19] Water and chemical fluxes from headwater streams were measured at several gauging stations operated by the Great Lakes Forest Centre[28] and at the lake outflows. These stations generally have continuous measurements of stream flow and approximately weekly measurements of chemical composition. The lakes were sampled bi-weekly in profile, at the deepest point.

Total aluminum (Al_t) samples were acidified with high-purity HNO_3 (1 ml/l) and quantitation achieved by graphite furnace AAS. Aluminum speciation was performed using the method of LaZerte,[29] in which the most biologically toxic species (i.e., inorganic monomeric forms, Al_i) are separated physically by 24-h dialysis using 1000 MWCO membrane. The aluminum is complexed with oxine and extracted into MIBK. Our procedure differed slightly from that of LaZerte[29] in that final aluminum species quantitation was achieved by back-extraction into 1% HNO_3 and measurement with a DC plasma emission spectrograph. See Chapter 1 for a critical discussion of these analytical methods.

In addition to the above analyses, results pertaining to Al_t in throughfall and soil waters[30] and ground water[31] will be presented in order to complete the "whole-watershed" picture.

B. St. Hippolyte Watersheds

In the Lac Pin Blanc watershed, shallow ground-water wells were located near the stream and pond. The water table height was measured daily and the water samples about every 5 d. Bulk rainfall was collected in polyethylene bottles using the system described by Likens et al.[32] Snowfall was collected in six plastic buckets (25-cm diameter) on an event basis; these samples were combined prior to chemical analysis. Continuous stream discharge was measured at a natural V-notch weir, whereas the water chemistry was determined in samples collected automatically several times a day. In the Hermine watershed, discharge measurement and stream sampling were performed automatically in 1985 and manually in 1986 at a V-notch weir.

In 1984, aluminum speciation was determined using the oxine method as discussed in Lalande and Hendershot.[33] Our research has shown that the first fraction (Al_1) includes mainly monomeric aluminum species, such as Al^{3+}, $AlOH^{2+}$, $Al(OH)_2^+$, and perhaps the dimer $Al_2(OH)_4^{2+}$. The second fraction (Al_2) measures the same forms as Al_1 plus monomeric aluminum bound to fluoride or organic ligands. The third fraction (Al_3) is considered to be the total reactive aluminum that, in addition to the forms measured in Al_2, includes polymeric forms of reactive aluminum that may be in either organic or inorganic form. (See also Chapter 1 for a discussion of aluminum fractionation.)

In 1985, aluminum speciation was performed using the PCV method[34,35] modified to give two forms of aluminum: (1) acid-extractable aluminum (Al_t), measured after acidifying the sample to 2% HNO_3; and (2) total monomeric aluminum (Al_{tm}), measured on an unacidified sample and believed to include Al^{3+} and simple complexes with OH, F, SO_4, and organic ligands.

In 1986, the Al_{tm} was separated analytically into charged and uncharged fractions by passing a solution through a cation exchange column. The exchangeable fraction is defined operationally to be inorganic aluminum (Al_i), whereas the nonexchangeable fraction is considered to be organically complexed aluminum (Al_o).[7]

IV. RESULTS AND DISCUSSION

A. Aluminum Variability Within a Watershed

Large variations in aqueous aluminum concentrations occur both within and between specific watershed compartments. For the purpose of this discussion, we shall separate aluminum variability into spatial and temporal components, realizing that, in reality, the two are often intimately related.

1. Spatial Variability

Table 1 presents average Al_t concentrations for various watershed compartments in the TLW. Total aluminum in the bulk deposition is low (31 to 33 μg/l) and only slightly elevated by passage through the deciduous canopy that predominates throughout the basin. The less than twofold increase in throughfall concentrations relative to the incident deposition appears to be influenced little by the growth "status" of the trees; i.e., there is little difference between throughfall concentrations in the summer growing season and the winter dormant season. Stemflow is further enriched in Al_t (two- to threefold greater than in the bulk deposition); but, in terms of the metal flux to the underlying components of the basin, stemflow is of minor importance.[30]

The highest Al_t values appear in soil waters. Samples collected under the surface organic horizons had values of 134 to 183 μg/l, while in the B horizon, concentrations were 357 to 380 μg/l. No consistent pattern between growing and dormant seasons is apparent. Leaf litter, which accounts for most of the material in the organic horizons, must contain a substantial quantity of aluminum that is readily leached, desorbed, or exchanged. The high concentrations found in the B horizon presumably reflect the high bulk soil concentrations of adsorbed aluminum that are present because of normal soil-formation processes. Since the B horizon acts as the primary site of aluminum demobilization, waters below this level have very much lower aluminum concentrations. The fact that there is a twofold decrease (102 to 45 μg/l) between the C horizon and slightly deeper ground water, however, may indicate that some aluminum mobilization from the B to the C horizon is occurring in response to an elevated, strong acid input from above.

The Al_t values are highly variable between headwater streams. In the TLW, streams at high elevation (e.g., streams F47, F49, and F50) generally occupy sub-basins with shallow surficial deposits. They also exhibit relatively higher — but variable — aluminum levels ranging from 94 to 197 μg/l. As will be developed more fully below, we ascribe high stream-water aluminum concentrations to geomorphological control of the flow pathways by which drainage waters reach the headwater stream channel. That is, we believe that a significant proportion of the water enters the stream by draining directly from the B horizon, the C horizon being excluded from the pathway by the physical constraint imposed by a shallow overburden. There is an obvious corollary to the situation in which only a small

Table 1
TOTAL ALUMINUM CONCENTRATIONS (μg/l) FOR VARIOUS COMPARTMENTS IN THE TURKEY LAKES WATERSHED

Watershed location	Sampling period	Total aluminum conc
Bulk deposition[a]	1983	31, 33
Throughfall[a]	1983	46, 51
Stemflow[a]	1983	90, 72
Oe Horizon (5—2 cm)[a]	1983	134, 183
B Horizon (60 cm)[a,b]	1983	380, 357
C Horizon (181 cm)[b]	1985	102
Ground water (280 cm)[b]	1985	45
Stream F47[c]	1984—1985	145
Stream F49[c]	1984—1985	94
Stream F50[c]	1984—1985	197
Stream F31[c]	1984—1985	38
Stream F34[c]	1984—1985	44
Batchawana Lake North	1983—1986	87
Batchawana Lake South	1983—1986	116
Wishart Lake	1983—1986	81
Little Turkey lake	1983—1986	63
Turkey Lake	1983—1986	33

Note: Data are median values except where noted.

[a] Foster and Nicolson[30]; two Al_t values correspond to unweighted mean concentrations for the dormant and growing seasons, respectively; all samples were collected at a low-elevation location within sub-basin F31.
[b] Chew et al.[31]; mean values for samples collected from piezometers at a high-elevation location near sub-basin F47; piezometer water at 67 cm had 701 μg/l total Al_t.
[c] J. A. Nicolson, unpublished data.

hydrological reservoir is available in the soil system; i.e., stream-flow generation will be dominated by quickflow mechanisms, yielding rapid variations in runoff and periods of low or "no" flow. All of the streams exhibit these characteristics. Clearly, flow pathway control existed long before the advent of acidic deposition; hence elevated contemporary Al_t levels in surface waters probably reflects a deposition-induced increase in the mobility of aluminum.

Table 2 presents the range and median for both Al_t and Al_i values obtained for the high-elevation streams (i.e., F47, F49, and F50) during the spring melt episodes of 1985 and 1986. In the TLW, more than 50% of the total annual flow occurs during spring melt.[28] Inorganic monomeric aluminum is the species generally considered to be most important from the standpoint of toxicity.[7] The difference between Al_t and Al_i represents soluble organic monomeric and polymeric species, as well as particulate materials. Since there was a fairly close correspondence in the magnitudes of Al_t and the nondialyzed, oxine-MIBK-defined species, the organic monomeric species almost always dominate the polymeric and particulate species in the TLW. As can be seen by the data in Table 2, there are also distinct differences in the aluminum species distributions among the headwater streams. Although the two most acidic streams (F47 and F50; median pH = 5.5 and 5.3, respectively) have the highest Al_t values, the proportion represented by Al_i was 69 and 34%, respectively. The DOC levels in F47 are also much lower than in F50 (see Table 2). The fact that the organic monomeric

Table 2
MEDIAN pH VALUE AND DOC (mg/l), AND RANGE AND MEDIAN TOTAL ALUMINUM AND INORGANIC MONOMERIC ALUMINUM (μg/l) OBSERVED DURING THE 1985 AND 1986 SPRING MELT EPISODES FOR SELECTED HIGH-ELEVATION STREAMS IN THE TURKEY LAKES WATERSHED

Stream	Median pH	Median DOC	Total aluminum		Inorganic monomeric aluminum	
			Range	Median	Range	Median
F47	5.5	2.4	108—420	281	68—371	195
F49	6.1	2.4	50—159	91	5—67	35
F50	5.3	4.1	156—227	193	23—226	66
Batchawana Lake outflow	5.8	4.2	72—159	116	3—244	55

component is much more important for F50 than for F47 implies that there is a difference in the character of the aluminum supply to these sub-basins. Of the two, F50 is by far the more steeply sloping and thus it is more likely that subsurface stormflow proceeds through organic soil horizons there as compared to sub-basin F47. Consistent with this hypothesis is the fact that Al_t in F50 is less than that in F47 (despite a lower pH value) and similar to that in the water draining from Oe horizons, as given in Table 1. Stream F49 exhibits higher pH values and lower Al_t and Al_i than either F47 or F50. Under these conditions, however, even the low DOC concentration (i.e., 2.4 mg/l) is sufficient to reduce the relative importance of Al_i to a proportion comparable to that of F50.

Streams draining low-elevation locations in the TLW (e.g., streams F31 and F34) have deeper overburden and a water-percolation pathway that probably includes a C horizon. Moreover, these tills contain a small amount (1 to 2%) of easily weathered $CaCO_3$, a geochemical factor which contributes to generally higher soil pH values at low elevation and, therefore, decreased aluminum mobility. The total aluminum in stream water at low elevation is very similar to the input precipitation concentrations.

Finally, whole-lake Al_t exhibits a gradient in value consistent with the stream levels discussed above. Batchawana Lake South has the highest concentration, not the "true" headwater (Batchawana Lake North). However, it receives comparatively more of its input as terrestrial basin runoff. All of the high-elevation, high-Al_t streams represented in Table 1 drain into Batchawana Lake South. Each of the succeeding downstream lakes has a lower average Al_t value. Figures 3 and 4 present isopleths of Al_t for the first 6 months of 1986 for Batchawana Lake South and Little Turkey Lake, respectively. The isopleths show not only the generally different Al_t levels present in the two lakes, but also the elevated concentrations that develop in bottom waters during periods of restricted lake circulation. Precipitation of aluminum species, or adsorption by lake seston followed by settling from the water column, are probably important removal processes for these lakes. This kind of removal process has been implicated in the co-removal of phosphorus and dissolved organic carbon that leads to the increased clarity and low-trophic status often associated with lakes affected by acidic deposition.[36-38]

2. Temporal Variability

Chemical variation in stream and lake waters of northern temperate watersheds is often intimately related to their hydrological variability. In the Turkey Lakes and St. Hippolyte watersheds, spring snowmelt is almost always the most pronounced hydrological event each year, often providing 30 to 50% (or more) of the total annual runoff.[28,39] However, heavy rainfall that often occurs during autumn months can also provide significant runoff volume.

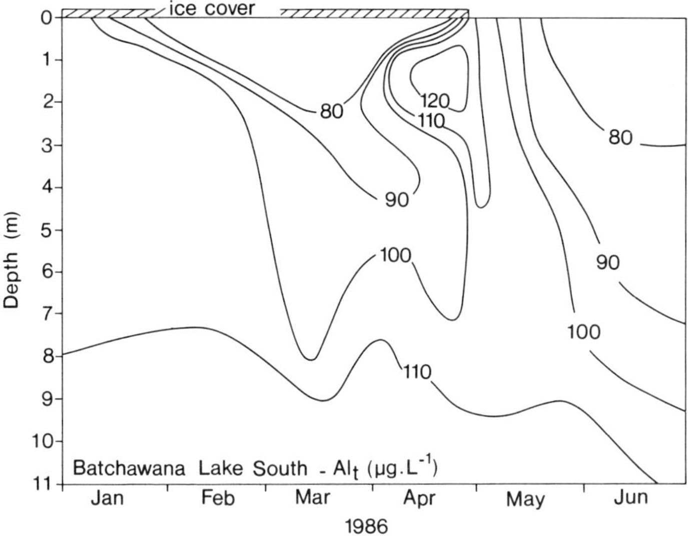

FIGURE 3. Total aluminum isopleths (μg/l) for January to June 1986, for Batchawana Lake South. Spring mixing occurred soon after ice disappearance in early May.

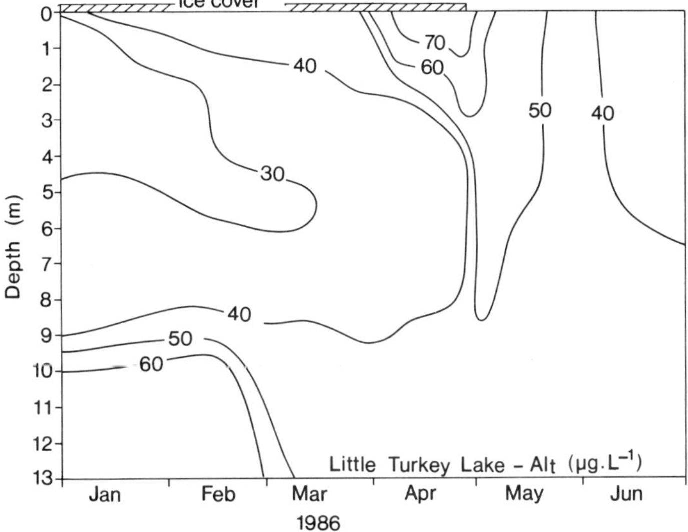

FIGURE 4. Total aluminum isopleths (μg/l) for January to June 1986, for Little Turkey Lake. Spring mixing occurred soon after ice disappearance in early May.

In either case, during the preceding period of low hydrological flux, the soluble material liberated by weathering of mineral or organic material accumulates in the soils where it can be flushed readily by a sudden, high-flow event.

During spring snowmelt, the preferential elution of pollutants from the snowpack during the early melt phase[40,41] and the flush of accumulated weathering products from terrestrial to aquatic ecosystems can cause a sharp drop in the pH values of lake or stream waters and a rise in the amount of aluminum.[29,42-44] Such chemical changes have been related to sudden fish kills, even where the water chemistry during the rest of the year poses no threat to

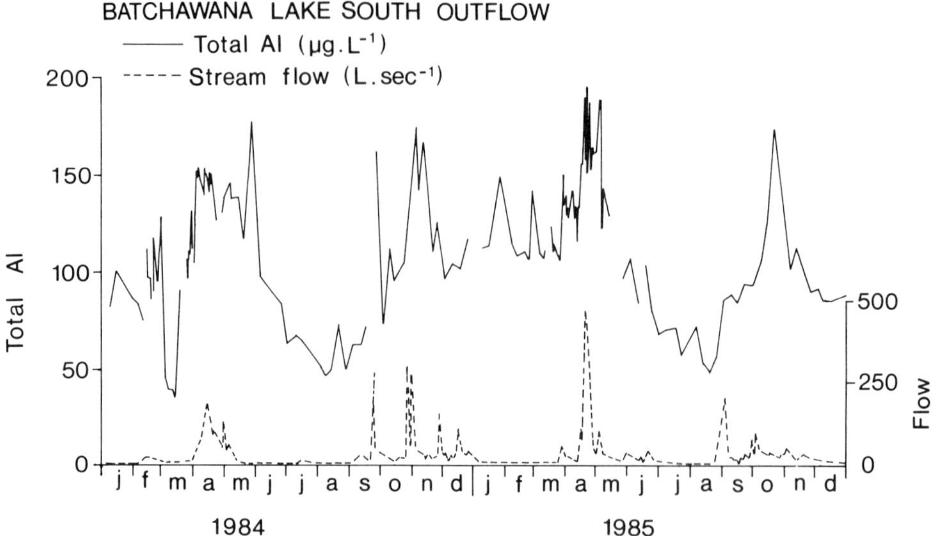

FIGURE 5. Temporal variation in Al_t (μg/l) and stream discharge (l/s) during 1984 and 1985 at the outflow of Batchawana Lake South.

fish.[10,45,46] Although the toxic effect of aluminum is dependent on its chemical species,[7] total concentrations as low as 100 to 200 μg/l have been associated with specific biological effects (see the review by Schindler[3]). Many streams and the headwater lake in the TLW exhibit episodically high values of both Al_t and Al_i during the spring melt period (see Table 2) that are within or above these critical levels.

Figure 5 presents the temporal variation observed in Al_t and discharge at the outflow of Batchawana Lake during 1984 and 1985. Although there is clearly substantial day-to-day variation even at this station, where flow through the lake would be expected to "damp" the influence of variable stream inputs, two general concentration maxima and minima can be identified each year. Maximum Al_t levels of about equivalent magnitude occur during the high-flow periods of each year, i.e., spring melt and autumn. Minimum Al_t occurs during the winter and summer months, with the latter being far more pronounced. The marked summer minima are similar to those observed in Adirondack watersheds.[47,48] They have been attributed to precipitation and within-lake retention of insoluble aluminum species in response to production-induced increases in water-column pH values. The sharp concentration minimum that occurred in March 1984 was not manifest in 1985. It may be attributed to the differing character of the spring melt episodes. In 1984, melting occurred abnormally early at the beginning of March; however, more than 2 weeks of extremely cold weather followed that reduced the runoff to its premelt intensity. The sharp minimum in Al_t in mid-March 1984, corresponds to this cold period. The small amount of early meltwater appears to have displaced relatively low-concentration lake water. This observation is in contrast to the usual runoff process, in which the cold, less dense meltwater flows across the lake immediately under the ice and exits via the lake water outflow. In the TLW, such meltwater tends to mix with only the top 1 to 2 m of lake water.[49]

The variability in Al_t within the lake water column during winter and spring is presented in Figures 3 and 4. High-aluminum stream water and direct seepage inputs to the lakes during snowmelt were manifest as a layer of higher-concentration lake water extending to a depth of approximately 2 m during April 1986. After ice disappeared from the lakes (at the end of April), a rapid increase in water temperature caused spring turnover — a water-column mixing that is typical of the dimictic lakes in eastern Canada. It should be noted that mixing in Batchawana Lake (see Figure 3) was incomplete, since the small peak in bottom water concentrations was not disrupted. After spring turnover, the small increase in Al_t with depth was reestablished.

The Al_i species present in the lakes during this time period were much lower in concentration, often approximately half of the Al_t values. Such Al_i values are probably not detrimental to biota. It is possible, however, that higher concentrations associated with direct subsurface seepage into the lakes may be locally important. Other studies in the TLW have shown that a significant portion of the spring runoff water enters the lakes as direct seepage and follows a flowpath within the top 1 m of the soil/till matrix.[50] High aluminum concentrations present in this water could affect fish-spawning habitats in near-shore gravel shoals. Gunn and Keller[51] have noted this phenomenon in other Ontario lakes.

The variations in Lac Pin Blanc stream discharge and water chemistry observed during snowmelt are presented in Figure 6. The figure covers the period March 21 to May 25, 1984. Stream flow (see Figure 6a), which was 7 l/s at the initiation of spring melt, rose sharply to form three major peaks on April 6 to 7, April 17, and April 25, with a smaller peak on May 9. Stream flow was influenced by both high temperatures and rainfall, causing average daily peak values to reach 160 to 180 l/s. Figure 7 shows comparable information for a ground-water well in the watershed. The water table position followed a pattern similar to that of stream discharge. The water table rose to a maximum during the first major melt event (April 5 to 7), then dropped to a minimum (April 11). A second maximum was observed on April 16 to 18, corresponding to the second major snowmelt event.

Figure 6b shows the variation of aluminum concentrations in the stream. Before April 4, the amount of total reactive aluminum (Al_3) was fairly high, around 125 μg/l. Beginning on April 5, there was a sharp rise in Al_3 concentration, to 180 μg/l, which corresponds to the first major peak in the discharge. During the same period, the concentration of Al_3 in well #7 rose to about 240 μg/l. The peak in Al_3 concentration was caused by the flushing of readily soluble or exchangeable material from the soils, aided by a rise in the water table and the generation of quickflow through surface horizons in low-lying areas. Once this first flush has passed, the positive correlation between stream water aluminum and discharge changes abruptly. Thereafter, stream flow peaks are matched by decreases in stream aluminum values.

We term this behavior of decreased aluminum with increased flow the "dilution effect", since it arises when a source of relatively pure water (snowmelt) is added to a solution flowing through the watershed.[52] The dilution effect occurs when the rate of water transport through the watershed is faster than the rate at which the basin can supply solutes. Although desorption or ion-exchange reactions are rapid, the absolute concentrations of most cations are controlled ultimately by primary mineral weathering (see Chapter 8). In small basins experiencing the high-water flux associated with spring melt, the water-soil contact time is normally too short for weathering to progress significantly. Clearly the pool of material available for dissolution, exchange, etc. during the latter melt stages is different in character from that which supplied the initial concentration peaks. The decrease in ion loading of the meltwater entering the soil over time also has the effect of decreasing the amount of exchangeable ions displaced from the soil-particle surfaces. The dilution of both surface and ground waters in the Lac Pin Blanc watershed was manifest by decreased electrical conductivity (EC) in both the stream and well water (see Figures 6c and 7c, respectively) as discharge peaked for the second and third time during the 1984 snowmelt episode. As the hydrological flux through the basin decreased following snowmelt, increasing water-soil contact time permitted a gradual increase in aluminum concentration (and EC) in both the stream and well water. Increasing soil temperature and biological activity may also increase the rate of formation of the labile aluminum pool in the soil.

Figure 6b shows that aquo- and hydroxy-monomeric aluminum (Al_1) and polymeric alu-

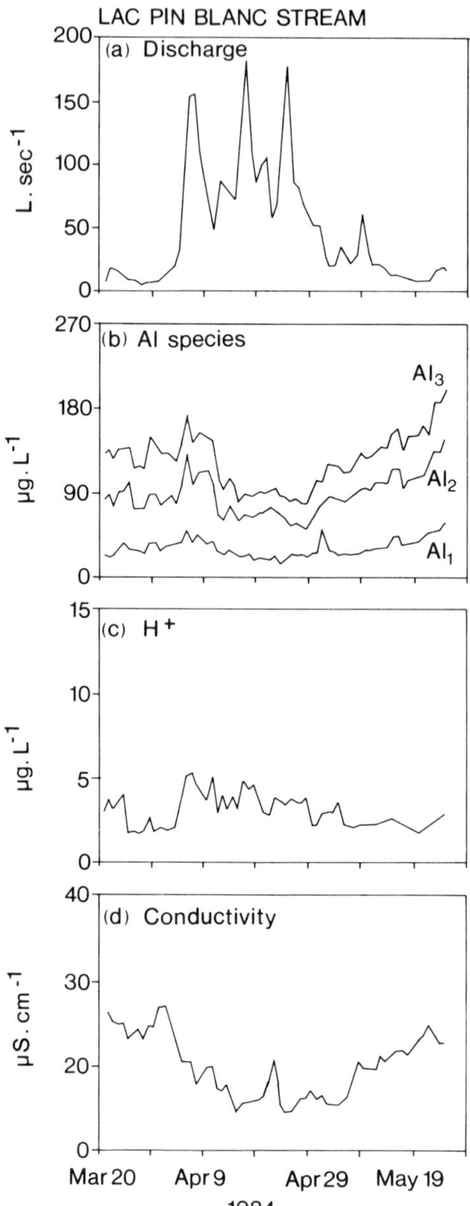

FIGURE 6. Temporal variation in: (a) discharge (l/s), (b) aluminum species (µg/l), (c) H^+ (µg/l), and (d) electrolytic conductivity (µS/cm) for March 21 to May 25, 1984 at the stream draining Lac Pin Blanc. See the text for details of the aluminum speciation.

minum (Al_3-Al_2) contributed about equally to the total reactive aluminum concentration throughout the melt period. Organically and fluoride-complexed monomeric aluminum (Al_2-Al_1) was more abundant than the other two forms, with its contribution increasing slightly after approximately May 4. The relative decrease in Al_1 was caused by a decrease in the amount of inorganic ligands capable of binding aluminum, and by the increase in pH value as the melt period proceeded. In general, Al_1 and Al_2-Al_1 co-vary, although there are periods

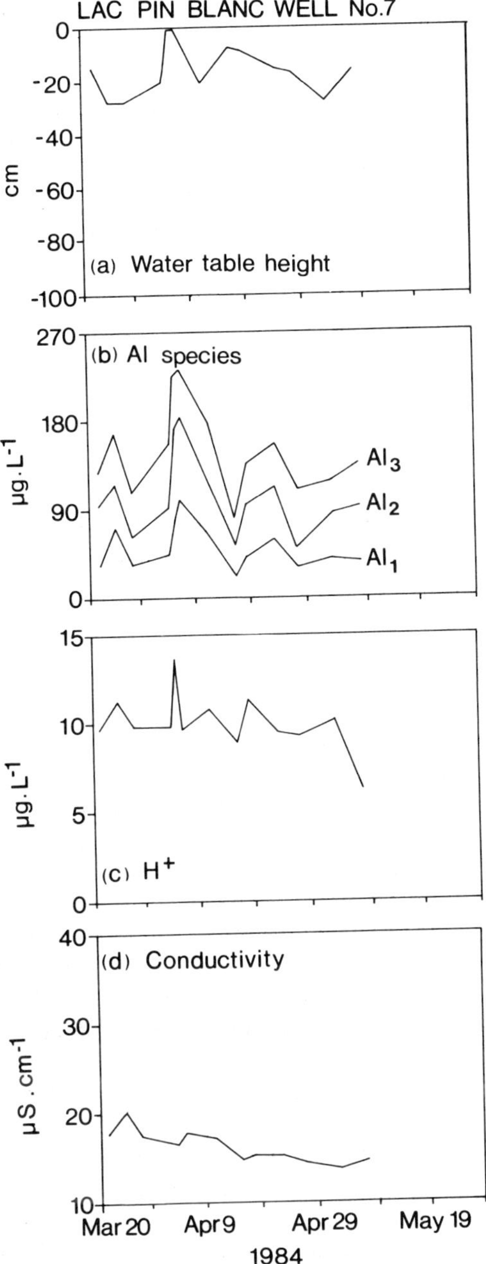

FIGURE 7. Temporal variation in: (a) water table height (cm), (b) aluminum species (μg/l), (c) H^+ (μg/l), and (d) electrolytic conductivity (μS/cm) for March 21 to May 25, 1984 in well #7 from the Lac Pin Blanc watershed. See the text for details of the aluminum speciation.

of exception. On the other hand, temporal variations in Al_3-Al_2 appear to be independent of Al_1 and Al_2-Al_1.

A major difference exists in the temporal variation of aluminum speciation between the stream and ground water (see Figures 6b and 7b, respectively). In the stream, peaks in total reactive aluminum (Al_3) were caused by increases in Al_1 and Al_2-Al_1 monomeric aluminum. In the wells, most of the variation is in Al_1 with a relatively smaller contribution from Al_2-Al_1. Since the most labile forms of aluminum are the most toxic to aquatic life, it is beneficial that a transformation of the most labile into less labile species occurs as the water moves from ground water to stream water.

Spring-melt studies conducted in the Hermine basin during 1985 yielded results (see Figure 8) in general agreement with the above discussion. The magnitude of the initial aluminum pulse that accompanied the first major stream-discharge peak was slightly more pronounced, whereas the degree of dilution of aluminum during subsequent periods of high flow was slightly less pronounced.

B. Aluminum Variability Between Watersheds

Table 3 presents Al_t values in various compartments from watersheds in North America and Europe. With the exception of Findley Lake, Washington, all of the watersheds receive moderate to heavy acidic deposition. Bulk deposition of Al_t is low at all sites except Solling, West Germany. Throughfall Al_t is always higher than in the incident precipitation, often by a factor of approximately two.

Soil water concentrations vary greatly among basins. Total aluminum in water collected from within or directly below organic soil horizons probably reflects the dominant influence of forest type; hence the highest concentrations are associated with watersheds having coniferous forest (Findley Lake, Lake Gårdsjön, and Solling) although the low value for McDonalds Branch is an exception. Like the TLW, Hubbard Brook, Laurel Hill, and The Integrated Lake-Watershed Acidification Study (ILWAS) (Adirondacks) are all deciduous forest. However, they receive comparatively higher levels of acidic deposition, which may account for their higher organic-horizon aluminum values. Findley Lake, which receives a very low acid input, has lower Al_t values in the B horizon than in the overlying organic horizon. In contrast, the heavily acidified European Gårdsjön and Solling basins exhibit very high B horizon Al_t levels, as does Laurel Hill in Pennsylvania. Both Hubbard Brook and ILWAS show a wide range in B-horizon concentrations, a reflection of the variability in soil characteristics found in these areas. The low value of Al_t in the TLW probably reflects the control over soil pH values (and therefore aluminum solubility) exerted by the $CaCO_3$ in the tills of the area.

Output stream concentrations represent the integrated watershed response to all deposition, vegetation, and geochemical factors. In the absence of an external mobilizing factor, very low Al_t values occur in stream water at Findley Lake, whereas geochemical factors influence the low to moderate levels observed in the TLW. The four basins in the northeastern U.S. exhibit a wide range in stream concentrations that is usually attributed to varying water flowpaths (see the references in Table 3). The stream water aluminum levels at Gårdsjön, Sweden are uniformly high; in fact, even ground water acidification has been reported in this area.[63]

C. Source of Aluminum in Soil Waters and Stream Waters

During the 1986 snowmelt in the Hermine basin, the stream hydrograph was modeled to assess the flow contribution from the solum (the soil horizons above the C horizon) and from the ground water (the material in or below the C horizon) using the approach of Pilgrim et al.[64] and Pinder and Jones.[65] Silicon was used as a tracer to calculate the relative contribution of water reaching the stream via the ground water or via the solum.[66] In the Hermine

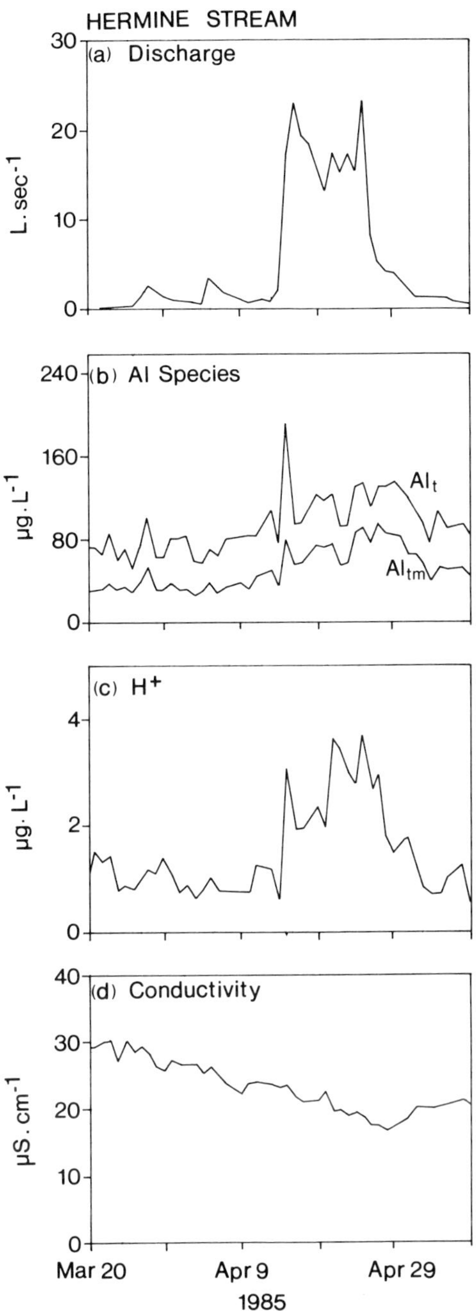

FIGURE 8. Temporal variation in: (a) discharge (l/s), (b) aluminum species (μg/l), (c) H^+ (μg/l), and (d) electrolytic conductivity (μS/cm) for March 20 to May 10, 1985 at the stream draining the Hermine watershed. See the text for details of the aluminum speciation. Note the differing axis scales as compared to Figures 6 and 7.

Table 3
TOTAL[a] ALUMINUM (μg/l) IN VARIOUS CATCHMENT COMPARTMENTS FROM SELECTED LOCATIONS

Location	Precipitation[c]	Through-fall	Soil water[b]			Ground water	Streams	Ref.
			Shallow	Intermediate	Deep			
Turkey Lakes Watershed, Ontario	32	49	134—183	357—380	102	45	38—197	30, 31, this study
St. Hippolyte, Quebec	20	—	100—600	47—650	—	—	50—220	52, 53
McDonalds Branch Watershed, New Jersey	12	46—182	452	604	728	300	14—502	54
Hubbard Brook, New Hampshire	tr	—	189—1010	251—850	—	—	150—710	55, 56
Laurel Hill, Pennsylvania	3	9	850	3590	—	—	81—900	57
Adirondacks, New York	—	—	420—790	220—1200	—	—	0—470	58
Findley Lake, Washington	30	60	720—790	580—740	—	—	20	59
Lake Gårdsjön, Sweden	tr	40	791—1740	2570—3110	—	—	583—828	60, 61
Solling, West Germany	190	310—500	560—1100	1700—1900	—	—	—	62

[a] "Total" aluminum may be either "acid extractable" or "total monomeric"; see individual references.
[b] Shallow = water collected within or below the organic horizons; intermediate = water collected within the B horizon; deep = water collected within the C horizon.
[c] tr = trace.

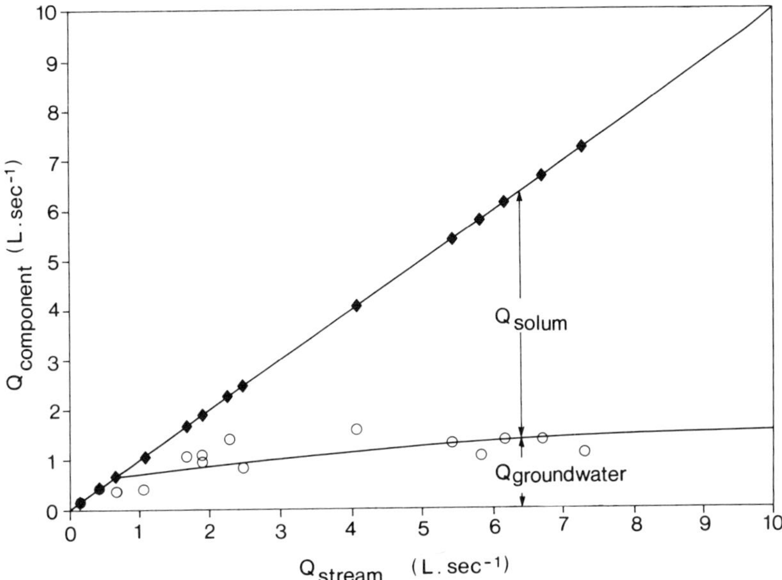

FIGURE 9. Plot of the best-fit relationship of the flow component model discussed in the text. Total stream flow is separated into ground water and solum input components in the Hermine watershed.

basin, the water flowing over the soil surface during the maximum flow events was observed to have a chemical composition similar to the water flowing in the solum, implying that there has been chemically effective contact between the overland flow water and the soil. As a result, we can assume a relatively simple system, wherein there are only two sources of water reaching the stream.[27] The model takes the form:

$$Q_tC_t = Q_gC_g + Q_sC_s \tag{1}$$

where Q_t = measured stream discharge [l/s], Q_g = ground water discharge [l/s], Q_s = solum discharge [l/s], C_t = measured stream Si concentration [mg/l], C_g = ground water Si concentration [mg/l] (assumed equal to the stream baseflow concentration), and C_s = measured solum Si concentration [mg/l]. The model was used to calculate Q_g and Q_s. The best-fit relationship to the 1986 data is shown in Figure 9.

According to this model, all of the discharge comes from the ground water when stream discharge is less than 0.6 l/s. At higher discharge, most of the increase is from flow through the solum. At a stream discharge of 20 l/s, only about 2 l/s are contributed by the ground water; the rest is coming from the solum. The large contribution from the solum in this watershed is explained by the dense C horizon that impedes infiltration and causes a temporary perched water table to develop during stormflow events. During melt events, therefore, most of the water reaching the stream passes through the solum, thereby accounting for the higher H^+ and aluminum concentrations measured in the stream during high flow (see Figure 8).

Figure 10 shows the relationship between pH value and Al^{3+} activity calculated with thermodynamic data compiled by Lindsay.[67] These calculations separate the inorganic monomeric aluminum into Al^{3+}, $AlOH^{2+}$, $Al(OH)_2^+$, $Al(OH)_3$, $Al(OH)_4^-$, AlF^{2+}, AlF_3, and $AlSO_4$ (see Chapter 8). Numerous authors have found that the Al^{3+} activity in soil waters appears to be controlled by the solubility of a gibbsite-like mineral. If gibbsite is the mineral controlling Al^{3+} activity in either soil or stream water, then the following equation describes the reaction:

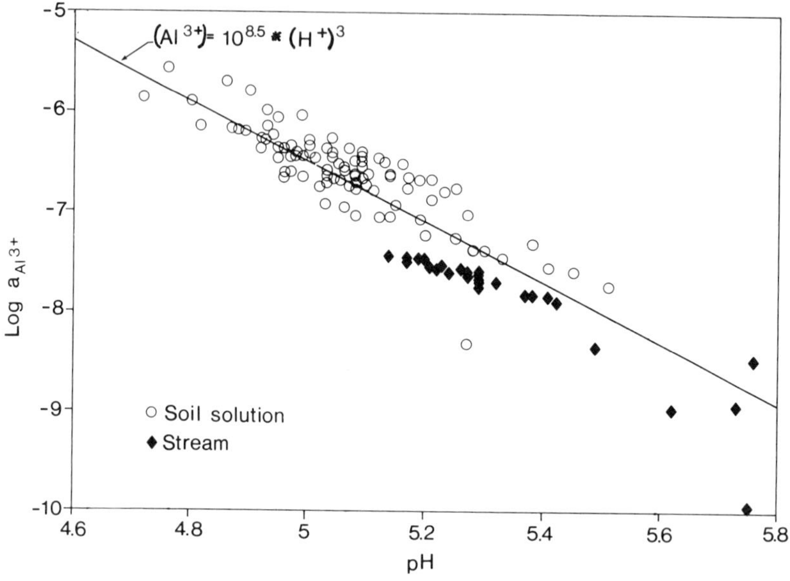

FIGURE 10. Plot of lot Al^{3+} activity vs. pH value for soil solution samples (circle) and stream samples (diamond) collected in the Hermine watershed during the spring melt episode of 1986. The theoretical gibbsite solubility line also is indicated.[67]

$$Al(OH)_3 + 3H^+ = Al^{3+} + 3H_2O \qquad (2)$$

For this reaction, when $\log[Al^{3+}]$ is plotted against pH values, the resulting line has a slope of -3, as shown in Figure 10 (see also Figure 1 in Chapter 8). In soil water from the Hermine watershed, the Al^{3+} activity closely follows the theoretical solubility relationship of a gibbsite-like mineral with $\log{}^*K_{so} = 8.5$ (Figure 10). However, in the podzolic soils of southern Québec, gibbsite has never been identified. In fact, none would be expected because of the high concentrations of dissolved organic matter and silicon that would interfere with the crystalization of this mineral (see Chapter 8). Another possibility is that the solid phase responsible for controlling the Al^{3+} activity is aluminum-interlayered vermiculite; however, this mineral is also absent from the clay fraction of the soils at St. Hippolyte. We conclude that the source of aluminum in the podzols is mainly amorphous material (organic or inorganic) that has a solubility similar to that of $Al(OH)_3$. For a comprehensive discussion of this problem, see Chapter 8.

The measured values of $\log[Al^{3+}]$ and pH in stream waters also are highly correlated, but the absolute slope of the line is greater than -3 — in contrast to the relationship observed for soil solutions (see Figure 10) — even during periods when most of the water is passing through the solum before reaching the stream. This lower slope is similar to what was observed at Hubbard Brook by Hooper and Shoemaker.[68] They hypothesized that the temporal variations in aluminum concentration observed during snowmelt are caused by "the depletion of a slowly forming, labile pool of Al in the soil." Furthermore, they noted that adjacent streams exhibit differing relationships, an observation similar to our findings in the TLW where two of the five headwater streams represented in Table 1 (F34 and F47, a high- and a low-pH value example, respectively) show a good correlation between Al_t and pH values, while the other three streams do not. Nordstrom and Ball[69] have observed a similar deviation from the theoretical slope of -3 at pH values below 4.6. However, in the study at St. Hippolyte, the deviation occurred below pH 5.4 and is probably not caused by the same mechanism. Sullivan et al.[70] investigated aluminum variation in response to episodic increases in flow in the Norwegian Birkenes catchment. Their observations generally agree with those at St. Hippolyte, although the peak aluminum concentrations associated with an initial discharge peak were primarily the result of elevated concentrations of organic aluminum species. Perhaps the difference in forest type (coniferous at Birkenes vs. deciduous at St. Hippolyte) can explain the differing species responses. In all cases, however, after the initial peak, aluminum concentrations remain relatively low in stream waters during subsequent increases in flow that occur before the soils have had time to increase the pool of readily soluble aluminum. The rate at which the labile aluminum pool in the soil forms and the factors influencing the rate are unknown and warrant further study (see Chapter 8). Observations from the present studies suggest that the time scale is on the order of weeks instead of days.

Clearly, the supply of aluminum to soil and surface waters is a complex hydrogeochemical process. Better knowledge of the flowpaths of water under differing water loading rates, of the chemical character of the labile aluminum pool in soils, and of the factors influencing its supply rate are necessary before it will be possible to predict aluminum behavior quantitatively in receiving surface waters.

ACKNOWLEDGMENTS

This paper could not have been completed without the willing participation of several individuals. Ray Semkin, Roy Neureuther, Marion Seymour, and Kay Palmer were involved in all aspects of the collection and analyses of samples from the TLW. John Nicolson kindly provided his unpublished stream concentration data so that the TLW picture presented in this Chapter would be complete. Claude Lapierre, Sylvain Savoie, Linda Mendes, Hélène Lalande, and François Courchesne collected and analyzed the samples from Québec province and helped to develop the concept of the importance of flowpath on stream water chemistry.

REFERENCES

1. **Cronan, C. S. and Schofield, C. L.,** Aluminum leaching response to acid precipitation: effects on high-elevation watersheds in the north-east, *Science,* 204, 304, 1979.
2. **van Breemen, N., Driscoll, C. T., and Mulder, J.,** Acidic deposition and internal proton sources in acidification of soils and waters, *Nature,* 307, 599, 1984.
3. **Dillon, P. J., Yan, N. D., and Harvey, H. H.,** Acidic deposition: effects on aquatic ecosystems, *Crit. Rev. Environ. Control,* 13, 167, 1984.
4. Tri-academy Committee, Acid deposition: effects on geochemical cycling and biological availability of trace elements, Report of the Tri-academy Committee on acid deposition (subgroup on metals), National Academy Press, Washington, D.C., 1985.
5. **Schindler, D. W.,** Effects of acid rain on freshwater ecosystems: a summary of recent research, *Science,* 239, 149, 1988.
6. **Hinrichsen, D.,** Multiple pollutants and forest decline, *Ambio,* 15, 258, 1986.
7. **Driscoll, C. T., Jr., Baker, J. P., Bisogni, J. J., Jr., and Schofield, C. L.,** Effect of aluminum speciation on fish in dilute acidified waters, *Nature,* 284, 161, 1980.
8. **Baker, J. P. and Schofield, C. L.,** Aluminum toxicity to fish in acidified waters, *Water Air Soil Pollut.,* 18, 289, 1982.
9. **Gunn, J. M. and Keller, W.,** Spawning site water chemistry and lake trout *(Salvelinus namaycush)* sac fry survival during spring snowmelt, *Can. J. Fish. Aquat. Sci.,* 41, 319, 1984.
10. **Henriksen, A., Skogheim, O. K., and Rosseland, B. O.,** Episodic changes in pH and aluminum-speciation kill fish in a Norwegian salmon river, *Vatten,* 40, 255, 1984.
11. **Hall, R. J., Driscoll, C. T., Likens, G. E., and Pratt, J. M.,** Physical, chemical, and biological consequences of episodic aluminum additions to a stream, *Limnol. Oceanogr.,* 30, 212, 1985.
12. **Dunne, T.,** Field studies of hillslope flow processes, in *Hillslope Hydrology,* Kirby, M. J., Ed., John Wiley & Sons, New York, 1978, 227.

13. **Freeze, R. A.,** Role of subsurface flow in generating surface runoff. II. Upstream source areas, *Water Resour. Res.*, 8, 1272, 1972.
14. **Atkinson, T. C.,** Techniques for measuring subsurface flow on hillslopes, in *Hillslope Hydrology*, Kirby, M. J., Ed., John Wiley & Sons, New York, 1978, 73.
15. **Mosley, M. P.,** Stream generation in a forested watershed, New Zealand, *Water Resour. Res.*, 15, 795, 1979.
16. **Beven, K. and German, P.,** Macropores and water flow in soils, *Water Resour. Res.*, 18, 1311, 1982.
17. **Horton, R. E.,** The role of infiltration in the hydrological cycle, *Trans, Am. Geophys. Union*, 14, 446, 1933.
18. **Price, A. G. and Hendrie, L. K.,** Water motion in a deciduous forest during snowmelt, *J. Hydrol.*, 64, 339, 1983.
19. **Semkin, R. G. and Jeffries, D. S.,** Bulk deposition of ions in the Turkey Lakes Watershed, *Water Pollut. Res. J. Can.*, 24, 474, 1986.
20. **Semkin, R. G. and Jeffries, D. S.,** Rock Chemistry in the Turkey Lakes Watershed, Turkey Lakes Watershed Unpublished Rep. No. 83-03, 1983.
21. **Kusmirski, R. T. and D. W. Cowell,** Mineralogy of Subsoil Samples, Turkey Lakes Watershed, Ontario, Turkey Lakes Watershed Unpublished Rep. No. 83-09, 1983.
22. **Craig, D. and Johnston, L. M.,** Turkey Lakes Groundwater Study: Aquifer Materials and Hydrogeological Instrumentation, Turkey Lakes Watershed Unpublished Rep. No. 83-22, 1983.
23. **Jeffries, D. S., Kelso, J. R. M., and Morrison, I. K.,** Physical, chemical, and biological characteristics of the Turkey Lakes Watershed, central Ontario, Canada, *Can. Sp. Publ. Fish. Aquat. Sci.*, 45 (Suppl. 1), 3, 1988.
24. **Wilson, C.,** Le climat du Québec, Service de la meteorologie du Canada, Etudes Climatiologiques No. 11, Ottawa, 1971.
25. **Hendershot, W. H., Dufresne, A., Lalande, H., and Wright, R. K.,** Aluminum speciation and movement in three small watersheds in the southern Laurentians, *Water Pollut. Res. J. Can.*, 19, 11, 1984.
26. **Dufresne, A.,** Mobilité de l'aluminum dans un bassin-versant des basses Laurentides, Mémoire de Maîtrise, Université de Montréal, Montréal, 1984.
27. **Savoie, S.,** Relationship Between the Chemistry of Soil Solutions and Stream Waters, St. Hippolyte, Québec, M.Sc. thesis, McGill University, Montréal, Québec, in preparation.
28. **Nicolson, J. A.,** Ion concentration in precipitation and streamwater in an Algoma maple-birch forest, in *Proc. Can. Hydrol. Symp. No. 15*, National Research Council Canada Publ. No. 24633, 1984, 123.
29. **Lazerte, B. D.,** Forms of aqueous aluminum in acidified catchments of central Ontario: a methodological analysis, *Can. J. Fish. Aquat. Sci.*, 41, 766, 1984.
30. **Foster, N. W. and Nicolson, J. A.,** Trace elements in the hydrologic cycle of a tolerant hardwood forest ecosystem, *Water Air Soil Pollut.*, 31, 501, 1986.
31. **Chew, H., Johnston, L. M., Craig, D., and Inch, K.,** Aluminum contamination of groundwater. II. Spring melt in the Chalk River and Turkey Lakes Watersheds—preliminary results, *Can. Sp. Publ. Fish. Aquat. Sci.*, 45 (Suppl. 1), 66, 1988.
32. **Likens, G. E., Bormann, F. H., Pierce, R. S., Eaton, J. S., and Johnson, N. M.,** *Biogeochemistry of a Forested Ecosystem*, Springer-Verlag, New York, 1977, chap. 2.
33. **Lalande, H. and Hendershot, W. H.,** Aluminum speciation in some synthetic systems: comparison of the fast oxine pH 5.0 extraction and dialysis methods, *Can. J. Fish. Aquat. Sci.*, 43, 231, 1986.
34. **Henriksen, A. and Bergman-Paulsen, I. M.,** An automatic method for determining aluminum in natural waters, *Vatten*, 4, 339, 1975.
35. **Rogeberg, E. J. S. and Henriksen, A.,** An automatic method for fractionation and determination of aluminum species in fresh-waters, *Vatten*, 41, 48, 1985.
36. **Dickson, W.,** Properties of acidified waters, in *Proc. Int. Conf. Ecol. Impact Acid Precip.*, Drablos, D. and Tollan, A., Eds., SNSF Project, Sandefjord, Norway, 1980, 75.
37. **Broberg, O.,** Phosphate removal in acidified and limed lake water, *Water Res.*, 18, 1273, 1984.
38. **Effler,S. W., Schafran, G. C., and Driscoll, C. T.,** Partitioning light attentuation in an acidic lake, *Can. J. Fish. Aquat. Sci.*, 42, 1707, 1985.
39. **Papineau, M.,** Composition chimique et bilan ionique de Lac Laflamme, (Forêt Montmorency), Québec, Environnement Canada, Direction des eaux intérieures, Région de Québec, 1984.
40. **Johannessen, M. and Henriksen, A.,** Chemistry of snow meltwater: changes in concentration during melting, *Water Resour. Res.*, 14, 615, 1978.
41. **Semkin, R. G. and D. S. Jeffries,** Storage and release of major ionic contaminants from the snowpack in the Turkey Lakes Watershed, *Water Air Soil Pollut.*, 31, 215, 1986.
42. **Jeffries, D. S., Cox, C. M., and Dillon, P. J.,** Depression of pH in lakes and streams in central Ontario during snowmelt, *J. Fish. Res. Board Can.*, 36, 640, 1979.

43. **Schofield, C. L. and Trojnar, J. R.,** Aluminum toxicity to brook trout *(Salvelinus fontinalis)* in acidified waters, in *Polluted Rain*, Rochester International Conference on Environmental Toxicity, Plenum Press, New York, 1979, 341.
44. **Seip, H. M., Muller, L., and Noas, A.,** Aluminum speciation: comparison of two methods in some acidic aquatic systems in southern Norway, *Water Air Soil Pollut.*, 23, 81, 1984.
45. **Grahn, O.,** Fishkills due to high aluminum concentration in lake water, in *Proc. Int. Conf. Ecol. Impact Acid Precip.*, Drablos, D. and Tollan, A., Eds., SNSF Project, Sandefjord, Norway, 1980, 310.
46. **Overrien, L. N., Seip, H. M., and Tollan, A.,** Acid precipitation effects on forest and fish, SNSF Final Report, Oslo, Norway, 1980.
47. **Driscoll, C. T. and Schafran, G. C.,** Short-term changes in the base neutralizing capacity of an acid Adirondack lake, New York, *Nature*, 310, 308, 1984.
48. **Schofield, C. L.,** Surface water chemistry in the ILWAS basins, in *The Integrated Lake-Watershed Acidification Study*, Vol. 4, Summary of Major Results, Goldstein, R. A., Ed., EPRI Rep. EA-3221, Lafayette, CA, 1984, chap. 6.
49. **Jeffries, D. S. and Semkin, R. G.,** Changes in snowpack, stream and lake chemistry in the Turkey Lakes Watershed, in *Proc. Int. Conf. Acid Precip.*, Lindau, West Germany; *VDI Ber. (Ver. Dtsch. Ing.)*, 500, 377, 1983.
50. **English, M. C., Jeffries, D. S., and Semkin, R. G.,** A preliminary examination of direct subsurface seepage into a Canadian Shield lake, *Can. Sp. Publ. Fish. Aquat. Sci.*, in press.
51. **Gunn, J. M. and Keller, W.,** Effects of acidic meltwater on chemical conditions at nearshore spawning sites, *Water Air Soil Pollut.*, 30, 545, 1986.
52. **Hendershot, W. H., Dufresne, A., Lalande, H., and Courchesne, F.,** Temporal variation in aluminum speciation and concentration during snowmelt, *Water Air Soil Pollut.*, 31, 231, 1986.
53. **Hendershot, W. H., Dufresne, A., Lalande, H., and Wright, R. K.,** Speciation of aluminum in different compartments of a drainage basin during snowmelt, in *Proc. Eastern Snow Conf.*, Montréal, Québec, 1985, 58.
54. **Turner, R. S., Johnson, A. H., and Wang, D.,** Biogeochemistry of aluminum in McDonalds Branch Watershed, New Jersey Pine Barrens, *J. Environ. Qual.*, 14, 314, 1985.
55. **Driscoll, C. T., van Breemen, N., and Mulder, J.,** Aluminum geochemistry in a forested Spodosol, *Soil Sci. Soc. Am. J.*, 49, 437, 1985.
56. **Johnson, N. M., Driscoll, C. T., Eaton, J. S., Likens, G. E., and McDowell, W. H.,** 'Acid rain', dissolved aluminum and chemical weathering at the Hubbard Brook Experimental Forest, New Hampshire, *Geochim. Cosmochim. Acta*, 45, 1421, 1981.
57. **Sharpe, W. E., DeWalle, D. R., Liebfried, R. T., Dinicola, R. S., Kimmel, W. G., and Sherwin, L. S.,** Causes of acidification of four streams on Laurel Hill in southwestern Pennsylvania, *J. Environ. Qual.*, 13, 619, 1984.
58. **Cronan, C. S.,** Vegetation and soil chemistry of the ILWAS watersheds, in *The Integrated Lake-Watershed Acidification Study*, Vol. 4, Summary of Major Results, Goldstein, R. A., Ed., EPRI Rep. EA-3221, Lafayette, CA, 1984, chap. 3.
59. **Ugolini, F. C., Minden, Dawson, H., and Zachara, J.,** An example of soil processes in the Abies Ambalis zone of central Cascades, Washington, *Soil Sci.*, 124, 291, 1977.
60. **Nilsson, S. I. and Bergkvist, B.,** Aluminum chemistry and acidification processes in a shallow podzol on the Swedish westcoast, *Water Air Soil Pollut.*, 20, 311, 1983.
61. **Nilsson, S. I.,** Budgets of aluminum species, iron and manganese in the Lake Gårdsjön catchment in SW Sweden, in *Lake Gårdsjön: An Acid Lake and its Catchment*, Ecological Bulletin No. 37, Andersson, F. and Olsson, B., Eds., Publishing House of the Swedish Research Councils, Stockholm, 1985, 120.
62. **Heinrichs, H. and Mayer, R.,** Distribution and cycling of major and trace elements in two central European forest ecosystems, *J. Environ. Qual.*, 6, 402, 1977.
63. **Hultberg, H. and Johansson, S.,** Acid groundwater, *Nordic Hydrol.*, 12, 51, 1981.
64. **Pilgrim, D. H., Huff, D. D., and Steele, T. D.,** Use of specific conductance and contact time relations for separating flow components in storm runoff, *Water Resour. Res.*, 15, 329, 1979.
65. **Pinder, G. F. and Jones, J. F.,** Determination of the groundwater component of peak discharge from the chemistry of total runoff, *Water Resour. Res.*, 5, 439, 1969.
66. **Hooper, R. P. and Shoemaker, C. A.,** A comparison of chemical and isotopic hydrograph separation, *Water Resour. Res.*, 22, 1444, 1986.
67. **Lindsay, W. L.,** *Chemical Equilibria in Soils*, John Wiley & Sons, New York, 1979, chap. 3.
68. **Hooper, R. P. and Shoemaker, C. A.,** Aluminum mobilization in an acidic headwater stream: temporal variation and mineral dissolution disequilibria, *Science*, 229, 463, 1985.
69. **Nordstrom, D. K. and Ball, J. W.,** The geochemical behavior of aluminum in acidified surface waters, *Science*, 232, 54, 1987.
70. **Sullivan, T. J., Christophersen, N., Muniz, I. P., Seip, H. M., and Sullivan, P. D.,** Aqueous aluminum chemistry response to episodic increases in discharge, *Nature*, 323, 324, 1986.

INDEX

A

B

C

D

E

H

I

N

O

P

Q

R

S

T

U

V

W

X

Z